Understanding Phylogenetics

Understanding Phylogenetics

Edited by Jesse Santos

SYRAWOOD
PUBLISHING HOUSE

New York

Published by Syrawood Publishing House,
750 Third Avenue, 9th Floor,
New York, NY 10017, USA
www.syrawoodpublishinghouse.com

Understanding Phylogenetics
Edited by Jesse Santos

International Standard Book Number: 978-1-68286-760-0 (Hardback)

Cataloging-in-Publication Data

Understanding phylogenetics / edited by Jesse Santos.
    p. cm.
Includes bibliographical references and index.
ISBN 978-1-68286-760-0
1. Phylogeny. 2. Evolution (Biology). I. Santos, Jesse.
QH367.5 .U53 2019
576.88--dc23

TABLE OF CONTENTS

PREFACE

Phylogenetics is a field of biology that studies the evolutionary history and relationship among individuals or groups of organisms. Phylogenetic inference methods, that evaluate observed heritable traits using studies of morphology or DNA sequences, are crucial in the development of a phylogenetic tree. Such studies are fundamental to the understanding of biodiversity, ecology, evolution and genomes. Phylogenetic inference involves computational techniques for implementing the criterion of optimality, methods of parsimony, maximum likelihood and Bayesian inference. This book is a valuable compilation of topics, ranging from the basic to the most complex advancements in the field of phylogenetics. It presents the complex subject of phylogenetics in the most comprehensible and easy to understand language. A number of latest researches have been included to keep the readers up-to-date with the global concepts in this area of study.

This book unites the global concepts and researches in an organized manner for a comprehensive understanding of the subject. It is a ripe text for all researchers, students, scientists or anyone else who is interested in acquiring a better knowledge of this dynamic field.

I extend my sincere thanks to the contributors for such eloquent research chapters. Finally, I thank my family for being a source of support and help.

Editor

The measure of success: geographic isolation promotes diversification in *Pachydactylus* geckos

Matthew P. Heinicke[1*], Todd R. Jackman[2] and Aaron M. Bauer[2]

Abstract

Background: Geckos of the genus *Pachydactylus* and their close relatives comprise the most species-rich clade of lizards in sub-Saharan Africa. Many explanations have been offered to explain species richness patterns of clades. In the *Pachydactylus* group, one possible explanation is a history of diversification via geographic isolation. If geographic isolation has played a key role in facilitating diversification, then we expect species in more species-rich subclades to have smaller ranges than species in less diverse subclades. We also expect traits promoting geographic isolation to be correlated with small geographic ranges. In order to test these expectations, we performed phylogenetic analyses and tested for correlations among body size, habitat choice, range sizes, and diversification rates in the *Pachydactylus* group.

Results: Both body size and habitat use are inferred to have shifted multiple times across the phylogeny of the *Pachydactylus* group, with large size and generalist habitat use being ancestral for the group. Geographic range size is correlated with both of these traits. Small-bodied species have more restricted ranges than large-bodied species, and rock-dwelling species have more restricted ranges than either terrestrial or generalist species. Rock-dwelling and small body size are also associated with higher rates of diversification, and subclades retaining ancestral conditions for these traits are less species rich than subclades in which shifts to small body size and rocky habitat use have occurred. The phylogeny also illustrates inadequacies of the current taxonomy of the group.

Conclusions: The results are consistent with a model in which lineages more likely to become geographically isolated diversify to a greater extent, although some patterns also resemble those expected of an adaptive radiation in which ecological divergence acts as a driver of speciation. Therefore, the *Pachydactylus* group may represent an intermediate between clades in which radiation is adaptive versus those in which it is non-adaptive.

Keywords: Biogeography, Systematics, Timetree, Allopatry, Radiation, Cladogenesis, Ancestral reconstruction, Phylogenetic comparative methods

Background

Discrete geographic regions, both continentally and on islands, often have biotas dominated by a relatively small number of species-rich lineages. The most obvious of these dominant groups are adaptive radiations in which a single ancestral species has given rise to descendants filling numerous niches, examples of which include Galapagos Finches (14 sp., 58% of breeding songbird species in the archipelago), Lake Victoria cichlids (169 sp., 67% of ray-finned fishes in the lake), and West Indies *Anolis* lizards (168 sp., 38% of lizards native to the islands) [1–3]. However, there are many other less-known examples of regionally prominent radiations. Among lizards, one of the most striking are geckos of the genus *Pachydactylus* (56 species) and its close relatives *Chondrodactylus* (6 species), *Colopus* (2 species), and *Elasmodactylus* (2 species). By species number, these geckos are the most successful radiation of lizards in southern Africa. Sixty four of 66 species occur in the southern African subcontinent, defined as that part of Africa south of the Zambezi

* Correspondence: heinicke@umich.edu
[1]Department of Natural Sciences, University of Michigan-Dearborn, 4901 Evergreen Rd., Dearborn, MI 48128, USA
Full list of author information is available at the end of the article

and Kunene rivers, and most are endemic to this region. These species occupy all major habitat types in southern Africa, and many species in the group display morphological novelties such as loss of adhesive toe pads or the evolution of interdigital webbing [4, 5]. Numerous other gecko genera are found in southern Africa, most of which are endemic or reach their peak diversity there, including *Afroedura*, *Afrogecko*, *Cryptactites*, *Goggia*, *Homopholis*, *Narudasia*, *Ramigekko*, and *Rhoptropus*, but *Pachydactylus* group species often dominate the gekkonid fauna, comprising, for example, 13 of 18 species in the Richtersveld of South Africa [6]. Likewise, none of these other genera approach *Pachydactylus* in its diversity of morphological or ecological variation.

Numerous possible causative factors have been posited or shown to explain the relative success of species-rich organismal groups. In classic adaptive radiations, the rapid evolution of morphological disparity may be the key process in spurring lineage accumulation [7, 8]. In other cases, the evolution of a novel trait may allow organisms possessing that trait to access underutilized resources, with utilization promoting ecological diversification and lineage accumulation. Examples include the evolution of antifreeze proteins in Antarctic icefishes and evolution of the pharyngeal jaw structures in parrotfishes [9, 10]. Sexual selection may also promote lineage accumulation, especially when this selection is for traits that serve as prezygotic isolating mechanisms such as male advertisement calls or color patterns serving as visual mate recognition systems [11, 12]. Finally, in some cases species-rich organismal groups may not actually exhibit a high diversification rate at all, but instead have a longer history of occupancy in a geographic region [13].

In the case of *Pachydactylus* and its relatives, none of these potential explanations are likely to fully account for the observed species diversity. *Pachydactylus* is divided into eight well-defined species groups [14, 15], each of which is believed to be monophyletic, and each of which is morphologically conservative. The degree of morphological disparity between these groups is not of a magnitude expected in a classic adaptive radiation [8]. This is not to say that there is no morphological variation within the genus. Some morphological novelties have evolved, including the previously mentioned foot characteristics as well as variation in scalation. However, such morphological novelties do not appear to be strong drivers of speciation in the group. For example, the most species-rich radiation of *Pachydactylus* geckos that has lost toe pads contains only three species, one of which still has toe pads [4].

Many geckos do have visual display systems or other traits which could theoretically promote diversification via divergent sexual selection. Examples include the

semaphore geckos (*Pristurus*), the dwarf geckos (*Sphaerodactylus*), and various day geckos (including *Cnemaspis*, *Lygodactylus*, and *Phelsuma*) [16–18]. In these genera, males are boldly patterned and use signaling behaviors to defend territories or attract mates. *Pachydactylus* and its relatives are strictly nocturnal, however, and typically have a drab pattern. Nor are any other prezygotic isolating mechanisms evident that could plausibly be hypothesized to be under sexual selection. Finally, the relative age of *Pachydactylus* and its relatives likely does not account for its diversity, either – the next closest relative of the *Pachydactylus* group is *Rhoptropus* [5], which also occurs mainly in southern Africa but includes only nine species.

Given that none of these explanations can fully account for the species diversity observed in *Pachydactylus* and its relatives, we hypothesize that geographic isolation leading to allopatric divergence plays a key role in lineage accumulation in *Pachydactylus* and its relatives. Populations in allopatry, if isolated for a sufficient period of time, can naturally speciate via genetic drift without requiring significant contributions from natural selection acting on divergent morphological traits or sexual selection promoting differentiation of mating systems among species [19, 20]. If geographic isolation does play a key role in diversification of *Pachydactylus* and its relatives, then traits promoting the formation of geographic isolation should affect both species' range sizes and diversification rate. The heritability of range size has been a matter of debate, but increasing numbers of studies demonstrate its heritability [21–25]. In at least some clades, including lizards, this heritability is associated with variable morphological or ecological traits [26, 27]. Likewise, numerous studies have reported trait-associated variation in diversification rates, especially since the development of BiSSE (binary-state speciation and extinction) and related models [28].

For *Pachydactylus* and its relatives, two variable traits of interest that may promote geographic isolation are body size and habitat preference. There is substantial body size variation in *Pachydactylus* and related genera, with the largest and smallest species having adult snout–vent lengths of 35 and 113 mm, respectively [29]. In many groups body size has been shown to be positively correlated with range size [30]. Habitat preference within *Pachydactylus* varies, with species showing preferences ranging from sand dunes to rocky cliffs to houses. In southern Africa, the periodic advance and retreat of Kalahari and Namib sands over geological time is linked to climatic variation [31–33]; this process has likely allowed intermittent connections to form between adjacent rocky habitats, but the prevailing pattern is that terrestrial habitats are relatively continuous while rocky habitats are more discontinuous. As a result, a preference for rocky habitats may be expected to be associated with geographic isolation and smaller range sizes. Such substrate specialization has been suggested to facilitate speciation in the *Pachydactylus* group [14, 34], but has never been explicitly tested.

We test whether body size or habitat preference is associated with the formation of geographic isolation in the *Pachydactylus* group in a phylogenetic context. We have generated a comprehensive time-calibrated multi-locus phylogeny of the group, and obtained body size and habitat preference trait data for all ingroup species. Geographic range size estimates are produced for all species, and the association between trait data and range size is quantified. We also estimate patterns of lineage accumulation through time and trait-associated estimates of diversification. Our data show that both body size and habitat preference affect range size, and that variation in these traits is also correlated with variation in diversification rate, suggesting that allopatric divergence following isolation has played a key role in speciation in the *Pachydactylus* group.

Methods

Phylogeny estimation

Previous studies have confirmed that the *Pachydactylus* is part of a monophyletic assemblage of morphologically similar geckos, also including genera *Chondrodactylus*, *Colopus*, and *Elasmodactylus* [4, 14, 35]. We sought to estimate a comprehensive phylogeny for this group, and obtained genetic samples from individuals of 55 of 56 *Pachydactylus* species, 6 of 6 *Chondrodactylus*, both *Colopus* species, and both *Elasmodactylus* species. These genera are part of a larger clade of geckos mainly distributed in Africa and Madagascar, and within this larger grouping they are most closely related to the genus *Rhoptropus* [5, 36]. As such, we included exemplars of 9 of 9 *Rhoptropus* species to serve as a near outgroup. An additional 18 gekkotan and 4 non-gekkotan taxa (*Anolis*, *Gallus*, *Python*, *Trachylepis*) were included as more distant outgroups, with outgroup species choice partially determined based on utility for molecular clock calibration. Nearly all ingroup sequences are associated with vouchered museum specimens. Sequences for four species (*Elasmdactylus tuberculosus*, *Pachydactylus namaquensis*, *P. tsodiloensis*, *P. visseri*) are exceptions, with sequences derived from genetic material obtained from captive-bred individuals; in these cases the live specimens were viewed by the authors to confirm identification and associated genetic material has been deposited in the Cryogenic Collection at the Museum of Comparative Zoology, Harvard University.

We constructed a sequence data set of nuclear and mitochondrial genes that evolve in a relatively clocklike fashion and have proven useful for determining relationships among species within gekkonid genera [37, 38]. The combined data set is 3443 bp (base pairs), including portions of the nuclear genes RAG1 (recombination activating gene 1; 1053 bp), KIF24 (kinesin family member 24; 592 bp) and PDC (phosducin; 395 bp), along with the complete mitochondrial ND2 gene (NADH dehydrogenase subunit 2; 1041 bp) and several adjacent tRNA genes (transfer RNA; 361 bp) (Table 1). All newly generated sequences were deposited in GenBank (accession numbers KY224166–KY224347).

For new sequences generated in this study, DNA was obtained from frozen or ethanol-preserved tissue samples using Qiagen DNeasy tissue kits under the manufacturer's protocol. PCR (polymerase chain reaction) amplification of fragments was performed in 25 µL reactions, under standard reaction conditions [39]. ND2, tRNA, RAG1, and PDC primers used in PCR and sequencing were the same as those used in [37]; KIF24 primers were derived from [40]. PCR purification was performed using AMPure magnetic beads, followed by cycle sequencing and purification using CleanSeq magnetic beads. Capillary electrophoresis was performed on an Applied Biosystems 3730xl sequencer. Sequence assembly was performed using BioEdit [41] or Geneious 5.1 [42], with alignment using Clustal [43]. Alignments of the protein-coding genes were edited manually to preserve reading frame and checked to ensure absence of premature stop codons, while those of the tRNAs were edited manually to preserve secondary structural features estimated in ARWEN [44].

Phylogenetic analyses were performed using maximum likelihood (ML) and Bayesian (BI) optimality criteria. For each analysis, model and partition choices were separately identified under the Bayesian Information Criterion using PartitionFinder [45]. In each case considered models of evolution were limited to those models that can be implemented by the programs used for phylogeny estimation. Greedy search schemes were employed and thirteen potential data blocks were considered: twelve data blocks corresponding to the three codon positions for each of the four protein-coding genes and the tRNA data comprising the thirteenth data block.

The ML analysis was performed using RAxML 8.2.4 [46]. One hundred independent searches were implemented on the original data set to identify the best tree, followed by 1,000 non-parametric bootstrap replicates to assess branch support. Based on the PartitionFinder results, the data were divided into eight partitions, each using one of two models: ND2 codon position 1, ND2 codon position 2, tRNAs, and (PDC position 1 + 2 + RAG1 position 1 + 2) used the GTR (general time reversible) + I + Γ model, while ND2 position 3, (PDC position 3 + RAG1 position 3), (KIF24 position 1 + 2) and KIF24 position 3 used the GTR + Γ model.

Table 1 Specimens and GenBank accession numbers of specimens used in this study

Species	ID Number	ND2	Rag1	PDC	KIF24
Anolis carolinensis	n/a	EU747728	AAWZ 02015549	AAWZ 02013979	NW_003338919
Chondrodactylus angulifer	MCZ R-184985	KY224209	KY224307	KY224257	
Chondrodactylus bibronii	CAS 201841	JN543886	JN543930	KY224258	KY224166
Chondrodactylus fitzsimonsi	MCZ R-185712	JN393945	KY224308	KY224259	KY224167
Chondrodactylus laevigatus	MCZ R-184819	KY224211	KY224310	KY224260	KY224168
Chondrodactylus pulitzerae	CAS 193828	KY224210	KY224309		
Chondrodactylus turneri	MCZ R-184410	KY224249	KM073525	KM073612	KM073800
Coleonyx variegatus	MVZ 161445	AB114446			
Coleonyx variegatus	CAS 205334		EF534777	EF534817	
Colopus kochii	CAS 214803	KY224212	KY224311	KY224261	KY224169
Colopus wahlbergii	NMZB 16974	JN569158	JN569191	JQ945366	
Correlophus ciliatus	AMS R-146595	JX024438	EF534778	EF534818	KU157544
Elasmodactylus tetensis	PEM R-5540	KY224213	KY224312	KY224262	KY224170
Elasmodactylus tuberculosus	MCZ:Cryo 3006	KY224214	KY224313	KY224263	KY224171
Euleptes europaea	no number	JN393941	EF534806	EF534848	KU157420
Gallus gallus	n/a	KT626857	NM_001031188	XM_004943303	NC_006127
Goggia braacki	PEM R-11911	KM073689	KM073528	KM073614	KM073802
Nephrurus levis	AMS 140561		GU459544	GU459746	KU157421
Nephrurus levis	SAMA R-19968	AY369018			
Oedura marmorata	SAMA R-34209	AY369015			
Oedura marmorata	AMS 143861		EF534779	EF534819	KU157428
Ophidiocephalus taeniatus	SAM R-44653	AY134601	HQ426303	HQ426214	KU157422
Pachydactylus acuminatus	MCZ R-185739	KY224215	KY224314	KY224264	KY224172
Pachydactylus affinis	PEM R-17545	KY224216	KY224315	KY224265	
Pachydactylus amoenus	AMB 8670	JN569163			
Pachydactylus angolensis	CAS 254887	KY224217	KY224316		
Pachydactylus atorquatus	MCZ R-184811	KY224218	KY224317	KY224266	
Pachydactylus austeni	LSUMZ H1629	KY224250	JQ945321	JQ945389	KY224173
Pachydactylus barnardi	MCZ R-184749	KY224219	KY224318	KY224267	
Pachydactylus bicolor	NMNW (AMB 7631)	JN543870+ KY224220	JN543911	KY224268	
Pachydactylus boehmei	MCZ R-184883	JN543906	JN543947	KY224270	KY224174
Pachydactylus capensis	MCZ R-184499	HQ165962	HQ165992	HQ165977	KY224175
Pachydactylus caraculicus	MCZ R-185767	JN543889	JN543933	KY224271	
Pachydactylus carinatus	LSUMZ 57293	KY224221	KY224319	KY224272	
Pachydactylus etultra	MCZ R-184978	HQ165959	HQ165989	HQ165974	KY224176
Pachydactylus fasciatus	MCZ R-185759	HQ165949	HQ165978	HQ165963	
Pachydactylus formosus	CAS 206715	KY224222	KY224320	KY224273	
Pachydactylus gaiasensis	MCZ R-184169	JN543891	KM073533	KM073615	KY224177
Pachydactylus geitje	PEM R-11226	JN543887	JN543931	KY224274	KY224178
Pachydactylus goodi	MCZ R-184783	KY224223	KY224321	KY224275	KY224179
Pachydactylus griffini	MCZ R-185741	KY224224	KY224322	KY224276	KY224180
Pachydactylus haackei	CAS 186341	KY224225	KY224323	KY224277	
Pachydactylus kladaroderma	PEM R-1253	KY224251	JQ945323	JQ945391	
Pachydactylus kobosensis	CAS 223904	KY224226	KY224324	KY224278	KY224181

Table 1 Specimens and GenBank accession numbers of specimens used in this study (Continued)

Pachydactylus labialis	MCZ R-184758	KY224227	KY224325	KY224279	KY224182
Pachydactylus latirostris	PEM R-16720	JN569141	JN569173	KY224280	KY224183
Pachydactylus macrolepis	PEM R-17668	JN569139	JN569170	KY224281	KY224184
Pachydactylus maculatus	CAS 186380	KY224228	KY224326	KY224282	KY224185
Pachydactylus maraisi	NMNW (JV 1856)	JN543871	JN543912	KY224269	
Pachydactylus mariquensis	NMB R10936	JN569157	JN569190	KY224283	KY224186
Pachydactylus mclachlani	MCZ R-185094	HQ165950	HQ165980	HQ165965	KY224187
Pachydactylus monicae	CAS 193418	HQ165952	HQ165982	HQ165967	KY224188
Pachydactylus montanus	MCZ R-184243	KY224229	KY224327	KY224284	KY224189
Pachydactylus namaquensis	MBUR 01770	KY224230			
Pachydactylus namaquensis	MCZ:Cryo 3007		KY224328	KY224285	KY224190
Pachydactylus oculatus	PEM R-1284	KY224231	KY224329	KY224286	
Pachydactylus oreophilus	MCZ R-185769	JN543892	JN543936	KY224287	KY224191
Pachydactylus oshaughnessyi	NMZB (DGB 611)	KY224232	KY224330	KY224288	KY224192
Pachydactylus otaviensis	MCZ R-184867	JN543893	JN543937	KY224289	KY224193
Pachydactylus parascutatus	CAS 214750	JN543894	JN543938	KY224290	KY224194
Pachydactylus punctatus	PEM R-12461	KY224233	KY224331	KY224291	
Pachydactylus purcelli	PEM R-16895	HQ165954	HQ165984	HQ165969	
Pachydactylus purcelli	MCZ R-184796				KY224195
Pachydactylus rangei	MCZ R-183725	JN543907	JN543948	JQ945392	
Pachydactylus reconditus	MCZ R-184856	KY224234	KY224332	KY224292	
Pachydactylus robertsi	NMNW R6697	KY224235	KY224333	KY224293	
Pachydactylus rugosus	CAS 201905	KY224252	JQ945325	JQ945393	
Pachydactylus sansteynae	CAS 214589	JN543898	KY224334	KY224294	
Pachydactylus scherzi	MCZ R-184938	KY224236	KY224335	KY224295	
Pachydactylus scutatus	MCZ Z37843	JN543901	JN543943	KY224296	KY224196
Pachydactylus serval	MCZ R-185989	HQ165956	HQ165986	HQ165986	KY224197
Pachydactylus tigrinus	NMB R10936	KY224237	KY224336	KY224297	KY224198
Pachydactylus tsodiloensis	MCZ:Cryo 3008	KY224238	KY224337	KY224298	KY224199
Pachydactylus vansoni	MCZ R-184434	KY224239		KY224299	KY224200
Pachydactylus vanzyli	NMNW (JV1761)	KY224253	JQ945326	JQ945394	KY224201
Pachydactylus visseri	MCZ:Cryo 3009	KY224240	KY224338	KY224300	KY224202
Pachydactylus waterbergensis	MCZ R-184751	KY224241	KY224339	KY224301	
Pachydactylus weberi	PEM R-12449	HQ165960	HQ165990	HQ165975	KY224203
Pachydactylus werneri	MCZ R-184960	KY224242	KY224340	KY224302	KY224204
Phelsuma inexpectata	JB 56	JN393939	JN393983	JN394016	
Phelsuma rosagularis	n/a	EU423292			
Phelsuma rosagularis	JB 109		HQ426306	HQ426217	
Phyllopezus pollicaris	MZUSP 92491	JX041417	EU293635		
Phyllopezus pollicaris	CENPAT12084				JQ827509
Phyllopezus pollicaris	JFBM 15822			HQ426225	
Pygopus nigriceps	ERP R29509	AY134604			
Pygopus nigriceps	SAMA R-23908		FJ571628		
Pygopus nigriceps	MVZ 197233			EF534823	
Python bivittatus	n/a		AEQU 010344888	AEQU 01027927	NW_006537073

Table 1 Specimens and GenBank accession numbers of specimens used in this study *(Continued)*

Python regius	n/a	AB177878			
Rhoptropus afer	MCZ R-183711	KY224254	KM073535	KM073616	KM073806
Rhoptropus barnardi	CAS 214658	KY224243	KY224341	KY224303	KY224205
Rhoptropus benguelensis	ANG_WC1834	KY224246	KY224346		
Rhoptropus biporosus	CAS 224030	KY224244	KY224342	KY224304	KY224206
Rhoptropus boultoni	CAS 214713	KY224256	EF534810	EF534852	KY224207
Rhoptropus bradfieldi	NMNW (to be accessioned)	KY224245	KY224343	KY224305	
Rhoptropus diporus	MCZ R-183737	KY224255	KY224344	KY224306	KY224208
Rhoptropus montanus	CAS 254867	KY224247	KY224345		
Rhoptropus taeniostictus	CAS 254908	KY224248	KY224347		
Sphaerodactylus nicholsi	CAS 198444	KU158020	EF534786	EF534826	KU157415
Sphaerodactylus roosevelti	CAS 198428	JN393943	EF534785	EF534825	
Sphaerodactylus torrei	JB 34	JX440519	EF534788	EF534829	KU157416
Teratoscincus microlepis	JFBM 15				KU157417
Teratoscincus microlepis	TG 00074	JX041451	EF534800	EF534842	
Teratoscincus roborowskii	CAS 171203	AF114252			
Teratoscincus roborowskii	TG 00070		EF534799	EF534841	
Teratoscincus roborowskii	JFBM 14				KU157418
Teratoscincus scincus	CAS 228808				KU157419
Teratoscincus scincus	JFBM 14252	JX041454	EF534801	EF534843	
Trachylepis varia	MCZ-R 184873		GU931671		
Trachylepis varia	TNHC 68769	GU931603			GU931534
Trachylepis varia	ZFMK 68413			KC345241	
Woodworthia maculata	RAH 292	GU459852	GU459449	GU459651	
Woodworthia maculata	RAH 92				KU157432

Specimen ID codes are as follows: AMB (Aaron M. Bauer field collection); AMS (Australian Museum, Sydney); ANG_WC (Werner Conradie field collection); CAS (California Acedemy of Sciences); CENPAT (Centro Nacional Patagónico, Puerto Madryn); DGB (Donald G. Broadley field collection); ERP (Eric R. Pianka field collection); JB (Jon Boone tissue collection); JFBM (James Ford Bell Museum of Natural History, University of Minnesota); JV (Jens Vindum field collection); LSUMZ (Louisiana State University Museum of Zoology); MBUR (Marius Burger field collection); MCZ (Museum of Comparative Zoology, Harvard University); MVZ (Museum of Vertebrate Zoology, University of California); MZUSP (Museum of Zoology, University of Sao Paulo); NMB (National Museum, Bloemfontein); NMNW (National Museum of Namibia, Windhoek); NMZB (National Museum of Zimbabwe, Bulawayo); PEM (Port Elizabeth Museum); RAH (Rod A. Hitchmough tissue collection); SAM/SAMA (South Australian Museum); TG (Tony Gamble tissue collection); TNHC (Texas Natural History Collection); ZFMK (Zoologisches Forschungsmuseum Alexander Koenig)

The BI analysis was implemented in BEAST 1.8.2 [47], using a Yule tree prior and uncorrelated lognormal relaxed clock. Based on the PartitionFinder results, the data were divided into ten partitions employing six distinct models: ND2 position 1, ND2 position 2, and tRNAs used the GTR + I + Γ model. ND2 position 3 used the GTR + Γ model. RAG1 position 1 + 2 used the TrN (Tamura-Nei) + I + Γ model. RAG1 position 3 and KIF24 position 1 + 2 used the HKY (Hasegawa-Kishino-Yano) + Γ model. PDC position 3 and KIF24 position 3 used the K80 (Kishino 1980) + Γ model. PDC position 1 + 2 used the TrNef + I + Γ model. Four replicate analyses were run for 50 million generations, sampled every 1000 generations. The first 5 million generations were discarded as burn-in. Effective sample sizes were estimated in Tracer 1.5 (>300 for all parameters in each run) to confirm the chain length was adequate.

BEAST 1.8.2 was also used to estimate divergence times simultaneously with phylogenetic relationships. The root prior (Lepidosauria-Archosauria divergence) was given a normal distribution (mean = 275 Ma [million years ago], SD = 15) encompassing the range of estimates for this divergence [48, 49]. Five constraints were also applied to internal nodes: most recent common ancestor (MRCA) of *Phelsuma rosagularis* and *P. inexpectata* (uniform prior; 0–8 Ma; [37]). MRCA of sampled *Sphaerodactylus* – *S. ocoae*, *S. roosevelti*, and *S. torrei* (exponential prior; mean = 3, offset = 15 Ma; [50, 51]). MRCA of *Woodworthia maculata* and *Oedura marmorata* (exponential prior; mean = 17, offset = 16; [52]). MRCA of *Ophidocephalus taeniatus* and *Pygopus nigriceps* (exponential prior; mean = 10, offset = 20; [53, 54]). MRCA of *Teratoscincus roborowskii* and *T. scincus* (exponential prior; mean = 3, offset = 10; [55]).

Trait data

Body size and habitat preference data were assigned to each species based on the authors' observations of specimens in the wild (62 of 66 ingroup species have been observed in-situ by the authors), supplemented by examination of vouchered museum specimens and information obtained from the literature [29, 56–58]. Maximum body size was treated in two ways depending on analysis. When possible, SVL (snout-vent length) was treated as a continuous character and the log-transformed maximum SVL was used. However, when treatment of size as a continuous character was not computationally feasible we instead treated size as a binary character. In *Pachydactylus* and related genera SVL is bimodal (Fig. 1). Those species with a maximum snout–vent length (SVL) < 70 mm comprised the "small" category, and those with a maximum SVL >75 mm comprised the "large" category. Habitat preference was divided into three categories. Those species that primarily shelter in burrows or under surface debris (logs, loose stones, aloe leaves, etc.), and forage actively on the ground, were classified as "terrestrial." Species that primarily shelter in rock cracks and forage on cliff faces or boulders were classified as "rupicolous." Finally, unspecialized species that both shelter and forage on a variety of surfaces (rock faces, tree trunks, buildings, etc.) were classified as "generalist climbers."

Range size estimates

Extent of occurrence (EOO) and area of occupancy (AOO) were defined as per the current International Union for the Conservation of Nature (IUCN) standards [1]. EOO was calculated as the area of the minimum convex polygon enclosing distribution records for each respective taxon. AOO was initially calculated as the sum of the total area of the quarter degree grid squares within which at least one record occurs. The final AOO was adjusted to an estimate of the actual suitable habitat within the occupied quarter degree squares based on the literature and the authors' field knowledge of each species. For all endemic South African species and for most species with a portion of their distribution occurring in South Africa, EOO and AOO values were previously estimated as part of the red list evaluation carried out in association with the Atlas and Red List of the Reptiles of South Africa, Lesotho and Swaziland [58]. Calculated EOO usually provides the broadest possible interpretation of the space used by a species, whereas the AOO represents a quite conservative estimate. However, for taxa known from single localities or several localities that are very close to one another, AOO as calculated above may yield a greater area than EOO. We used EOO or AOO, which ever was the greater, as our estimates of species' ranges. These values were log transformed when used in analyses.

Phylogenetic comparative methods

We performed a variety of comparative analyses to investigate the relationship among phylogeny, divergence times, trait data, and range sizes. All comparative analyses were completed in replicate on both the BEAST maximum clade-credibility tree and on 1000 post-burnin trees randomly sampled from the BEAST posterior distribution. These trees were pruned to remove outgroups (for which we have incomplete taxon sampling and no trait data). The package Phytools [59], implemented in R 3.2.2 [60] was used compute phylogenetic signal of range size using both the K and λ statistics [61, 62]. We also used phytools to plot lineages through time and test for constancy of lineage accumulation through time using the γ statistic of Pybus and Harvey [63].

State dependent diversification of range size on trait data (SVL or habitat) were tested using OUwie [64], which allows for tests of correlation between a multistate vs. continuous trait in a phylogenetic context. Because body size evolution and habitat choice may be coupled [65], we also tested for auto-correlation between these two traits, for a total of three analyses: (1) SVL vs. range size, (2) habitat vs. range size, and (3) SVL vs. habitat. SVL was treated as a multistate (binary) trait in test (1), but was treated as a continuous trait in test (3) to facilitate analysis. In all three cases, we tested the hypothesis that the optimum continuous trait value, θ, differed depending on the identity of the multistate trait value, i.e. whether large- and small-bodied species differ in range

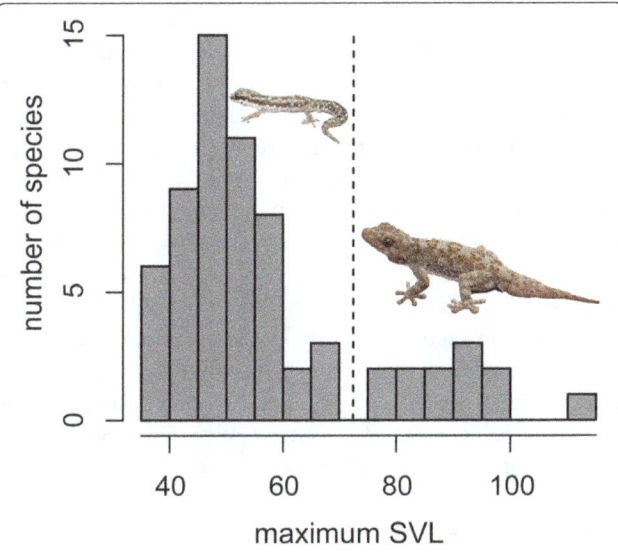

Fig. 1 Maximum snout–vent length for species in the genera *Pachydactylus*, *Chondrodactylus*, *Colopus*, and *Elasmodactylus*. Values are based on literature sources along with observations of field-collected and museum-preserved specimens. The small-bodied species *P. geitje* and large-bodied *P. namaquensis* are illustrated

size (test 1), whether species differing in habitat use differ in range size (test 2), or whether species differing in habitat use differ in body size (test 3). We performed these tests by estimating ancestral states of the multi-state character on the phylogeny, and then fitting values of θ (trait optimum), α (pull toward optimum), and σ (rate of change of trait) for the continuous trait under two model regimes. The null Brownian Motion (BM) model regime estimated single values of θ, α, and σ that did not depend on the state of the multistate character. This was tested against a more complex Ornstein-Uhlenbeck (OU) model in which there were multiple θ parameters, one per multistate character state. We used ΔAICc (corrected Akaike Information Criterion) values to identify which model provided a better fit for the data. All three tests were performed on 1000 trees randomly sampled from the post-burnin BEAST posterior distribution, and on each of these 1000 trees we performed 100 ancestral reconstructions of the multistate trait using stochastic character mapping [66] implemented in phytools, resulting in a total of 100,000 model fits per test, each with a unique combination of phylogeny and ancestral state estimate.

We also estimated trait-associated rates of speciation (λ), extinction (μ), and transition rate (q) under a BiSSE [28] model for SVL data or a multiple-state speciation and extinction (MuSSE) [67] model for habitat data implemented in Diversitree. Because hypothesis testing in a BiSSE or MuSSE framework can have a high Type I error rate [68–70] and low statistical power when data sets contain fewer than several hundred terminal taxa [71], we refrain from explicitly testing the statistical significance of character-associated variation in model parameter estimates. Instead, we fit models in which each trait was given individual λ, μ, and q parameters strictly to determine estimates of these model parameters. Model fitting was performed in an Markov chain Monte Carlo (MCMC) framework with runs lasting 1100 generations and the first 100 discarded as burn-in. These estimates were obtained for each of 1000 trees randomly sampled from the BEAST posterior distribution, resulting in each parameter estimate being obtained from 1,000,000 observations.

Results
Phylogeny and divergence times
The phylogenies estimated in both the ML and BI analyses are very similar (Fig. 2), and most branches receive strong support. As expected, the grouping of *Pachydactylus*, *Chondrodactylus*, *Colopus*, and *Elasmodactylus* is monophyletic (BI/ML support values 1.0/97) and these are in turn most closely related to *Rhoptropus* (support values 100/1.0). Within the ingroup, the topology resembles that estimated by Bauer and Lamb [14], which

included 26 fewer ingroup taxa and was estimated from ~1,600 fewer nucleotide sites, but there are some notable differences. Most notably, both the genera *Colopus* and *Elasmodactylus* are recovered as non-monophyletic. One species of *Colopus*, *C. kochi*, is embedded in *Pachydactylus* and is most closely related to the *Pachydactylus mariquensis* group (support values 1.0/93), a set of four species represented by a single taxon in [14]. The other *Colopus* species, *C. wahlbergii*, is also embedded in *Pachydactylus*, but there is not strong support for any set of *Pachydactylus* species being its closest relatives (support values 0.76/54), although there is strong support for its association with *Pachydactylus* to the exclusion of *Chondrodactylus* and *Elasmodactylus* (100/1.0). The two *Elasmodactylus* species are outside a group containing all *Pachydactylus*, *Colopus*, and *Chondrodactylus* species, with *E. tuberculosus* being more closely related, but with poor support (0.37/46). Within *Pachydactylus*, recognized species groups [14, 15, 35, 72–74] are recovered as monophyletic with strong support as are many of the species-level relationships within these groups. However, within the speciose *serval/weberi* and northwestern groups, in which many new taxa have been added, species relationships are more highly modified. In the first of these, the basal division into reciprocally monophyletic *serval* and *weberi* groups is not supported, and the former makes the latter paraphyletic. Relationships among species groups in *Pachydactylus* remain unresolved, with most groups connected by exceptionally short internodes. There are two exceptions. The *serval/weberi* group and *capensis* group are closest relatives, as are the *geitje* and *rugosus* groups.

The divergences between *Rhoptropus* and *Pachydactylus* + *Chondrodactylus* + *Colopus* + *Elasmodactylus* occurred in the early Cenozoic (66–43 Ma). This is a similar pattern as observed in other gekkonids, in which relatively species-rich regional radiations undergo initial diversification in the early Cenozoic (e.g. [5, 37, 75]). The short internodes connecting *Pachydactylus* species groups are indicative of a relatively high diversification rate in the mid-Cenozoic ~30–35 Ma. The lineage through time (LTT) plot shows that the rate of lineage accumulation remains steady or slowly increases to this point after which there is a noticeable decline (Fig. 3). The overall trend is of significantly decreasing lineage accumulation through time (mean γ value = −5.8, p < 1 x 10^{-5} for all 1000 sampled trees).

Comparative analyses
Ancestral reconstruction of body size in the *Pachydactylus* group suggests that being large-bodied is ancestral for the group (Fig. 4). A shift to small body size occurred once early in the evolutionary history of the group, and there have been only two reversals. Reconstruction of

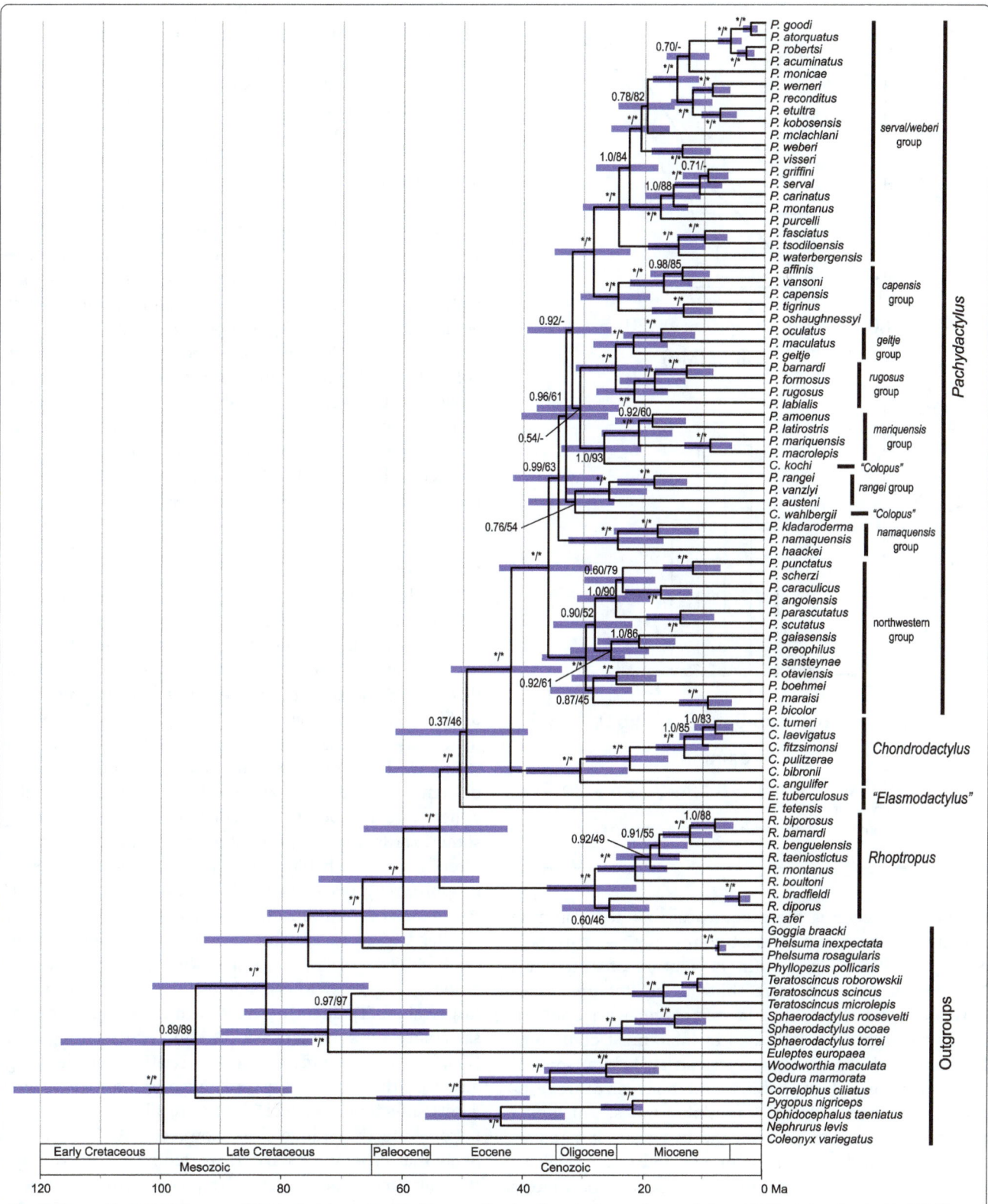

Fig. 2 Time-calibrated phylogeny of *Pachydactylus* and related genera. The topology is the maximum clade credibility tree estimated in BEAST with non-gekkotan outgroups cropped for clarity. Support values (Bayesian posterior probabilities/ML bootstrap) are given at nodes; asterisks indicate nodes with Bayesian support values = 1.0 and ML bootstrap values > 95. Named species groups and genera are given to the right. Geologic epochs and eras are indicated on the timescale; post-Miocene epochs (Pliocene, Pleistocene, Holocene) are not labeled

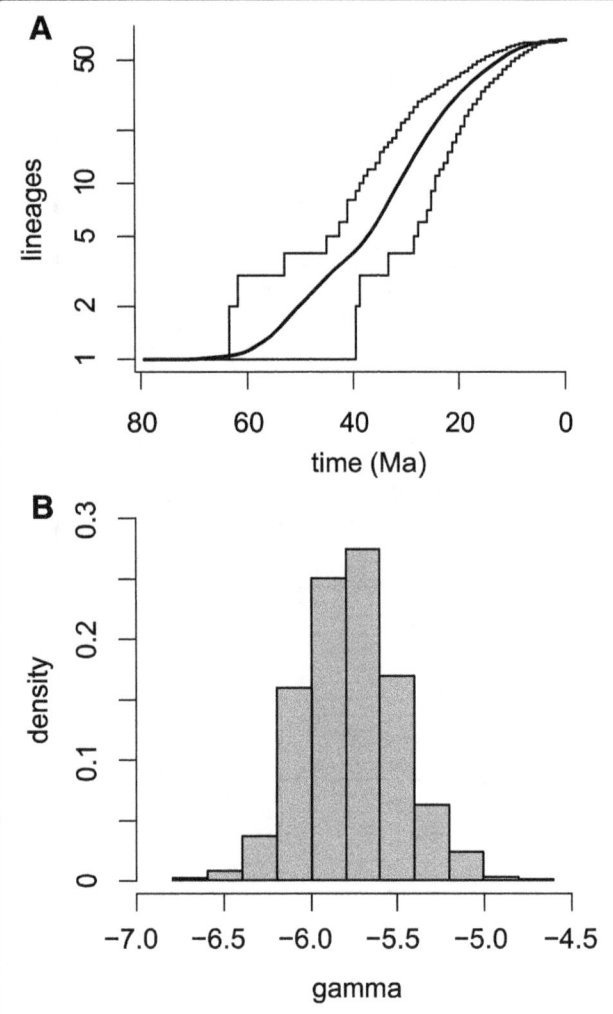

Fig. 3 a Lineage accumulation in the *Pachydactylus* group. The plot depicts LTT curves for 1000 trees randomly sampled from the BEAST posterior distribution. **b** Histogram of γ statistic estimates for the 1000 LTT curves depicted in (**a**)

Pagel's λ ($\lambda = 0.46$, p = 0.13), but the estimate of K is slightly non-significant ($K = 0.57$, p = 0.077). The estimated global optimum range size (θ) for small-bodied species is $10^{4.5}$ km^2, approximately one order of magnitude smaller than large-bodied species ($\theta = 10^{5.4}$ km^2). When comparing habitat preference, rock-dwelling species have the smallest estimated global optimum extent of occurrence ($\theta = 10^{4.1}$ km^2), followed by terrestrial species ($\theta = 10^{5.1}$ km^2) with generalized climbers having the largest geographic ranges ($\theta = 10^{5.8}$ km^2). Trait-associated estimates of speciation and extinction rates are less variable (Fig. 6). Small-bodied species are estimated to have slightly higher speciation (mean λ[small-bodied] = 0.055; mean λ[large-bodied] = 0.040) and lower extinction rates, but there is extensive overlap. Habitat-associated estimates of diversification rate also overlap, especially between terrestrial species and generalized climbers, although rock-dwelling species are estimated to have speciated at somewhat higher rates (mean λ[generalized climber] = 0.032; mean λ[terrestrial] = 0.012; mean λ[rock-dwelling] = 0.065).

Discussion

While the heritability of range size has been demonstrated for many lineages, possible mechanistic explanations have varied, and include niche breadth [27], dispersal ability [76, 77], and morphological characteristics [26, 78] of lineages, as well as the geographic limits of biomes, landmasses, or hydrological basins [79]. In many cases, these factors may be interlinked. In this study, we focus on two traits, body size and habitat requirements, that were expected to affect dispersal ability either directly because smaller organisms, including some lizards, may disperse shorter distances [80], or indirectly, because habitat patchiness can restrict dispersal if appropriate dispersal corridors are not available [81]. As expected, within the *Pachydactylus* group the smaller-bodied species occupying more patchily distributed habitats are the species with the smallest geographic ranges. Other studies that have measured dispersal ability directly have shown that reduced dispersal ability does not always lead to reduced range size [77], but in *Pachydactylus* and its relatives our data suggest that dispersal ability and range size are correlated. Traits affecting dispersal ability are likely not the only factors affecting range size, however. Minimally, it is likely that geographic barriers, including major river systems and mountain ranges, also play a significant role in restricting the ranges occupied by individual species. For example, the species *P. austeni* and *P. goodi* are known only from south of the Orange River even though suitable habitats for each of these species also exist to the north [58].

habitat preference is more equivocal, but the common ancestor of the group is most commonly reconstructed as a generalized climber (in 80% of reconstructions). What is clear is that more shifts in habitat preference have occurred than shifts in body size, with approximately 26 transitions indicated in total, most commonly between rock-dwelling and terrestrial habitat preferences. Although both habitat and body size are estimated to have shifted multiple times, including reversals, correlation between the two traits is not particularly strong based on fits of BM and OU models — out of 100,000 model fits, the BM model incorporating only a single global SVL optimum was favored according to the AIC 31% of the time (Fig. 5a).

In contrast, both body size and habitat preference are strongly correlated with range size (Fig. 5b, c). Range size displays significant phylogenetic signal based on

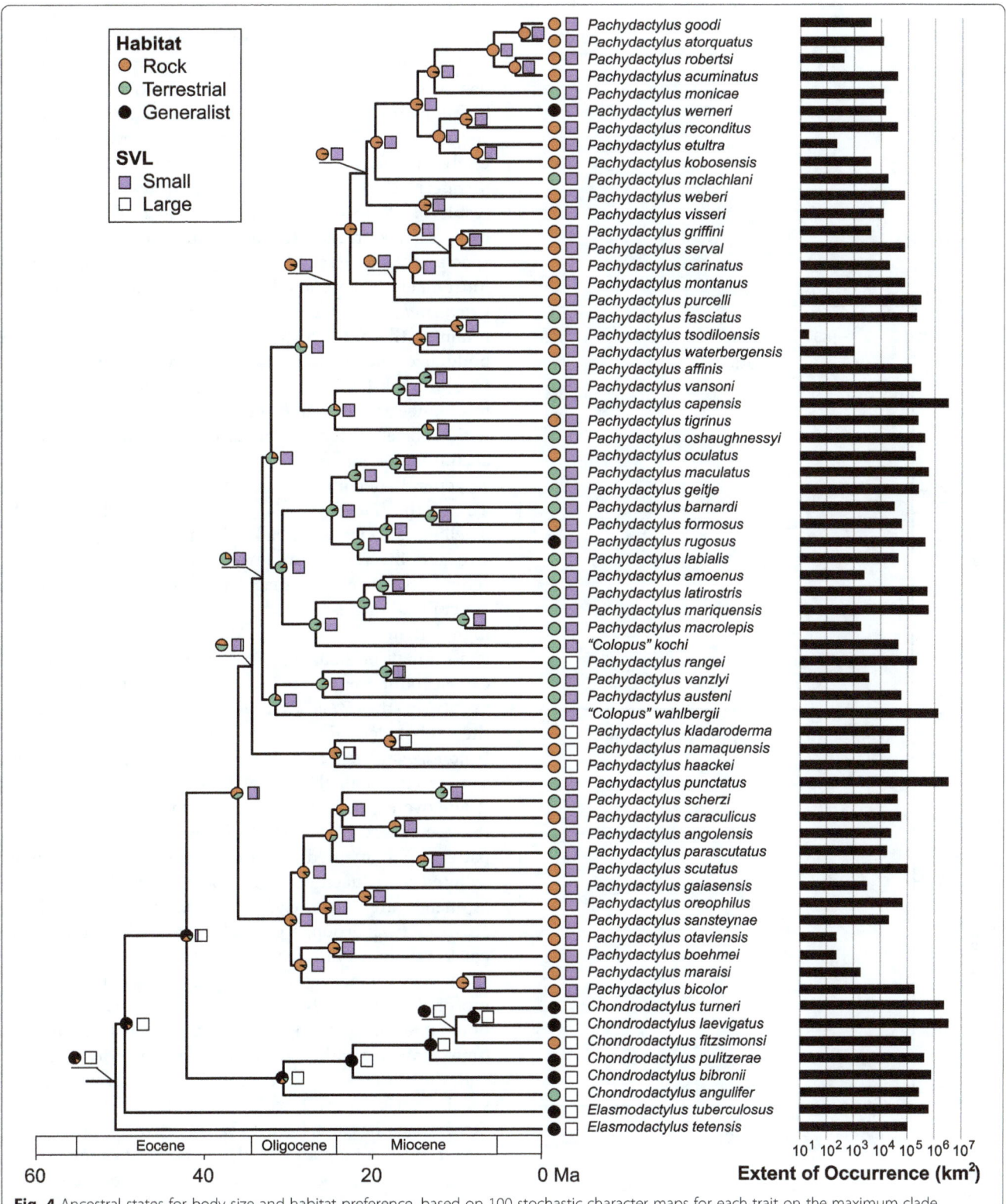

Fig. 4 Ancestral states for body size and habitat preference, based on 100 stochastic character maps for each trait on the maximum clade credibility tree. Range size values for each species are given to the right of each terminal branch

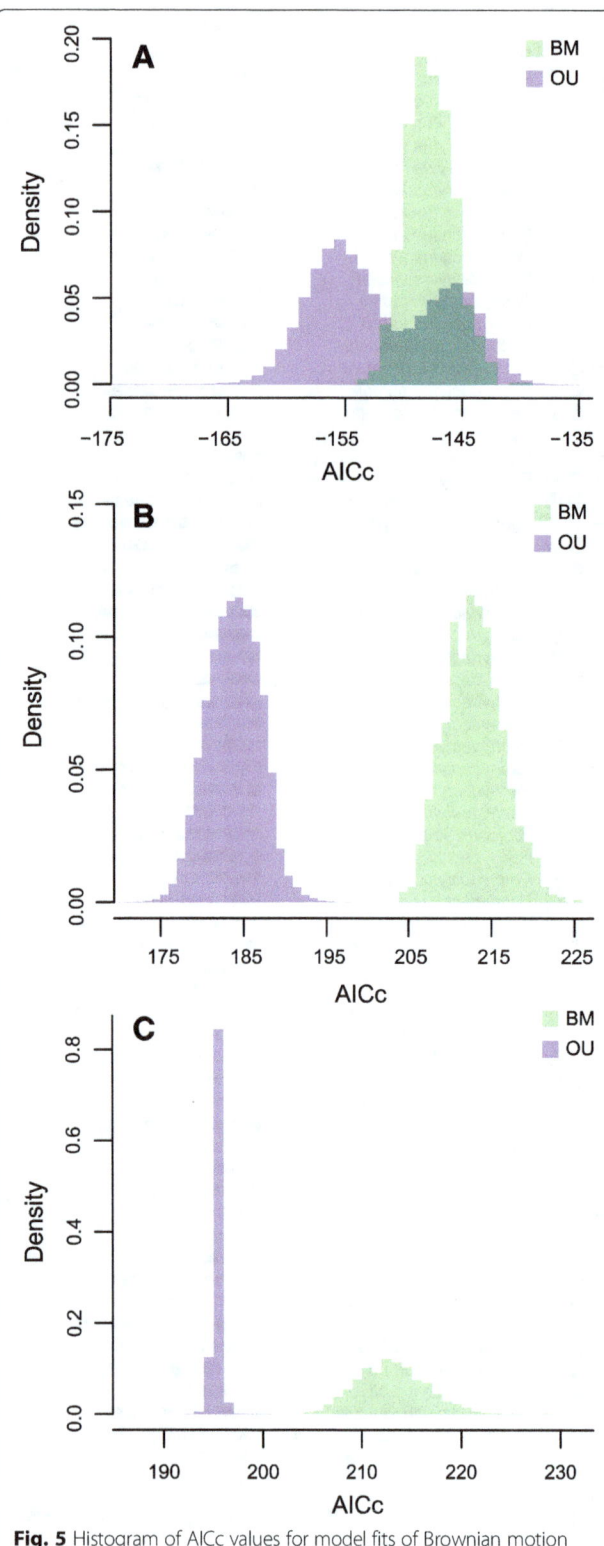

Fig. 5 Histogram of AICc values for model fits of Brownian motion (BM) and Ornstein-Uhlenbeck (OU) models of trait diversification estimated in OUwie. **a** habitat preference vs. SVL. **b** habitat preference vs. range size. **c** SVL vs. range size

Taken as a whole, the observed patterns of trait evolution, range size, and diversification are consistent with an evolutionary scenario in which diversification has been dominated by geographic isolation followed by allopatric speciation. Based on our analyses, we suggest that geographic isolation has developed more easily in *Pachydactylus* + *Colopus* than it has in *Elasmodactylus* or *Chondrodactylus*, at least partly as a result of *Pachydactylus* + *Colopus* species being more likely to have traits promoting this isolation. Ancestral species in the *Pachydactylus* group as a whole were most likely large-bodied habitat generalists, and most *Chondrodactylus* and *Elasmodactylus* species have retained these traits to the present. We infer small body size and habitat specialization (for either terrestrial or rock-dwelling lifestyles) to appear in the common ancestor of *Pachydactylus* + *Colopus*, coincident with a brief observed increase in the rate of lineage accumulation in the *Pachydactylus* group, followed by a general decline in diversification rate measured across the *Pachydactylus* group as a whole. Rock-dwelling species especially differ strikingly in range size and diversification rate, having extents of occurrence two orders of magnitude smaller than habitat generalists and estimated rates of diversification 2–4X higher than other species. Allopatric speciation of isolated small-bodied, rock-dwelling lineages therefore can account for much of the observed taxonomic diversity in the *Pachydactylus* group. Not surprisingly, the subclades that have retained ancestral traits (*Chondrodactylus* and *Elasmodactylus*) are much less species-rich than those that have not.

The overall decline in diversification rate through time that we observe in the *Pachydactylus* group is similar to patterns documented in many lineages that are often attributed to reduced ecological opportunity through time as niches are filled (e.g. [82–85]). In the case of the *Pachydactylus* group, a general pattern of morphological conservatism within species groups, exemplified by the small number of shifts in body size (Fig. 2) and digital morphology [4, 5] through time is in line with expectations if ecological opportunity has decreased through time. However, shifts in habitat use are more frequent, and the number of co-occurring *Pachydactylus* group species varies from 1 to 13, suggesting that ecological niche space has not been exhausted. An alternative explanation that may also partly explain the observed rate slowdown is a geographic model as described above. In clades dominated by allopatric speciation, diversification rates may decline as vicariance events affect fewer species as species' geographic ranges decline through time [86, 87]. The relatively low species diversity of *Chondrodactylus*, which includes mostly large-bodied habitat generalists (i.e., species with large geographic ranges),

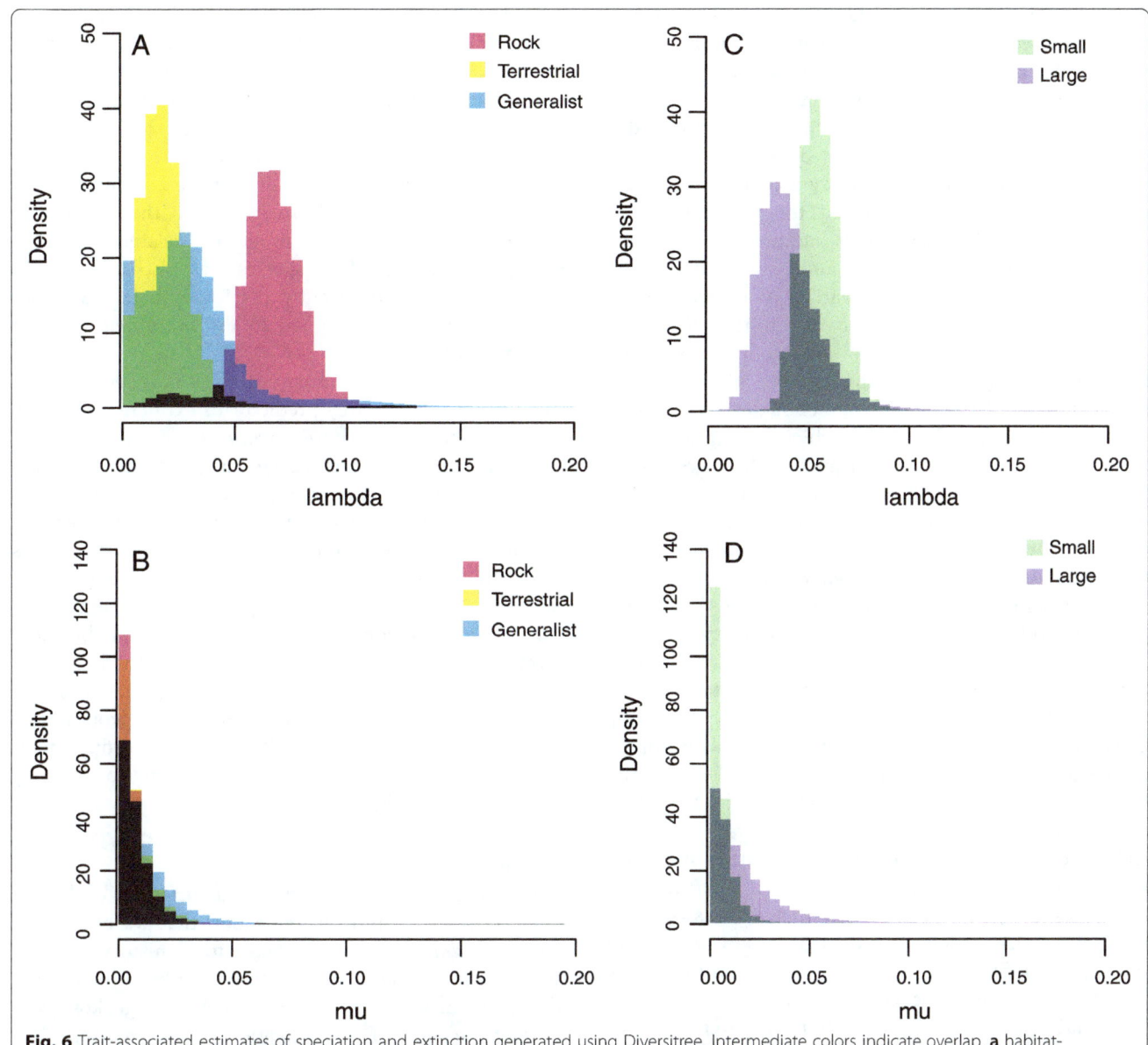

Fig. 6 Trait-associated estimates of speciation and extinction generated using Diversitree. Intermediate colors indicate overlap. **a** habitat-associated speciation rate. **b** habitat-associated extinction rate. **c** body size-associated speciation rate. **d** body-size associated extinction rate

compared to *Pachydactylus*, which includes mostly small-bodied habitat specialists, supports this model.

As indicated above, a jump in lineage accumulation coincident with the appearance of habitat-specialist clades in the mid-Cenozoic ~30–35 Ma is contrary to the general pattern of declining diversification rate through time. It is possible that climatic or geomorphic processes active at this time were especially favorable for isolating lineages, resulting in increased speciation. Major periods of tectonic uplift in eastern and southern Africa did not commence until approximately the Oligocene-Miocene boundary (23 Ma) [88–90], making large-scale geomorphological change incompatible with the observed rate

increase. However, a major climatic regime shift did occur at the Eocene-Oligocene boundary: a global cooling associated with Antarctic glaciation [91, 92]. In Africa, this shift resulted in aridification and greater environmental heterogeneity, including reduction in forest cover [93, 94], and would have greatly increased the available habitat for arid-adapted *Pachydactylus* group geckos, potentially facilitating rapid radiation. A similar pattern occurs in forest-adapted chameleons, where rapid radiation is coincident with wide availability of suitable habitat, in the case of chameleons during the Eocene [95].

In performing this study, we have attempted to minimize confounding factors and methological biases.

For example, we collected data for nearly all target taxa and integrated all comparative analyses across a sample of 1000 credible trees to avoid sampling or phylogenetic biases. Even so, our interpretations should be treated cautiously. The observed relationships between trait data and geographic range extent are based on correlation. While we chose to focus on body size and habitat use specifically because we expected them to affect geographic range, it is possible that one or more other factors co-varying with body size and habitat use are the actual drivers of range size variation among species. Trait-associated measurements of diversification rate utilized the BiSSE model. Even though our ingroup phylogeny was comprehensive, the number of taxa in our data set may not have been large enough to avoid inadequacies of the model [69, 71], which is why we refrain from ascribing statistical significance to these results. Alternate methods of trait-dependant diversification (e.g. [70]) likewise are best suited to larger data sets. One possible way to increase data set size is to incorporate taxa from across the Afro-Malagasy clade of geckos, but the interrelationships of genera within this large radiation are still poorly resolved and relevant trait data are missing for many species. Finally, range size estimates are based on known collection localities and a correct interpretation of species-level taxonomy in the group. Collecting effort varies greatly by country, with, for example, less than 10,000 amphibian and reptile collection records in Angola, 38,000 in Namibia, and >100,000 in South Africa [58, 96]. Some species also vary phenotypically and have named subspecies that further study may reveal to warrant specific status (e.g., *Pachydactylus punctatus*; [97]). However, given that trait-associations with range size varied by orders of magnitude, we do not expect refinement of species' range limits or taxonomy to strongly influence our results.

Beyond interpretation of evolutionary patterns, the results of this study also have significant implications for taxonomy and conservation. The phylogenetic results indicate that *Elasmodactylus* and *Colopus* are not monophyletic, and both species of *Colopus* are nested in *Pachydactylus*. Although we recover *Elasmodactylus* as non-monophyletic, its monophyly cannot be wholly discounted given the poor support for the node joining *E. tuberculosus* with *Chondrodactylus* + *Colopus* + *Pachydactylus*. Performing a Shimodaira-Hasegawa (SH) test also shows that the likelihood of our best-scoring tree (lnL −97035.168404) is not significantly higher than the likelihood of a tree in which *Elasmodactylus* is constrained to be monophyletic (lnL −97036.513963; *p* > 0.05). These species also share morphological traits rare or absent in other *Pachydactylus* group species, including preanal pores and easily broken skin [14]. Thus, we suggest that taxonomic decisions regarding these species

be delayed until each species' phylogenetic position is better established. We refer both *Colopus* species to *Pachydactylus*. *Colopus wahlbergii* is morphologically divergent and in our analyses its position in *Pachydactylus* is equivocal, with moderate support for an association with the *rangei* group. Thus, we refer it to no species group within *Pachydactylus*. In contrast, *Colophs kochi* is deeply nested in *Pachydactylus*, and there is strong support for its placement as closely related to the *P. mariquensis* group. This species was also included in *Pachydactylus* until recently [14]. We therefore advise that *C. kochi* be re-assigned to the *mariquensis* group within *Pachydactylus*.

One important determinant of rarity is range size, and small range size is a key predictor of extinction risk [98, 99]. Thus, species in the *Pachydactylus* group inheriting traits promoting smaller ranges also inherit traits promoting greater rarity. However, our analyses show these same traits to be associated with higher rates of diversification. Given the difficulty in estimating extinction rates from phylogenies [100], it is unclear if this higher diversification rate is observed despite a higher extinction rate, or if extinction rates do not depend on the measured traits in the *Pachydactylus* group. Notwithstanding this difficulty, these results stress the importance of defining a frame of reference when measuring evolutionary "success." In the case of the *Pachydactylus* group, more widespread, common species tend to belong to relatively species-poor subclades.

Conclusions

The relationships among morphological and ecological traits, range size, and diversification that we observe in the *Pachydactylus* group points to a history of geographic isolation contributing significantly to the group's species richness compared to other African geckos. Even so, some aspects of diversification in the *Pachydactylus* group, including early evolution of divergent traits within the group, are consistent with patterns observed in classic adaptive radiations. In this sense, the process of diversification of *Pachydactylus* group geckos may be considered intermediate between a true adaptive radiation on one hand and a non-adaptive radiation (as observed in plethodontid salamanders; [101, 102]) on the other. It is likely that many other species-rich groups share this same intermediate pattern.

Abbreviations
AICc: Corrected Akaike information criterion; AOO: Area of occupancy; BI: Bayesian inference; BiSSE: Binary-state speciation and extinction; BM: Brownian motion; bp: Base pairs; EOO: Extent of occurrence; GTR: General time reversible model; HKY: Hasegawa-Kishino-Yano model; IUCN: International union for the conservation of nature; K80: Kimura 1980 model; KIF24: Kinesin family member 24; LTT: Lineages through time; Ma: Million years ago; MCMC: Markov chain Monte Carlo; ML: Maximum likelihood; MRCA: Most recent common ancestor; MuSSE: Multiple-state

speciation and extinction; ND2: NADH dehydrogenase subunit 2; OU: Ornstein-Uhlenbeck; PCR: Polymerase chain reaction; PDC: Phosducin; RAG1: Recombination activating gene 1; SH: Shimodaira-Hasegawa; SVL: Snout-vent length; TrN: Tamura-Nei model; tRNA: Transfer RNA

Acknowledgements
We thank the late Donald G. Broadley, Mirko Barts, Jon Boone, William R. Branch, Marius Burger, and Krystal Tolley for providing access to some specimens used in the study. Johan Marais, Bill Branch, Werner Conradie, Trip Lamb, Don Broadley, Randy Babb, Paul Moler, Ross Sadlier, Glenn Shea, Jens Vindum, and many Villanova University Masters students participated in collecting.

Funding
This project was funded by the University of Michigan (MPH), NSF DEB-0515909 (AMB, TRJ), DEB-0844523 (AMB, TRJ), DEB-1019443 (AMB), DEB-1556255 (AMB), DEB-1556585 (MPH), and EF-1241885 (subaward 13–0632) (AMB). Funding bodies played no role in study design or implementation.

Authors' contributions
MPH and AMB designed the project. Data were collected by MPH, TRJ, and AMB. Data analysis was performed by MPH. MPH wrote the paper with input from TRJ and AMB. All authors read and approved the final manuscript.

Competing interests
The authors declare that they have no competing interests.

Author details
[1]Department of Natural Sciences, University of Michigan-Dearborn, 4901 Evergreen Rd., Dearborn, MI 48128, USA. [2]Department of Biology, Villanova University, 800 Lancaster Avenue, Villanova, PA 19085, USA.

References
1. IUCN. The IUCN Red List of Threatened Species. Version 2015–4. http://www.iucnredlist.org. Accessed 20 Mar 2016.
2. Froese R, Pauly D. FishBase, version (10/2015). http://www.fishbase.org. Accessed 20 Mar 2016.
3. Hedges, SB. CaribHerp: West Indian Amphibians and Reptiles. http://www.caribherp.org. Accessed 20 Mar 2016.
4. Lamb T, Bauer AM. Footprints in the sand: independent reduction of subdigital lamellae in the Namib–Kalahari burrowing geckos. Proc R Soc Lond B Biol Sci. 2006;273:855–64.
5. Gamble T, Greenbaum E, Jackman TR, Russell AP, Bauer AM. Repeated origin and loss of adhesive toepads in geckos. PLoS One. 2012;7:e39429.
6. Bauer AM, Branch WR. The herpetofauna of the Richtersveld national park and the adjacent northern Richtersveld, Northern cape Province, Republic of South Africa. Herpetological Nat Hist. 2001;8:111–60.
7. Schluter D. The ecology of adaptive radiation. Oxford: Oxford University Press; 2000.
8. Losos JB, Mahler DL. Adaptive radiation: the interaction of ecological opportunity, adaptation, and speciation. In: Bell MA, Futuyma DJ, Eanes WF, Levinton JS, editors. Evolution since Darwin: the first 150 years. Sunderland: Sinauer; 2010. p. 381–420.
9. Alfaro ME, Brock CD, Banbury BL, Wainwright PC. Does evolutionary innovation in pharyngeal jaws lead to rapid lineage diversification in labrid fishes? BMC Evol Biol. 2009;9:1.
10. Rutschmann S, Matschiner M, Damerau M, Muschick M, Lehmann MF, Hanel R, Salzburger W. Parallel ecological diversification in Antarctic notothenioid fishes as evidence for adaptive radiation. Mol Ecol. 2011;20:4707–21.
11. Deutsch JC. Colour diversification in Malawi cichlids: evidence for adaptation, reinforcement or sexual selection? Biol J Linn Soc. 1997;62:1–4.
12. Masta SE, Maddison WP. Sexual selection driving diversification in jumping spiders. Proc Natl Acad Sci. 2002;99:4442–7.
13. Wiens JJ, Graham CH, Moen DS, Smith SA, Reeder TW. Evolutionary and ecological causes of the latitudinal diversity gradient in hylid frogs: treefrog trees unearth the roots of high tropical diversity. Am Nat. 2006;168:579–96.
14. Bauer AM, Lamb T. Phylogenetic relationships of southern African geckos in the Pachydactylus group (Squamata: Gekkonidae). Afr J Herpetol. 2005;54:105–29.
15. Bauer AM, Heinicke MP, Jackman TR, Branch WR. Systematics of the Pachydactylus mariquensis group of geckos (Reptilia: Squamata: Gekkonidae): Status of P. mariquensis latirostris, P. m. macrolepis and P. amoenus. Navorsinge van die Nasionale Museum Bloemfontein. 2011;27:85–107.
16. Arnold EN. Historical changes in the ecology and behaviour of semaphore geckos (Pristurus, Gekkonidae) and their relatives. J Zool. 1993;229:353–84.
17. Ikeuchi I, Mori A, Hasegawa M. Natural history of Phelsuma madagascariensis kochi from a dry forest in Madagascar. Amphibia-Reptilia. 2005;26:475–83.
18. Regalado R. Does dichromatism variation affect sex recognition in dwarf geckos? Ethol Ecol Evol. 2015;27:56–73.
19. Coyne JA. Genetics and speciation. Nature. 1992;355:511–5.
20. Coyne JA, Orr HA. Speciation. Sunderland: Sinauer Associates; 2004.
21. Jablonski D. Heritability at the species level: analysis of geographic ranges of Cretaceous mollusks. Science. 1987;238:360–3.
22. Gaston KJ. Species-range size distributions: products of speciation, extinction and transformation. Philos Transact A Math Phys Eng Sci. 1998; 353:219–30.
23. Webb TJ, Gaston KJ. On the heritability of geographic range sizes. Am Nat. 2003;161:553–66.
24. Waldron A. Null models of geographic range size evolution reaffirm its heritability. Am Nat. 2007;170:221–31.
25. Vamosi SM, Vamosi JC. Perspective: Causes and consequences of range size variation: the influence of traits, speciation, and extinction. Front Biogeography. 2012;4:168–77.
26. Lee MS, Skinner A, Camacho A. The relationship between limb reduction, body elongation and geographical range in lizards (Lerista, Scincidae). J Biogeogr. 2013;40:1290–7.
27. Slatyer RA, Hirst M, Sexton JP. Niche breadth predicts geographical range size: a general ecological pattern. Ecol Lett. 2013;16:1104–14.
28. Maddison WP, Midford PE, Otto SP. Estimating a binary character's effect on speciation and extinction. Syst Biol. 2007;56:701–10.
29. Branch WR. Field guide to snakes and other reptiles of southern Africa. 3rd ed. Struik: Cape Town; 1998.
30. Gaston KJ, Blackburn TM. Range size-body size relationships: evidence of scale dependence. Oikos. 1996;479–85.
31. Lancaster N. Late Quaternary paleoenvironments in the southwestern Kalahari. Palaeogeogr Palaeoclimatol Palaeoecol. 1989;70:367–76.
32. Partridge TC. The evidence for Cainozoic aridification in southern Africa. Quat Int. 1993;17:105–10.
33. Thomas DS, Shaw PA. The evolution and characteristics of the Kalahari, southern Africa. J Arid Environ. 1993;25:97–108.
34. Bauer AM. Evolutionary scenarios in the Pachydactylus group geckos of southern Africa: new hypotheses. Afr J Herpetol. 1999;48:53–62.
35. Lamb T, Bauer AM. Phylogenetic relationships of the large-bodied members of the African lizard genus Pachydactylus (Reptilia: Gekkonidae). Copeia. 2002;2002:586–96.
36. Bauer AM. Phylogeny and biogeography of the geckos of southern Africa and the islands of the western Indian Ocean: a preliminary analysis. In: Peters G, Hutterer R, editors. Vertebrates in the Tropics. Bonn: Zoologisches Forschungsinstitut und Museum A. Koenig; 1990. p. 274–84.
37. Heinicke MP, Greenbaum E, Jackman TR, Bauer AM. Phylogeny of a trans-Wallacean radiation (Squamata, Gekkonidae, Gehyra) supports a single early colonization of Australia. Zool Scr. 2011;40:584–602.
38. Heinicke MP, Greenbaum E, Jackman TR, Bauer AM. Evolution of gliding in Southeast Asian geckos and other vertebrates is temporally congruent with dipterocarp forest development. Biol Lett. 2012;8:994–7.
39. Heinicke MP, Daza JD, Greenbaum E, Jackman TR, Bauer AM. Phylogeny, taxonomy and biogeography of a circum-Indian Ocean clade of leaf-toed geckos (Reptilia: Gekkota), with a description of two new genera. Syst Biodivers. 2014;12:23–42.
40. Portik DM, Bauer AM, Jackman TR. The phylogenetic affinities of Trachylepis sulcata nigra and the intraspecific evolution of coastal melanism in the western rock skink. Afr Zool. 2010;45:147–59.
41. Hall TA. BioEdit: a user-friendly biological sequence alignment editor and analysis program for Windows 95/98/NT. Nucleic Acids Symp Ser. 1999;41:95–8.

42. Kearse M, Moir R, Wilson A, Stones-Havas S, Cheung M, Sturrock S, Buxton S, Cooper A, Markowitz S, Duran C, Thierer T. Geneious Basic: an integrated and extendable desktop software platform for the organization and analysis of sequence data. Bioinformatics. 2012;28:1647–9.

43. Larkin MA, Blackshields G, Brown NP, Chenna R, McGettigan PA, McWilliam H, Valentin F, Wallace IM, Wilm A, Lopez R, Thompson JD. Clustal W and Clustal X version 2.0. Bioinformatics. 2007;23:2947–8.

44. Laslett D, Canbäck B. ARWEN, a program to detect tRNA genes in metazoan mitochondrial nucleotide sequences. Bioinformatics. 2008;24:172–5.

45. Lanfear R, Calcott B, Ho SY, Guindon S. PartitionFinder: combined selection of partitioning schemes and substitution models for phylogenetic analyses. Mol Biol Evol. 2012;29:1695–701.

46. Stamatakis A. RAxML version 8: a tool for phylogenetic analysis and post-analysis of large phylogenies. Bioinformatics. 2014;30:1312–3.

47. Drummond AJ, Suchard MA, Xie D, Rambaut A. Bayesian phylogenetics with BEAUti and the BEAST 1.7. Mol Biol Evol. 2012;29:1969–73.

48. Kumar S, Hedges SB. A molecular timescale for vertebrate evolution. Nature. 1998;392:917–20.

49. Reisz RR, Müller J. Molecular timescales and the fossil record: a paleontological perspective. TRENDS Genet. 2004;20:237–41.

50. Kluge AG. Cladistic relationships of sphaerodactyl lizards. Am Mus Novit. 1995;3139:1–23.

51. Iturralde-Vinent MA, MacPhee RD. Age and paleogeographical origin of Dominican amber. Science. 1996;273:1850.

52. Lee MS, Hutchinson MN, Worthy TH, Archer M, Tennyson AJ, Worthy JP, Scofield RP. Miocene skinks and geckos reveal long-term conservatism of New Zealand's lizard fauna. Biol Lett. 2009;5:833–7.

53. Hutchinson MN. The first fossil pygopodid (Squamata, Gekkota) and a review of mandibular variation in living species. Memoirs Queensland Museum. 1997;41:355–66.

54. Lee MS, Oliver PM, Hutchinson MN. Phylogenetic uncertainty and molecular clock calibrations: a case study of legless lizards (Pygopodidae, Gekkota). Mol Phylogenet Evol. 2009;50:661–6.

55. Macey JR, Wang Y, Ananjeva NB, Larson A, Papenfuss TJ. Vicariant patterns of fragmentation among gekkonid lizards of the genus Teratoscincus produced by the Indian collision: a molecular phylogenetic perspective and an area cladogram for Central Asia. Mol Phylogenet Evol. 1999;12:320–32.

56. Loveridge A. Revision of the African lizards of the family Gekkonidae. Bull Mus Comp Zool. 1947;98:1–469.

57. Alexander GJ, Marais J. A guide to the reptiles of southern Africa. Cape Town: Struik; 2007.

58. Bates MF, Branch WR, Bauer AM, Burger M, Marais J, Alexander GJ, De Villiers MS (eds). Atlas and red list of the reptiles of South Africa, Lesotho and Swaziland. Pretoria: South African National Biodiversity Institute; 2014.

59. Revell LJ. Phytools: an R package for phylogenetic comparative biology (and other things). Methods Ecol Evol. 2012;3:217–23.

60. R Core Team. R: A language and environment for statistical computing. Vienna, Austria: R Foundation for Statistical Computing; 2015.

61. Blomberg SP, Garland T, Ives AR. Testing for phylogenetic signal in comparative data: behavioral traits are more labile. Evolution. 2003;57:717–45.

62. Pagel M. Inferring the historical patterns of biological evolution. Nature. 1999;401:877–84.

63. Pybus OG, Harvey PH. Testing macro-evolutionary models using incomplete molecular phylogenies. Proc R Soc Lond B Biol Sci. 2000;267:2267–72.

64. Beaulieu JM, Jhwueng DC, Boettiger C, O'Meara BC. Modeling stabilizing selection: expanding the Ornstein–Uhlenbeck model of adaptive evolution. Evolution. 2012;66:2369–83.

65. Collar DC, Schulte II, James A, Losos JB. Evolution of extreme body size disparity in monitor lizards (Varanus). Evolution. 2011;65:2664–80.

66. Huelsenbeck JP, Nielsen R, Bollback JP. Stochastic mapping of morphological characters. Syst Biol. 2003;52:131–58.

67. FitzJohn RG. Quantitative traits and diversification. Syst Biol. 2010;59:619–33.

68. FitzJohn RG. Diversitree: comparative phylogenetic analyses of diversification in R. Methods Ecol Evol. 2012;3:1084–92.

69. Rabosky DL, Goldberg EE. Model inadequacy and mistaken inferences of trait-dependent speciation. Syst Biol. 2015;64:340–55.

70. Rabosky DL, Huang H. A robust semi-parametric test for detecting trait-dependent diversification. Syst Biol. 2016;65:181–93.

71. Davis MP, Midford PE, Maddison W. Exploring power and parameter estimation of the BiSSE method for analyzing species diversification. BMC Evol Biol. 2013;13:1.

72. Bauer AM, Lamb T. Phylogenetic relationships among members of the Pachydactylus capensis group of southern African geckos. Afr Zool. 2002;37: 209–20.

73. Bauer AM, Lamb T, Branch WR. A revision of the Pachydactylus serval and P. weberi groups (Reptilia: Gekkota: Gekkonidae) of Southern Africa, and with the description of eight new species. Proc Calif Acad Sci. 2006;57:595–709.

74. Lamb T, Bauer AM. Relationships of the Pachydactylus rugosus group of geckos (Reptilia: Squamata: Gekkonidae). Afr Zool. 2000;35:55–67.

75. Wood PL, Heinicke MP, Jackman TR, Bauer AM. Phylogeny of bent-toed geckos (Cyrtodactylus) reveals a west to east pattern of diversification. Mol Phylogenet Evol. 2012;65:992–1003.

76. Böhning-Gaese K, Caprano T, van Ewijk K, Veith M. Range size: disentangling current traits and phylogenetic and biogeographic factors. Am Nat. 2006; 167:555–67.

77. Lester SE, Ruttenberg BI, Gaines SD, Kinlan BP. The relationship between dispersal ability and geographic range size. Ecol Lett. 2007;10:745–58.

78. Olifiers N, Vieira MV, Grelle CE. Geographic range and body size in Neotropical marsupials. Glob Ecol Biogeogr. 2004;13:439–44.

79. Machac A, Zrzavý J, Storch D. Range size heritability in Carnivora is driven by geographic constraints. Am Nat. 2011;177:767–79.

80. Sinervo B, Calsbeek R, Comendant T, Both C, Adamopoulou C, Clobert J. Genetic and maternal determinants of effective dispersal: the effect of sire genotype and size at birth in side-blotched lizards. Am Nat. 2006;168:88–99.

81. Beier P, Noss RF. Do habitat corridors provide connectivity? Conserv Biol. 1998;12:1241–52.

82. Rabosky DL, Lovette IJ. Density-dependent diversification in North American wood warblers. Proc R Soc Lond B Biol Sci. 2008;275:2363–71.

83. Gavrilets S, Losos JB. Adaptive radiation: contrasting theory with data. Science. 2009;323:732–7.

84. Burbrink FT, Pyron RA. How does ecological opportunity influence rates of speciation, extinction, and morphological diversification in New World ratsnakes (tribe Lampropeltini)? Evolution. 2010;64:934–43.

85. Mahler DL, Revell LJ, Glor RE, Losos JB. Ecological opportunity and the rate of morphological evolution in the diversification of Greater Antillean anoles. Evolution. 2010;64:2731–45.

86. Pigot AL, Phillimore AB, Owens IP, Orme CD. The shape and temporal dynamics of phylogenetic trees arising from geographic speciation. Syst Biol. 2010;59:660–73.

87. Moen D, Morlon H. Why does diversification slow down? Trends Ecol Evol. 2014;29:190–7.

88. Partridge TC. Of diamonds, dinosaurs and diastrophism: 150 million years of landscape evolution in southern Africa. S Afr J Geol. 1998;101:167–84.

89. Sepulchre P, Ramstein G, Fluteau F, Schuster M, Tiercelin J, Brunet M. Tectonic uplift and eastern Africa aridification. Science. 2006;313:1419–23.

90. Cowling RM, Proches S, Partridge TC. Explaining the uniqueness of the Cape flora: incorporating geomorphic evolution as a factor for explaining its diversification. Mol Phylogenet Evol. 2009;51:64–74.

91. Zachos J, Pagani M, Sloan L, Thomas E, Billups K. Trends, rhythms, and aberrations in global climate 65 Ma to present. Science. 2001;292:686–93.

92. Liu Z, Pagani M, Zinniker D, DeConto R, Huber M, Brinkhuis H, Shah SR, Leckie RM, Pearson A. Global cooling during the Eocene-Oligocene climate transition. Science. 2009;323:1187–90.

93. Davis CC, Bell CD, Fritsch PW, Mathews S. Phylogeny of Acridocarpus-Brachylophon (Malpighiaceae): implications for Tertiary tropical floras and Afroasian biogeography. Evolution. 2002;56:2395–405.

94. Kissling WD, Eiserhardt WL, Baker WJ, Borchsenuis F, Couvreur TLP, Balslev H, Svenning J. Cenozoic imprints on the phylogenetic structure of palm species assemblages worldwide. Proc Natl Acad Sci. 2012;109:7379–84.

95. Tolley KA, Townsend TM, Vences M. Large-scale phylogeny of chameleons suggests African origins and Eocene diversification. Proc R Soc B. 2013;280: 20130184.

96. JRS Biodiversity Foundation grant. Digitizing southwestern-African herpetological collections. http://jrsbiodiversity.org/grants/university-of-florida/. Accessed 1 Aug 2016.

97. Hewitt J. Some new forms of batrachians and reptiles from South Africa. Rec Albany Mus. 1935;4:283–357.

98. Gaston KJ. Rarity. London: Chapman & Hall; 1994.

99. Harris G, Pimm SL. Range size and extinction risk in forest birds. Conserv Biol. 2008;22:163–71.

2

The conquering of North America: dated phylogenetic and biogeographic inference of migratory behavior in bee hummingbirds

Yuyini Licona-Vera and Juan Francisco Ornelas[*] ⓘ

Abstract

Background: Geographical and temporal patterns of diversification in bee hummingbirds (Mellisugini) were assessed with respect to the evolution of migration, critical for colonization of North America. We generated a dated multilocus phylogeny of the Mellisugini based on a dense sampling using Bayesian inference, maximum-likelihood and maximum parsimony methods, and reconstructed the ancestral states of distributional areas in a Bayesian framework and migratory behavior using maximum parsimony, maximum-likelihood and re-rooting methods.

Results: All phylogenetic analyses confirmed monophyly of the Mellisugini and the inclusion of *Atthis*, *Calothorax*, *Doricha*, *Eulidia*, *Mellisuga*, *Microstilbon*, *Myrmia*, *Tilmatura*, and *Thaumastura*. Mellisugini consists of two clades: (1) South American species (including *Tilmatura dupontii*), and (2) species distributed in North and Central America and the Caribbean islands. The second clade consists of four subclades: Mexican (*Calothorax*, *Doricha*) and Caribbean (*Archilochus*, *Calliphlox*, *Mellisuga*) sheartails, *Calypte*, and *Selasphorus* (incl. *Atthis*). Coalescent-based dating places the origin of the Mellisugini in the mid-to-late Miocene, with crown ages of most subclades in the early Pliocene, and subsequent species splits in the Pleistocene. Bee hummingbirds reached western North America by the end of the Miocene and the ancestral mellisuginid (bee hummingbirds) was reconstructed as sedentary, with four independent gains of migratory behavior during the evolution of the Mellisugini.

Conclusions: Early colonization of North America and subsequent evolution of migration best explained biogeographic and diversification patterns within the Mellisugini. The repeated evolution of long-distance migration by different lineages was critical for the colonization of North America, contributing to the radiation of bee hummingbirds. Comparative phylogeography is needed to test whether the repeated evolution of migration resulted from northward expansion of southern sedentary populations.

Keywords: Bee hummingbirds, Biogeography, Mellisugini, Molecular phylogeny, Migration, North America

Background

Bird migration is one of the most extraordinary behaviors found in nature. The voyage for migration involves a fascinating suite of characters including navigational systems, physiological specializations and the seasonal timing of events [1, 2]. Our knowledge of several ecological aspects of migration has become impressive over time [3, 4]; however, much remains to be learned on how long-distance seasonal migration repeatedly evolved in a wide variety of bird lineages and about the selection pressures underlying the evolution of migration [5–8]. In particular, the origin and geographical directionality of long-distance seasonal migration has been widely debated in the literature (e.g., [7]), centered in two prominent ideas originated on the examination of current distributions of migratory species and their presumed sister species: the 'southern-home' and the 'northern-home' hypotheses. The

* Correspondence: francisco.ornelas@inecol.mx
Departamento de Biología Evolutiva, Instituto de Ecología, A.C., Carretera Antigua a Coatepec No. 351, El Haya, Xalapa, 91070 Veracruz, Mexico

'southern-home' hypothesis posits that the breeding migratory species from the temperate regions are returning to their tropical ancestral ranges during the winter, whereas the 'northern-home' hypothesis postulates that the ancestral temperate range of migratory species becomes harsh for survival and depart to the novel tropics during the winter, and then returning to their ancestral home for breeding [6, 9]. In the 'southern-home' hypothesis, it is assumed that migration should evolve from sedentary ancestors to migratory descendants in response to ecological change or vice versa in the 'northern-home' hypothesis [10].

Escaping from intraspecific competition and the environmental seasonality with low food availability during the breeding season has been interpreted as being crucial for the evolution of migration [6, 11–13]. However, other factors such as increased harshness of climatic conditions and variation in resource availability during the non-breeding season, predation or parasitism would also make species to shift their breeding ranges and become migrants [6, 12]. Likely, migrant populations originating from southern tropical regions might have shifted their ranges northwards through long-distance dispersal coupled with climatic cycles [14], assuming competition in the tropical breeding ranges or the use of seasonally abundant resources in temperate regions as the driving forces for the northward expansion and evolution of migration [2, 6, 9, 15]. Several authors have envisioned scenarios for the transition from a sedentary to a migratory species over evolutionary time [9, 11, 12, 16]. As a result of the differential effects of intraspecific and interspecific competition, increasing seasonality of climate or by certain patterns of climatic change during the Tertiary and Pleistocene glaciations, Cox [11] proposed that migration evolves from changing the initial sedentary condition to that of a partial migrant, having with both permanent sedentary populations and populations migrating into seasonally favorable adjacent areas. Partial migrants then evolve further through extinction of sedentary populations and expansion into derived forms with separate or disjunct seasonal ranges. In contrast, Levey and Stiles [16] developed a scenario where temporal and spatial variation of resources, especially for fruit- and nectar-feeding birds, led to altitudinal intra-tropical migration, predisposing these birds to migrate out of the tropics.

Despite the appeal of intraspecific competition and variation in resource availability as being the first step for the evolution of migration, these scenarios have several shortcomings (reviewed in [6]) including those that have shown how in a small fraction of recently-expanded populations migratory behavior can increase rapidly when favored by selection (e.g., [17, 18]). Therefore, the repeated evolution of long-distance seasonal

migration within bird lineages linked to the occurrence of relatively fast range expansions to take advantage of abundant resources could be the result of selective pressures occurring throughout several climatic cycles affecting resource availability in seasonally changing environments.

More recently, Somveille and collaborators [19–22] examined global spatial patterns in the diversity of migratory birds, and found strong support for the hypothesis that seasonality is the main force driving bird migration worldwide (see also [23]). Whereas the previous studies attempt to explain the ecological mechanisms driving the higher diversity of migratory species in the Northern Hemisphere [22, 23], Rolland et al. [8] used molecular phylogenies that included most extant bird species to infer that sedentary behavior is ancestral and that migratory behavior evolved independently multiple times during the evolutionary history of birds. They also found that seasonal migration increases diversification via sedentary populations arising from migratory populations (asymmetrical speciation), in which speciation of ancestral species into one sedentary and one migratory species was more frequent in migratory species than sedentary. Their results suggest that the evolution of seasonal migration in birds has facilitated diversification through the divergence of migratory subpopulations that become sedentary, and illustrate asymmetrical diversification as a mechanism by which diversification rates are decoupled from species richness.

Hummingbirds (Trochilidae) are one of the largest bird families, with ca. 338 species distributed in the Americas [24]. The most recent molecular phylogeny suggests that hummingbirds split from their sister group, swifts and treeswifts, ca. 42 million years ago (MYA) in Eurasia and that the age of the common ancestor of hummingbirds in South America is ca. 22 MYA [25]. Given the gap between these two events and the absence of relevant fossils in the Americas, McGuire et al. [25] hypothesized that hummingbirds reached North America by dispersal across Beringia. After that, hummingbirds dispersed to the South American continent and may have become extinct both in Europe and North America [25]. Hummingbirds have diversified into nine clades (Topazes, Hermits, Mangoes, Brilliants, Coquettes, *Patagona*, Mountain Gems, Bees, Emeralds), seven of which rapidly diversified in South America in conjunction with the Andean uplift. The common ancestor of the other two clades, Bees and Mountain Gems, recolonized North America ca. 12 MYA [25], before the formation of the Central American land bridge and closure of the Isthmus of Panama. While hummingbird diversification probably increased in conjunction with the Andean uplift according to divergence dating using substitution rate priors (rather than fossil calibrations) [25], other divergence-dating analyses using both fossil calibrations and substitution rate priors retrieved older divergence splits between Bees and Mountain Gems (20–25 MYA; [26]), suggesting that North American hummingbirds are not recent colonizers and may have only become extinct in Europe [26, 27].

The 'bee' hummingbirds (Mellisugini tribe; [24]) comprise an assemblage of 16 genera and 36 small species distributed throughout the Americas, from southern Canada to South America [28]. Although some species are geographically widespread (e.g., *Archilochus* spp.; [28]), other have very restricted distributions such as the smallest bird of the world (*Mellisuga helenae*) endemic to Cuba. The most extensive molecular phylogeny of hummingbirds to date [25], with at least one representative species for each genus in the Mellisugini, estimated its relatively recent origin (ca. 5 MYA), revealed a high rate of diversification (0.57 species/MYA), as compared to other hummingbird clades. This phylogeny retrieved Mellisugini as composed of two main clades: one clade included species informally named "woodstars" distributed in South America and *Tilmatura dupontii* with distribution in Central and North America, and the second clade contained species arranged as in two subclades: (1) *Calypte*, *Selasphorus* and *Atthis* species, and (2) "sheartails" (*Doricha eliza* and *Calothorax lucifer*), *Archilochus* (*A. colubris* and *A. alexandri*), *Calliphlox evelynae* and *Mellisuga minima*, in which phylogenetic relationships between the "sheartails" and the other species within the subclade are not supported. Besides the high rate of diversification, Mellisugini species are distinguished by the dimorphic tail morphology, which in males the rectrices are unusual in shape to produce sounds and acrobatic courtship displays during the breeding season (e.g., [29–32]).

Mellisugini is the only group of hummingbirds with long-distance seasonal migration and, therefore, an interesting study group from a biogeographic perspective. Most of the species in Canada and the USA are obligate, long-distance seasonal migrants, which vacate their entire breeding range to winter mainly in Mexico [28]. Several aspects of hummingbird long-distance seasonal migration are particularly remarkable, with journeys across the Gulf of Mexico by *Archilochus colubris* or those of more than 6000 km by *Selasphorus rufus*, breeding in western United States and Canada and overwintering in Mexico [33]. However, the origin and evolution of migratory behavior and the impact on hummingbird diversification has not been studied. The evolution of hummingbird migration is a complex phenomenon to address because it is thought to evolve rapidly in response to selection [9, 34–36]. Previous phylogenetic hypothesis [25] suggests that migratory behavior is not evolutionarily constrained, as both sedentary and long-distance migratory species seem to have evolved repeatedly within the Mellisugini. Understanding

the evolution of migratory behavior within the Mellisugini is important, particularly because they are susceptible to rapid evolutionary change, i.e. their high rate of net diversification with species accumulation during their brief 5 MYA history [25], and because they can change their migratory behavior to escape from increased harshness of climatic conditions during the Pleistocene glacial cycles [37], and from seasonal changes in the phenology and availability of nectar floral resources by current global climate changes [38]. Unfortunately, the lack of a wider geographic sampling and the absence of some North American representative species from previous phylogenetic analyses, has not allowed having a fully resolved phylogeny of the group to understanding the evolution of long-distance seasonal migration and timing of diversification and colonization patterns.

The objectives of our study were to: (1) reconstruct the phylogenetic relationships among bee hummingbirds increasing both geographical and intraspecific sampling, (2) estimate divergence times between species and genera, and (3) reconstruct the ancestral range at each divergence event, and subsequent temporal and geographical shifts on migration in bee hummingbirds. The suite of morphological and behavioral characters coupled with the wide variety of environments where they live, including the most xeric environments tolerated within hummingbirds, have been linked to their relatively rapid radiation with highest rate of species' accumulation [25]. Thus, the Mellisugini present a useful model for exploring hidden biodiversity due to its wide distribution in both North and South American continents and recent biogeographic origin, and for understanding the potential impact of shifts between sedentary and long-distance migratory behavior on diversification of bee hummingbirds because migratory and non-migratory species, and species with partial migration (migratory and non-migratory populations) occur only in the North American continent.

Methods
Sampling and laboratory methods
The data set included 116 samples of bee hummingbirds from North America and the Caribbean Islands and 1–2 samples of bee hummingbirds from South America (*n* = 16 samples), representing all 16 genera of bee hummingbirds (32 of the 36 extant species, 89%). Tissue samples were unavailable for four species: *Chaetocercus astreans* (Colombia), *C. berlepschi* (Ecuador), *C. heliodor* (Colombia, Venezuela and Ecuador), and *Mellisuga helenae* (Cuba). Most of these species are endemic and range-restricted; *Chaetocercus berlepschi* is threatened by habitat loss [39, 40]. We include new sequence data for 60 individuals from the genera *Archilochus*, *Atthis*, *Calothorax*, *Doricha*, *Calypte*, *Selasphorus* and *Tilmatura*

to supplement the data set in McGuire et al. [25] and Feo et al. [31]. Additionally, we included a single individual of each of 15 species of mountain gems and emeralds to be used for sequence alignment and as outgroups. Samples were obtained from vouchered tissue collections (see Acknowledgements) and from our collecting efforts in Mexico.

DNA was extracted from tissue or tail feathers with the DNeasy Tissue extraction kit (Qiagen, Valencia, CA, USA) using the standard protocol. We amplified and sequenced six gene regions, two mitochondrial protein coding genes—1041 base pairs (bp) of nicotinamide dehydrogenase subunit 2 (*ND2*) and 807 bp of nicotinamide dehydrogenase subunit 4 (*ND4*), and four nuclear loci—1085 bp of fibrinogen beta chain intron (*FBG I7*), 551 bp of adenylate kinase 1 intron 5 (*AK1 I5*), 577 bp of ornithine decarboxylase 1 introns 6 and 7 intervening exon (*ODC1*), and 635 bp of Z-linked muscle, skeletal, receptor tyrosine kinase intron 3 (*MUSK I3*) using specific primers (Additional file 1). Protocols for PCR reactions and for sequencing the PCR products are described elsewhere [41]. The products were read on a 310 automated DNA sequencer (Applied Biosystems) at the INECOL's sequencing facility. Finally, assembled sequences were edited and checked for quality, pre-aligned using MAFFT v7 (http://mafft.cbrc.jp/alignment/server/), and then manually aligned using PhyDE [42]. Newly generated sequences have been submitted to GenBank (Accession nos. *ND2*: KX855335– KX855393; *ND4*: KX855394– KX855450; *AK1 I5*: KX855451– KX855509; *MUSK I3*: KX855568– KX855624; *ODC1*: KX855510– KX855567; *FBG I7*: KX855625– KX855637; Additional file 2). The alignments supporting the results of this article are available in the Dryad Digital Repository (http://dx.doi.org/10.5061/dryad.68fn0) as Licona-Vera and Ornelas [43].

Phylogenetic reconstruction

The phylogeny was reconstructed using Bayesian inference (BI), maximum-likelihood (ML) and maximum parsimony (MP). We performed BI comparative phylogenetic analyses using MrBAYES v3.2.2 [44] and the CIPRES Science Gateway [45] on the following data sets: (1) only mitochondrial genes ('unpartitioned mtDNA'), (2) only mitochondrial genes as two partitions ('partitioned mtDNA'), (3) only nuclear genes ('unpartitioned nuDNA'), (4) only nuclear genes as four partitions ('partitioned nuDNA'), (5) combined loci data set with a single model ('concatenated'), (6) each DNA region as one partition ('mtDNA + nuDNA'), and (7) with a set partition-specific DNA evolution models of each gene ('6-partitions'). We used jMODELTEST v2.1.7 [46] to select an appropriate model of nucleotide substitution for each locus and the concatenated data set. GTR + I + G (*ND2*), TrN + I + G

(*ND4*), K80 + G (*AK1 I5*), HKY (*MUSK I3*), HKY (*ODC1*), HKY + G (*FBG I7*), GTR + I + G (mtDNA data set), GTR + G (nuDNA data set), and TrN + I + G (concatenated) were selected as the best fitting models and incorporated as prior information in the Bayesian analyses. For each data set, two parallel Markov chain Monte Carlo (MCMC) analyses were executed simultaneously for 30 million generations, sampling every 10,000 generations. Output parameters were visualized using TRACER v1.6 (http://tree.bio.ed.ac.uk/software/tracer/). A 25% burn-in was used, and a majority rule consensus tree was calculated and visualized in FIGTREE v1.4.3 (http://tree.bio.ed.ac.uk/software/figtree/). We computed Bayes factors with the harmonic means [47] to determine whether applying partition-specific models for the combined data sets significantly improved the explanation of the data.

The ML analysis for the concatenated data set was run using RAxML v8.2.9 [48] with a GTRGAMMA model for each partition. Node support for the ML tree was estimated with 1000 bootstrap replicates.

The MP analysis was run for the concatenated data set in NONA v2.0 [49] using WINCLADA [50], with nucleotide characters treated as equally weighted and unordered. We ran 1000 iterations, holding 10 trees per iteration with 10%of the nodes constrained, and all the parameters set to default. Branch support was assessed using bootstrap resampling, 1000 bootstrap-resampled pseudo-replicate matrices were each analyzed using 100 random addition sequences (multi*100). Ten trees were retained during TBR swapping after each search initiation (hold/10).

Divergence time estimation

A Bayesian relaxed-clock analysis was performed in BEAST v2.4.4 [51, 52] to assess species divergence times using the six genes. We constrained Trochilidae and the hummingbird clades used as outgroups (emeralds and mountain gems) as monophyletic based on McGuire et al. [25]. Divergence times were estimated using an uncorrelated lognormal relaxed clock model across all genes, with the trees linked and the substitution models for each partition unlinked [53]. We calibrated our divergence-dating analyses using a Yule speciation model and three calibration strategies for divergence time estimation: (1) incorporating a separate normally distributed substitution rate calibration priors for *ND2*, *ND4*, *AK1*, *FGB*, and *ODC* using the mean substitution rates proposed by Lerner et al. [54] to model the tree prior, allowing the substitution rate prior for *MUSK I3* to be calculated by BEAST because no substitution rate was available; (2) using as secondary calibration the age of the split between mountain gems and bee hummingbirds (normal, mean 12.0 MYA, SD ± 1, range of 13.9–10.3

MYA) according to McGuire et al. [25] to calibrate the root of the tree; and (3) using both strategies, secondary calibration + substitution rates. This strategy was also used for divergence time estimation using a reduced data set, which includes one individual for each of the species to contrast results of divergence time estimation from single vs. multiple-individuals data sets.

Two independent chains of MCMC were run with 50 million generations, sampling every 5000 generations. Results were visualized in TRACER v1.6 (http://tree.bio.ed.ac.uk/software/tracer/) to confirm appropriate burn-in, adequate effective samples sizes (ESS > 200) of the posterior distribution for all parameters, and to assess convergence among runs by comparing likelihoods of parameters. The three independent runs were combined with LOGCOMBINER v2.4.4 [51, 52] and the resulting maximum clade credibility tree and 95% highest posterior (HPD) distributions of each estimated node annotated using TREEANNOTATOR v2.4.4 [51, 52] and visualized in FIGTREE v1.4.3 (http://tree.bio.ed.ac.uk/).

Ancestral areas of bee hummingbirds

We reconstructed ancestral geographic ranges using Bayesian methods with BBM (Bayesian Binary MCMC) analyses implemented in RASP v3.2.1 [55]. This method determines the probability of each ancestral geographical region for each node averaged over the collection of trees derived from a Bayesian MCMC analysis [56, 57]. To reconstruct the ancestral areas, we loaded 6000 trees from the Bayesian Inference analyses using MrBAYES. The breeding distributions of each sample was obtained from del Hoyo et al. [28] and crossed with the status and distribution information compiled by the Cornell Laboratory of Ornithology as input (www.allaboutbirds.org/guide). We coded each individual in the data set as occurring in one or more of the following areas: A = western North America, B = eastern North America, C = southeastern Mexico and Central America, D = West Indies, and E = South America (Additional file 3). These regions were based on a modified map of the ecoregions (http://maps.tnc.org/gis_data.html) proposed by Blair and Sánchez-Ramírez [58]. The posterior probabilities for nodes in the phylogeny with >0.90 were estimated to incorporate information from most nodes of the tree but minimizing phylogenetic 'noise' from poorly supported relationships. The maximum number of areas in ancestral ranges was constrained to three, *Amazilia rutila* assigned as outgroup using the 'custom' option, and the ancestral areas for nodes visualized on the condensed tree. Analyses were run for 50,000 iterations, sampling every 100 generations, the first 25% of which were discarded as burn-in, with the JC + gamma model of state transitions used as input.

Evolution of migratory behavior

Ancestral state reconstruction was used to map migratory behavior onto the resulting molecular phylogeny. The evolution of migratory behavior was reconstructed using maximum parsimony (MP) and maximum-likelihood (ML) methods. We traced the evolution of migratory behavior over the molecular phylogeny using two topologies: one with all samples and the other with one sample per species. The first topology corresponds to the best estimate of Mellisugini phylogenetic relationships using the 6-partitions data set (see Results), a Bayesian 50% majority rule consensus tree of 132 samples of bee hummingbirds. We used the 18,000 post-burnin trees from the BEAST analysis to account for the phylogenetic uncertainty in the ancestral state reconstruction. The second topology was a Bayesian 50% majority rule consensus tree using one sample per species (see Results). This tree was obtained from a BEAST analysis using the same parameters and the 6-partitions strategy described on methods section of phylogenetic reconstruction.

For ancestral state reconstruction we used three different coding schemes mainly based on information in del Hoyo et al. [28] and Malpica and Ornelas ([37]; Additional files 3 and 4). In coding Scheme 1 species that migrate seasonally between different latitudinal geographical breeding and wintering ranges were coded as migrants (i.e., obligate, long-distance migration; [59]) and non-migratory species were coded as sedentary. For this coding scheme, we also considered as migratory those species with partial migration, in which some individuals or populations are fully migratory across their range and other individuals or populations are sedentary (*Selasphorus platycercus* and *Calothorax lucifer*; [60]). The migratory state does not include tropical hummingbird species that may undertake altitudinal or short-distance migration at the fringes of their northern ranges in the northern temperate region (e.g., *Amazilia violiceps*, *Eugenes fulgens*, *Heliomaster constantii*; [59]). In coding Schemes 2 and 3, species with partial migration were coded as sedentary or polymorphic, respectively, to test the robustness of our conclusions to potential ambiguities in character state coding [60]. Coding species with a simple binary codification (sedentary or migrant) and pruning trees to species probably mask or confuse the ancestral character reconstruction of species susceptible to rapid evolutionary change of migratory behavior [34–36]. Thus, we conducted ancestral state reconstruction using the data set with multiple individuals for a given species to compare results with those obtained for species-level analyses (single-individual data set). Also, insights might be gained from sampling several individuals if these provide the signal at the phylogenetic level of when the shifts from migratory to sedentary (or vice versa) occurred between populations in species with both sedentary and migratory populations. Here, single individuals of *C. lucifer* and *S. platycercus* with migratory and sedentary populations were classified as either migrant or sedentary based on data on Malpica and Ornelas [37] and because samples of single individuals were collected during the breeding season from known allopatric migratory or sedentary populations. Species names, English names and distributional range for the Mellisugini species used in this study are provided in Additional file 5.

MP and ML based ancestral state reconstruction were conducted in Mesquite v3.11 [61] using each of the coding schemes of migratory behavior described above. To account for topological uncertainty we used the 'trace character over trees' option, which summarizes the ancestral state reconstruction over a series of trees. All reconstructions were integrated over the last 18,000 post burn-in of the Bayesian analysis and the ancestral states were summarized using the 'Count trees with uniquely best states' option on the maximum credibility tree using Mesquite. A reconstruction is regarded as equivocal when there are two or more equally parsimonious states inferred at a particular node. For the parsimony ancestral character reconstructions character-state changes were set as unordered, with other parameters as default. In the ML approach, the character state for each ancestral node was reconstructed using the Markov k-state 1 parameter model (Mk1), which specifies an equal probability of any state change and considers the rate of change the only parameters. We conducted ancestral state reconstruction with more than one method because each of the two methods described above suffers from certain advantages and limitations [62–66].

Ancestral states of migratory behavior in the Mellisugini were also reconstructed using the re-rooting method of Yang et al. [67] as implemented in PHYTOOLS v0.5 [68] in R v3.3.0 [69]. This method re-roots the phylogeny at every node and calculates the phylogenetically independent contrast for the root node, taking advantage of the fact that this value is the maximum likelihood estimate for that node [70, 71]. We used the Mk1 model for reconstruction of the character state for each ancestral node, assuming equal rate of evolution. Since the likelihood approach is not applicable for polymorphic characters this reconstruction was performed using only the parsimony approach.

Results

Phylogeny

Bayesian analyses using the entire gene data set (nuDNA + mtDNA) resulted in a well-supported phylogeny of the Mellisugini tribe and close relatives (Fig. 1). The summary of MP and ML bootstrap values and the Bayesian posterior probabilities are presented on the branches of the BI 50% majority-rule consensus tree (Fig. 1). Results of other

Bayesian analyses using partition or unpartitioned nuDNA or mtDNA data sets are given in Additional file 6. Given the stronger support between clades in the analyses of the nuDNA + mtDNA data set and Bayes Factors, we considered the MrBayes results of concatenated mtDNA + nuDNA genes (with 6-partitions strategy) to be our best estimate of phylogenetic relationships in the Mellisugini. We rely on this tree in our ancestral state reconstructions and discussion of the evolution of migration and of biogeography. Changes to previous phylogenetic topologies of the Mellisugini (e.g. [25]) are indicated in Additional file 7.

Divergence dating and ancestral areas of bee hummingbirds

The topology of the BEAST time-tree using the third calibration strategy (secondary calibration + substitution rates (Fig. 2) was concordant with those derived from other reconstruction methods (Fig. 1, Additional file 6). The BEAST analyses indicated the most recent common ancestor (MRCA) for the Mellisugini originated approximately 9.93 MYA (95% HPD 11.94–7.92 MYA) in mainland North America ~6 million years before the final closure of the Isthmus of Panama (Fig. 2). The ancestor originated in either western North America or southern Mexico and Central America, with relatively high support for nodes A (separating bee hummingbirds and mountain gems) and B (separating South and North American bee hummingbirds) yielded by the ancestral area reconstruction (AC, 97% and 79%, respectively; Fig. 2). By the end of the Miocene, the bee hummingbirds first reached western North America (node C; A, 60%). Accordingly, subsequent major nodes (nodes D, E, F, K, J) were reconstructed as nearly 100% western North America (A). Although it is uncertain where the ancestor of bee hummingbirds was distributed in the region, the analysis suggests that western North America was colonized during the early diversification of the group with dispersals into other regions of the Northern Hemisphere. The BEAST analyses also showed a mid-to-late Miocene split separating South American woodstars from the other bee hummingbirds (node B; Fig. 2), divergence of the Mexican sheartails from other North American bee hummingbirds in the late Miocene (node C), and that the diversification of the South American woodstars (node L), Caribbean sheartails (node H) and the split between the *Calypte-Selasphorus* subclades (node E) occurred in the early Pliocene (Fig. 2). Details of ages for other nodes of interest are shown in Table 1.

Evolution of migratory behavior

The results of ancestral state reconstruction of long-distance migratory behavior on the Bayesian 50% majority-rule consensus tree of a reduced 32 bee

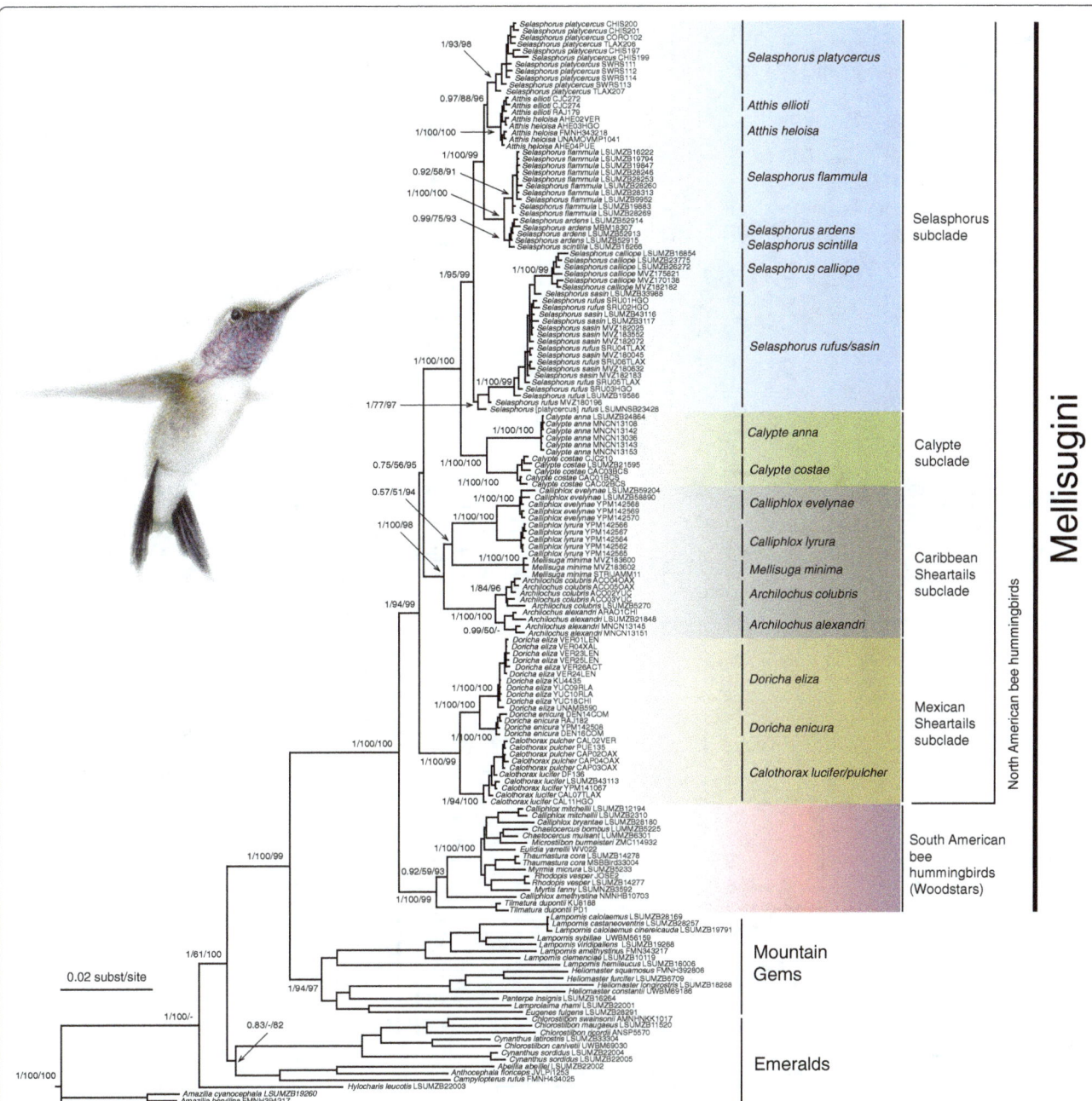

Fig. 1 Phylogenetic 50% majority-rule consensus tree of the Mellisugini hummingbird from the Bayesian analysis of the combined NADH dehydrogenase subunit 2 (*ND2*) and subunit 4 (*ND4*) mitochondrial protein coding genes, and fibrinogen beta chain intron (*FBG I7*), adenylate kinase 1 intron 5 (*AK1 I5*), ornithine decarboxylase 1 introns 6 and 7 intervening exon (*ODC1*), and Z-linked muscle, skeletal, receptor tyrosine kinase intron 3 (*MUSK I3*) nuclear loci. Partitioning considerably improved mean −lnL values in the Bayesian analyses, with unpartitioned arithmetic mean −lnL = −35,190.36, compared with −34,147.09 for two partitions and −3341.47 for six partitions. Bayes factor comparison also indicated that the 6-partitioned analysis provided better explanations than other data analyses: 2lnB (6-partitions/unpartitioned) = 3697.78, and 2lnB (6-partitions/2-partitions) = 1611.24 significantly above the threshold value of 10. Bayesian posterior probabilities (PP) followed by bootstrap values (ML and MP, respectively) are shown above the branches (only bootstrap values above 50 and PP values above 0.5 are shown for the main clades) for the partitioned analyses. Note that the ID of the only sample of *Selasphorus platycercus* (LSUMNSB23428) included in the phylogeny presented by McGuire et al. [25] is likely incorrect. Painting by Marco Pineda (courtesy of Juan Francisco Ornelas) showing *Calothorax lucifer* (male)

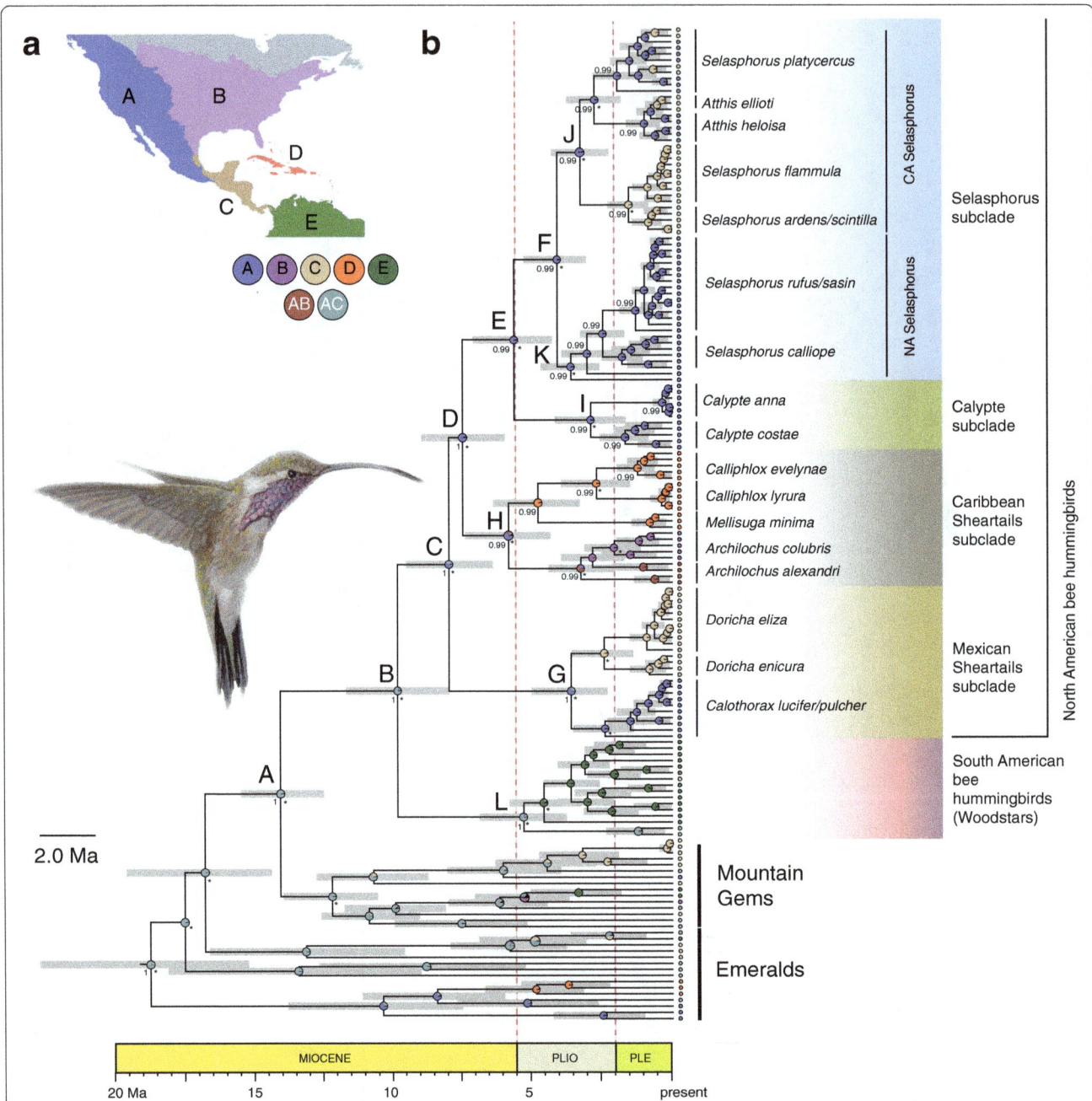

Fig. 2 a Biogeographic regions used in the RASP analysis: A = western North America, B = eastern North America, C = southeastern Mexico and Central America, D = West Indies, and E = South America. **b** Chronogram of the Mellisugini lineages based on the third calibration method (secondary calibration + substitution rates) with a Yule speciation model for the combined *ND2* and *ND4* mitochondrial genes and *FBG I7, AK1 I5, ODC1,* and *MUSK I3* nuclear loci data set. Purple bars indicate 95% Highest Posterior Density (HPD) intervals for selected nodes. The pink dotted vertical lines denote the time span of the Pliocene. Results using Bayesian methods with BBM (Bayesian Binary MCMC) analyses implemented in RASP [55] are drawn on this topology. The ancestral origin for each taxon, as delimited in (a), is shown on the terminal lineages. Pie charts at nodes represent the probabilities of the ancestral distributions. These probabilities account for the phylogenetic uncertainty in the rest of the tree and the biogeographic uncertainty at each node. Asterisks next to main nodes refer to posterior probability support for each node (* > 0.8 posterior probability). Painting by Marco Pineda (courtesy of Juan Francisco Ornelas) showing *Calothorax pulcher* (male)

hummingbird species data set using only one individual per species and the 6-partitions strategy (Additional file 8) are shown in Fig. 3. The MP and ML reconstructions of migratory behavior provided similar results with high certainty in the bee hummingbirds, in which the phylogenetic position of migratory species indicates multiple independent origin of long-distance migratory behavior (Fig. 3a–c). ML, MP, and the re-rooting method of Yang

Table 1 Divergence dates (MYA) of bee hummingbirds for various nodes estimated with a Bayesian uncorrelated lognormal relaxed-clock approach using a Yule speciation tree prior as implemented in BEAST

| Node | Several-individuals | | | | | | Single-individual | |
| | Secondary calibration + substitution rates | | Secondary calibration | | Substitution rates | | Secondary calibration + substitution rates | |
	PP	Mean (95% HPD)	PP	Mean (95% HPD)	PP	Mean (95% HPD)	PP	Mean (95% HPD)
Node A: Bee hummingbirds/Mountain Gems	1.0	14.08 (15.54–12.51)	1.0	11.83 (13.79–9.85)	1.0	19.97 (24.81–15.64)	0.99	12.49 (13.58–11.37)
Node B: SA/NA bee hummingbirds	1.0	9.93 (11.94–7.92)	0.99	7.68 (9.81–5.60)	1.0	12.50 (16.04–9.11)	1.0	5.92 (6.59–5.24)
Node C: NA crown (Mex. sheartails/other)	1.0	8.03 (9.82–6.44)	0.99	6.32 (8.24–4.51)	1.0	9.56 (12.17–7.05)	1.0	4.98 (5.60–4.41)
Node D: Caribbean Sheartails/other	1.0	7.51 (9.14–5.95)	1.0	5.95 (7.83–4.27)	1.0	8.90 (11.31–6.59)	1.0	4.77 (5.36–4.23)
Node E: *Calypte* subclade/other	0.99	5.72 (7.20–4.31)	0.99	4.12 (5.62–2.73)	1.0	6.71 (9.08–4.87)	0.99	3.11 (3.61–2.62)
Node F: CA *Selasphorus*/NA *Selasphorus*	0.99	4.14 (5.27–3.04)	0.99	3.1 (4.29–1.97)	1.0	4.79 (9.08–4.87)	0.99	2.18 (2.61–1.80)
Node G: Mexican Sheartails crown	1.0	3.61 (5.11–2.39)	1.0	3.09 (4.66–1.70)	1.0	4.39 (6.17–2.71)	1.0	2.20 (2.76–1.69)
Node H: Caribbean sheartails crown	0.99	5.75 (7.35–4.21)	0.99	4.61 (6.21–3.13)	1.0	6.93 (9.28–4.86)	0.99	4.04 (4.61–3.44)
Node I: *Calypte* crown	0.99	3.09 (2.58–0.90)	0.99	2.56 (3.79–1.40)	1.0	3.46 (5.22–1.96)	0.99	2.14 (2.61–1.69)
Node J: CA *Selasphorus* crown	0.99	3.31 (4.42–2.29)	0.99	2.23 (3.21–1.33)	1.0	3.82 (5.12–2.57)	0.99	1.43 (1.77–1.10)
Node K: NA *Selasphorus* crown	0.99	3.66 (4.74–2.60)	0.99	2.64 (1.61–0.48)	1.0	4.24 (5.81–2.86)	0.99	1.16 (1.53–0.79)
Node L: South American crown	1.0	5.42 (7.04–3.81)	0.99	5.44 (7.38–3.67)	1.0	6.44 (8.72–4.32)	1.0	4.30 (4.94–3.65)

Posterior probabilities (PP) given for each node; estimates are given as mean ages (in millions of years) with 95% Highest Posterior Density (HPD) intervals in parentheses. The divergence time (mean 12.0 MYA, SD ± 1, range of 13.9–10.3 MYA) between mountain gems and bee hummingbirds [25] was used for temporal calibration of the root node of the tree and the mean substitution rates proposed by Lerner et al. [54] to model the tree prior. Node letters correspond to those in Figs 2–3. NA North America, SA South America, CA Central America

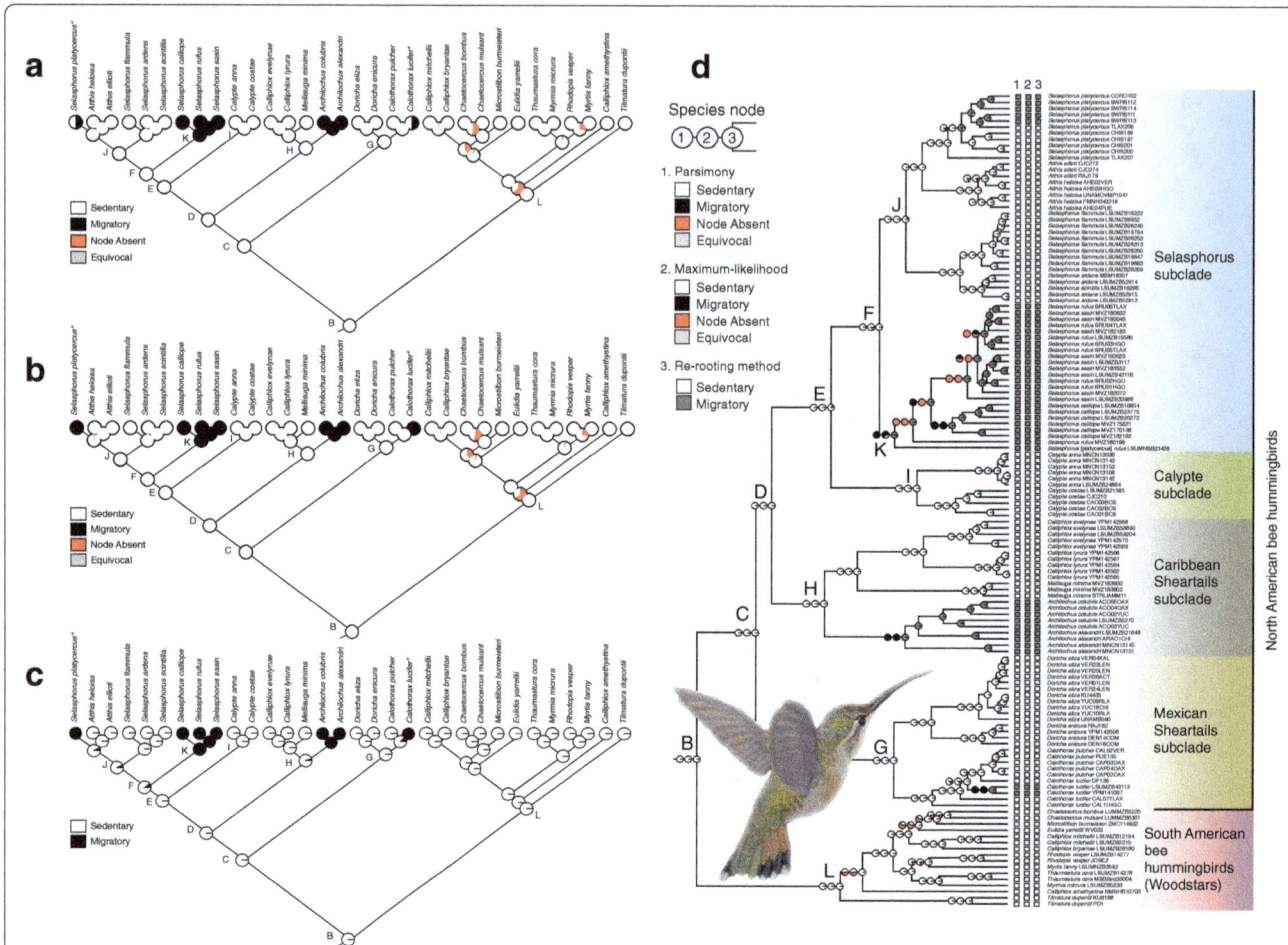

Fig. 3 Ancestral state reconstructions across the set of 18,000 post burn-in BEAST trees for the Mellisugini based on parsimony in which species with migratory and sedentary populations (*Selasphorus platycercus* and *Calothorax lucifer*) were coded as polymorphic (**a**) or migratory (**b**), and estimation of ancestral states of migratory behavior carried out with the tree obtained from the Bayesian analysis under the ML criterion and the MK1 model using all samples (**c**). Ancestral state reconstructions results obtained for the MP, ML and re-rooting methods using the data set with several samples (**d**). Each square at the tips of the tree represents the state of each extant taxon and the pie charts at each node represent the probability of the state of the common ancestor present at that node. Dark gray squares indicate fully migrant individuals and white squares indicate sedentary individuals. Asterisks next to species names indicate species with migratory and sedentary populations. Painting by Marco Pineda (courtesy of Juan Francisco Ornelas) showing *Calothorax lucifer* (female)

et al. [67] ancestral-state reconstructions supported a sedentary ancestral mellisuginid (bee hummingbird) regardless of the coding scheme used for migratory behavior (Fig. 3a–c). Because the results using the various coding schemes were largely the same, we only present the results based on one of the coding schemes for each of the ancestral state reconstructions: parsimony and species with partial migration (*C. lucifer* and *S. platycercus*) coded as polymorphic (Fig. 3a), and maximum likelihood and re-rooting with *C. lucifer* and *S. platycercus* coded as migratory (Fig. 3b–c). Basal nodes of all North American subclades (incl. Monophyly of Caribbean sheartails) were reconstructed with strong support for the sedentary state with MP, ML and re-rooting methods (Fig. 3a–c), and migratory behavior was gained four times during the evolution of the Mellisugini: once in the Mexican sheartails

subclade (*Calothorax lucifer*) and the Caribbean sheartails subclade (*Archilochus* species), and twice in the *Selasphorus* subclade (*Selasphorus rufus*, *S. sasin* and *S. calliope* group and *S. platycercus*). Similar results were obtained for the MP, ML and re-rooting ancestral state reconstructions using the data set with several samples (Fig. 3d), except that migratory behavior was been lost at least once in *S. platycercus* (see also [37]).

Discussion

Phylogeny of bee hummingbirds

Our phylogenetic analyses recovered a monophyletic South American clade sister to other bee hummingbirds in Central America, the Caribbean islands and North America. Despite this, the backbone of our tree topology is not entirely consistent with that for previous studies with fewer taxa and individual samples ([25, 72]; see Additional file 7). According to our analyses, Mexican sheartails split very soon after the split between South American and North American bee hummingbirds, and sister to remaining subclades. The *Selasphorus* subclade is composed of two groups according to our results, the North American *Selasphorus* (*S. calliope*, *S. sasin*, *S. rufus*) and the Central American *Selasphorus* arranged by geography, *S. flammula*, *S. scintilla* and *S. ardens* from Costa Rica and Panama and *S. platycercus*, *Atthis heloisa* and *A. ellioti* mainly from Mexico and Guatemala.

Our DNA sequence data set (including 162 samples and 2–6 loci for each sample) contains 8.8% missing data, and this incompleteness is unlikely to have negatively impacted the accuracy of phylogenetic reconstruction because the number of characters in the analysis is large [73, 74]. Perhaps adding more taxa and more samples per taxon improved the accuracy of our phylogenetic reconstruction, in which monophyly of the Caribbean sheartails (*Archilochus*, *Mellisuga*, *Calliphlox*) is strongly supported. Lastly, our dense individual sampling within *S. platycercus* indicated that this species is nested within the group of Central American *Selasphorus* and *Atthis* species, the *Selasphorus* subclade, and did leverage our phylogenetic and ancestral state reconstructions from ID errors in single-individual species representation of previous phylogenetic reconstructions.

Divergence dating and ancestral areas of bee hummingbirds

According to our results, the crown age of the Mellisugini, ca. 9.93 MYA using the 'several-individuals' data set and the 'secondary calibration + substitution rates' calibration strategy, is older than those estimated by McGuire et al. [25] and Abrahamczyk and Renner [72], 5.3 and 7.22–6.71 MYA, respectively. When using the 'single-individuals' data set and the 'secondary calibration + substitution rates' calibration strategy, the crown age of the Mellisugini is similar, ca. 6.59–5.24 MYA (Table 1). These obvious differences across studies are likely due to the different calibration strategies and taxon sampling employed. McGuire et al. [25] included 27 species and 61 samples of bee hummingbirds and Abrahamczyk and Renner [72] 25 species and 26 samples, whereas our study included 32 species and 132 samples. A discrepancy in age estimates is observed in the comparison of the posterior mean age estimates between the 'single-individual' species samples and the 'several-individuals' species samples (taxon sampling variation) for the 6-partitions data sets (Table 1). In this case, we know that all sampling and data conditions are identical between the two analyses except for the density of taxon sampling within species. The age estimates for the Mellisugini (nodes B–L) differ on average by about 2 MYA, with the all sampling consistently yielding older

ages leading to very different biogeographic conclusions. Although the difference is greater for the deeper nodes in the tree, the impact becomes minor as one moves to the tips of the tree (Fig. 2, Table 1). Our results indicated that different sampling strategies have yielded different estimates and potential errors in molecular dating likely due to sampling bias for recent evolutionary radiations. In discussing age estimates in the subsequent discussion, we refer primarily to the posterior mean point estimated obtained with the 'several-individuals' species samples for the 6-partitions data set with more precision in divergence time estimation of shallow nodes.

The early divergences within the Mellisugini are estimated to have occurred after the mid-Miocene climatic optimum (nodes B, C and D of Fig. 2; [75]), but the majority of divergence events occurred much later, from the Pliocene to the mid-late Pleistocene (Fig. 2). Based on these results, we propose that the formation of the mountain systems in Mexico and Central America from mid-Miocene to the Pliocene was critical in providing favorable habitats and climatic conditions for the divergence of bee hummingbirds in the region. The bee hummingbirds dispersed to North America in the mid-Miocene, and then its history was probably marked by a period of expansion to xeric environments and segregation into xeric and moist temperate forests directly associated with a global decrease in temperature and humidity during the Late Miocene [76] and desert formation in North America [77]. Divergences of the Mexican sheartails and the *Calypte* and *Archilochus* species from their ancestors (Fig. 2) dated at 8.03, 5.72 and 5.76 MYA, respectively, coincides with the Miocene peak in speciation rates in some plants characteristic of xeric environments in Mexico, including *Agave* [78] and cacti [79], through a region of the country similar to that through which *Calothorax* is currently distributed and feed upon. The radiation of the Mellisugini in Central America, Caribbean islands and North America must have been relatively rapid. During the Late Miocene, the lineage would already be possibly occupying the main mountain ranges in Central America, and occurring in Mexico, in both montane and dry environments. The divergences of *Calothorax*, *Archilochus*, and the NA *Selasphorus* from their sedentary ancestors are dated at 3.09, 5.75, 4.14 MYA, respectively, suggesting that these divergences occurred in the Pliocene. These divergences also coincide with the second peak in speciation rates in *Agave* sensu *lato* dated at 3–2.5 MYA influenced by nectar feeding bats [78], and the transition from bee-pollination to hummingbird-pollination in Mexico in the *Opuntia-Nopalea* cacti clade dated at 5.73 MYA (95% HPD 8.7–3.42 MYA; [80]). In contrast, divergence of migratory *S. platycercus* from its sedentary ancestor is dated at 1.54 MYA (95% HPD 2.27–0.93 MYA), suggesting that the evolution of migration in *S. platycercus* occurred in the Pleistocene as shown by ancestral state reconstruction (Fig. 3).

The Late Miocene ages of hummingbird-dependent plant clades in North America (9–5 MYA; [72] coincide with the timings of divergence events of bee hummingbirds during the Pliocene and mid-late Pleistocene (Table 1). Our interpretation is also supported by the age of the oldest hummingbird-adapted group in North America, *Lonicera* (Caprifoliaceae), with a stem age of 9.2 MYA and a crown age of 7.0 MYA [81], the age of hummingbird-pollinated *Psittacanthus* mistletoes in Mexico with a stem age of 9.68 MYA and a crown age of 7.43 MYA [82, 83], and by the Pleistocene origin of *Penstemon* in the Rocky Mountains with subsequent migration and radiation to the Cascade–Sierra Nevada cordillera and then into southwestern North America and throughout eastern North America [84]. Interestingly, range expansions of bee hummingbirds in North America during the Pliocene seem to correspond to Pliocene divergences within the hummingbird-pollinated *Psittacanthus* mistletoes apparently linked to habitat shifts [82, 83]. For instance, the ages of the *Calothorax* sheartails, with a stem age of 3.6 MYA (95% HPD 5.11–2.39 MYA) and a crown age of 2.4 MYA (95% HPD 3.61–1.41 MYA), coincide with the timing of divergence events of the *Psittacanthus* mistletoes they currently pollinate (*P. auriculatus* distributed in the xeric areas of Oaxaca and *P. calyculatus* distributed in pine-oak forests along the Trans-Mexican Volcanic Belt; [85–87]), with a stem age of 3.1 MYA and a crown age of 1.8 MYA [82, 83]. The ages of *Calypte costae*, with a stem age of 3.1 MYA (95% HPD 4.49–1.78 MYA) and a crown age of 1.7 MYA (95% HPD 2.58–0.91 MYA), coincide with the timing of divergence events of the *Psittacanthus* mistletoes they currently pollinate in the Sonoran Desert, *P. sonorae*, with a stem age of 4.8 MYA and a crown age of 0.3 MYA [82, 83].

Like the hummingbirds and their coevolved plants in North America, the earliest divergence within the Caribbean sheartails (5.7 MYA; Table 1) indicates that they were also contemporaneous with lineages of hummingbird-adapted flowers [72, 88]. Overall, the results of our divergence-dating analysis seem to indicate that the Pliocene range expansions of bee hummingbirds are connected with the biogeography of their host plants and provide interesting insights on how range expansions into North America via habitat changes facilitated the evolution of migration in this group.

Evolution of migratory behavior

Our study is the first to provide phylogenetic evidence for the repeated evolution of long-distance migratory behavior in the radiation of the Mellisugini, with the crown ancestors of the main clades and North American subclades reconstructed as sedentary. Our results suggest that long-distance seasonal migration arose independently four times in the Mellisugini: once in the Mexican sheartails subclade (*Calothorax lucifer*) and the Caribbean sheartails subclade (*Archilochus* species), and twice in the *Selasphorus* subclade (*Selasphorus rufus*, *S. sasin* and *S. calliope* group and *S. platycercus*). Our study also showed that migratory lineages are generally more closely related to non-migratory lineages than to other migratory lineages, and that long-distance seasonal migration arose at different times. For instance, the split between migratory *Archilochus* species and sedentary species endemic to the Caribbean islands (*Mellisuga* and *Calliphlox* species) occurred at 5.7 MYA, whereas genetic differentiation between migratory *C. lucifer* and *S. platycercus* and their sedentary ancestors seem to have started during the late Pleistocene.

Our ancestral area reconstruction appears to explain the migration patterns of earliest *Archilochus* species, earliest *A. colubris* breeding in eastern NA (USA and Canada) and migrant to mainly eastern Mexico, Central America and some Caribbean islands, and *A. alexandri* breeding in western NA (USA and Canada) and migrant to western Mexico. A phylogeographic approach accompanied with paleodistributional modeling would be needed to test whether the differences in migratory patterns between *Archilochus* sister species were influenced by Pleistocene climate change and their range shifts occurred earlier. Using ecological niche modeling and phylogeographic data, Malpica and Ornelas [37] showed first that *S. platycercus* is a niche tracker and then that the climate conditions associated with modern obligate migrants in the USA were not present during the LIG, which provides indirect evidence for recent migratory behavior in *S. platycercus* on the temporal scale of glacial cycles. Their study also revealed that the evolution of migration within *S. platycercus* produced no significant genetic structure using nuclear microsatellites (nSSRs), migratory and sedentary groups of populations form an admixed population. The fact that they detected no significant genetic differentiation between migratory populations of *S. platycercus* and sedentary populations of the species in central Mexico (*platycercus* subspecies) is surprising because these hummingbirds inhabit different breeding areas of the USA and Mexico, and no evidence of sympatry at overwintering sites in Mexico has been noted. However, phylogeographic analyses and population genetic methods revealed that both migratory populations in the USA and sedentary populations in Mexico of the *platycercus* subspecies form one admixed population, and that sedentary populations from southern Mexico and Guatemala (*guatemalae* subspecies) diverged earlier (0.75 MYA) and undertook independent evolutionary trajectories [37].

Several studies have explicitly outlined a similar frame-work for addressing the evolution of bird migration between North and Central America, in particular for species with migratory populations in Canada and USA and sedentary populations in Mexico and Central America [10, 89]. For example, Milá et al. [90] examined the evolution of migration in the chipping sparrow (*Spizella passerina*) with sedentary populations in Mexico and the southern USA and migratory populations in the northern USA and Canada, and found evidence that migration was driven by northern range expansion from sedentary populations following glacial episodes.

Our study provides phylogenetic evidence for a seden-tary origin for the Mellisugini, but certainly is not the first to deal with this question in birds of the Northern Hemisphere (e.g., [60, 89–98]). Studies that have reconstructed the ancestral state of migration in a phylogenetic context have found either equivocal results [93, 99, 100] or results in favor of a migratory [10, 60, 101] or a sedentary ancestor [89, 92, 98, 102–104]. Accord-ing to a well-supported molecular phylogeny of *Cath-arus* thrushes *sensu lato* (incl. *Hylocichla mustelina*), long-distance seasonal migration is reconstructed as the ancestral condition at most basal nodes when put-ting character changes as close to the root of the tree as possible (ACCTRAN resolving option), and north of Mexico is reconstructed as the ancestral area with the origin of the clade at 8 MYA, diversification of *Catharus* from *Hylocichla* occurring at 6.9 MYA, and further lineage divergence within *Catharus* starting in the early Pliocene at 4.7 MYA [10]. Within *Catharus*, migratory behavior was lost after the first speciation event in the genus and was geographically and tempor-ally correlated with Central American distributions and the final closure of the Central American Seaway. Subsequently, migratory behavior was re-gained twice in *Catharus* and was geographically and temporally correlated with a re-colonization of North America in the late Pleistocene [10]. Counter to our results for the Mellisugini, the ancestral wood-warbler (Parulidae) was reconstructed as migratory using the well-supported molecular phylogeny of Lovette et al. [97], with losses of migration as prevalent as gains throughout the evolutionary history of Parulidae [60]. These results suggest that extant sedentary tropical radiations in the Parulidae represent losses of long-distance seasonal mi-gration and colonization of the tropics from temperate regions [60]. However, many derived non-migratory clades descended from non-migratory ancestors, sup-porting the notion that the ancestor of the Parulidae was a non-migrant [98]. Using a phylogenetic model of the joint evolution of breeding and non-breeding, wintering ranges to infer the biogeographic history in the emberizoid passerine birds, Winger et al. [101]

found that seasonal migration between breeding ranges in North America and winter ranges in the Neotropics evolved primarily via shifts of winter ranges toward the tropics from ancestral ranges in North America.

In the Mellisugini, it seems that migration evolved out of the tropics through the northern extension of ancestral tropical or subtropical breeding ranges into temperate regions ('southern home-theory'; [6, 9]). According to results of the Rolland et al.'s [8] study that included most extant bird species, we infer that seden-tary behavior is ancestral and migratory behavior evolved several times during the evolutionary history of the Mellisugini. Testing increased diversification rates in the Mellisugini with the evolution of migration is hampered by the lack of statistical power (see [8] for further discussion). Nonetheless, the divergence of a migratory species into two migratory daughter species tend to be less frequent that the divergence of a sedentary species into two sedentary daughter species, consistent with the findings of Rolland et al. [8] and predictions of Helbig [5] and Claramunt et al. [105] that genetic differentiation is reduced in migratory species with high dispersal capacity. The results of Rolland et al.'s [8] study suggest that the mobility of migratory species promotes the colonization of new areas and, if adapted to the new habitat, populations can become sedentary and diverge from the founding migratory species.

Several factors would have influenced the way bee hum-mingbirds colonized the northern portion of North Amer-ica in the mid-Pliocene. Following the first dispersal of ancestral *Archilochus* hummingbirds from the Caribbean islands to the northeast and northwest of North America, either along the coastal slope or across the Gulf of Mexico, it is possible that range changes in the Pleistocene caused multiple populations to lose migration and stay restricted to the Caribbean Islands with subsequent speci-ation. Successful dispersal of North American *Selasphorus* hummingbirds occurred from Central America and southern Mexico to the northwest of the continent. The evolution of migratory populations from ancestral seden-tary populations in southern Mexico of *S. platycercus* occurred later, likely due to Pleistocene climate changes (see also [90]). These scenarios are consistent with the idea that geographic isolation during the Late Pleistocene account for intraspecific and sister-species-level diver-gence largely based on habitat shifts influenced by climate change (e.g., [102, 106–108]), particularly shifts subse-quent to the LGM produced distinct migratory pathways and further genetic differentiation [93, 109]. Therefore, it seems that divergence of a migratory species into two mi-gratory daughter species is linked to a seemly rare event of changing migratory trajectories widely documented in some songbirds (e.g., [93, 109–116]). Further study using a comparative phylogeographic approach accompanied with ecological niche modeling and more or faster molecular markers (e.g., SNPs, SSRs) should provide finer resolution to the history of migration in the Mellisugini, particularly for contrasts between shallow lineages with sedentary and migratory species (e.g., *Calothorax lucifer*/ *C. pulcher*). For testing the effects of cyclical glacial changes on producing seasonally unstable habitats, and driving northward expansion and the evolution of seasonal migration and contraction into southern sedentary populations, and increased sampling of South American woodstars, further study will be needed to test whether the evolution of long-distance seasonal migration in North American bee hummingbirds has facilitated diversification in the Mellisugini through the divergence of migratory subpopulations that become sedentary [8, 60].

We cannot eliminate the possibility that long-distance migratory behavior evolved relatively early in the evolution of the Mellisugini. If this was the case, a migratory ancestor lost migration multiple times. Under this hypothesis colonization of North America and the Caribbean Islands would be more likely, despite that the phylogenetic evidence of that migratory ancestor is now lost, as temperate niches remained relatively open with ephemeral resources with subsequent losses of migratory at later times environments became less seasonal. If this model were statistically supported, its results would suggest that long-distance migratory behavior evolved once in the base of the Mellisugini tree, with several subsequent losses towards the end of the Pliocene. In coding SA bee hummingbirds and mountain-gems as migratory assuming that a migratory ancestor lost migration multiple times, ancestral character state reconstruction yielded equivocal results for the node of the Mellisugini and further ancestral nodes within the tribe were reconstructed as sedentary. When forcing only SA bee hummingbirds to be migratory, both the node of the Mellisugini and further ancestral nodes for main clades within the tribe were reconstructed as sedentary (Results not shown).

Cox's [11] model predicts that migratory species will be derived from sedentary species within the seasonal subtropics, and that migratory behavior is a derived character state. We believe that the Mellisugini lineage fits this model in many ways, particularly because most migrant species are closely related to the sedentary species found in the seasonal highlands of Mexico and Central America. An important result of our study is that long-distance migratory species do not form a monophyletic group. The relationships between migrant and resident species within the Mexican sheartails, Caribbean sheartails, and *Selasphorus* subclades are more complex than we expected. Also, the repeated gains of migration occurred during the Late Pliocene and this suggests that potential responses, i.e. the temporal evolution of migratory

behavior, can be linked to historical, climatic and ecological events on a phylogeny [7, 10]. However, we cannot ignore the possibility that long-distance migratory behavior in the Mellisugini was the ancestral state with several drop-offs of migration. Given the high degree of lability of the trait [34–36] and assuming that the phylogenetic signal of long-distance migratory behavior in the Mellisugini is an artifact of phylogenetic inertia in biogeographic range (including latitude and temperature seasonality), these questions seem unanswerable, making long-distance seasonal migration non tractable over substantial evolutionary time until comparative genomic data sets for migratory/sedentary closely related species pairs and for migratory and non-migratory populations of species with partial migration become available.

Conclusions

Pliocene's mountain building in Mexico and Pleistocene climate changes were the primary feature that structured diversity in the Mellisugini. These results are consistent with Cox's [12] idea that the Mexican Plateau and arid southwestern United States have acted as staging areas for the evolution of hummingbird migration. Range expansions of early lineages of the Mellisugini seem to be connected with the biogeography of their host plants and provide interesting insights on how range expansions into North America via habitat changes facilitated the evolution of migration in this group. Recently evolved lineages in all subclades of the Mellisugini appear to have undergone long-distance seasonal migration, albeit in different directions. This history of repeated evolution of migration within the Mellisugini allowed for divergence across common biogeographic regions spanned by North American bee hummingbirds. It is likely that, without repeated evolution of migration in different directions, diversification of the Mellisugini would have decelerated towards the present [25]. Thus, molecular patterns of diversification within the Mellisugini reflect a dynamic history of divergence, the main lineages during the Pliocene linked to the formation of the mountain systems in Mexico and Central America and further divergence by the evolution of seasonal migration during the Pleistocene.

Additional files

Additional file 1: Primers employed in this study. (DOC 39 kb)

Additional file 2: Species names, voucher information, locality, and GenBank accession numbers for specimens sequenced in this study. (DOC 102 kb)

Additional file 3: Species names, distributional codes and migratory status for ancestral state reconstruction analyses of the Mellisugini species used in this study. A = western North America, B = eastern North America, C = eastern Mexico and Central America, D = West Indies, E = South America; M = migratory, S = sedentary (binary character codification). (DOC 169 kb)

Additional file 4: Migratory status of the Mellisugini species for ancestral state reconstruction analysis used in this study. M = migratory, S = sedentary. (DOC 57 kb)

Additional file 5: Species names, English names and distributional range for the Mellisugini species used in this study. (DOC 75 kb)

Additional file 6: Bayesian 50% majority rule consensus trees of 132 representatives of bee hummingbirds (32 of the 36 extant species, 89%), 15 of mountain gems and 15 of emeralds. The trees are based on data sets of (a) only mitochondrial genes ('unpartitioned mtDNA data set'), (b) only mitochondrial genes as two partitions ("partitioned mtDNA data set"), (c) only nuclear genes ('unpartitioned nuDNA data set"), and (d) only nuclear genes as four partitions ("partitioned nuDNA data set"). Posterior probabilities (PP) > 0.5 are shown. (PDF 930 kb)

Additional file 7: Comparison of backbone tree topologies of the Mellisugini. (a) McGuire et al. [25], (b) Abrahamczyk & Renner [72], and (c) Bayesian 50% majority rule consensus tree of 32 bee hummingbird species of this study in Additional file 7. Asterisks denote nodes with 1.0 posterior probability (PP) support. Numbers at nodes reflect posterior probabilities less than 1.0. Support values for nodes of phylogeny in (b) are not provided in Abrahamczyk & Renner [72]. (PDF 425 kb)

Additional file 8: Bayesian 50% majority rule consensus tree of 32 bee hummingbird species and representatives of mountain gems and emeralds used as outgroups. The tree is based on a combined data set of all available fragments of *ND2*, *ND4*, *AK1 I5*, *MUSK I3*, *ODC1* and *FBG I7* and partition-specific DNA evolution models of each gene ('6-partitions data set'). Posterior probabilities (PP) > 0.5 are shown. (PDF 404 kb)

Acknowledgements
We thank Cristina Bárcenas, Antonio Acini Vásquez, Andrés Ortíz-Rodríguez, Clementina González, Flor Rodríguez-Gómez, Eduardo Ruiz-Sánchez, María José Pérez-Crespo and Andreia Malpica for field and lab assistance; and Cristina González-Rubio (CIBNOR), Borja Milá (MNCN-C SIC) and Rosa Alicia Jiménez (MC: Escuela de Biología, USAC) for providing tissue samples essential to this work. The samples collected in Mexico were conducted with the permission of the Secretaría de Medio Ambiente y Recursos Naturales, Instituto de Ecología, Dirección General de Vida Silvestre (permit numbers: INE: SEMARNAP, D00-02/3269, INE SGPA/DGVS/02038/07, 01568/08, 02517/09, 07701/11, 13528/14, 02577/15, 06448/16). Borja Milá provided useful comments on previous versions of the manuscript. This work constitutes partial fulfillment of Y.L.V's doctorate in Biodiversity and Systematics at INECOL.

Funding
This project was funded by the Departamento de Biología Evolutiva, Instituto de Ecología, A.C. (INECOL) awarded to J.F.O. (20030/10563). Y.L.V. was supported by a doctoral scholarship (262561) from CONACyT. The publication costs were financed by the Dirección General of the INECOL (20029/60813).

Authors' contributions
The authors of this paper have a general interest in the evolutionary history of hummingbirds. For this paper, YLV was involved in collecting most samples and obtaining the molecular data, and together with JFO in performing the phylogenetic and dating analyses, writing the manuscript and interpreting the molecular and phylogenetic data. Both authors read and approved the final manuscript.

Competing interests
The authors declare that they have no competing interests.

References
1. Berthold P. Control for bird migration. London: Chapman and Hall; 1996.
2. Berthold P. Bird migration. 2nd ed. Oxford: Oxford University Press; 2001.
3. Faaborg J, Holmes RT, Anders AD, Bildstein KL, Dugger KM, Gauthreaux SA Jr, et al. Recent advances in understanding migration systems of new world land birds. Ecol Appl. 2010;80(1):3–48.
4. Supp SR, La Sorte FA, Cormier TA, Lim MCW, Powers DR, Wethington SM, et al. Citizen-science data provides new insight into annual and seasonal variation in migration patterns. Ecosphere. 2015;6(1):15.
5. Helbig AJ. Evolution of bird migration: a phylogenetic and biogeographic perspective. In: Avian migration (eds P Berthold, E Gwinner, E Sonnenschein). Berlin, Germany: Springer; 2003. p. 3–20.
6. Salewski V, Bruderer B. The evolution of bird migration–a synthesis. Naturwissenschaften. 2007;94(4):268–79.
7. Zink RM. The evolution of avian migration. Biol J Linn Soc. 2011;104(2):237–50.
8. Rolland J, Jiguet F, Jønsson KA, Condamine FL, Morlon H. Settling down of seasonal migrants promotes bird diversification. Proc R Soc B. 2014; 281(1784):20140473.
9. Rappole JH. The ecology of migrant birds: a Neotropical perspective. Washington, D.C.: Smithsonian Institution Press; 1995.
10. Voelker G, Bowie RCK, Klicka J. Gene trees, species trees and earth history combine to shed light on the evolution of migration in a model of avian system. Mol Ecol. 2013;22(12):333–3344.
11. Cox G. The role of competition in the evolution of migration. Evolution. 1968;22(1):180–92.
12. Cox G. The evolution of avian migration systems between temperate and tropical regions of the new world. Am Nat. 1985;126(4):451–74.
13. Lack D. Bird migration and natural selection. Oikos. 1968;19(1):1–9.
14. Safriel UN. The evolution of Palearctic migration—the case for southern ancestry. Isr J Zool. 1995;41(3):417–31.
15. Alerstam T, Hedenström A, Åkesson S. Long-distance migration: evolution and determinants. Oikos. 2003;103(2):247–60.
16. Levey DJ, Stiles FG. Evolutionary precursors of long-distance migration: resource availability and movement patterns in the Neotropical landbirds. Am Nat. 1992;140(3):447–76.
17. Terrill SB, Berthold P. Ecophysiological aspects of rapid population growth in a novel migratory blackcap (*Sylvia atricapilla*) population: an experimental approach. Oecologia. 1990;85(2):266–70.
18. Helbig AJ. Inheritance of migratory direction in a bird species: a cross-breeding experiment with SE- and SW-migrating blackcaps (*Sylvia atricapilla*). Behav Ecol Sociobiol. 1991;28(1):9–12.
19. Somveille M, Manica A, Butchart SHM, Rodrigues ASL. Mapping global diversity patterns for migratory birds. PLoS One. 2013;8(8):e70907.
20. Ehlers J, Gibbard PL. The extent and chronology of Cenozoic global glaciation. Quat Int. 2007;164–165:6–20.
21. Greenberg R, Kozlenko A, Etterson M, Dietsch T. Patterns of density, diversity, and the distribution of migratory strategies in the Russian boreal forest avifauna. J Biogeogr. 2008;35(11):2049–60.
22. Somveille M, Rodrigues ASL, Manica A. Why do birds migrate? A macroecological perspective. Glob Ecol Biogeogr. 2015;24(6):664–74.
23. Somveille M. The global ecology of bird migration: patterns and processes. Front Biogeography. 2016;8(3):e32694.
24. McGuire JA, Witt CC, Remsen JV Jr, Dudley R, Altshuler DL. A higher-level taxonomy for hummingbirds. J Ornithol. 2009;150:155–65.
25. McGuire JA, Witt CC, Remsen JV Jr, Corl A, Rabosky DL, Altshuler DL, et al. Molecular phylogenetics and the diversification of hummingbirds. Curr Biol. 2014;24(8):910–6.
26. Ornelas JF, González C, de los Monteros JA E, Rodríguez-Gómez F, García-Feria LM. In and out of Mesoamerica: temporal divergence of *Amazilia* hummingbirds pre-dates the orthodox account of the completion of the Isthmus of Panama. J Biogeogr. 2014;41(1):168–81.
27. Pacheco MA, Battistuzzi FU, Lentino M, Aguilar RF, Kumar S, Escalante AA. Evolution of modern birds revealed by mitogenomics: timing the radiation and origin of major orders. Mol Biol Evol. 2011;28(6):1927–42.
28. del Hoyo J, Elliott A, Sargatal J. (eds.) Handbook of the Birds of the World, Vol. 5, Barn-owls to hummingbirds. Barcelona: Lynx Editions; 1999. p. 537–680.
29. Clark CJ, Elias DO, Prum RO. Aeroelastic flutter produces hummingbird feather songs. Science. 2011;333(6048):1430–3.
30. Clark CJ, Feo JT, Van Dongen WFD. Sounds and courtship displays of the Peruvian Sheartail, Chilean Woodstar, oasis hummingbird, and a hybrid male Peruvian Sheartail x Chilean Woodstar. Condor. 2013;115(3):558–75.
31. Feo TJ, Musser MJ, Bery J, Clark CJ. Divergence in morphology, calls, song, mechanical sounds, and genetics supports species status for the Inaguan hummingbird (Trochilidae: *Calliphlox "evelynae" lyrura*). Auk. 2014;132(1):248–64.
32. Licona-Vera Y, Ornelas JF. Genetic, ecological and morphological divergence between populations of the endangered Mexican Sheartail hummingbird (*Doricha eliza*). PLoS One. 2014;9(7):e101870.

33. Arizmendi MC, Berlanga H. Colibríes de México y Norteamérica. México, D.F: Comisión Nacional para el Conocimiento y Uso de la Biodiversidad (CONABIO); 2014.

34. Berthold P, Pulido F. Heritability of migratory activity in a natural bird population. Proc R Soc Lond B, Bot Sci. 1994;257(1350):311–5.

35. Pulido F, Berthold P, Mohr G, Querner U. Heritability of the timing of autumn migration in a natural bird population. Proc R Soc Lond B, Bot Sci. 2001;268(1470):953–9.

36. Pulido F, Berthold P. Current selection for lower migratory activity will drive the evolution of residency in a migratory bird population. Proc Natl Acad Sci USA. 2010;107(6):7341–6.

37. Malpica A, Ornelas JF. Postglacial northward expansion and genetic differentiation between migratory and sedentary populations of the broad-tailed hummingbird (Selasphorus platycercus). Mol Ecol. 2014;23(2):435–52.

38. McKinney AM, CaraDonna PJ, Inouye DW, Barr B, Bertelsen CD, Waser NM. Asynchronous changes in phenology of migratory broad-tailed hummingbirds and their early- season nectar resources. Ecology. 2012;93(9):1987–93.

39. BirdLife International. Chaetocercus berlepschi. IUCN Red List of Threatened Species. Version 2013.2. International Union for Conservation of Nature. http://dx.doi.org/10.2305/IUCN.UK.2016-3.RLTS.T22688279A93190225.en. Retrieved 2 June 2017.

40. IUCN. The IUCN Red List of Threatened Species. Version 3.1. Available: http://www.iucnredlist.org . Accessed 23 Nov 2013.

41. McGuire JA, Witt CC, Altshuler DL, Remsen JV Jr. Phylogenetic systematics and biogeography of hummingbirds: Bayesian and maximum likelihood analyses of partitioned data and selection of an appropriate partitioning strategy. Syst Biol. 2007;56(5):837–56.

42. Müller K, Müller J, Neinhuis C, Quandt D. PhyDE – Phylogenetic Data Editor, v0.995. Available at: http://www.phyde.de. 2006.

43. Licona-Vera Y, Ornelas JF. Data from: the conquering of North America: dated phylogenetic and biogeographic inference of migratory behavior in bee hummingbirds. 2017. Dryad Digital Repository. http://dx.doi.org/10.5061/dryad.68fn0 .

44. Ronquist F, Huelsenbeck J. MrBayes 3: Bayesian phylogenetic inference under mixed models. Bioinformatics. 2003;19(12):1572–4.

45. Miller MA, Pfeiffer W, Schwartz T. Creating the CIPRES science Gateway for inference of large phylogenetic trees, Proceedings of the Gateway computing environments workshop (GCE), 14 Nov. 2010, New Orleans, LA. Washington, DC: Institute of Electrical and Electronics Engineers (IEEE); 2010. p. 1–8.

46. Darriba D, Taboada GL, Doallo R, Posada D. JModelTest 2: more models, new heuristics and parallel computing. Nat Methods. 2012;9:772.

47. Nylander JAA, Ronquist F, Huelsenbeck JP, Nieves-Aldrey JL. Bayesian phylogenetic analysis of combined data. Syst Biol. 2004;53(1):47–67.

48. Stamatakis A. RAxML Version 8: a tool for phylogenetic analysis and postanalysis or large phylogenies. Bioinformatics. 2014;30(9):1312–3.

49. Goloboff PA. NoName (NONA), Version 2.0. Program and documentation. Tucumán: Fundación Instituto Miguel Lillo; 1997.

50. Nixon KC. WinClada, Version 1.00.08. Program and Documentation. Ithaca (NY): Cornell University Press; 2002.

51. Drummond AJ, Suchard MA, Xie D, Rambaut A. Bayesian phylogenetics with BEAUti and the BEAST 1.7. Mol Biol Evol. 2012;8(29):1969–73.

52. Bouckaert R, Heled J, Kühnert D, Vaughan T, Wu C-H, Xie D, et al. BEAST 2: a software platform for Bayesian evolutionary analysis. PLoS Comput Biol. 2014;10(4):e1003537.

53. Drummond AJ, Ho SY, Phillips MJ, Rambaut A. Relaxed phylogenetics and dating with confidence. PLoS Biol. 2006;4(5):e88.

54. Lerner HRL, Meyer M, James HF, Hofreiter M, Fleischer RC. Multilocus resolution of phylogeny and timescale in the extant adaptive radiation of Hawaiian honeycreepers. Curr Biol. 2011;21(21):1838–44.

55. Yu Y, Harris AJ, He XJ. RASP (reconstruct ancestral state in phylogenies): a tool for historical biogeography. Mol Phylogenet Evol. 2015;87:46–9.

56. Nylander JAA, Olsson U, Alström P, Sanmartín I. Accounting for phylogenetic uncertainty in biogeography: a Bayesian approach to dispersal-vicariance analysis of the thrushes (Aves: Turdus). Syst Biol. 2008; 57(2):257–68.

57. Harris AJ, Xiang QY. Estimating ancestral distributions of lineages with uncertain sister groups: a statistical approach to dispersal–Vicariance analysis and a case using Aesculus L. (Sapindaceae) including fossils. J Syst Evol. 2009;47(5):349–68.

58. Blair C, Sánchez-Ramírez S. Diversity-dependent cladogenesis throughout western Mexico: evolutionary biogeography of rattlesnakes (Viperidae: Crotalinae: Crotalus and Sistrurus). Mol Phylogenet Evol. 2016;97:145–54.

59. Boyle WA. Altitudinal bird migration in North America. Auk. 2017;134(2):443–65.

60. Winger BM, Lovette IJ, Winker DW. Ancestry and evolution of seasonal migration in the Parulidae. Proc R Soc B. 2012;279(1728):610–8.

61. Maddison WP, Maddison DR. Mesquite: a modular system for evolutionary analysis. Version 3.11. 2009–2016 [http://mesquiteproject.wikispaces.com].

62. Cunningham CW, Omland KE, Oakley TH. Reconstructing ancestral character states: a critical reappraisal. Trends Ecol Evol. 1998;13(9):361–6.

63. Cunningham CW. Some limitations of ancestral character-state reconstruction when testing evolutionary hypotheses. Syst Biol. 1999;48(3):665–74.

64. Ronquist F. Bayesian inference of character evolution. Trends Ecol Evo. 2004;19(9):475–81.

65. Ekman S, Andersen HL, Wedin M. The limitations of ancestral state reconstruction and the evolution of the ascus in the Lecanorales (lichenised Ascomycota). Syst Biol. 2008;57(1):141–56.

66. Schäffer S, Koblmüller S, Pfingstl T, Sturmbauer C, Krisper G. Ancestral state reconstruction reveals multiple independent evolution of diagnostic morphological characters in the "higher Oribatida" (Acari), conflicting with current classification schemes. BMC Evol Biol. 2010;10:246.

67. Yang Z, Kumar S, Nei M. A new method of inference of ancestral nucleotide and amino acid sequences. Genetics. 1995;141(4):1641–50.

68. Revell LJ. Two new graphical methods for mapping trait evolution on phylogenies. Methods Ecol Evol. 2013;4(8):754–9.

69. R Development Core Team. R: A language and environment for statistical computing. R foundation for statistical computing, Vienna, Available at: http://www.r-project.org/). 2012.

70. Felsenstein J. Phylogenies and the comparative method. Am Nat. 1985; 125(1):1–15.

71. Pagel M. The maximum likelihood approach to reconstructing ancestral character states of discrete characters on phylogenies. Syst Biol. 1999;48(3):612–22.

72. Abrahamczyk S, Renner SS. The temporal build-up of hummingbird/plant mutualisms in North America and temperate South America. BMC Evol Biol. 2015;15:104.

73. Wiens JJ. Missing data and the design of phylogenetic analyses. J Biomed Inform. 2006;39(1):34–42.

74. Wiens JJ, Moen DS. Missing data and the accuracy of Bayesian phylogenetics. J Syst Evol. 2008;46(3):307–14.

75. Cerling TE, Harris JM, MacFadden BJ, Leakey MG, Quade J, Eisenmann V, et al. Global vegetation change through the Miocene/Pliocene boundary. Nature. 1997;389:153–8.

76. Zachos JC, Shackleton NJ, Revenaugh JS, Pälike H, Flower BP. Climate response to orbital forcing across the Oligocene-Miocene boundary. Science. 2001;292(5515):274–8.

77. Riddle B, Hafner D. A step-wise approach to integrating phylogeographic and phylogenetic biogeographic perspectives on the history of a core north American warm deserts biota. J Arid Environ. 2006;66(3):435–61.

78. Good-Avila SV, Souza V, Gaut BS, Eguiarte LE. Timing and rate of speciation in Agave (Agavaceae). Proc Natl Acad Sci USA. 2006;103(24):9124–9.

79. Arakaki M, Christin P-C, Nyffeler R, Lendel A, Eggli U, Ogburn RM, et al. Contemporaneous and recent radiations of the world's major succulent plant lineages. Proc Natl Acad Sci USA. 2011;108(20):8379–84.

80. Hernández-Hernández T, Brown JW, Schlumpberger BO, Eguiarte LE, Magallón S. Beyond aridification: multiple explanations for the elevated diversification of cacti in the new Wold succulent biome. New Phytol. 2014;202(4):1382–97.

81. Smith SA, Donoghue MJ. Combining historical biogeography with niche modelling in the Caprifolium clade of Lonicera (Caprifoliaceae, Dipsacales). Syst Bot. 2010;59(3):322–41.

82. Ornelas JF, Gándara E, Vásquez-Aguilar AA, Ramírez-Barahona S, Ortiz-Rodriguez AE, González C, et al. A mistletoe tale: postglacial invasion of Psittacanthus schiedeanus (Loranthaceae) to Mesoamerican cloud forests revealed by molecular and species distribution modeling. BMC Evol Biol. 2016;16:78.

83. Pérez-Crespo MJ, Ornelas JF, González-Rodríguez A, Ruiz-Sanchez E, Vásquez-Aguilar AA, Ramírez-Barahona S. Phylogeography and population differentiation in the Psittacanthus calyculatus (Loranthaceae) mistletoe: a complex scenario of the climate-volcanism interaction along the trans-Mexican Volcanic Belt. J Biogeogr. 2017;00(0):00.

84. Wolfe AD, Randle CP, Datwyler SL, Morawetz JJ, Arguedas N, Díaz J. Phylogeny, taxonomic affinities, and biogeography of *Penstemon* (Plantaginaceae) based on ITS and cpDNA sequence data. Am J Bot. 2006;93(11):1699–713.

85. Azpeitia F, Lara C. Reproductive biology and pollination of the parasitic plant *Psittacanthus calyculatus* (Loranthaceae) in central Mexico. J Torrey Bot Soc. 2006;133(3):429–38.

86. Díaz Infante S, Lara C, Arizmendi MC, Eguiarte LE, Ornelas JF. Reproductive ecology and isolation of *Psittacanthus calyculatus* and *P. auriculatus* mistletoes (Loranthaceae). PeerJ. 2016;4:e2491.

87. Pérez-Crespo MJ, Lara C, Ornelas JF. Uncorrelated mistletoe infection patterns and mating success with local host specialization in *Psittacanthus calyculatus* (Loranthaceae). Evol Ecol. 2016;30(6):1061–80.

88. Abrahamczyk S, Souto-Vilarós D, McGuire JA, Renner SS. Diversity and clade ages of West Indian hummingbirds and the largest plant clades dependent on them: a 5–9 Myr young mutualistic system. Biol J Linn Soc. 2015;114(4):848–59.

89. Outlaw DC, Voelker G, Mila B, Girman DJ. The evolution of migration in, and historical biogeography of the *Catharus* thrushes: a molecular phylogenetic approach. Auk. 2003;120(2):299–310.

90. Milá B, Smith TB, Wayne RK. Postglacial population expansion drives the evolution of long-distance migration in a songbird. Evolution. 2006;60(11):2403–9.

91. Klicka J, Voelker G, Spellman GM. A molecular phylogenetic analysis of the "true thrushes" (Aves: Turdinae). Mol Phylogenet Evol. 2005;34(3):486–500.

92. Outlaw DC, Voelker G. Phylogenetic tests of hypotheses for the evolution of avian migration: a case study using the Motacillidae. Auk. 2006;123(2):455–66.

93. Ruegg KC, Hijmans RJ, Moritz C. Climate change and the origin of migratory pathways in the Swainson's Thrush*catharus ustulatus*. J Biogeogr. 2006;33(7):1172–82.

94. Winker K, Pruett CL. Seasonal migration, speciation, and morphological convergence in the genus *Catharus* (Turdidae). Auk. 2006;123(4):1052–68.

95. Voelker G, Light JE. Palaeoclimatic events, dispersal and migratory losses along the afro-European axis as drivers of biogeographic distribution in *Sylvia* warblers. BMC Evol Biol. 2011;11:163.

96. Topp CM, Pruett CL, McCracken KG, Winker K. How migratory thrushes conquered northern North America: a comparative phylogeographic approach. PeerJ. 2013;1:e206.

97. Lovette IJ, Pérez-Emán JL, Sullivan JP, Banks RC, Fiorentino I, Córdoba-Córdoba S, et al. A comprehensive multilocus phylogeny for the wood-warblers and a revised classification of the Parulidae (Aves). Mol Phylogenet Evol. 2010;57(2):753–70.

98. Simpson RK, Johnson MA, Murphy TG. Migration and the evolution of sexual dichromatism: evolutionary loss of female coloration with migration among wood-warblers. Proc R Soc Lond B, Bot Sci. 2015;282(1809):20150375.

99. Joseph L, Lessa E, Christidis L. Phylogeny and biogeography in the evolution of migration: shorebirds of the *Charadrius* Complex. J Biogeogr. 1999;26(2):329–42.

100. Chesser R. Evolution in the high Andes: the phylogenetics of *Muscisaxicola* ground-tyrants. Mol Phylogenet Evol. 2000;15(3):369–80.

101. Winger BM, Barker FK, Ree RH. Temperate origins of long-distance seasonal migration in new world songbirds. Pro Natl Acad Sci USA. 2014;111(33):12115–20.

102. Cicero C, Johnson N. Molecular phylogeny and ecological diversification in a clade of new world songbirds (genus *Vireo*). Mol Ecol. 1998;7(10):1359–70.

103. Kondo B, Omland K. Ancestral state reconstructions of migration: multistate analysis reveals rapid changes in new world orioles (*Icterus* spp.). Auk. 2007;124(2):410–9.

104. Rheindt F, Christidis L, Norman J. Habitat shifts in the evolutionary history of a Neotropical flycatcher lineage from forest and open landscapes. BMC Evol Biol. 2008;8:193.

105. Claramunt S, Derryberry EP, Remsen JV Jr, Brumfield RT. High dispersal ability inhibits speciation in a continental radiation of passerine birds. Proc R Soc Lond B, Bot Sci. 2012;279(1733):1567–74.

106. Johnson NY, Cicero C. New mitochondrial DNA data affirm the importance of Pleistocene speciation in north American birds. Evolution. 2004;58(5):1122–30.

107. Weir J, Schluter D. Ice sheets promote speciation in boreal birds. Proc R Soc Lond B, Bot Sci. 2004;271(1551):1881–7.

108. Lovette IJ. Glacial cycles and the tempo of avian speciation. Trends Ecol Evol. 2005;20(2):57–9.

109. Ruegg KC, Smith TB. Not as the crow flies: a historical explanation for circuitous migration in Swainson's thrush (*Catharus ustulatus*). Proc R Soc Lond B, Bot Sci. 2002;269(1498):1375–81.

110. Chamberlain CP, Bensch S, Feng X, Åkesson S, Andersson T. Stable isotopes examined across a migratory divide in Scandinavian willow warblers (*Phylloscopus trochilus trochilus* and *Phylloscopus trochilus acredula*) reflect their African winter quarters. Proc R Soc Lond B, Bot Sci. 2000;267(1438):43–8.

111. Bensch S, Grahn M, Muller N, Gay L, Åkesson S. Genetic, morphological, and feather isotope variation of migratory willow warblers show gradual divergence in a ring. Mol Ecol. 2009;18(14):3087–96.

112. Rolshausen G, Segelbacher G, Hobson KA, Schaefer M. Contemporary evolution of reproductive isolation and phenotypic divergence in sympatry along a migratory divide. Curr Biol. 2009;19(24):2097–101.

113. Procházka P, Stokke BG, Jensen H, Fainová D, Bellinvia E, Fossøy F, et al. Low genetic differentiation among reed warbler *Acrocephalus scirpaceus* populations across Europe. J Avian Biol. 2011;42(2):103–13.

114. Bairlein F, Norris DR, Nagel R, Bulte M, Voigt CC, Fox JW, et al. Cross-hemisphere migration of a 25 g songird. Biol Lett. 2012;8(4):505–7.

115. Delmore KE, Hübner S, Kane NC, Schuster R, Andrew RL, Câmara F, et al. Genomic analysis of a migratory divide reveals candidate genes for migration and implicates selective sweeps in generating islands of differentiation. Mol Ecol. 2015;24(8):1873–88.

116. Delmore KE, Toews DPL, Germain RR, Owens GL, Irwin DE. The genetics of seasonal migration and plumage color. Curr Biol. 2016;26(16):2167–73.

3

Phylogenetic and paleobotanical evidence for late Miocene diversification of the Tertiary subtropical lineage of ivies (*Hedera* L., Araliaceae)

V. Valcárcel[1][*] (iD), B. Guzmán[2], N. G. Medina[3], P. Vargas[2] and J. Wen[4]

Abstract

Background: *Hedera* (ivies) is one of the few temperate genera of the primarily tropical Asian Palmate group of the Araliaceae, which extends its range out of Asia to Europe and the Mediterranean basin. Phylogenetic and phylogeographic results suggested Asia as the center of origin and the western Mediterranean region as one of the secondary centers of diversification. The bird-dispersed fleshy fruits of ivies suggest frequent dispersal over long distances (e.g. Macaronesian archipelagos), although reducing the impact of geographic barriers to gene flow in mainland species. Genetic isolation associated with geographic barriers and independent polyploidization events have been postulated as the main driving forces of diversification. In this study we aim to evaluate past and present diversification patterns in *Hedera* within a geographic and temporal framework to clarify the biogeographic history of the genus.

Results: Phylogenetic (biogeographic, time divergence and diversification) and phylogeographic (coalescence) analyses using four DNA regions (*nrl*TS, *trn*H-*psb*A, *trn*T-*trn*L, *rpl*32) revealed a complex spatial pattern of lineage divergence. Scarce geographic limitation to gene flow and limited diversification are observed during the early-mid Miocene, followed by a diversification rate increase related to geographic divergence from the Tortonian/Messinian. Genetic and palaeobotanical evidence points the origin of the *Hedera* clade in Asia, followed by a gradual E-W Asian extinction and the progressive E-W Mediterranean colonization. The temporal framework for the E Asia - W Mediterranean westward colonization herein reported is congruent with the fossil record. Subsequent range expansion in Europe and back colonization to Asia is also inferred. Uneven diversification among geographic areas occurred from the Tortonian/Messinian onwards with limited diversification in the newly colonized European and Asian regions. Eastern and western Mediterranean regions acted as refugia for Miocene and post-Miocene lineages, with a similar role as consecutive centers of centrifugal dispersal (including islands) and speciation.

Conclusions: The Miocene Asian extinction and European survival of *Hedera* question the general pattern of Tertiary regional extinction of temperate angiosperms in Europe while they survived in Asia. The Tortonian/Messinian diversification increase of ivies in the Mediterranean challenges the idea that this aridity period was responsible for the extinction of the Mediterranean subtropical Tertiary flora. Differential responses of *Hedera* to geographic barriers throughout its evolutionary history, linked to spatial isolation related to historical geologic and climatic constraints may have shaped diversification of ivies in concert with recurrent polyploidy.

Keywords: Eastern and western Mediterranean, Tertiary refuge, Centrifugal dispersal, Climate-driven spatial speciation

* Correspondence: virginia.valcarcel@uam.es
[1]Department of Biology (Botany), Universidad Autónoma de Madrid, Madrid, Spain
Full list of author information is available at the end of the article

Background

Hedera (ivies) is an Old World root-climber genus that extends from N Africa to Europe and S Asia [1–4]. The main diagnostic characters for species identification and recognition are morphological features from foliar trichomes and vegetative leaves [5]. However, ploidy level and geographic distribution provide fundamental information for species delimitation [5, 6] (Fig. 1). For example, ploidy level was essential for the identification of two morphologically similar species that were traditionally considered as the same species (*H. helix*: 2×, *H. hibernica*: 4×) [7], or for the segregation of two N African endemics (*H. algeriensis*: 4×, *H. maroccana*: 2×) [8–10]. In addition, geographic isolation helped distinguish two closely related species (*H. iberica*, SW Iberian Peninsula; *H. maderensis*, Madeira) [11], or disclose incipient speciation processes (*H. nepalensis* var. *nepalensis*, Himalaya; var. *sinensis*, E & SE China) [12, 13]. The combination of morphological and cytogenetic variation together with geographic information, help characterize 12 species (14 taxa): 6 diploid species (3 island endemics), 2 tetraploid species, 4 hexaploid taxa (2 island endemics), and 1 octoploid species (Fig. 1). The numerous island endemics (five) and the strong geographic structure detected in the DNA sequence variation [4, 11, 14–16] are interpreted as an imprint of the geographic barriers in the diversification process of *Hedera*. However, the endozoochorus dispersal syndrome of ivies, mainly mediated by birds [17, 18], together with the winter ripping of their fleshy fruits when food is scarce for animals, suggests that small geographic obstacles might not be such effective barriers to gene flow.

Ivies usually occupy shaded and humid understories of temperate and subtropical woodlands and riparian vegetation. However, they can also live in extremely dry environments such as fully sun-exposed rocks [19, 20]. Not all species are equally tolerant to both deep shaded and drought environments. Indeed, whereas *H. helix* occurs under the above-described contrasted environments, other species show very strict environmental requirements. For example, *H. iberica* is restricted to warm humid places ("Canuto") in southern Iberia [21], where remnants of the Tertiary flora also occur [22, 23]. Another example is *H. canariensis* that only occurs in humid, warm and shaded understories of the Macaronesian subtropical woodland 'Laurisilva' [24]. The fact that some species display the subtropical affinity that characterized the Asian Palmate group of the Araliaceae while others show a strict temperate tendency, may be suggesting that climate might have also contributed to the speciation in *Hedera*.

Different phylogenetic studies have described a very complex evolutionary history for this genus [4, 6, 13–16, 25]. The origin of the *Hedera* clade has been estimated in the Oligocene in Asia during the deep radiation of the Asian Palmate group of the Araliaceae [25, 26]. In previous phylogeographic studies conducted within *Hedera*, Asia was also suggested as the ancestral area of the extant species of ivies [16, 25]. It has been hypothesized that extinctions in Asia together with Mediterranean colonization and diversification have led to the present distribution. In this scenario, current Asian species of the genus would be the result of a re-colonization from the Mediterranean [15].

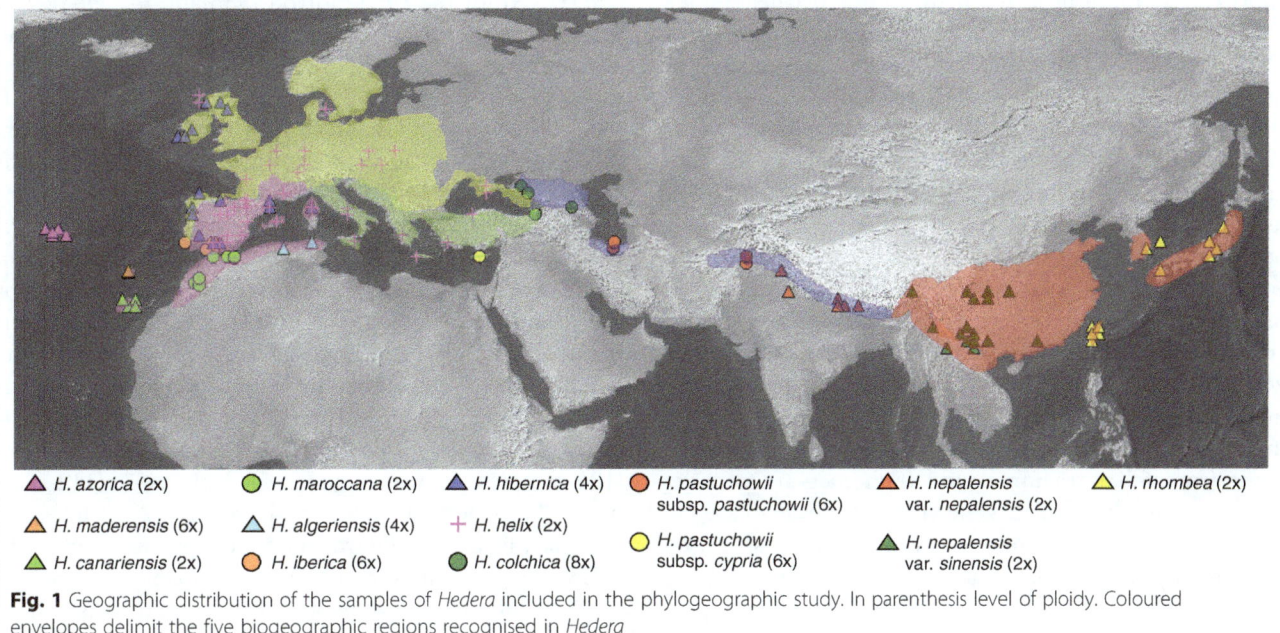

△ *H. azorica* (2x)   ○ *H. maroccana* (2x)   △ *H. hibernica* (4x)   ○ *H. pastuchowii* subsp. *pastuchowii* (6x)   △ *H. nepalensis* var. *nepalensis* (2x)   △ *H. rhombea* (2x)

△ *H. maderensis* (6x)   △ *H. algeriensis* (4x)   ✚ *H. helix* (2x)   ○ *H. pastuchowii* subsp. *cypria* (6x)   △ *H. nepalensis* var. *sinensis* (2x)

△ *H. canariensis* (2x)   ○ *H. iberica* (6x)   ● *H. colchica* (8x)

Fig. 1 Geographic distribution of the samples of *Hedera* included in the phylogeographic study. In parenthesis level of ploidy. Coloured envelopes delimit the five biogeographic regions recognised in *Hedera*

In this study we aim to reconstruct ancient and recent diversification patterns in *Hedera* under the working hypothesis that geographic barriers have determined the main patterns of diversification by promoting speciation. Under this working hypothesis, divergence events in *Hedera* would be expected to occur preferentially between areas after a colonization event. The resulting geographic isolation would have led to a strong geographic structure in the genetic variation. To evaluate this hypothesis we analyzed three plastid DNA spacer/intron regions (*trn*H-*psb*A, *trn*T-*trn*L, and *rpL*12) and the nuclear ribosomal ITS (nrDNA) region. First, a *nr*ITS dated phylogeny was reconstructed and used as a starting point to conduct biogeographic and diversification analyses. Phylogenetic results based on *nr*ITS data were examined together with those obtained from the phylogeographic analysis of the plastid dataset. To achieve our ultimate goal of clarifying the biogeographic and phylogeographic history of *Hedera*, the following specific objectives were addressed, the: (1) study of the Mediterranean and Asian areas using a geographically-balanced targeted sampling, (2) reconstruction of past and present diversification patterns, and (3) evaluation of the importance of geographic barriers in promoting isolation.

Methods
Taxon sampling and sequencing
Phylogenetic sampling
A phylogenetic-based study was performed to provide a temporal context to conduct the biogeographic and diversification analyses for reconstructing the evolution of the *Hedera* clade. Wide sampling of outgroup is needed for biogeographic inferences when the nodes of interest approximate the root of the ingroup tree because node's estimates partly rely on the optimization of their stems [27]. Therefore, all the generic-lineages of the Asian Palmate group have been included, as well as the putative sister-group of the Asian Palmate group (the *Aralia-Panax* group). The phylogenetic-based analyses (biogeographic, diversification and divergence age analyses) used the *nr*ITS region because: (1) there is a large number of available sequences of Araliaceae, (2) it is more variable than the fastest evolving plastid regions and provides more resolved tree topologies, (3) it better complements the evolutionary history of the genus where nuclear and plastid incongruence has been previously reported due to hybridization [15, 25], and (4) main diagnostic characters in the taxonomy of *Hedera* come from foliar trichomes and trichomes are genetically controlled by nuclear genes [28, 29]. In any case, all the analyses were also done with the plastid dataset used for the phylogeographic study (see below) but not included in the study because the lack of branch support may have resulted in inaccurate interpretations of the

biogeographic, divergence age and diversification results. The *nr*ITS dataset included 34 samples representing 12 species of *Hedera*, 44 of the other 20 generic-lineages of the Asian Palmate group, and 12 other Araliaceae genera (Additional file 1). *Harmsiopanax ingens* was used as the outgroup. All the 90 *nr*ITS sequences were obtained from previous studies [6, 15, 25, 26] and downloaded from GenBank (http://www.ncbi.nlm.nih.gov, Additional file 1).

Phylogeographic sampling
A phylogeographic study was conducted to reconstruct the geographic pattern of genetic diversity within *Hedera*, including 153 samples representing the 12 species (14 taxa) recognised (Fig. 1 and Additional file 2). The number of samples per species varied between 5 and 40, except for *H. algeriensis* (endemic to N Algeria and N Tunisia) for which only two samples were available. Sampling effort was more intensive on the two most widespread species, leading to the inclusion of 40 samples of the European *H. helix* and 32 of the Asian *H. nepalensis*. Samples were selected to represent the whole geographic range of each species with an emphasis on the areas considered as Tertiary refugia both in the Mediterranean [30] and in China [31]. To investigate the geographic origin of *Hedera*, *Kalopanax septemlobus* was included as the outgroup (Additional file 2) [25]. Additionally, to evaluate the potential impact of the uncertainty on the sister group of *Hedera* [25], six other Asian Palmate genera were also included (Additional file 2).

Three plastid DNA regions were analyzed (*rpL*32, *trn*H-*psb*A and *trn*T-*trn*L) for this part of the study. The primers used for the amplifications were as follows: (1) trn a and trn b for *trn*T-L spacer [32], (2) rpL32F and *trn*L(UAG) for the *rpL*32 intron [33], and (3) *trn*HR and *psb*AF for the *trn*H-*psb*A spacer [34]. Amplifications and sequencing protocols followed Valcárcel et al. [15] for the *trn*T-*trn*L region, Mitchell et al. [26] for the *trn*H-*psb*A spacer and Shaw et al. [33] for the *rpL*32 intron. As a result 270 sequences were newly generated in this study (76 for *rpL*32, 89 for *trn*T-*trn*L, and 111 for *trn*H-*psb*A). The three plastid DNA regions of *K. septemlobus* were taken from Li et al. [35] and downloaded from GenBank (http://www.ncbi.nlm.nih.gov), as well as for the other six genera of Araliaceae included. The sampling for the *trn*T-*trn*L spacer was completed by the addition of 44 sequences from our previous phylogeographic studies [15, 16]. Three DNA matrices were compiled using only *Kalopanax* as the outroup: *trn*T-*trn*L (134 samples, 89 new sequences), *trn*H-*psb*A (112 samples, 111 new sequences), and *rpL*32 (77 samples, 76 new sequences). Alignments were carried out with MUSCLE [36] followed by manual revision in Geneious v9.0.5 (http://www.geneious.com). Sequences were concatenated into a fourth matrix with the program

Sequence Matrix [37], only including samples with the three DNA regions sequenced (66 samples). A fifth matrix was additionally built to check for the impact of the different rootings of the *Hedera* network on the geographic interpretations, including 65 samples of *Hedera* plus 7 different genera of Araliaceae.

Phylogenetic-based analyses
Divergence age estimates
Divergence age estimates were inferred from the *nr*ITS matrix through a relaxed molecular clock implemented in Beast v.1.7.5 [38]. The substitution rate variation was modeled using an uncorrelated lognormal distribution and a Birth-Death process was applied to model speciation. The best evolutionary model for each of the DNA regions was selected by jModeltest setting a threshold of 3 ΔAIC (Additional file 3) [39]. The analyses were run in the absence of topological constraints, except for the calibration nodes. Two MCMC analyses were run for 100 million generations sampled every 10,000 generations. Convergence, mixing and effective sample size (ESS) of model parameters were assessed using Tracer 1.5 [40]. Samples from the two independent runs were pooled after removing a 25% burn-in using Log Combiner 1.7.5 [38]. Trees were summarized in a maximum clade credibility (MCC) tree obtained in TreeAnotator 1.7.5. Seven leaf macrofossils and two pollen grain fossils have been recorded in *Hedera* (Table 1). The taxonomy of ivies is mainly based on foliar trichomes that are not well preserved in fossils. Also, pollen grains do not show any morphological variation between the extant species of ivies [3]. Therefore, certainty on the phylogenetic placements of these fossils is limited. Only the oldest fossil found (Oligocene, Table 1) can be placed with certainty as a calibration point at the stem of *Hedera*, as inferred from the age recovered for the lineage of *Hedera* in previous studies [between 36.6 Mya and 51.55 Mya; 25]. However, we decided not to use this node as a calibration point in the final analyses because

the stem of *Hedera* represents an uncertain node in the phylogeny [25]. Instead, two previous divergence time estimates obtained from plastid DNA were employed as secondary calibration points [25]. This secondary calibration approach is more conservative since the age estimates in which it is based were obtained from a fossil-based calibration with a certain placement of fossils on robust nodes [25]. Accordingly, the crown groups of the Asian Palmate and *Hedera* clades were set as normal distributions of 72.55 ± 9.0 Myr and 7.65 ± 3.5 Myr, respectively. Calibration accuracy was tested by comparing divergence times herein estimated to the ages of reliable *Hedera* fossils [41–50] (Table 1). A second Beast analysis was performed using the oldest fossil of *Hedera* as the minimum age for the stem of *Hedera* to double check the posterior age recovered for the crown of *Hedera* in the secondary calibration estimate.

Biogeographic range estimation
Estimation of spatial patterns of geographic diversification in *Hedera* was conducted using a model-based likelihood method (Lagrange) and the *nr*ITS dataset. A Dispersal-Extinction-Cladogenesis analysis was performed over multiple trees using a script provided by Richard Ree (*pers. com.*). A multi-tree approach was essential to account for the impact of phylogenetic uncertainty on the ancestral range estimate of the stem group of *Hedera*, due to its ambiguous sister-group relationship [25]. For this purpose, 1000 post-burnin trees from the *nr*ITS Beast analysis were randomly selected with the R-package ape [51] and used as input trees for Lagrange. The geographic range of the Araliaceae was divided into eight regions based on floristic endemicity with special emphasis on *Hedera* distribution: (A) tropical Africa, (B) Neotropics, (C) Australia, (D) western Mediterranean region (including Macaronesia, hereafter W Mediterranean), (E) eastern Mediterranean region (hereafter E Mediterranean), (F) Europe, (G) western Asia (hereafter W Asia), and (H) eastern Asia (hereafter

Table 1 Fossil records of *Hedera* (entries arranged in chronological order)

Taxon	Locality	Biogeographic region	Age	Size class	Reference
Hedera sp.	Pongsan, Korea	E Asia	Oligocene (39.9–23 Mya)	Macrofossil	[44, 46]
Hedera cf. *multinervis*	Abkhazia, Georgia	W Asia	Miocene	Macrofossil (leaf)	[47]
Hedera cf. *multinervis*	Vegora, Greece	E Mediterranean	Miocene	Macrofossil (leaf)	[41, 48]
Hedera orbiculata	Silesia, Poland	Europe	Langhian (16.0–11.6 Mya)	Microfossil (pollen)	[41]
Hedera cf. *multinervis*	Cerdanya, Spain	W Mediterranean	Tortonian (11.6–11.3 Mya)	Macrofossil (leaf)	[42, 45]
Hedera sp.	Iberian Peninsula	W Mediterranean	Upper Miocene (11.7–5.3 Mya)	Microfossil (pollen)	[45]
Hedera cf. *helix*	Italy	W Mediterranean	Messinian (7.2–5.3 Mya)	Macrofossil (leaf)	[49]
Hedera sp.	NW Portugal	Europe	Pliocene (5.3–2.6 Mya)	Microfossil (pollen)	[50]
Hedera orbiculata	Thuringia, Germany	Europe	Piacenzian (3.6–2.6 Mya)	Macrofossil	[43]

E Asia). The codification of areas for each sample is provided in Additional file 1. In range constraints, adjacency of areas was allowed only between areas that share the edge (i.e., between W Mediterranean and Europe, and between E Mediterranean and W Asia). Maximum range size was set to two areas. Ranges allowed in the analysis included all possible combinations within those imposed by adjacency and maximum range size. Because we were only interested in recovering the early evolution of *Hedera*, results were only computed for the most internal nodes of *Hedera* with support (posterior probability (PP) >0.95; Fig. 2). Particularly, eight nodes were analyzed: the stem and crown groups of the *Hedera* clade (nodes 0 and 1; Fig. 2), two main clades (nodes 2 and 3; Fig. 2) and two main subclades (nodes 4 and 5; Fig. 2). The output file obtained from the Lagrange analysis of 1000 trees was read and parsed with a new R script herein designed for parsing multi-tree Lagrange results (Additional file 4). This script makes automatic parsing Lagrange multi-tree results, which simplifies the process of summarizing results saving time. The results were summarized as the mean of probabilities estimated by Lagrange for the posterior trees analyzed. The specific ancestral areas of the descendants of a given node are provided when a congruent biogeographic pattern is consistently recovered over the multi-trees analyzed (e.g., nodes 0 and 1; Fig. 2). If the estimated ancestral areas of the descendants of a given node *i* resulted in incongruent biogeographic patterns when the multi-trees results were analyzed together, only the ancestral areas of node *i* are provided with no specification to the descendant lineages. The ancestral area for a given node *i* is assumed to be the combination of the ancestral areas estimated for its descendants. For example, if the most probable biogeographic patterns for a given node are (1) E|D at a mean probability of 0.39, (2) D|E at a mean probability of 0.16, and (3) ED|D at a mean probability of 0.14; then, a simplification of the results are shown by providing the ancestral area for the node as ED with a mean probability of 0.69.

Diversification analyses

The 6500 post-burnin Beast trees were pruned to only consider the clade of *Hedera* and one tip per species, except for non-monophyletic species for which one tip per species-lineages was kept (18 tips, 14 taxa). We decided to only analyze the clade of *Hedera* because the independent analysis of a particular clade is recommended to isolate its diversification pattern from the heterogeneous diversification patterns of other clades in the phylogeny [52]. The stem of *Hedera* was included in the analysis because ignoring long branches before crown nodes may result in inaccurate interpretations of the diversification pattern within the crown group [53, 54].

The resulting 6500 post-burnin pruned trees were used as the inputs for the diversification analyses. The log-transformed number of extant taxa was plotted against time (LTT plot) for the 6500 post-burnin pruned trees using the R-package ape [51]. Fitness to speciation models with one, two or three diversification rates was also tested as implemented in the R-package LASER [55]. The best evolutionary model was selected for each of the 6500 post-burnin pruned trees based on the AIC using a ΔAIC of 4. Two contingency table tests with one dimension and two levels were performed over the results of the 6500 trees using chi-square goodness of fit [56] in R [57]: (1) number of trees with constant vs. variable rate models and (2) number of trees with Yule two rates vs. Yule three rates variable models.

The Phylogenetic Diversity (PD) index of Faith [58] measures the length of evolutionary pathways that connect a given set of taxa as the sum of branch lengths connecting taxa in a given area. In this study, Faith's index was used to account for the amount of PD of *Hedera* represented in each of the five major endemicity areas delimited for ivies in the biogeographic analyses (W Mediterranean, Europe, E Mediterranean, W Asia, E Asia). The aim of this analysis was to evaluate whether the most species-rich endemicity areas also hold the greatest evolutionary diversity. Faith's PD was estimated as implemented in the R-package Picante [59]. To assess if regions have significantly higher or lower PDs than at random expectations for a given number of species, PDs were calculated over the 1000 randomized posterior trees used for the Lagrange analysis but pruned to only consider the clade of *Hedera* and one tip per species-lineages. Two-tailed test was used to compare the observed PDs to the null distribution of the 1000 random replicates (significance level of 0.05).

Network reconstructions based on the coalescence [58] analysis of plastid haplotypes were performed on the five-plastid DNA matrices. Statistical Parsimony (SP) was applied as implemented in TCS 1.13 [60]. The 95% probability limit of parsimonious connections was applied and gaps were coded as missing data. Predictions from coalescent theory were applied to deals with homoplasy [61, 62]. To test the hypothesis that interior and tip haplotypes are equally frequent we used a contingency table test with one dimension and two levels using chi-square goodness of fit [56] in R [57].

Results
Estimates of divergence times
The early divergence age estimates of the *Hedera* clade (Fig. 2, Additional file 5) fit the timing set by the fossil

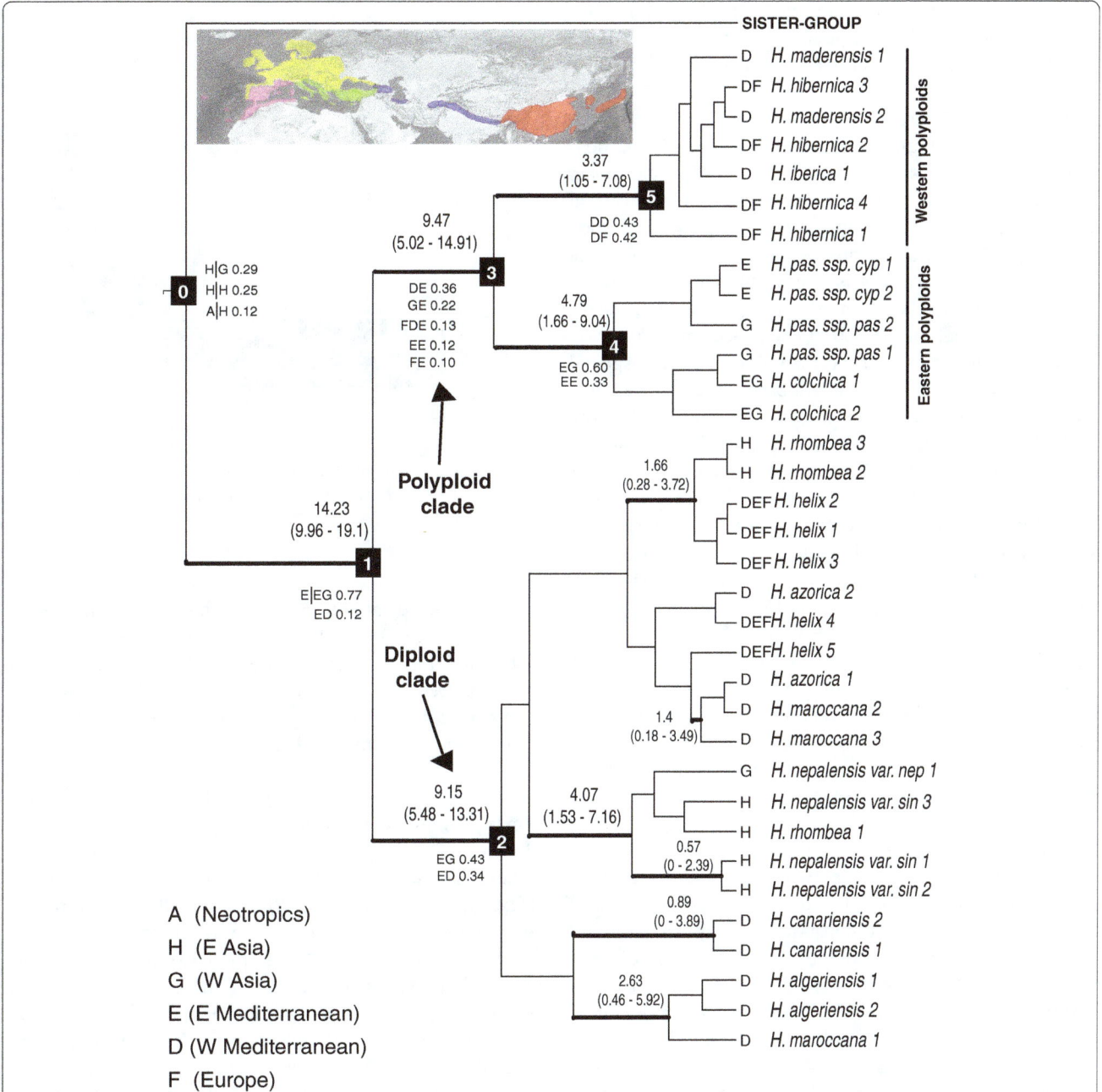

Fig. 2 Beast Maximum Clade Credibility chronogram of the *Hedera* clade from the *nr*ITS dataset. Mean ages and 95% CI are only represented for clades with >0.95 Posterior Probability support. Nodes of interest are labelled. Ancestral areas and estimated probabilities obtained from the Lagrange multi-tree analysis are provided only for the nodes of interest. The complete Asian Palmate group MCC tree including the 90 sequences of Araliaceae is shown in Additional file 5. Branch lengths were modified from the original tree (Additional file 5) to better fit the biogeographic results

record (Table 1). The two main clades of *Hedera* (diploid and polyploid) diverged in parallel during the late Miocene (9.47 / 9.15 Mya, 5.02–14.91 / 5.48–13.31 Mya 95% CI; nodes 2 / 3, Fig. 2), five million years after the crown age divergence of the *Hedera* clade in the early-mid Miocene (14.23, 9.96–19.1 Mya 95% CI; node 1; Fig. 2). The divergence time estimate for the eastern polyploid subclade is 4.79 Mya (1.66–9.04 Mya 95% CI;

node 4), whereas 3.37 Mya (1.05–7.08 Mya 95% CI; node 5) is recovered for the western polyploid subclade. The diploid clade displays a large basal polytomy that prevents from any further divergence time or biogeographic estimates (Fig. 2). The posterior estimates on the calibration node 1 lay outside the bound of the calibration, which is interpreted as low influence of this calibration prior on the posterior. The MCC Beast tree obtained after removing this secondary calibration point and including the oldest fossil of *Hedera* as the calibration point at the stem of *Hedera* revealed similar posterior node estimates (Additional file 6).

Ancestral range inference

Results from the multi-tree Lagrange analysis (1000 posterior random trees) plotted on the *nr*ITS MCC Beast tree of *Hedera* posit Asia as the most likely ancestral area for the stem of the genus (|H or |G, *P* = 0.66; node 0; Fig. 2). The most probable range inferred for the crown group of *Hedera* is the combination of the E Mediterranean and W Asia (*P* = 0.77; node 1). This result suggests that an extinction event occurred in E Asia between *Hedera's* stem and crown along with a dispersal event to the E Mediterranean. Subsequently, two equally plausible biogeographic scenarios are inferred for the diploid clade (from node 1 to 2; Fig. 2): (1) persistence in the ancestral area (EG, *P* = 0.43; node 2) or (2) W Asian extinction and dispersal to the W Mediterranean (ED, *P* = 0.34; node 2). The most probable biogeographic scenario for the polyploid clade (from node 1 to 3; Fig. 2) is the E Mediterranean persistence (E, *P* = 0.93; node 3), W Asian extinction (≠ G, *P* = 0.71), and the W Mediterranean colonization (D, *P* = 0.49). The W Mediterranean is recovered as part of the ancestral area for the divergence of the western polyploid subclade (node 5; D, *P* = 0.85; Fig. 2), whereas divergence of the eastern polyploid subclade may have occurred in the E Mediterranean and W Asia (EG, *P* = 0.66; node 4) or in the E Mediterranean (EE, *P* = 0.33).

Diversification analyses

Rate variable models (Yule two rates and Yule three rates) are more frequently selected as the best fitting speciation process than constant ones (pureBirth and birth-death) among the 6500 posterior pruned trees (4003 vs. 2497, respectively; *U* = 359.5319, *p*-value <0.001; Additional file 7). The Yule two rates model estimates an initial diversification of 0.04 per lineage per unit of time (−0.14–0.23 95% CI) that increased to 0.29 (0.024–0.55 95% CI) at around 7.82 Mya (5.94–9.71 95% CI). The equally probable Yule three rates (*U* = 0.89, *p*-value = 0.4) recovered an initial diversification of 0.04 (−0.29–0.37 95% CI) that increased to 0.94 (−2.96–4.45 95% CI) at around 10.19 Mya (7.42–12.97 95% CI) and decreased to 0.36 (−1.11–1.84 95% CI) at 3.20 Mya (0.35–6.05 95% CI). The LTT plot of the 6500 posterior pruned trees describes a flat initial diversification rate period followed by an ever-increasing diversification pattern from the Miocene onwards for the *Hedera* clade (Fig. 3). This graphical representation is congruent with the recovery of Yule two rates and Yule three rates as the best fitting speciation models. The E Mediterranean and W Asia are the only regions that reveal greater PDs

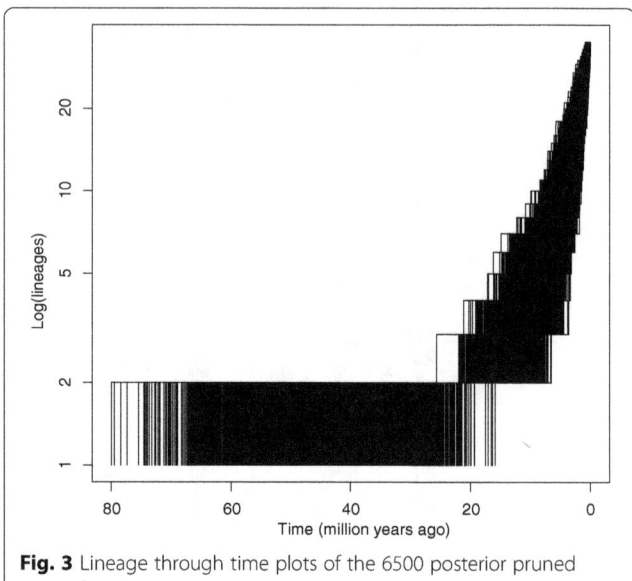

Fig. 3 Lineage through time plots of the 6500 posterior pruned trees of the *Hedera* clade from the *nr*ITS dataset

than expected at random for the given number of species (Table 2). The largest value of Faith's PD index is estimated for W Mediterranean (72.05, Table 2), while the smallest value is computed for E Asia (32.54). The remaining three main biogeographic areas display similar low values of PD index (Table 2).

Phylogeographic networks

Twenty-four haplotypes are detected within *Hedera* when the three-plastid DNA regions are concatenated (Fig. 4, Additional file 8). The highest number of haplotypes is detected in the Mediterranean region with 15 haplotypes, 57% of them exclusive to E Asia. Seven haplotypes are detected in Asia, 67% of them exclusive. Finally, five haplotypes occur in Europe, with only the 2 of them exclusive. The most frequent haplotype (Hp 8) is detected in 14 Asian samples (9 from E Asia and 5 from W Asia) of two species (*H. colchica*, *H. nepalensis*), followed by two Euro-Mediterranean haplotypes (Hp 22: 8 Mediterranean samples, 3 European; Hp 16: 4 European samples, 2 Mediterranean). Comparable geographic patterns of haplotype diversity are obtained when individual plastid matrices are analyzed (Additional files 9, 10 and 11). The W Mediterranean is consistently recovered as the geographic region with the highest number of haplotypes (*rpL*32: 6 Hps, *trn*H-*psb*A: 5 Hps, *trn*T-*trn*L: 7 Hps; Additional files 9, 10 and 11). The area with the second highest number of haplotypes is E Asia according to *trn*H-*psb*A (4 Hps; Additional file 10) and *trn*T-*trn*L (8 Hps; Additional

Table 2 Summary of different diversity indicators in *Hedera* according to the five biogeographic regions recognized

Biogeographic region	N_{spp}	Ploidy levels	Interior Hps	Tip Hps	Observed PD	Lower PD	Upper PD
W Mediterranean	8	2×, 4×, 6×, 8×	6	7	72.05***	73.51	75.12
E Mediterranean	3	2×, 6×, 8×	0	2	44.95*	38.73	41.77
Europe	2	2×, 4×	2	3	43.53$^{n.s.}$	31.29	34.64
W Asia	3	2×, 6×, 8×	1	2	44.34***	38.92	41.98
E Asia	2	2×	1	4	32.54***	38.31	41.47

Nspp: Number of species, *HPs:* Haplotypes, *Observed PD:* Faith's Phylogenetic Diversity Index [48, 52], Lower PD and Upper PD: Lower and upper bounds of the null distribution of the empirical randomization of PD. Level of significance is indicated as followed: ***$P \leq 0.001$, *$P \leq 0.05$, n.s. $P > 0.05$. Interior and tip haplotypes are according to Fig. 4

file 11) while Europe according to *rpL*32 (4 Hps; Additional file 9).

The number of interior haplotypes is 10 (6 unambiguous, 29 samples; 4 ambiguous, 4 samples) while 14 are tip haplotypes (23 samples). Five of the six unambiguous interior haplotypes are from the W

Mediterranean (5) and Europe (2) while only one from Asia (Table 2, Fig. 4). The mean number of sequences per haplotype varies from 4.8 for unambiguous interior haplotypes and 1.64 for tip haplotypes. Also the ratio of single-sample haplotyes (singleton) vs. non-singleton haplotypes varies between 0.2 for interior

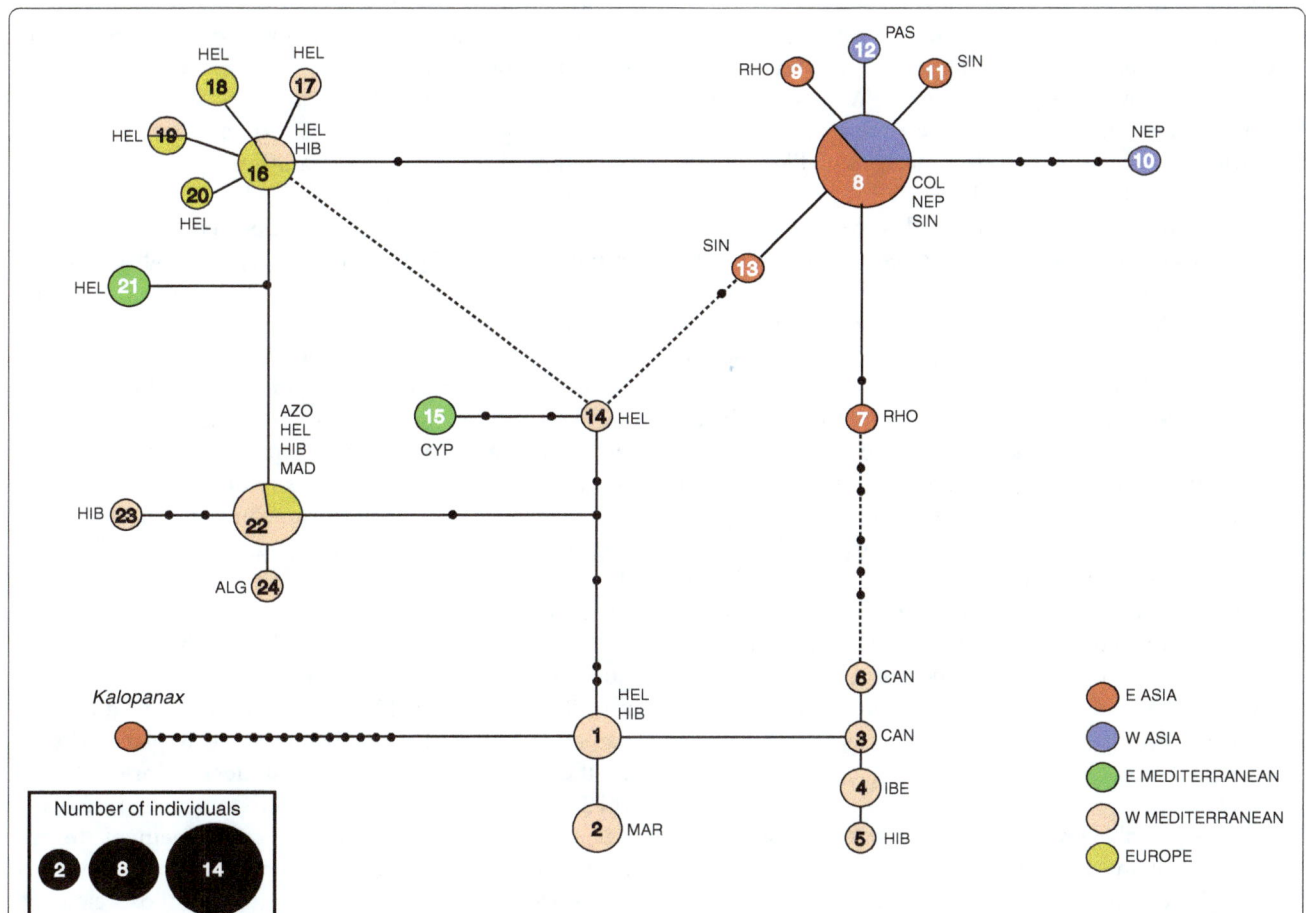

Fig. 4 Phylogeographic network of the *Hedera* plastid matrix (*rpL*32, *trn*H-*psb*A, *trn*T-*trn*L) including *Kalopanax septemlobus* as outgroup. Haplotype numbering is according to Additional file 8. Circle dimensions are proportional to the number of samples displaying each haplotype as indicated at the bottom. Lines indicate a single nucleotide substitution and dots (●) represent extinct or not-detected haplotypes. Dashed lines indicate ambiguities resolved under predictions of the Coalescent Theory [65]. Abbreviation names of the taxa displaying each haplotype are as follows: ALG, *H. algeriensis*; AZO, *H. azorica*; CAN, *H. canariensis*; COL, *H. colchica*; CYP, *H. pastuchowii* subsp. *cypria*; HEL, *H. helix*; HIB, *H. hibernica*; IBE, *H. iberica*; MAD, *H. maderensis*; MAR, *H. maroccana*; NEP, *H. nepalensis* var. *nepalensis*; PAS, *H. pastuchowii* subsp. *pastuchowii*; RHO, *H. rhombea*; SIN, *H. nepalensis* var. *sinensis*

haplotypes and 0.6 for tip haplotypes. Interior haplotypes are more frequent than tip ones ($U = 14.087$, p-value <0.001). This result together with the application of Templeton's rules allows us to resolve the network uncertainties (Fig. 4).

Limited taxonomic congruence is detected in the haplotype network, with the widespread Euro-Mediterranean *H. helix* and *H. hibernica* and the E Asian *H. rhombea* displaying between two and eight unrelated haplotypes (Fig. 4 and Additional file 8). However, strong geographic structure and environmental affinity is detected (Fig. 4). The W Mediterranean haplotypes are scattered through the network, whereas the majority of the European and Asian haplotypes are organized in two star-like groups. The Euro-Mediterranean and Asian start-like groups are connected to each other through their most widespread haplotypes (Hp 16 and Hp 8, respectively; Fig. 4). The haplotypes of all the samples included from the southwest Mediterranean are connected with no missing haplotype needed (hereafter "relict SW Mediterranean haplotypes"). The independent analysis of each plastid region consistently recovers the group of relict SW Mediterranean haplotypes and the Asian star-like group connected to the widespread Euro-Mediterranean star-like group (Additional files 9, 10 and 11).

The *Hedera* network is connected to *Kalopanax* (outgroup) through the relict SW Mediterranean Hp 1 with 16 missing haplotypes needed for connection. Similarly, a relict SW Mediterranean haplotype is connected to an Asian outgroup when the seven genera of Araliaceae are included (Additional file 12). Connection to outgroup slightly differs when plastid datasets are analyzed independently (Additional files 9, 10 and 11). The biogeographic connection to the outgroup is inferred through a widespread Euro-Mediterranean haplotype in *rpL*32 (Additional file 9) or through E Asian haplotypes in *trn*H-*psb*A (Additional file 10) and *trn*T-*trn*L (Additional file 11). The Taiwanese sample that connects to the outgroup in the *trn*H-*psb*A network (Additional file 10) is recovered as a tip haplotype when the three DNA regions are analyzed together (Hp 7; Fig. 4).

Discussion
Uneven geographic diversification of *Hedera*: E Asia as both ancestral and sink area

Eastern Asia is inferred as the most likely ancestral area for the *Hedera* clade (Fig. 2), which is congruent with the E Asian location of the oldest fossil record of *Hedera* (Table 1). This geographic context is also consistent with the fact that all the remaining 20 generic-lineages of the Asian Palmate group occur in E Asia. Indeed, tropical and subtropical SE Asian environments are inferred to be the center of diversification of the Asian Palmate

group and some of its genera [26, 63–69]. This E Asian placement of the ancestor of the temperate *Hedera* clade is congruent with the Asian origin of numerous lineages of the flora of Europe [70–72]. Westward dispersal and E Asia extinction appear to have occurred early in the evolution of the clade of *Hedera* leading to an initial divergence during the Lower-Middle Miocene in W Asia and the E Mediterranean (Fig. 2). The fossil record also supports the timing of this westward migration, since reliable fossils of *Hedera* are found for the first time in W Asia (Georgia) [47] and the E Mediterranean (Greece) [48] during the Lower-Middle Miocene (Table 1). The two early descendant lineages (polyploid and diploid clades) persisted in the E Mediterranean with a probable and independent W Mediterranean colonization coupled with W Asian extinction in the polyploid clade. The low diversification rate found for this period in the *Hedera* clade ($r_0 = 0.03–0.04$, >10.19 Mya - >7.82 Mya; Fig. 3, Additional file 7) is better explained by a high extinction rate in Asia. This Asian extinction scenario supports the previous biogeographic hypothesis of Asian extinction and Mediterranean differentiation of *Hedera* [15, 16], but challenges the general pattern of Tertiary regional extinction of temperate genera in Europe and survival in Asia or North America [73–75]. The E-W gradual extinction in Asia is not only supported by the biogeographic analysis (Fig. 2), but also by the high number of missing haplotypes that are needed to connect the relict SW Mediterranean group to the Asian outgroup in the network (Fig. 4, Additional file 12). Soon after the westward migration and Asian extinction, a diversification rate increase is detected for *Hedera* starting in the Tortonian/Messinian (10.19 or 7.82 Mya; Fig. 3, Additional file 7), when ivies were already established at both sides of the Mediterranean as inferred from biogeographic reconstruction (Fig. 2) and fossil record (Table 1). The increase in aridity that synchronously started across the whole basin by that time [76–78] has been proposed to be responsible for the extinction of most of the sub-tropical Tertiary plants that previously inhabited this area [79–81]. The diversification rate increase of *Hedera* is intriguing since it places the intensification of diversification under an arid climate, which is an unpredicted scenario for a group of plants that need humid environments. Confinement to wet habitats in locally isolated environments could have promoted isolation and thus speciation for those elements of the sub-tropical Tertiary flora that were also tolerant to arid conditions. This could be the case of ivies that are considered part of the dry sub-tropical Tertiary flora [69, 82, 83]. Indeed, certain Mediterranean traits have been observed in *Hedera* species: low stomatal density primarily on leaf underside (Virginia Valcárcel *pers. obs.*), high density of indumentum on young shoots (Hugh A.

McAllister *pers. obs.*), and flowering only in sun-exposed branches (Pablo Vargas *pers. obs.*). Independent colonization events from the Mediterranean to other areas of Europe and independent back colonizations to Asia are inferred coinciding with the highest diversification periods (10.19–3.20 Mya or 7.82 Mya - present, Additional file 7). Interestingly, the intense diversification that accompanied this post-Miocene geographic range expansion did not occur evenly in all areas. As a result, an unbalanced diversity pattern is currently observed with the ancestral area (E Asia) operating as a sink region for the diversity of extant ivies (Table 2) and the distant Mediterranean basin acting as its source of re-colonization.

E and W Mediterranean as refugia in the Miocene and Pliocene

Two main regions are considered as Tertiary and Quaternary refugia for temperate and subtropical plant lineages in Eurasia, the Mediterranean basin [30] and SW China [73, 84]. The fact that Pliocene and Pleistocene extinctions were more intense in Europe than in Asia [75] has been explained by several factors, such as climatic and topographic changes, complex geological history, or the antiquity of the flora in Asia ([84], but see [85]). As a whole, there is a greater diversity of temperate genera in Asia than in Europe [73]. After a thorough full sampling of *Hedera* in the Mediterranean and Asian Tertiary refugia (Fig. 1), the Mediterranean region is not only still inferred as the main center of diversification for ivies [16] but also as a cumulative lineage refugium. The E and W Mediterranean are interpreted as consecutive refugia with differential roles in the evolution of the genus. Whereas the E Mediterranean acted as the refugium for *Hedera* during the dramatic changes of the Miocene, the W Mediterranean is identified as a more recent refugium for ivies during the climatic changes that shaped the Mediterranean flora from the Pliocene onwards. Our biogeographic analysis suggests that the E Mediterranean was part of the ancestral area back during the early differentiation of *Hedera* in the Miocene (nodes 2–3; Fig. 2), which is also supported by the time of the first fossils of *Hedera* in the Mediterranean that are from the eastern side [48] (Table 1). This pattern reveals that the E Mediterranean was a phylogenetic center of diversification in the past that served as a Miocene refugium for the ancestors of the lineages of extant ivies. This scenario is also supported by the fact that the E Mediterranean has a greater Faith's PD than expected by chance (Table 2). The reason why the E Mediterranean currently shows low diversity might be due to higher extinction rates in this area and higher levels of differentiation in the W Mediterranean. Currently, the E Mediterranean only has three species, three ploidy levels, two-plastid tip haplotypes and a relatively low PD index

(45, Table 2). The limited diversity of extant ivies in the E Mediterranean contrasts the results of some studies showing this side of the basin as a center of diversification for different groups of organisms [86–92]. From the ancestral refugium in the E Mediterranean a centrifugal dispersal resulted in the colonization of the W Mediterranean in the first place and the back colonization of W Asia in more recent times. A similar centrifugal dispersal from the E Mediterranean has been described for *Anthemis* also in the Miocene [93]. Despite the relevance of the E Mediterranean in the early evolution of the genus, the W Mediterranean is the region that accounts for the highest diversity in ivies. Indeed, the W Mediterranean displays the greatest number of extant species (eight), ploidy levels (four), plastid haplotypes (13, Additional files 7), and Faith's PD (72, Table 2). This is consistent with the pattern observed in the phylogeographic study where the W Mediterranean samples are scattered through the network displaying internal and tip haplotypes both in relict and derived groups (Fig. 4). Also, seven out of the 10 internal haplotypes and six out of the 13 tip haplotypes are widespread or unique to western Mediterranean samples (Fig. 4). All these results support the W Mediterranean as a secondary center of diversification [6], which might also be the reason for this area to have the greatest PD index but with this index lower than the one expected by chance (Table 2). All sources of evidence agree with the W Mediterranean as a source area of dispersal in the post-Miocene colonization of Europe and E Asia. Indeed, the star-like organization of the Asian and Euro-Mediterranean haplotypes (Fig. 4) suggest a recent and fast colonization from the W Mediterranean.

The relative importance of the E and W Mediterranean regions as refugia for the Mediterranean flora varies among groups [see 30 for details] with a general trend towards considering the W Mediterranean as the main Tertiary refugium (33 cases vs. 19 in E Mediterranean) [30]. It seems reasonable to assume that the role of the E Mediterranean as an ancestral refugium might have been overlooked because the more recent diversification processes might have erased footprints of earlier events, as in the case of *Hedera*.

Geographic isolation related to new habitats as the driver for *Hedera* evolution

The pattern of endemicity in *Hedera*, with 10 endemic taxa (five island endemics) and three widespread species (*H. helix*, *H. hibernica* and *H. nepalensis*), can only be explained by a key role of geographic isolation in speciation. However, geographic isolation seems less likely for a bird-dispersed fleshy-fruited plant group, like *Hedera*, due to the expected long-distance dispersability of endozoochorous dispersal syndromes mediated by birds [94]. Successful dispersal is supported by the occurrence of three species of *Hedera* in three archipelagoes of

Macaronesia. Nevertheless, complex and different geographic patterns of genetic variation have been suggested for fleshy-fruited plants [14, 95, 96]. Evidence points to additional external factors that may alter the efficiency of dispersal mediated frugivory, such as site availability (spatial limitation) [97]. The geographic structure inferred from the biogeographic and phylogeographic reconstructions of *Hedera* is not spatially or temporally constant. Geographic isolation is interpreted for the recent past of the genus from the geographic congruence of main clades diverged during the Pliocene (nodes 4, 5; Fig. 2). In contrast, back to the early evolution of *Hedera*, diversification occurred within the same geographic area (nodes 1–3, Fig. 2). This may be explained by a combination of historical and contemporary geological and ecological factors. The early divergence of the genus that led to the diploid and polyploid clades most likely took place in the E Mediterranean and W Asia during the Lower-Middle Miocene (Fig. 2, Table 1). By that time the Arabian microplate collided with Eurasia resulting in a major change to a colder climate and the uplift of E Mediterranean basin (~16 Mya) [98]. This may have promoted regional or habitat-dependent isolation that cannot be observed at the scale of biogeographic analyses. The independent range expansion to the W Mediterranean that occurred in the diploid and polyploid clades before the end of the Miocene (Fig. 2) does not seem to result in the observation of any divergence pattern either. During that period the Mediterranean region was under seasonal subtropical climate, which may have been crucial for *Hedera* establishment and dispersal across the basin, hindering geographic isolation. Given that the diploid and polyploid clades persisted in the E Mediterranean, the similarity in the diversification pattern detected could be related to the short compressional event that took place in Eastern Aegean 9–8 Mya [99] that may have promoted a secondary contact between the two clades. Such contact provided a reliable context for the inter-lineage hybridization already proposed in *Hedera* [6, 15] that might have eventually resulted in genome duplication [100]. Also, it provides a likely explanation for the incongruence detected between plastid and nuclear results [15] (cf. Figs. 2 and 4). The evolution of the diploid clade remains unclear, whereas a vicariant event is found for the divergence of the polyploid clade leading to the eastern and western polyploid subclades. Interestingly the time of lineage differentiation for eastern polyploids appears to be older than that of the western one, which was unexpected given that the W Mediterranean basin is much older [98]. This may be the result of a geographic filter due to the closer proximity of the E Mediterranean to the ancestral range of ivies (E Asia).

Conclusions

The consequences of the geological changes occurred in the Mediterranean area during Lower-Middle Miocene were dramatic for the Mediterranean Tertiary plant lineages [79, 81]. However, the subtropical *Hedera* clade survived and diversified in an arid and relatively constant climate. Indeed, the diversification rate increase of ivies since the Tortonian/Messinian under the increasing aridity in the Mediterranean, suggests that a climate-driven spatial limitation (i.e. habitat availability) may have enabled geographic speciation. Cumulative Miocene and post-Miocene refugia are detected in E and W Mediterranean, respectively, that also acted as consecutive dispersal centers.

Additional files

Additional file 1: List of the studied material included in the phylogenetic-based analyses. Localities, geographic area codification and GenGank accession numbers of the nuclear Internal Transcribed Spacer are provided. Areas abbreviation are as follows: A, Neotropics; B, Tropical Africa; C, Asutralia: D, W Mediterranean; E, E Mediterranean; F, Europe; G, W Asia; H, E Asia. Papers of reference are provided as superscript as follows: (1) Li et al. [35], (2) Mitchell et al. [26], (3) Mitchell et al. [66], (4) Valcárcel et al. [25], (5) Vargas et al. [7], and (6) Valcárcel et al. [16]. (DOCX 141 kb)

Additional file 2: GenBank accession numbers of the studied material included in the phylogeographic study. (DOCX 171 kb)

Additional file 3: Summary of DNA sequences variation and evolutionary models best fitting the nuclear and plastid matrices. (DOCX 41 kb)

Additional file 4: R script for parsing Lagrange results from multi-tree analyses. (R 22 kb)

Additional file 5: Beast Maximum Clade Credibility chronogram of the *nr*ITS dataset of Araliaceae with the secondary calibration approach. Legend: Mean ages and 95% CI are only represented for clades with >0.5 Posterior Probability support. (PDF 255 kb)

Additional file 6: Beast Maximum Clade Credibility chronogram of the *nr*ITS dataset of Araliaceae with the fossil calibration approach. Mean ages and 95% CI are only represented for clades with >0.5 Posterior Probability support. (PDF 236 kb)

Additional file 7: Fitness to speciation models. Summary of the results of the fitness to speciation models estimated from 6500 posterior pruned trees of the *Hedera* clade. N indicates the number of trees that recover a given evolutionary model. Mean values and 95% CI are provided for AIC and each of the parameters of the models. (DOCX 49 kb)

Additional file 8: List of the haplotypes detected in the samples included in the phylogeographic study. Taxa name and general distribution is provided. Locality and voucher is specified for each sample as well as the number of haplotype detected individually for the three regions (*rpL*32, *trn*H-*psbA*, *trn*T-*trn*L). Last column indicates the number of haplotype when combining the three-plastid regions *rpL*32, *trn*H-*psbA* and *trn*T-*trn*L (HP³). (DOCX 106 kb)

Additional file 9: *Hedera* phylogeographic network obtained from the Statistical Parsimony analysis of the *rpL*32 plastid region. Haplotype numbering is according to Additional file 8. Circle dimensions are proportional to the number of samples displaying each haplotype. *Kalopanax septembolus* is used as outgroup. (PDF 96 kb)

Additional file 10: *Hedera* phylogeographic network obtained from the Statistical Parsimony analysis of the *trn*H-*psb*A plastid region. Haplotype numbering is according to Additional file 8. Circle dimensions are proportional to the number of samples displaying each haplotype. *Kalopanax septembolus* is used as outgroup. (PDF 93 kb)

Additional file 11: *Hedera* phylogeographic network obtained from the Statistical Parsimony analysis of the *trn*T-*trn*L plastid region. Haplotype numbering is according to Additional file 8. Circle dimensions are proportional to the number of samples displaying each haplotype. *Kalopanax septembolus* is used as outgroup. (PDF 108 kb)

Additional file 12: *Hedera* phylogeographic network obtained from the Statistical Parsimony analysis of the three-plastid regions (*trn*H-*psb*A, *trn*T-*trn*L, *rpL*32) and using seven Araliaceae genera as outgroup. Circle dimensions are proportional to the number of samples displaying each haplotype. Abbreviation names of the taxa displaying each haplotype are as follows: ALG, *H. algeriensis*; AZO, *H. azorica*; CAN, *H. canariensis*; COL, *H. colchica*; CYP, *H. pastuchowii* subsp. *cypria*; HEL, *H. helix*; HIB, *H. hibernica*; IBE, *H. iberica*; MAD, *H. maderensis*; MAR, *H. maroccana*; NEP, *H. nepalensis* var. *nepalensis*; PAS, *H. pastuchowii* subsp. *pastuchowii*; RHO, *H. rhombea*; SIN, *H. nepalensis* var. *sinensis*. (PDF 188 kb)

Abbreviations

My: Million years; Mya: Million years ago; PD: Phylogenetic index; PP: Posterior probability

Funding

Part of the data was obtained with the financial support of a grant to V. Valcárcel from Junta de Andalucía.

Authors' contributions

VV obtained half of the data, performed the biogeographic, diversification and phylogeographic analyses, helped developing the R-script, interpreted all results and lead the writing process. BG obtained half of the data, performed the divergence age analysis, wrote the methodology for this analysis in the manuscript and contributed to the results interpretation and writing of the manuscript. NGM developed the R-script for summarizing the multi-tree results from biogeographic analysis and contributed to the writing of the manuscript. PV and JW led the elaboration of the hypothesis, provided funding for obtaining the data and were both major contributors to help interpret the results and improve the writing of the manuscript. All authors read and approved the final manuscript.

Competing interests

The authors declare that they have no competing interests.

Author details

[1]Department of Biology (Botany), Universidad Autónoma de Madrid, Madrid, Spain. [2]Department of Biodiversity and Conservation, Real Jardín Botánico, CSIC, Madrid, Spain. [3]Department of Botany, Faculty of Science, University of South Bohemia, Ceske Budejovice, Czech Republic. [4]Department of Botany/MRC 166, Smithsonian Institution, Washington, DC, USA.

References

1. Meusel H, Jäger E, Weinert E. Vergleichende Chorologie Der Zentraleuropäischen Flora. Jena: Veb Gustav Fischer Verlag; 1965.
2. Mabberley DJ. The plant-book. Cambridge: Cambridge University Press; 1997.
3. Valcárcel V: Taxonomy, systematics and evolution of *Hedera* L. (Araliaceae). Universidad Pablo de Olavide; 2008.
4. Green AF, Ramsey TS, Ramsey J. Phylogeny and biogeography of ivies (*Hedera* spp., Araliaceae), a polyploid complex of woody vines. Syst Bot. 2011;36:1114–27.
5. Rutherford A, McAllister HA, Mill RR. New ivies from the Mediterranean area and Macaronesia. The Plantsman. 1993;15:115–28.
6. Vargas P, Mcallister HA, Morton C, Jury SL, Wilkinson MJ. Polyploid speciation in *Hedera* (Araliaceae): phylogenetic and biogeographic insights based on chromosome counts and ITS sequences. Plant Syst Evol. 1999;219:165–79.
7. McAllister HA, Rutherford A. *Hedera helix* L. and *H. hibernica* (Kirchner) bean (Araliaceae) in the British isles. Watsonia. 1990;18:7–15.
8. Rutherford A. The history of the Canary Islands ivy and its relatives. Ivy J. 1984;10:13–8.
9. Rutherford A. The ivies of Andalusia (southern Spain). Ivy J. 1989;15:7–17.
10. McAllister HA. Canary and Algerian ivies. The Plantsman. 1988;10:27–9.
11. Ackerfield J, Wen J. Evolution of *Hedera* (the ivy genus, Araliaceae): insights from chloroplast DNA data. Int J Plant Sci. 2003;164:593–602.
12. Ackerfield J. Trichome morphology in *Hedera* (Araliaceae). Edinb J Bot. 2001;58:259–67.
13. Ackerfield J, Wen J. A morphometric analysis of *Hedera* L. (the ivy genus, Araliaceae). Adansonia Sér. 2002;324:197–212.
14. Grivet D, Petit RJ. Phylogeography of the common ivy (*Hedera* sp.) in Europe: genetic differentiation through space and time. Mol Ecol. 2002;11:1351–62.
15. Valcárcel V, Fiz O, Vargas P. Chloroplast and nuclear evidence for multiple origins of polyploids and diploids of *Hedera* (Araliaceae) in the Mediterranean basin. Mol Phylogenet Evol. 2003;27:1–20.
16. Valcárcel V, Vargas P. Phylogenetic reconstruction of key traits in the evolution of ivies (*Hedera* L.). Plant Syst Evol. 2013;229:447–58.
17. Ridley HN. The dispersal of plants throughout the world. L. Reeve & Co: Kent; 1930.
18. Guitián J. Dispersal of ivy *Hedera helix* seeds by birds: time spent in the plant and seed removal efficiency. Ardeola. 1987;34:25–35.
19. Kollmann J, Grubb PJ. Recruitment of fleshy-fruited species under different shrub species: control by under-canopy environment. Ecol Res. 1999;14:9–21.
20. Sack L, Grubb PJ, Maranon T. The functional morphology of juvenile plants tolerant of strong summer drought in shaded forest understories in southern Spain. Plant Ecol. 2003;168:139–63.
21. Valcárcel V, Rutherford A, Miller R, McAllister HA. *Hedera* L. in *Flora iberica*. Vol. X. Araliaceae-Umbelliferae. Edited by Nieto Feliner G. Madrid: Departamento de publicaciones del CSIC; 2003.
22. McAllister HA. New work on ivies. Int Dendrol Soc Yearb. 1981;1981:106–9.
23. Rodríguez-Sánchez F, Hampe A, Jordano P, Arroyo J: Past tree range dynamics in the Iberian Peninsula inferred through phylogeography and palaeodistribution modelling: A review. Rev Palaeobot Palynol 2010, 162: 507–521. [Iberian Floras through Time: Land of Diversity and Survival].
24. Bramwell D, Bramwell Z. Flores Silvestres de Las Islas Canarias. Editorial Rueda: Madrid, Spain; 1990.
25. Valcárcel V, Fiz O, Wen J. The origin of the early differentiation of ivies (*Hedera* L.) and the radiation of the Asian Palmate group (Araliaceae). Mol Phylogenet Evol. 2014;70:492–503.
26. Mitchell A, Li R, Brown JW, Schönberger I, Wen J. Ancient divergence and biogeography of Raukaua (Araliaceae) and close relatives in the southern hemisphere. Aust Syst Bot. 2012;25:432–46.
27. Ronquist F. Dispersal-vicariance analysis: a new approach to the quantification of historical biogeography. Syst Biol. 1997;45:195–203.
28. Marks MD. Molecular genetic analysis of trichome development in *Arabidopsis*. Annu Rev Plant Physiol Plant Mol Biol. 1997;48:137–63.
29. Ishida T, Kurata T, Okada K. Wada: a genetic regulatory network in the development of trichomes and root hairs. Annu Rev Plant Biol. 2008;59:365–86.
30. Médail F, Diadema K. Glacial refugia influence plant diversity patterns in the Mediterranean Basin. J Biogeogr. 2009;36:1333–45.
31. López-Pujol J, Zhang F-M, Sun H-Q, Ying T-S, Ge S. Centres of plant endemism in China: places for survival or for speciation? J Biogeogr. 2011;38:1267–80.
32. Taberlet P, Gielly L, Pautou G, Bouvet J. Universal primers for amplification of three non-coding regions of chloroplast DNA. Plant Mol Biol. 1991;17:1105–9.
33. Shaw J, Lickey EB, Schilling EE, Small RL. Comparison of whole chloroplast genome sequences to choose noncoding regions for phylogenetic studies in angiosperms: the tortoise and the hare III. Am J Bot. 2007;94:275–88.
34. Sang T, Crawford D, Stuessy T. Chloroplast DNA phylogeny, reticulate evolution, and biogeography of Paeonia (Paeoniaceae). Am J Bot. 1997;84:1120.
35. Li R, Ma P-F, Wen J, Yi T-S. Complete sequencing of five Araliaceae chloroplast genomes and the phylogenetic implications. PLoS One. 2013;8:e78568.
36. Edgar RC. MUSCLE: multiple sequence alignment with high accuracy and high throughput. Nucleic Acids Res. 2004;32:1792–979.
37. Vaidya G, Lohman DJ, Meier R. SequenceMatrix: concatenation software for the fast assembly of multi-gene datasets with character set and codon information. Cladistics. 2010;27:171–80.
38. Drummond AJ, Rambaut A. BEAST: Bayesian evolutionary analysis by sampling trees. BMC Evol Biol. 2007;7:214.

39. Posada D. jModelTest: phylogenetic model averaging. Mol Biol Evol. 2008;25:1253–6.

40. Rambaut A, Drummond AJ: *Tracer v1.4*. 2007.

41. Szafer W. Miocenska Flora ze Starych Gliwic na Slasku (Miocene Flora of stare Gliwice in upper Silesia). Pr Inst Geol. 1961;33:1–205.

42. Barrón E, Postigo-Mijarra JM, Diéguez C. The late Miocene macroflora of the Cerdanya Basin (eastern Pyrenees, Spain): towards a synthesis. Paleontogr Abt B Paleobotany - Paleophytology. 2014;291:85–129.

43. Mai DH, Walther H. Die pliozaenen Floren von Thueringen, Deutsche Demokratische Republik. Quartaerpalaeontologie. 1988;7:55–297.

44. Rim KH. Fossils of North Korea. Pyongyang: Science and Technology Press; 1994.

45. Müller J. Fossil pollen records of extant angiosperms. Bot Rev. 1981;47:1–142.

46. Kong WS. The vegetational and environmental history of the pre-Holocene period in the Korean Peninsula. Korean J Quat Res. 1992;6:12.

47. Kolakovskii AA, Shakryl AK. Kimmeriyskaya flora Gul'ripsha (Bagazhishta). Tr Sukum Bot Sada. 1978;24:134–56.

48. Kvaček Z, Velitzelos D, Velitzelos E. Late Miocene Flora of Vegora Macedonia N. Greece: University of Athens, Greece; 2002.

49. Kovar-Eder J, Kvaček Z, Martinetto E, Roiron P: Late Miocene to Early Pliocene vegetation of southern Europe (7–4 Ma) as reflected in the megafossil plant record. *Palaeogeogr Palaeoclimatol Palaeoecol* 2006, 238: 321–339. [*Late Miocene to Early Pliocene Environment and Climate Change in the Mediterranean Area*].

50. Vieira M, Poças E, Pais J, Pereira D. Pliocene flora from S. Pedro da Torre deposits (Minho, NW Portugal). Geodiversitas. 2011;33:71–85.

51. Paradis E, Claude J, Strimmer K. APE: analyses of phylogenetics and evolution in R language. Bioinformatics. 2004;20:289–90.

52. Ricklefs RE. History and diversity: explorations at the intersection of ecology and evolution. Am Nat. 2007;170:S56–70.

53. Crisp MD, Cook LG. Explosive radiation or cryptic mass extinction? Interpreting signatures in molecular mhylogenies. Evolution. 2009;63:2257–65.

54. Fiz-Palacios O, Valcárcel V. From Messinian crisis to Mediterranean climate: a temporal gap of diversification recovered from multiple plant phylogenies. Perspect Plant Ecol Evol Syst. 2013;15:130–7.

55. Rabosky DL. LASER: a maximum likelihood toolkit for detecting temporal shifts in diversification rates from molecular phylogenies. Evol Bioinforma. 2006;2:257–60.

56. Crawley MJ. The R book. Chichester: Wiley; 2007.

57. R Development Core Team: *R: A Language and Environment for Statistical Computing*. Vienna, Austria; 2011.

58. Faith DP. Conservation evaluation and phylogenetic diversity. Biol Con. 1992;61:1–10.

59. Kembel SW, Cowan PD, Helmus MR, Cornwell WK, Morlon H, Ackerly DD, et al. Picante: R tools for integrating phylogenies and ecology. Bioinformatics. 2010;26:1463–4.

60. Clement M, Posada D, Crandall KA. TCS: A computer program to estimate gene genealogies. Mol Ecol. 2000;9:1657–9.

61. Crandall KA, Templeton AR. Empirical tests of some predictions from coalescent theory with applications to intraspecific phylogeny reconstruction. Genetics. 1993;134:959–69.

62. Posada D, Crandall KA. Intraspecific gene genealogies: trees grafting into networks. Trends Ecol Evol. 2001;16:37–45.

63. Li R, Wen J. Phylogeny and biogeography of Asian *Schefflera* (Araliaceae) based on nuclear and plastid DNA sequences data. J Syst Evol. 2014;52:431–49.

64. Li R, Wen J. Phylogeny and diversification of Chinese Araliaceae based on nuclear and plastid DNA sequence data. J Syst Evol. 2016;54:453–67.

65. Li R, Wen J. Phylogeny and biogeography of *Dendropanax* (Araliaceae), an Amphi-Pacific Disjunct genus between tropical/subtropical Asia and the Neotropics. Syst Bot. 2013;38:536–51.

66. Mitchell A, Wen J. Phylogeny of *Brassaiopsis* (Araliaceae) in Asia based on nuclear ITS and 5S-NTS DNA sequences. Syst Bot. 2005;30:872–86.

67. Plunkett GM, Wen J, Ii PPL. Infrafamilial classifications and characters in Araliaceae: insights from the phylogenetic analysis of nuclear (ITS) and plastid (*trnL-trnF*) sequence data. Plant Syst Evol. 2004;245:1–39.

68. Wen J, Plunkett GM, Mitchell AD, Wagstaff SJ. The evolution of Araliaceae: a phylogenetic analysis based on ITS sequences of nuclear ribosomal DNA. Syst Bot. 2001;26:144–67.

69. Wen J, Nie Z-L, Soejima A, Meng Y. Phylogeny of Vitaceae based on the nuclear *GAI1* gene sequences. Can J Bot. 2007;85:731–45.

70. Axelrod D. Evolution and biogeography of Madrean–Tethyan sclerophyll vegetation. Ann Mo Bot Gard. 1975;62:280–334.

71. Harris AJ, Xiang QY, Thomas DT: Phylogeny, origin, and biogeographic history of *Aesculus* L. (Sapindales) an update from combined analysis of DNA sequences, morphology and fossils. Taxon 2009, 58:1–19.

72. Harris A, Wen J. Xiang Q-Y (Jenny): inferring the biogeographic origins of inter-continental disjunct endemics using a Bayes-DIVA approach. J Syst Evol. 2013;51:117–33.

73. Wen J. Evolution of eastern Asian and eastern North American disjunct pattern in flowering plants. Annu Rev Ecol Syst. 1999;30:421–55.

74. Latham RE, Ricklefs RE. Global patterns of tree species richness in moist forests: energy-diversity theory does not account for variation in species richness. Oikos. 1993;67:325–33.

75. Qian H, Ricklefs RE. Large-scale processes and the Asian bias in temperate plant species diversity. Nature. 2000;407:180–2.

76. Krijgsman W. The onset of the Messinian salinity crisis in the eastern Mediterranean (Pissouri Basin, Cyprus). Earth Planet Sci Lett. 2002;194:299–310.

77. Ivanov D, Ashraf AR, Mosbrugger V, Palamarev E. Palynological evidence for Miocene climate change in the Forecarpathian Basin (central Paratethys, NW Bulgaria). Palaeogeogr Palaeoclimatol Palaeoecol. 2002;178:19–37.

78. Van Dam JA. Geographic and temporal patterns in the late Neogene (12.3 ma) aridification of Europe: the use of small mammals as paleoprecipitation proxies. Palaeogeogr Palaeoclimatol Palaeoecol. 2006;238:190–218.

79. Thompson JD. Plant evolution in the Mediterranean. Oxford: Oxford University Press; 2005.

80. Rodríguez-Sánchez F, Pérez-Barrales R, Ojeda F, Vargas P, Arroyo J: The Strait of Gibraltar as a melting pot for plant biodiversity. *Quat Sci Rev* 2008, 27: 2100–2117. [*The Coastal Shelf of the Mediterranean and Beyond: Corridor and Refugium for Human Populations in the Pleistocene*].

81. Jiménez-Moreno G, Fauquette S, Suc JP. Miocene to Pliocene vegetation reconstruction and climate estimates in the Iberian Peninsula from pollen data. Rev Paleobot Palynol. 2010;162:410–5.

82. Herrera C. Historical effects and sorting processes as explanations for contemporary ecological patterns: character syndromes in Mediterranean woody plants. Am Nat. 1992;140:421–46.

83. Verdú M, Dávila P, García-Fayos P, Flores-Hernández N, Valiente-Banuet A. "convergent" traits of Mediterranean woody plants belong to pre-Mediterranean lineages. Biol J Linn Soc. 2003;78:415–27.

84. Qiu YX, Fu CX, Comes HP. Plant molecular phylogeography in China and adjacent regions: tracing the genetic imprints of quaternary climate and environmental change in the world's most diverse temperate flora. Mol Phylogenet Evol. 2011;59:225–44.

85. Qian H. A comparison of the taxonomic richness of temperate plants in East Asia and North America. Am J Bot. 2002;89:1818–25.

86. Torrecilla P, López-Rodríguez JA, Catalán P. Phylogenetic relationships of *Vulpia* and related genera (Poeeae; Poaceae) based on analysis of ITS and trnL-F sequences. Ann Mo Bot Gard. 2004;91:124–58.

87. Oberprieler C. Temporal and spatial diversification of Circum-Mediterranean Compositae-anthemideae. Taxon. 2005;54:951–66.

88. Inda LA, Torrecilla P, Catalán P, Ruiz-Zapata T. Phylogeny of *Cleome* L. and ITS close relatives *Podandrogyne* Ducke and *Polanisia* Raf. (Cleomoideae, Cleomaceae) based on analysis of nuclear ITS sequences and morphology. Plant Syst Evol. 2008;274:111–26.

89. Mansion G, Rosenbaum G, Schoenenberger N, Bacchetta G, Rosselló JA, Conti E. Phylogenetic analysis informed by geological history supports multiple, sequential invasions of the mediterranean basin by the angiosperm family Araceae. Syst Biol. 2008;57:269–85.

90. Micó E, Sanmartín I, Galante E. Mediterranean diversification of the grass-feeding Anisopliina beetles (Scarabaeidae, Rutelinae, Anomalini) as inferred by bootstrap-averaged dispersal-vicariance analysis. J Biogeogr. 2013;36:546–60.

91. Ree RH, Sanmartín I. Prospects and challenges for parametric models in historical biogeographical inference. J Biogeogr. 2009;36:1211–20.

92. Sanmartín I. Dispersal vs. vicariance in the Mediterranean: historical biogeography of the Palearctic Pachydeminae (Coleoptera, Scarabaeoidea). J Biogeogr. 2003;30:1883–97.

93. Lo Presti RM, Oberprieler C. Evolutionary history, biogeography and eco-climatological differentiation of the genus *Anthemis* L. (Compositae, Anthemideae) in the circum-Mediterranean area. J Biogeogr. 2009;36:1313–32.

94. Costa JM, Ramos JA, da Silva LP, Timoteo S, Araújo PM, Felgueiras MS, et al. Endozoochory largely outweighs epizoochory in migrating passerines. J Avian Biol. 2014;45:59–64.

95. Hampe A, Arroyo J, Jordano P, Petit R. Rangewide phylogeography of a bird-dispersed Eurasian shrub: contrasting Mediterranean and temperate glacial refugia. Mol Ecol. 2003;12:3415–26.

96. Dubreuil M, Riba M. González–Martínez SC, Vendramin GG, Sebastiani F, Mayol M. Genetic effects of chronic habitat fragmentation revisited: strong genetic structure in a temperate tree, *Taxus baccata* (Taxaceae), with great dispersal capability. Am J Bot. 2010;97:303–10.

97. García C, Jordano P, Arroyo JM, Godoy JA. Maternal genetic correlations in the seed rain: effects of frugivore activity in heterogeneous landscapes. J Ecol. 2009;97:1424–35.

98. Krijgsman W. The Mediterranean: Mare nostrum of earth sciences. Earth Planet Sci Lett. 2002;205:1–12.

99. Ring U, Laws S, Bennet M. Structural analysis of a complex nappe sequence and late-orogenic basins from the Aegean Island of Samos, Greece. J Struct Geol. 1999;21:1575–601.

100. Escudero M, Martín-Bravo S, Mayrose I, Fernández-Mazuecos M, Fiz-Palacios O, Hipp AL, et al. Karyotypic changes through dysploidy persist longer over evolutionary time than polyploid changes. PLoS One. 2014;9:e85266.

4

Dated tribe-wide whole chloroplast genome phylogeny indicates recurrent hybridizations within Triticeae

Nadine Bernhardt[1]* (iD), Jonathan Brassac[1], Benjamin Kilian[1,2] and Frank R. Blattner[1,3]

Abstract

Background: Triticeae, the tribe of wheat grasses, harbours the cereals barley, rye and wheat and their wild relatives. Although economically important, relationships within the tribe are still not understood. We analysed the phylogeny of chloroplast lineages among nearly all monogenomic Triticeae taxa and polyploid wheat species aiming at a deeper understanding of the tribe's evolution. We used on- and off-target reads of a target-enrichment experiment followed by Illumina sequencing.

Results: The read data was used to assemble the plastid locus *ndh*F for 194 individuals and the whole chloroplast genome for 183 individuals, representing 53 Triticeae species and 15 genera. We conducted Bayesian and multispecies coalescent analyses to infer relationships and estimate divergence times of the taxa. We present the most comprehensive dated Triticeae chloroplast phylogeny and review previous hypotheses in the framework of our results. Monophyly of Triticeae chloroplasts could not be confirmed, as either *Bromus* or *Psathyrostachys* captured a chloroplast from a lineage closely related to a *Bromus*-Triticeae ancestor. The most recent common ancestor of Triticeae occurred approximately between ten and 19 million years ago.

Conclusions: The comparison of the chloroplast phylogeny with available nuclear data in several cases revealed incongruences indicating past hybridizations. Recent events of chloroplast capture were detected as individuals grouped apart from con-specific accessions in otherwise monopyhletic groups.

Keywords: Hybridization, Whole chloroplast genome, Phylogeny, Next-generation sequencing, Divergence times, Triticeae, Wheat, *Triticum*, *Aegilops*, *Psathyrostachys*

Background

The economically important grass tribe Triticeae Dumort. consists of approximately 360 species and several subspecies in 20-30 genera. Triticeae taxa occur in temperate and dry regions of the World and harbour the important cereals bread wheat (*Triticum aestivum*), barley (*Hordeum vulgare*), rye (*Secale cereale*) and their wild relatives [1, 2]. Yet there is no good understanding of the relationships among Triticeae taxa, although a multitude of molecular phylogenies have been produced [3–11]. The acceptance levels of taxa vary greatly among authors on the genus-level and below (for recent reviews see [1, 12, 13]). One important reason is the complex

mode of evolution within Triticeae. The majority of species are allopolyploids and many of them likely have originated repeatedly, involving genetically different parent species [14–19]. Bread wheat is the most prominent polyploid and evolved via consecutive hybridizations of three diploids and thereby combines three related genomes (named **A**, **B** and **D**) [7, 20]. In Triticeae and many other crops such genomes were defined through cytogenetic characterization of chromosomes together with the analysis of their pairing behaviour in interspecific and intergeneric crosses (for reviews see [1, 12, 21]). It is assumed that diploid species and monogenomic taxa are the basic units within Triticeae and that the heterogenomic polyploids form a second level of taxonomic entities [22, 23].

Triticeae are known to have low barriers against hybridization, which result in mixed or even recombinant

---

* Correspondence: bernhardt@ipk-gatersleben.de
[1]Leibniz Institute of Plant Genetics and Crop Plant Research (IPK), Gatersleben, Germany
Full list of author information is available at the end of the article

phylogenetic signals from nuclear data [10, 20]. In contrast, phylogenetic analyses of plastid sequences provide clear information on maternal lineages, as organelles are mostly uniparentally inherited and non-recombining in angiosperms [24], although chloroplast capture [25] can result in deviating phylogenetic hypotheses. Yet, plastid sequence data is limited for Triticeae. Studies based on a tribe-wide taxon sampling are rare and focused on single or few plastid markers [9, 26–28]. To date, the number of whole plastid genome sequences is increasing [29–34], however, entire chloroplast genomes are mainly available for the domesticated taxa and their closest relatives. These previous studies provide only limited insight in the maternal phylogeny of Triticeae, as only one to few accessions per taxon were included and often support values for the taxonomic units are low [26, 28, 35].

Here we present phylogenetic analyses of chloroplast sequences based on a comprehensive set of monogenomic Triticeae species plus allopolyploid representatives of the wheat group (i.e. taxa belonging to the genera *Aegilops* and *Triticum*). For each species we included multiple individuals to sample part of the intraspecific variation. We performed a target-enrichment and next-generation sequencing (NGS) approach that, among nuclear loci (which will be published elsewhere), targeted the chloroplast *ndh*F gene. Since chloroplasts occur in high copy number in the plant cell, they represent a large fraction of the off-target reads when sequencing reduced complexity libraries, which can be used to assemble almost complete chloroplast genomes [36]. Our dataset was complemented by chloroplast genomes stored in the GenBank database. Multispecies coalescent (MSC) analyses based on *trn*K-*mat*K, *rbc*L and *ndh*F were used for dating of the major splits within the evolution of the tribe and to reconsider the monophyly of the Triticeae chloroplast lineages.

Methods

Plant materials

Aiming at a good representation of taxa for phylogenetic inference we analysed 194 individuals representing approximately 53 species belonging to 15 genera (depending on taxonomic treatment applied) of the grass tribe Triticeae and included *Bromus* and *Brachypodium* accessions as outgroup taxa (Table 1, Additional file 1: Table S1). The accessions were acquired from the International Center for Agricultural Research in the Dry Areas (ICARDA), the seed bank of the Leibniz Institute of Plant Genetics and Crop Research (IPK), the National Small Grain Collection of the US Department of Agriculture (USDA), the Czech Crop Research Institute, and the Laboratory of Plant Genetics (Kyoto University). Additional seed material was collected during field trips. Multiple accessions per species and intra-specific entities

were selected if possible. All materials were grown from seed and identified based on morphological characters if an inflorescence was produced. Plant material obtained from germplasm repositories that was found to be in conflict with its taxonomic affiliation was only included in the analyses if the taxon could be unequivocally determined. Vouchers of the morphologically identified materials (Additional file 1: Table S1) were deposited in the herbarium of IPK (GAT).

Laboratory work

Flow-cytometric measurements were conducted to determine the ploidy level for all accessions. All analyses followed the protocol of Doležel et al. [37] on a CyFlow Space flow cytometer (Partec). At least 7500 nuclei were counted. Only measurements with a coefficient of variation (CV) for sample and standard peak <4% were accepted. Samples that recurrently produced CV values >4% were repeated in Galbraith's buffer containing 1% polyvinylpyrrolidon (vol/vol) and 0.1% Triton X-100 (vol/vol). At least three measurements per species were carried out. If only a single accession of a species could be retrieved from a seed bank, its ploidy level was estimated three times. Samples of the same species were processed on at least two different days to account for instrument drifts.

Genomic DNA was extracted either from 10 mg silica-dried leaves using the DNeasy Plant Mini Kit (Qiagen) or from 5 g of freeze-dried leaves using the cetyltrimethyl-ammonium bromide (CTAB) method [38]. DNA quantifications were done using the Qubit dsDNA BR Assay (Life Technologies) or the Quant-iT PicoGreen dsDNA Assay Kit (Invitrogen) on a Tecan Infinite 200 microplate reader according to the manufactures instructions. The LE220 Focused-Ultrasonicator (Covaris) was used to shear 3 μg genomic DNA in 130 μl TE buffer for every sample into fragments having an average length of 400 bp with the following settings: instantaneous ultrasonic power (PIP) 450 W, duty factor (df) 30%, cycles per burst (cpb) 200. The treatment was applied for 100 s. The sheared DNA was used in a sequence-capture approach (SureSelect^{XT} Target Enrichment for Illumina Paired-End Sequencing, Agilent Technologies) targeting at 450 nuclear single-copy loci aiming for 0.01–0.02% of a Triticeae genome. Baits complementary to chloroplast *ndh*F based on 628 bp of available *Hordeum vulgare, Aegilops tauschii, Pseudoroegneria spicata, Triticum urartu* (identical to EF115541.1, JQ754651.1, KJ174105.1 and AF056180.1, respectively) and 2073 bp of *Brachypodium distachyon* (identical to AF251451.1) sequences were designed as well. The pairwise sequence identity was larger then 99% among Triticeae taxa and 96% when comparing the Triticeae taxa with *Brachypodium*. Baits were designed to cover the entire 2073 bp of *ndh*F as well as

Table 1 Overview of Triticeae and outgroup taxa considered

Species	Genome	Ploidy (N)	Distribution area
Aegilops bicornis Jaub. & Spach	S*	2× (4)	SE Mediterranean
Aegilops biuncialis Vis.	UM	4× (4)	SW-SE Europe, N Africa, SW Asia
Aegilops columnaris Zhuk.	UM	4× (2)	SW Asia
Aegilops comosa Sm.	M	2× (4)	Balkans
Aegilops crassa Boiss.	DM/DDM	4× (1)/6× (2)	SW Asia
Aegilops cylindrica Host	DC	4× (2)	SE Europe, W Asia
Aegilops geniculata Roth	MU	4× (3)	E Europe, W Asia, Macaronesia
Aegilops juvenalis Eig	DMU	6× (2)	SW Asia,
Aegilops kotschyi Boiss.	S*U	4× (1)	SW Asia, NE Africa
Aegilops longissima Schweinf. & Muschl.	S*	2× (5)	E Mediterranean
Aegilops markgrafii (Greuter) K. Hammer	C	2× (5)	NE Mediterranean
Aegilops neglecta Req. ex Bertol.	UM/UMN	4× (2)/6× (2)	Mediterranean to SW Asia
Aegilops peregrina (Hack.) Maire & Weiller	SU	4× (1)	SW Asia, N Africa
Aegilops searsii Feldman & Kislev	S*	2× (5)	E Mediterranean
Aegilops sharonensis Eig	S*	2× (1)	Israel, Lebanon
Aegilops speltoides Tausch	S	2× (6)	E Mediterranean
Aegilops tauschii Coss.	D	2× (4)	SW–C Asia
Aegilops triuncialis L.	UC	4× (2)	Mediterranean to SW Asia
Aegilops umbellulata Zhuk.	U	2× (3)	SE Europe, SW Asia
Aegilops uniaristata Vis.	N	2× (3)	SE Europe
Aegilops ventricosa Tausch	DN	4× (2)	SW Europe, N Africa
Agropyron cristatum (L.) Gaertn.	P	2× (2)/4× (4)	S Europe, NECW Asia
Amblyopyrum muticum (Boiss.) Eig	T	2× (6)	Turkey
Australopyrum retrofractum (Vickery) A. Löve	W	2× (4)	SE Australia
Dasypyrum villosum (L.) P. Candargy	V	2× (5)	SW–SE Europe, Caucasus
Eremopyrum bonaepartis (Spreng.) Nevski	Ft/Xe/FtXe	2×/4× (5)	SE–E Europe, WC Asia
Eremopyrum triticeum (Gaertn.) Nevski	Ft	2× (3)	SE–E Europe, WC Asia
Henrardia persica (Boiss.) C.E. Hubb.	O	2× (4)	SE Europe, SW Asia
Heteranthelium piliferum Hochst. ex Jaub. & Spach	Q	2× (4)	SE Europe, SW Asia
Hordeum bulbosum L.	I	4× (1)	Mediterranean to C Asia
Hordeum marinum Huds.	Xa	2× (1)	Mediterranean
Hordeum murinum L.	Xu	2× (1)	Mediterranean to C Asia
Hordeum pubiflorum Hook. f.	I	2× (1)	S Argentina
Hordeum vulgare L.	H	2× (2)	SW Asia
Psathyrostachys juncea (Fisch.) Nevski	Ns	2× (6)	E Europe, NC Asia
Pseudoroegneria cognata (Hack.) A. Löve	St	6× (1)	SW Asia, West Himalaya
Pseudoroegneria spicata (Pursh) A. Löve	St	2× (2)/6× (1)	NW of Northern America
Pseudoroegneria stipifolia (Czern. ex Nevski) A. Löve	St	2× (1)/4×(2)	E Europe, N Caucasus
Pseudoroegneria strigosa (M. Bieb.) A. Löve	St	2× (2)/6× (2)	Balkans, Crimea
Pseudoroegneria tauri (Boiss. & Balansa) A. Löve	St	2× (5)	E Mediterranean, S Caucasus
Secale cereale L.	R	2× (4)	Turkey
Secale strictum C. Presl	R	2× (4)	S Europe, SW Asia, N Africa
Taeniatherum caput-medusae (L.) Nevski	Ta	2× (6)	S Europe, SW Asia, N Africa
Thinopyrum distichum (Thunb.) A. Löve	E	4× (2)	S Africa

Table 1 Overview of Triticeae and outgroup taxa considered *(Continued)*

Thinopyrum spp. Löve	E	6× (1)/8× (2)	SE Europe, SW Asia, N Africa
Triticum aestivum L.	BAD	6× (6)	Caucasus, Iran
Triticum monococcum L.	A	2× (10)	Turkey
Triticum timopheevii (Zhuk.) Zhuk.	GA	4× (7)	SW Asia
Triticum turgidum L.	BA	4× (10)	Lebanon
Triticum urartu Thumanjan ex Gandilyan	A	2× (5)	E Mediterranean, Caucasus
Triticum zhukovskyyi Menabde & Ericzjan	GAA	6× (1)	Caucasus
Brachypodium distachyon (L.) P. Beauv.		4× (1)	S Europe, SW Asia, N Africa
Brachypodium pinnatum L.) P. Beauv.		4× (1)	Europe, NCW Asia, NE Africa
Bromus inermis Leyss.		4× (1)	SW Asia, Caucasus
Bromus tectorum L.		4× (1)	Europe, SW Asia, N Africa

The genome, determined ploidy levels, number of included accessions (N), and the main native distribution for all taxa sequenced in this study is given. The genomes names of allopolyploid *Aegilops* taxa are follwing Kilian et al. [74] and Li et al. [84] for S*. Genome denominations for *Hordeum* follow Blattner [107], and Bernhardt [12] for the remaining taxa. Different seed banks adopt different taxonomic treatments that may vary in the number of (sub)species recognized. More comprehensive information about the used accessions, including the species names used in the donor seed bank and the country of origin is provided in Additional file 1: Table S1

each polymorphism between the reference sequences at least five times. After the enrichment procedure all samples were barcoded and pooled (following [39, 40]) at equimolar ratios. Capture libraries were sequenced on the Illumina HiSeq 2000 or MiSeq. The flowcells were loaded aiming for a sequencing coverage of at least 40X.

Data assembly

We used the captured *ndh*F and the off-target read fraction (i.e. reads for which no capture probes were designed in the target-enrichment experiment) to assemble whole chloroplast genomes. GENEIOUS versions R8–R10 (Biomatters Ltd.) were used for quality control and downstream analyses. Read pairs were set with an average insert size of 300 bp and bases with an error probability above 5% were trimmed. Chloroplast genomes were assembled in a two-step procedure consisting of (1) the generating of a species-specific reference sequences followed by (2) the creation of individual-based chloroplast assemblies.

In the first step we assembled species-specific chloroplast sequences by combining reads of multiple accessions of a single species. This increased the coverage for a species-specific chloroplast genome compared to the usage of data of an individual sample only. In a few cases single accessions were found to contribute a large amount of variation to these assemblies. These accessions were excluded from species-specific assemblies (Additional file 1: Table S1). The reads were either mapped to GenBank sequences of conspecific or closely related taxa (for *Aegilops*, *Hordeum* and *Triticum* species), or to *Hordeum vulgare* (EF115541), a well-studied basal organism in Triticeae, for taxa lacking conspecific chloroplast genomes in GenBank. One inverted repeat

was cleaved off the GenBank sequences as no sequence variation has been found between the inverted repeats of the same chloroplast genome. A careful comparison of Triticeae chloroplast genomes available in GenBank showed a large amount of insertions and deletions (indels) among the sequences from single species. In case several chloroplast genomes per species were retrieved from GenBank, those were aligned and an annotated consensus was created as reference to check for intraspecific indels. Then a stringent read mapping approach was used that considered only reads with mates mapping in proper distance according to the insert size (±50%). This was done to avoid the inclusion of chimerical Illumina reads, which have been reported to occur frequently [41]. All read mappings were performed using the GENEIOUS mapper with five iterations, allowing at maximum 15% of mismatches per read and a maximum gap size of 1000 bp to encompass large deletions. The assembly results were compared and manually checked for inconsistencies (i.e. indels the assembler was unable to resolve). Consensus sequences were called using the 50% majority rule. Up to five rounds of mapping and inspection were performed, each time using the contig obtained previously.

In the second step, for each sequenced individual chloroplast sequences were assembled by mapping all reads to their species-specific consensus sequence generated in step (1). Read mappings were performed using the GENEIOUS mapper with five iterations, allowing at maximum 10% of mismatches per read and a maximum gap size of 100 bp. The assembly results were manually checked for inconsistencies. No global coverage threshold was applied as the read coverage for single accessions were relatively low. However, single nucleotide polymorphisms (SNPs) compared to the reference covered by a single read were masked. Finally consensus sequences were called using the 'Highest Quality' option, which is able to resolve conflicts between reads because it takes the relative residue quality into account. 'N' were called for positions without coverage. Whole chloroplast sequences with more than 50% missing data were excluded from further analyses.

A multiple sequence alignment of the whole chloroplast genomes generated in step (2) plus a set of GenBank-derived sequences (Additional file 1: Table S1) was generated using MAFFT 7 (http://mafft.cbrc.jp/alignment/software; accessed in November 2016; [42]) applying the auto algorithm in combination with the 'nwildcard' option. The alignment was manually curated. The sequences generated in the scope of this study were annotated by comparing them to the annotations of the GenBank accession number KJ592713 [43] in GENEIOUS. All sequences were submitted to GenBank (accession numbers KX591961-KX592154 and KY635999-KY636181). The number of parsimony-informative positions was inferred using PAUP* 4.0b10 [44].

Phylogenetic analyses

We performed a Bayesian phylogenetic analysis for ndhF, as the sequence of this locus could be retrieved for all individuals without any missing data. First, unique ndhF haplotypes were identified using TCS 1.2.1 [45]. The best-suited model of sequence evolution was identified on the data matrix of unique haplotypes with JMODELTEST 2.1.4 [46] using the default parameters. The Bayesian information criterion (BIC; [47]) was selected for model choice because of its high accuracy [46] and its tendency to favour simpler models than the Akaike information criterion (AIC; [48]). Bayesian inference (BI) was performed in MRBAYES 3.2.6 [49] using the model inferred by JMODELTEST. BI consisted of four independent analyses each running for 20 million generations and sampling a tree every 1000 generations.

BI of the whole chloroplast genome alignment were run with MRBAYES 3.2.6 on the CIPRES (Cyberinfrastructure for Phylogenetic Research) Science Gateway 3.3 [50] for two datasets: (1) the complete alignment and (2) one alignment with positions having more than 50% of missing data being masked in GENEIOUS version R10. The best-fitting models of sequence evolution were estimated by making the Monte Carlo Markov chain (MCMC) sampling across all substitution models ([51]; 'lset nst = mixed'). For each dataset we performed three analyses, testing the impact of different rate settings, i.e. (1) a gamma-distributed rate variation, (2) a proportion of invariable sites and (3) with both combined to be able to identify the best-suited substitution model by comparing the posterior probabilities with AIC through MCMC (AICM; [52]), which is less computing intensive

though not as accurate as the application of Bayes factors [53], in TRACER. Each analysis was performed with two independent Metropolis coupled MCMC analyses each with four sequentially heated chains (temperature set to 0.05) until the standard deviation of split frequencies reached 0.009, a maximum of 10 million generations or the maximum runtime of CIPRES. Trees were sampled every 500 generations. For all Bayesian analyses conducted Brachypodium distachyon (EU325680) was set as outgroup and the convergence of the runs was assessed in TRACER v. 1.6 [54]. A consensus tree was computed after deleting a burn-in of the first 25% of trees.

Additionally, a Bayes factor (BF; [55]) analysis was carried out for the ndhF dataset to further evaluate the monophyly of Triticeae chloroplasts. Mean marginal log-likelihoods were computed using the stepping-stone sampling [56] in MRBAYES 3.2.6 [49] for monophyletic and polyphyletic relationships of Triticeae and the substitution model as identified in JMODELTEST. Each analysis consisted of two million generations with four independent runs of four parallel chains. The BF was evaluated using ten as a cut-off value [57].

Estimating divergence times using trnK-matK, rbcL and ndhF

We inferred a calibrated phylogeny for the three plastid loci trnK-matK, rbcL and ndhF. First, we tested the robustness of the calibration of the most recent common ancestor (MRCA) of Brachypodium and Triticeae when increasing the sampling for Triticeae from 12 to 37 species compared to Marcussen et al. [20]. For this a Bayesian coalescence analysis based on trnK-matK, rbcL and ndhF for the subfamily Pooideae was performed. The same GenBank sequences were assembled to form a contiguous sequence as described and used in Marcussen et al. [20]. This set of GenBank accessions was complemented with sequences assembled in this study whenever additional taxa or more sequence information from a certain taxon could be added. Following Marcussen et al. [20] we restricted ourselves to one sequence per species. We used the species-specific sequences from step (1) of the sequence assembly procedure, over the selection of a single accession per taxon, comparable to Pelser et al. [58]. This allowed us to employ all phylogenetic information available for a taxon and to overcome stretches of missing data. Conspecific sequences used for consensus inference showed 99.96 – 100% of identical sites. The best partitioning schemes and DNA substitutions models were inferred with PARTITIONFINDER [59, 60] comparing all possible partitioning schemes. The analysis was carried out using the combination of age priors for analyses 2, 4, 6, 10 and 17 as published in Marcussen et al. [20] in BEAST 2.4.1 [61]. For each setting one replicate was performed. Priors on the root age were estimated as stem node ('use originate'). We found the divergence time of *Brachypodium* and Triticeae as inferred by Marcussen et al. [20] to be robust. Second, we performed a multispecies coalescent (MSC) analysis using it as the secondary calibration point in million years ago (Ma) as normally distributed priors for the root of *Brachypodium*-Triticeae (mean 44.44 Ma ± 3.53) on *trn*K-*mat*K, *rbc*L and *ndh*F for all Triticeae accessions. We excluded gene sequences of *trn*K-*mat*K and *rbc*L if they showed more than 50% of missing data and sequences of all polyploid wheat accessions. Sequences of *Zea mays*, *Oryza sativa*, *Brachypodium distachyon* and two *Bromus* species were included as outgroup taxa. The taxa *Triticum monococcum* and *T. boeoticum*, *Secale cereale* and *S. vavilovii*, *Pseudoroegneria tauri* and *Ps. libanotica*, *Taeniatherum caput-medusae* and *Tae. crinitum*, *Agropyron cristatum* and *Agr. cimmericum* were each subsumed under the same species name (Additional file 1: Table S1), as no pronounced genetic variation were detected in the analysis of whole chloroplast sequences. Hereby we were following existing taxonomic treatments, which already unify these taxa under a single species name (see, e.g. [62]).

We performed MSC analyses for a dataset including *Psathyrostachys* and another one without it to evaluate the impact of this taxon on divergence times. Monophyly of Triticeae was not enforced for either analysis as suggested by the Bayes Factor analysis. For each dataset, first, the best partitioning schemes and DNA substitution models were inferred with PARTITIONFINDER searching all partitioning schemes. The analysis was run with the substitution models being linked, the Yule species tree prior, as well as the piecewise linear and constant root population model. Since the rate constancy was systematically rejected for all loci by the likelihood-ratio test [63], an uncorrelated lognormal clock model ([64]; uniform ucld.mean: min 0, max 0.01) was used. Trees were sampled every 5000 generations. Four independent analyses were performed and each was run for 600 million generations. All MSC analyses were run using the BEAGLE library [65]. Effective sample sizes (ESS) and convergence of the analyses were assessed using TRACER v. 1.6 [54]. An appropriate burn-in was estimated from each trace file, and the four analyses were combined with LOGCOMBINER as part of the BEAST package. Then a maximum clade credibility (MCC) tree was summarised with TREEANNOTATOR and visualized with FIGTREE 1.4.2 [66].

Results

Ploidy levels

Flow cytometric measurements were performed for all accessions to be able to distinguish between different ploidy levels for accessions of the same species (Additional file 1: Table S1). We identified di- and tetraploid accession

for *Agropyron cristatum*, *Eremopyrum bonaepartis*, *Pseudoroegneria stipifolia* and *Ps. strigosa*, and detected tetra- and hexaploid cytotypes for *Aegilops crassa* and *Ae. neglecta*. Flow cytometric measurements were used as additional information to confirm species affiliations [67]. For example, comparing of the genome sizes measured for the diploid species *Thinopyrum bessarabicum* and *Th. elongatum* to the data from the Kew Angiosperm DNA C-values database revealed that the analysed accessions actually represent polyploids instead of diploids. For more information on problematic material from seed banks see Additional file 1: Table S1.

Sequence assembly

The target-enrichment protocol and Illumina sequencing were applied to 194 accessions, covering 53 species of 15 genera (dependent on the applied classificatory system) and three outgroup species (i.e. *Bromus* and *Brachypodium*, Table 1, Additional file 1: Table S1). Whole chloroplast genomes were assembled in a two-step procedure via (1) an intermediate step of generating a species-specific reference if there was none available in GenBank and (2) the assembly of the chloroplast of each accession via read mapping to sequences from step (1).

The average coverage of the chloroplast genome varies largely between single samples and depends mainly on the actual sequencing depth. Between approximately 50% and 90% of the reads mapping to the chloroplast mapped to *ndh*F (Additional file 2: Table S2), which was included in the bait design. Thus, the *ndh*F gene could be assembled for all accession without missing data. We identified 64 unique haplotypes when comparing the *ndh*F gene data plus the sequences downloaded from GenBank (Additional file 1: Table S1). The alignment of these 64 haplotypes had a total length of 2232 bp with 186 (8.3%) parsimony-informative sites.

The entire alignment of whole chloroplast genome sequences comprised 222 sequences, 39 of them were downloaded from GenBank. This alignment ranged from *psb*A in the large single-copy region to partial *ndh*H in the small single-copy region and had a total length of 123,531 bp. It had 9064 (7.3%) parsimony-informative positions. The data matrix included 9.3% of missing data ('N'). These randomly distributed stretches of missing data occur in the alignment in regions where the sequencing coverage was insufficient. Additionally the matrix revealed 7.5% of gaps due to length variation between taxa. In several cases taxa showed long indels in intergenic regions, thus the same 900 bp deletion was found between *rpl*23 and *ndh*B in *Pseudoroegneria*, *Thinopyrum* and *Dasypyrum*. Many short indels (3–40 bp) were found in introns of coding genes (e.g. *ysf3*) and intergenic spacers. A variant of this alignment, having regions with 50% of missing data being removed, had a total length of 114,788 bp. In this case 8717 (7.6%) positions of the alignment were parsimony informative, while 9.2% of the characters were constituted by N's and 0.8% of gaps. Alignment masking mainly excluded regions of length variation due to short repeat motives in intergenic regions. With only few substitutions per chloroplast intraspecific variation was generally very low.

The alignment revealed insertions unique to some GenBank sequences whose true occurrence could not be confirmed by our data: no reads from our analysed individuals mapped to these insertions. Moreover, BLAST searches of these regions returned mitochondrial and/ or nuclear genomic data as best hit (e.g. KC912690, KC912692, KC912693, KC912694) indicating assembly artefacts. Those GenBank sequences were excluded from further analyses (Additional file 1: Table S1).

Phylogenetic analyses

We performed a BI analysis on the set of 64 unique *ndh*F haplotypes with the model of sequences substitution set to GTR + G [68, 69], as identified by JMODELTEST. The phylogenetic tree obtained from *ndh*F (Fig. 1) shows Triticeae to be paraphyletic, as the lineage of *Psathyrostachys* appears to have diverged before the lineage of *Bromus*, although the position of *Bromus* is with a posterior probability (pp) of 0.88 not well supported. The branch lengths for the *Bromus* group are considerably longer compared to *Psathyrostachys*. The topology shows that individuals of most species and/or genera form monophyletic groups. However, *Eremopyrum bonaepartis* is polyphyletic, as the diploid plastid type of *E. bonaepartis* groups as sister to *Henrardia persica*, while the haplotypes of all tetraploid *E. bonaepartis* and diploid *E. triticeum* form a clade with *Agropyron cristatum*. A common maternal ancestor can be hypothesized for *Agropyron*, *Australopyrum*, *Eremopyrum* and *Henrardia* as these taxa form a well supported clade, which is sister to the clade of *Hordeum* species. The clades formed by the genera *Heteranthelium*, *Secale* and *Taeniatherum* are placed on a polytomy together with a clade formed by taxa having a **B**, **G** or **S** genome [i.e. *Aegilops speltoides* (**S**) and all polyploid *Triticum* taxa (**B/G**)], the clade of taxa with an **E**, **St** or **V** genome (i.e. *Thinopyrum*, *Pseudoroegneria* and *Dasypyrum*), and the clade of all remaining *Aegilops*, *Amblyopyrum* and diploid *Triticum* taxa. *Pseudoroegneria* appears paraphyletic, as *Dasypyrum* and *Thinopyrum* haplotypes group within this clade. The backbone of this clade represents a polytomy. Notably the placement of the otherwise monophyletic *Dasypyrum* is not supported. Several different haplotypes can be distinguished for various species of *Pseudoroegneria* itself (e.g. *Ps. spicata*, *Ps. strigosa*, *Ps. tauri*, *Ps. stipifolia*). Furthermore, the two **A**-genome

species *Triticum urartu* and *T. monococcum* are monophyletic. Also all **D**-genome species (i.e. *Ae. tauschii*, *Ae. cylindrica* and *Ae. ventricosa*) form a clade. Both genomic groups are located on a polytomy together with the remaining *Aegilops* species and *Amblyopyrum*. *Aegilops crassa* and *Ae. juvenalis* (**D'**) group apart from the other **D** taxa and show a *ndh*F haplotype with less nucleotide differences to **S*** than to **D** chloroplast lineages (i.e. one SNP difference to **S*** vs. three and five SNPs to **D**). All diploid and polyploid **S*** species sequenced in the scope of this study share the same *ndh*F haplotype. *Aegilops comosa* (**M**) and *Ae. uniaristata* (**N**) are sister species. All **U**-genome taxa fall into the same clade together with *Aegilops geniculata* (**M°**) and *Amblyopyrum muticum* (**T**). *Aegilops triuncialis* accessions possess **U** as well as **C** haplotypes.

Sometimes, single accessions of a species group within the otherwise monophyletic clade of another species. Thus, the accession AE_1831 of *Aegilops markgrafii* (**C**) falls into the clade of *Amblyopyrum muticum* (**T**) while KP_2012_119 of *Aegilops biuncialis* (**U**) falls within *Ae. geniculata* (**M°**). The accession AE_586 of *Aegilops neglecta* (**U**) groups together with *Ae. markgrafii* (**C**). Further, intraspecific variation within *ndh*F was found in several cases, for example, for *Aegilops comosa*, *Ae. speltoides*, *Amblyopyrum muticum*, and *Dasypyrum villosum*. With a score of 36.4, BF strongly favours Triticeae chloroplasts as paraphyletic (Additional file 3: Table S3) when *Psathyrostachys* is included in the analysis.

As the resolution of the phylogenetic tree from the *ndh*F dataset is not sufficient to distinguish between more recently diverged taxa, the whole chloroplast genome dataset was phylogenetically analysed by BI using an alignment of the entire chloroplast genomes and a variant of it were positions having more then 50% of missing data have been masked. In both cases MRBAYES revealed a *gtrsubmodel* in combination with gamma-distributed rate variations as best-suited substitution model. The topologies (Fig. 2, Additional file 4: Figure S1) returned from both analyses are mainly congruent to each other and to the *ndh*F tree. However, nodes of deep splits supported moderately for the complete plastid data matrix show higher support in the dataset where low-covered regions have been masked. This is, for example, the case for the split of the ancestor of *Bromus*. The branch length differences between *Bromus* and *Psathyrostachys* are in agreement with the *ndh*F tree. In contrast to the *ndh*F dataset, the whole chloroplast phylogenies are able to provide a hypothesis of the relationships between all major genomic groups. They suggest that the **E**, **St**, and **V** clade (i.e. *Thinopyrum*, *Pseudoroegneria* and *Dasypyrum*) diverged before *Heteranthelium*, which in turn split before *Secale* and *Taeniatherum*. *Pseudoroegneria spicata* forms its own clade that diverged first

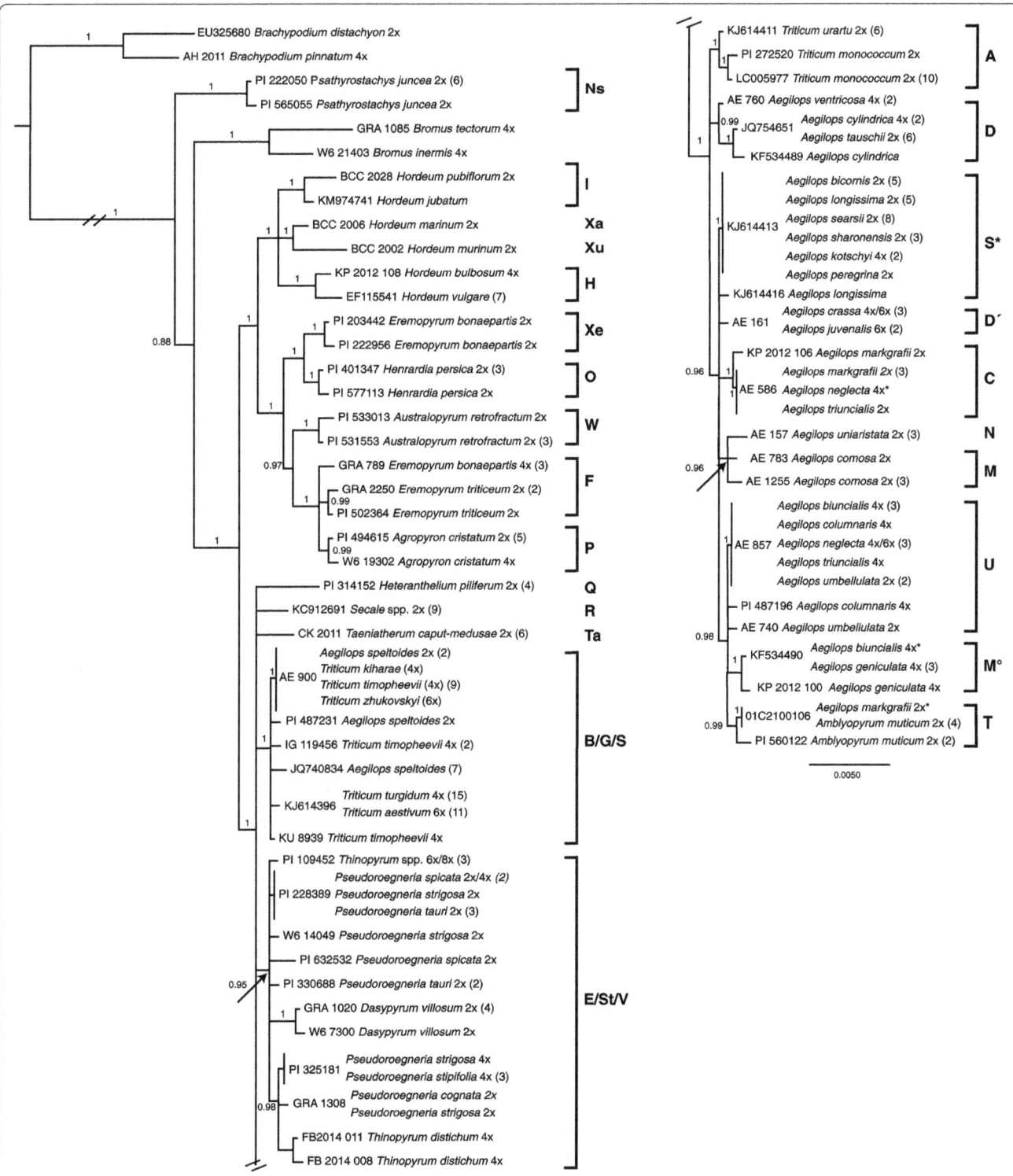

Fig. 1 Phylogenetic tree derived from 2232 bp of the chloroplast locus *ndh*F and Bayesian inference. The multiple sequence alignment consisted of 64 unique haplotypes that originated from 194 accessions sequenced in the scope of this study and 41 sequences retrieved from GenBank. *Brachypodium distachyon* was set as outgroup taxon. Posterior probabilities (pp) for the main clades are depicted next to the nodes if they were higher then 0.75. Each unique haplotype is named with a distinctive identifier. For detailed information which accession possesses which haplotype and species synonyms see Additional file 1: Table S1. The ploidy level is indicated behind taxon labels. If there are multiple accessions per taxon sharing the same haplotype, the number of accessions is provided behind the taxon label. Single accessions grouping apart from other accessions of their taxon are marked with an *asterisk*. To the right the genomic groups are shown. *Arrows* with support values indicate the nodes they refer to

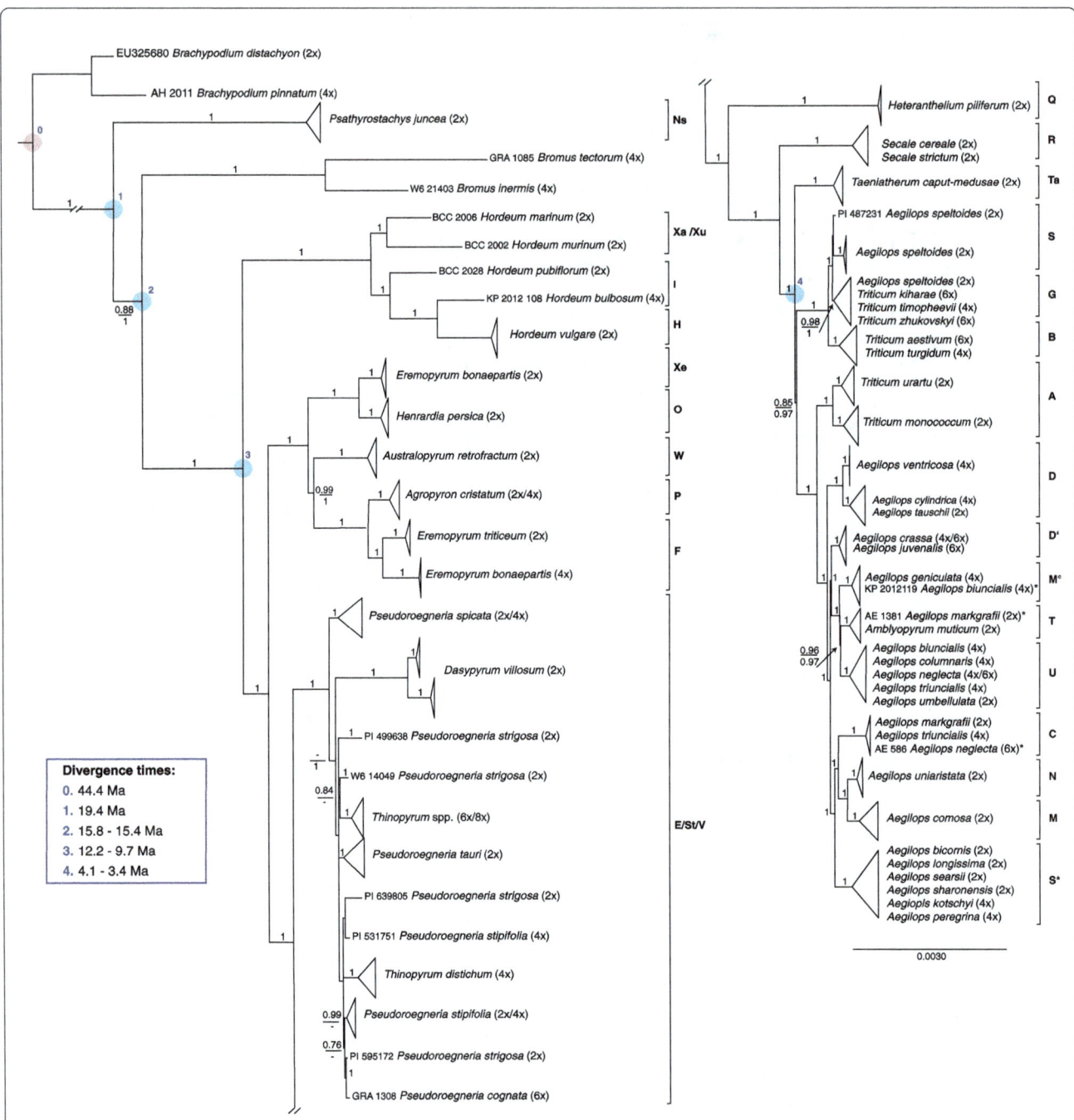

Fig. 2 Phylogenetic tree derived from an alignment of whole genome chloroplast sequences via Bayesian phylogenetic inference. The multiple sequence alignment comprised 183 genomes assembled in the present study and 39 genomes that were downloaded from GenBank. *Brachypodium distachyon* was defined as outgroup taxon. The tree shown corresponds to an analysis based on the complete alignment of 123,531 base pairs (bp). Clades were collapsed into triangles to reflect the main groupings. The area of the *triangles* reflects the genetic variation contained in a certain clade. Posterior probabilities (pp) for the main clades are depicted next to the nodes if they were higher then 0.75. Support values of a second Bayesian phylogenetic analysis based on 114,788 bp of whole chloroplast genomes, where alignment positions with more than 50% of missing data were masked, are shown below the values of the corresponding nodes in the complete chloroplast analysis if the values differed between analyses. Ploidy levels are provided in brackets after the taxon labels. Single accessions grouping apart from other accessions of their taxon are highlighted with an *asterisk*. To the *right* the genomic groups are indicated. The *red circle* represents the secondary calibration point from Marcussen et al. [20] used for node calibrations in multispecies coalescent analyses (MSC). Major nodes are shown in *blue* and their estimated ages in million years are given in the box. Two age values for the same node correspond to the analysis with (first value) and without (second value) the inclusion of *Psathyrostachys*. For more information on the results of the MSC analyses see Additional file 5: Figure S2 and Additional file 6: Figure S3. For the full representation of the tree showing the grouping of all single accessions see Additional file 4: Figure S1. For species synonyms see Additional file 1: Table S1. Arrows with support values indicate the nodes they refer to

from *Dasypyrum* and the remaining taxa within this clade. However, the *Dasypyrum* chloroplast genomes are characterized by rather long branches compared to other taxa in this clade. Furthermore, *Dasypyrum* comprises two well-differentiated haplotypes. *Aegilops speltoides* and the polyploid wheat species form three groups: (1) most *Ae. speltoides* accession form a clade of their own (**S**), (2) some *Ae. speltoides* accessions group together with *Triticum timopheevii, T. zhukovskyi* and the artificially synthesized wheat *T. kiharae* (**G**) and (3) all accessions of *T. turgidum* and *T. aestivum* share the same haplotype (**B**). The not supported placement of one *Ae. speltoides* accession (PI_48721) close to the **S** group, shifts to a supported position in the **G** group when regions with an high extent of missing data were masked. Additionally, the usage of entire chloroplast genomes resolves that diploid *Triticum* species (**A**) diverged before the **D**-genome taxa and the remaining *Aegilops* species and *Amblyopyrum*. The phylogeny also indicates that **D'** is closely related but distinct from **D**. Further, **M°**, **T** and **U** taxa form a clade, that diverged before the split of taxa having a **C, N, M** or **S*** genome. Within this clade the sister species relationship of *Aegilops comosa* and *Ae. uniaristata* is confirmed. *Aegilops comosa* (**M**) groups distinct from the other **M°** plastid type. The species *Ae. searsii, Ae. bicornis, Ae. longissima, Ae. sharonensis* form a clade together with the polyploid *Ae. kotschyi* and *Ae. peregrina* (**S***) indicating only very little sequence variation. Concordant to the *ndh*F tree, one sequence each of *Ae. markgrafii* (AE_1831), *Ae. biuncialis* (KP_2012_119) and *Ae. neglecta* (AE_586) group apart from the other sequences of their respective taxon.

Ages of clades

Divergence times were estimated based of *trn*K-*mat*K, *rbc*L and *ndh*F sequences for each accession included in the study and using an uncorrelated lognormal clock model and a secondary calibration on the MRCA of *Brachypodium distachyon* and Triticeae in *BEAST. Different ages for the split of Triticeae and *Bromus* were obtained depending on the in- or exclusion of the genus *Psathyrostachys*. Including *Psathyrostachys*, Triticeae are paraphyletic and the ages are slightly older but with larger and overlapping 95% highest-posterior densities (HPD) compared to the dataset that does not comprise *Psathyrostachys* (Additional file 5: Figure S2, Additional file 6: Figure S3). In the analysis including *Psathyrostachys* the most recent common ancestor (MRCA) of Triticeae and *Bromus* occurred approximately 19.44 Ma (95% HPD = 12.66-27.20). The split of *Bromus* and the remaining Triticeae (termed "core Triticeae") occurred approximately 15.77 Ma (95% HPD = 9.38-22.75). The age of this split

does not seem affected by the absence of *Psathyrostachys* (15.41 Ma, 95% HPD = 10.72-20.83). However, the MRCA of the core Triticeae occurred approximately 12.17 Ma (95% HPD = 7.65-17.44) including *Psathyrostachys* and nearly 2.5 million years later (9.68 Ma, 95% HPD = 7.42-12.21) in the analysis omitting this early diverging lineage. The MRCA of *Aegilops, Triticum* and *Amblyopyrum* (plus *Taeniatherum*) occurred around 4.14 Ma (95% HPD = 2.48-6.44) including *Psathyrostachys* and 3.38 Ma (95% HPD = 2.35-4.47), when omitting it.

Discussion
Plant materials
The analysed accessions were mainly acquired from several seed banks (i.e. ICARDA, IPK, USDA, the Czech Crop Research Institute) but additional material was collected during field trips. Multiple accessions per species and intra-specific entities were selected to be able to detect intraspecific genetic variability.

The performance of genome size measurements allowed the distinction of ploidy level differences for accessions of the same species. Our finding of different ploidy levels within *Agropyron cristatum, Eremopyrum bonaepartis, Pseudoroegneria strigosa, Aegilops crassa* and *Ae. neglecta* are in agreement with previous work [70–74]. For the first time we report the occurrence of different ploidy levels for *Pseudoroegneria stipifolia*.

Few accessions have been found having unexpected genome sizes, like in *Thinopyrum*. Concerns about the condition of seed bank material have been raised in other studies and are related to the fact that it is often maintained under conditions that permit open pollination over several rounds of seed replication [75, 76]. As Triticeae show species-specific genome sizes [67, 77, 78] the performance of flow cytometric measurements is a good strategy to detect problematic material, especially in the case of perennial Triticeae where inflorescences for morphological species determination cannot always be obtained within the timeframe of a research project. Also in this study, a few selected accessions needed to be excluded due to deviations in genome size or morphological characters. However, the vast majority of the material did not reveal any peculiarities and samples directly collected in the wild always grouped with other samples of the same species.

Sequence assembly
In this study we assembled the chloroplast *ndh*F gene and complete chloroplast genomes using for the latter off-target sequence reads of a target-enrichment approach and NGS sequencing for a comprehensive set of Triticeae taxa. The *ndh*F gene could be assembled for 194 accessions representing 53 Triticeae and three

outgroup species without missing data, as it was included in the bait design for sequence enrichment. We obtained a set of 183 whole chloroplast genome sequences that provide new plastid genomes of 36 Triticeae species out of 15 genera for which so far no such sequence was available. From these data we estimated the maternal relationships within Triticeae. In previous studies off-target reads have been successfully analysed in diverse organism groups [36, 79–82]. Because the chloroplast occurs in high copy number in the cells, it constitutes the main fraction of off-target reads in target-enrichment approaches in plants. Therefore the majority of reads identified as chloroplast DNA originated most probably from this genome and not from parts that were transferred from the chloroplast to the nuclear genome, which should be rare in off-target reads.

The pooling of samples from multiple conspecific individuals allowed us to overcome the low coverage for individual samples and to assemble chloroplast genomes to be used as taxon-specific reference for the assembly of individual chloroplast genomes for accessions for which no conspecific reference was available in GenBank. Stretches of missing data remain in the final individual-based assemblies of the plastid genomes. As these stretches occur randomly along the chromosome, they do not influence the detection of structural differences (indels) between chloroplast genomes of species and/or genera. Generally, indels and base substitutions occur mostly in spacer regions of the Triticeae chloroplast genomes. An increase in sequencing depth may have allowed assembling the chloroplast genomes of all individuals without any missing data. However, the comparison of accessions sequenced with different depths shows that overall higher sequencing coverage will not guarantee a complete chloroplast sequence, as off-target regions are randomly (or not) retained during the enrichment process. The most problematic part in assembling the reads was to reach confidence about the detected indel positions, as the short read length of 2×100 bp of the Illumina platform did not always cover such regions completely. The whole genome sequences we provide were carefully checked manually and compared to available sequences in GenBank. Comparable to other studies (e.g. [32, 43]) we were not able to confirm all parts of GenBank-derived sequences obtained from whole-genome shotgun sequencing. It might be that they contain some non-identified assembly errors. With the now available longer Illumina paired-end reads of 2×250 bp these problems should become less severe in future studies. Finally, the topologies validated our assembly procedure, as previously published GenBank sequences always grouped in their respective clades irrespective of the small differences found.

Maternal phylogeny of Triticeae

In this work we aimed for a molecular phylogeny of the chloroplast lineages in Triticeae. The results from *ndh*F and whole chloroplast genome phylogenetic analyses are mainly in agreement with hypotheses previously published for groups within the tribe [9, 26, 83] and with respect to the domesticated wheats and their close relatives [30, 31, 84]. Compared to these latter publications a better understanding was obtained, particularly because of the comprehensive taxon sampling, the usage of whole chloroplast genomes, and the inclusion of multiple individuals per species.

The tribe Triticeae is generally accepted to be monophyletic [22, 23, 85–87] with *Bromus*, the only genus in the tribe Bromeae, being the sistergroup to all Triticeae [88, 89]. However, based on our data, but also previously published chloroplast data [26, 35, 90], the monophyly of Triticeae was either rejected or not supported. As morphology [23] and also phylogenies based on nuclear data place *Psathyrostachys* at the base of Triticeae close to *Hordeum* ([10]; own unpublished data), we see two possibilities to explain the chloroplast phylogeny. Thus, either *Psathyrostachys* obtained the chloroplast of a close and nowadays extinct relative belonging to the ancestral Triticeae-Bromeae gene pool, or vice versa an ancestor belonging to the *Bromus* stem group obtained a chloroplast from early Triticeae. In any case, a chloroplast phylogeny including *Bromus* and *Psathyrostachys* might not reflect Triticeae relationships very well, at least for its basal groups, and will also influence the outcome of molecular dating approaches (see below).

The retrieved chloroplast phylogeny indicates a common maternal ancestor for the genera *Australopyrum*, *Eremopyrum*, *Agropyron* and *Henrardia*, with *Eremopyrum*, *Agropyron* and *Henrardia* currently having overlapping distribution areas in southern Europe and western Asia. The monogenomic genus *Australopyrum* (**W**) and all allopolyploid taxa possessing a **W** genome (*Stenostachys* - **HW**, *Anthosachne* - **StYW**, *Connorochloa* – **StYHW**; taxa not sampled) are endemic to dry and temperate Australasia [91]. This supports speciation in allopatry after long-distance dispersal of an *Australopyrum* progenitor and likely recurrent formation of allopolyploid taxa involving numerous other Triticeae species in Australasia. A sister relationship between the species of *Agropyron* and *Eremopyrum* has also been proposed by other studies. However, when *Eremopyrum bonaepartis* was included, *Eremopyrum* became polyphyletic with the diploid cytotype being sister to *Henrardia*. This is in agreement with earlier findings [10, 92, 93].

Similar to Mason-Gamer [83] we found that *Pseudoroegneria*, *Dasypyrum* and *Thinopyrum* form a monophyletic clade indicating that they belong to the same maternal lineage. A sister relationship of *Pseudoroegneria* and *Dasypyrum* has been proposed recently by Escobar et al. [10] based on nuclear data. In our dataset *Dasypyrum* groups however within *Pseudoroegneria*. Within *Dasypyrum*, accessions from Bulgaria and Italy cluster together, while material from Turkey and Greece form another sub-clade. Hence, this pattern may indicate some recent local differentiation. The polyphyletic grouping of *Thinopyrum* within this clade can be explained either by incomplete lineage sorting (ILS) or because *Thinopyrum* repeatedly captured different plastid types of *Pseudoroegneria*. A close relationship to the *Aegilops-Triticum-Amblyopyrum* group has been reported for *Thinopyrum* based on nuclear data [3, 83, 93–95]. This incongruence might be explained by the fact that *Thinopyrum*, but also *Dasypyrum* and *Pseudoroegneria* are outcrossing taxa [10, 96], which seems to increase the chance of chloroplast capture via hybridization and back-crossing [25]. Moreover, most taxa have overlapping distribution areas in the Caucasus region, also facilitating hybridization. Our results revealed no major sequence variation among chloroplast genomes of *Secale strictum* and *S. cereale/S. vavilovii*. This points to an only recent diversification within this genus.

It is well known that the species of *Triticum*, *Aegilops* and *Amblyopyrum muticum* are closely related and of rather recent origin [7, 10, 20, 26]. To date, there is no general agreement on how taxa within this species complex are related to each other, even at the diploid level. There is an on-going dispute if *Aegilops* and *Triticum* should be merged into one genus, and if *Amblyopyrum muticum* should be included into *Aegilops* [74, 84, 97–99]. In agreement with Bordbar et al. [9], the chloroplast phylogeny revealed that *Am. muticum* possesses a chloroplast genome similar to the **M** and **U** genome groups, although based on nuclear data *Am. muticum* appears to be sister to all *Aegilops* and *Triticum* species [7]. The *Aegilops*-like chloroplast genome of *Am. muticum* might be explained by the existence of a common ancestor and therefore a chloroplast genome already shared before divergence of these lineages. Alternatively, it may indicate that it captured the chloroplast from one of these species or their MRCA, which is geographically possible, as distribution areas overlap in Turkey and Armenia.

Polyploid *Triticum* species and *Aegilops speltoides* formed a clade supporting that *Ae. speltoides* is the maternal donor of polyploid wheat genomes. The relationships within this clade corroborate the hypothesis that two different *Ae. speltoides* lineages were involved in their formation [30, 74, 100, 101]. The direct maternal donor for *Triticum timopheevii* and *T. zhukovskyi* (**G**) could be identified, as they share the chloroplast haplotype of three *Ae. speltoides* accessions originating from Iraq and Syria. However the donor remains uncertain for *Triticum turgidum*

and *T. aestivum* (**B**), indicating that either our sampling of *Ae. speltoides* was not sufficient to cover the species diversity or pointing to a nowadays extinct donor lineage. Alternatively, Gornicki et al. [30] suggested, that tetraploidisation within this clade predates the one of *T. timopheevii*.

All taxa of the genus *Triticum* s.str. Fall into one clade together with *Aegilops* and *Amblyopyrum*. *Triticum* taxa that were elevated to species rank by Dorofeev et al. [102] could not be distinguished on the basis of their chloroplast haplotypes, which supports the taxonomic treatment of van Slageren [97] subsuming them under the same species name (Additional file 1: Table S1).

Based on chloroplast data and supported by the findings of Petersen et al. [7] and Li et al. [84], *Ae. speltoides* (**S**) appears to be the species that diverged earliest from all other *Aegilops* species. Generally the wheat group is characterized by short branch lengths and plastid haplotypes shared by multiple species. This is most probably due to the only recent divergence of these species.

Chloroplast capture as indicator of hybridization events

The exchange of chloroplasts among closely related plant species has been reported in diverse plant groups and the effect of hybridization on Triticeae taxa is a matter of discussion. For example, a homoploid hybrid origin of the **D**-genome lineage involving the **A**- and **B**-genome lineages is the subject of a recent dispute [20, 84, 98, 99]. However, our and previous studies [30, 31, 84] revealed three independent but closely related chloroplast lineages with plastids of the **A**-genome lineage being more closely related to the ones of the **D** genome, which can be explained by consecutive divergence. Hence, if such a hybridization event occurred it only affected the nuclear genome.

Although recent publications agree that the detection of hybridization events depends mainly on taxon sampling [19], so far all postulated hypotheses for Triticeae are based on a limited choice of taxa. In our study, three possible cases of ancient chloroplast captures were identified, i.e. for (1) *Bromus/Psathyrostachys*, (2) *Thinopyrum* and (3) *Amblyopyrum*, as the chloroplast phylogeny looks considerably different from phylogenies retrieved from nuclear data [7, 83]. More recent events of chloroplast captures were identified for single accessions of the species *Aegilops biuncialis*, *Ae. markgrafii*, *Ae. neglecta* and *Ae. triuncialis* that grouped within clades of other closely related species. We assume such hybridization events to occur frequently between various taxa of the wheat group due to incomplete reproductive isolation among these young species.

Ages of clades

To obtain dated phylogenies of Triticeae we used the split of *Brachypodium* and Triticeae as secondary calibration point [20] based on *trn*K-*mat*K, *rbc*L and *ndh*F sequences. Pros and cons of using chloroplast data for the estimation of divergence times were already discussed by Middleton et al. [31] who argued that splits of chloroplast lineages might be older than the respective species, resulting in overestimated taxon ages for medium-aged and young clades. For dating in Triticeae we see an additional concern using chloroplast data. Due to mostly low substitution rates in plastid genomes [103] also underestimation of ages is possible in young clades, as fixation of mutations occur as a stochastic process [30, 104, 105] that might be slower than species diversification. In these cases already well-diverged taxa might still possess very similar or identical chloroplast haplotypes [106], resulting in lower age estimations in comparison to nuclear data. This might be the case for many nodes of our tree, although the divergence times retrieved for the main splits are generally about 1 million years older than the ones obtained by Middleton et al. [31]. Our analyses suggest the occurrence of a MRCA for the *Aegilops/Triticum* group at approximately 4 Ma, while divergence times of this complex were proposed to date back to approximately 3 Ma [31] or 6.55 Ma based on a dataset of five nuclear and one plastid gene [20].

Another critical topic regarding chloroplast-based dating in Triticeae results from the chloroplast data of *Psathyrostachys*. Our results support the hypothesis that the chloroplast of either *P. juncea* or a *Bromus* ancestor was obtained through chloroplast capture from a taxon belonging the *Bromus*/Triticeae stem lineage, resulting in *P. juncea* clearly falling outside the otherwise monophyletic Triticeae. We strongly favour an event of chloroplast capture over ILS as the cause for the observed relationships. The pronounced sequence variation between *Bromus*, *Psathyrostachys* and the remaining Triticeae for entire chloroplast genomes is best explained by strong and independent sequence divergence of *Bromus* and *Psathyrostachys* compared to the remaining Triticeae. Moreover, in case ILS represents the reason for the observed relationships our coalescent-analyses should have returned the same age for the MRCA of Triticeae-Bromeae with and without the inclusion of *Psathyrostachys*. However, we obtained age estimations that differed by approximately 4 million years.

As the direction of chloroplast capture remains unknown, we estimate the MRCA of all Triticeae to an age of between 10 and 19 million years. When comparing in- vs. exclusion of *P. juncea* the age estimations for all clades are robust, as they fall generally within the 95%

HPD (Additional file: 5: Figure S2, Additional file 6: Figure S3).

Conclusions

We assembled chloroplast sequence data of a large set of monogenomic Triticeae and polyploid wheats by combining on- as well as off-target reads of a sequence-capture approach coupled with Illumina sequencing. This approach allowed us to produce a set of 183 Triticeae chloroplast genomes. These sequences provide new plastid genomes for 39 Triticeae, two *Bromus* and one *Brachypodium* species. Moreover, the data was used to estimate the most comprehensive hypothesis of relationships among Triticeae chloroplast lineages to date.

We infer that an early event of chloroplast capture was involved in the evolution of *Psathyrostachys* or *Bromus*. Either *Psathyrostachys* or *Bromus* obtained a chloroplast from a taxon closely related to a common ancestor of the Triticeae-Bromeae lineage that diverged approximately 19.44 Ma, as the *Psathyrostachys* chloroplast haplotype groups at a deeper node than *Bromus* in our whole-genome phylogeny. We can, however, not safely determine the direction of chloroplast exchange in this case, as this would need the inclusion of much more Bromeae species.

We identified taxa that share the same maternal lineage (e.g. *Agropyron*, *Eremopyrum* and *Heteranthelium*; *Pseudoroegneria* and *Dasypyrum*). Conflicts to nuclear phylogenies (i.e. the grouping of *Thinopyrum*, *Amblyopyrum*) likely indicate old events of chloroplast introgression, while some cases of pronounced intraspecific variation could be attributed to recent events of hybridization, as foreign chloroplast types grouped within otherwise monophyletic species groups (i.e. *Ae. biuncialis* and *Ae. markgrafii*, *Ae. neglecta*).

As plastids are maternally inherited in these grasses, they provide supplementary information to nuclear data. For example, the plastid data indicate the polyphyly of *Eremopyrum*. Moreover, the possession of an *Aegilops*-like chloroplast type of *Amblyopyrum* might reject a taxonomic treatment completely separate from *Aegilops*. Hence, plastid data can facilitate understanding Triticeae evolution, which in turn is crucial on the way to a robust taxonomic system for the entire tribe of Triticeae. However, plastid phylogenies will never be able to infer all hybridization events involved in speciation, e.g. when nuclear genomes got introgressed while chloroplast lineages remains unaffected.

Additional files

Additional file 1: Table S1. Accessions considered in the study. Overview of the material considered in this study. For all materials, the GenBank identifier, the accession and species name as used in this study (Species) as well as their species synonyms used in the donor seed banks or in the NCBI GenBank (Material source/Reference) are provided. The genome symbol, and the country of origin, where the material was originally collected are given. The ploidy level measured in the scope of this study and the information if a herbarium voucher could be deposited in the herbarium of IPK Gatersleben (GAT) is given. Genomic formulas of tetraploids and hexploids are given as "female x male parent". The genomes of *Aegilops* taxa follow Kilian et al. [74] and Li et al. [84]. Genome denominations for *Hordeum* follow Blattner [107] and Bernhardt [12] for the remaining taxa. (XLS 84 kb)

Additional file 2: Table S2. Read numbers mapping to the complete chloroplast sequences and *ndh*F. Number of reads mapping and mean coverage for the entire chloroplast genome and *ndh*F after the removal of duplicated reads. Also the proportions of all reads mapping to the chloroplast that mapped to *ndh*F are given. (XLS 66 kb)

Additional file 3: Table S3. Marginal likelihoods and Bayes factor evaluation of Triticeae chloroplast relationships. Stepping-stone estimates of marginal likelihoods calculated with MrBayes 3.2.6 on the *ndh*F dataset and Bayes factor estimated as $2(H_1-H_2)$, where H_1 enforces monophyly and H_2 enforces polyphyly of Triticeae chloroplasts. $BF_{12} < -10$ indicates strong support for model 2. (DOC 27 kb)

Additional file 4: Figure S1. Full representation of the Bayesian phylogenetic tree based on whole chloroplast genome sequences. The multiple sequence alignment comprised 183 genomes assembled in the present study and 39 genomes that were downloaded from GenBank. *Brachypodium distachyon* was used as outgroup taxon. The tree shown is based on the complete alignment of 123,531 base pairs (bp). Posterior probabilities (pp) for the main clades are depicted next to the nodes if they were higher then 0.75. Support values of a second Bayesian analysis based on 114,788 bp of whole chloroplast genomes were alignment positions with more than 50% of missing data were masked are shown below the values of the corresponding nodes in the complete chloroplast analysis if the values differed between analyses. For clades comprising multiple taxa, the taxon affiliation of single accession is indicated by the same symbols behind accession and taxon name (e.g. ':", *). The ploidy level is provided in brackets after the taxon label. Single accessions grouping apart from other accessions of their taxon are shown in bold. To the right the genomic groups are indicated. The red circle represents the secondary calibration point from Marcussen et al. [20] used for node calibrations in multispecies coalescent analyses (MSC). Major nodes are shown in blue. Their estimated ages in million years are given in the box. Two age values for the same node correspond to the analysis with *Psathyrostachys* (first value) and without it (second value). For more information on the results of the MSC analyses see Additional file 5: Figure S2 and Additional file 6: Figure S3. For the full representation of the tree showing the grouping of all single accessions see Additional file 4: Figure S1. For species synonyms see Additional file 1: Table S1. Arrows with support values indicate the nodes they refer to. (PDF 555 kb)

Additional file 5: Figure S2. Calibrated species trees based on *trn*K-*mat*K, *rbc*L, and *ndh*F including *Psathyrostachys*. Calibrated multispecies coalescent derived from three chloroplast loci *trn*K-*mat*K, *rbc*L and *ndh*F of all Triticeae accessions (excluding polyploid wheats). Sequences of *Brachypodium distachyon*, *Oryza sativa* and *Zea mays* were included as outgroups. Posterior probability values are given for all nodes. Divergence time estimates were inferred using the secondary calibration points from Marcussen et al. [20] for the *Brachypodium*-Triticeae split (mean 44.44 million years ago). Node bars indicate the age range with 95% interval of the highest probability density. For the analysis *Triticum monococcum* and *T. boeoticum, Secale cereale* and *S. vavilovii, Pseudoroegneria tauri* and *Ps. libanotica, Taeniatherum caput-medusae* and *Tae. crinitum, Agropyron cristatum* and *Agr. cimmericum* were each subsumed under a single species name (Additional file 1: Table S1). (JPEG 1085 kb)

Additional file 6: Figure S3. Calibrated species trees based on *trn*K-*mat*K, *rbc*L, and *ndh*F omitting *Psathyrostachys*. Calibrated multispecies coalescent derived from three chloroplast loci *trn*K-*mat*K, *rbc*L and *ndh*F considering all genomic Triticeae groups covered in the study but omitting *Psathyrostachys* and polyploid wheats. Sequences of *Brachypodium distachyon, Oryza sativa* and *Zea mays* were included as outgroups. Posterior probability

values are given for all nodes. Divergence time estimates were inferred using the secondary calibration points from Marcussen et al. [20] for the *Brachypodium*-Triticeae split (mean 44.44 million years ago). Node bars indicate the age range with 95% interval of the highest probability density. For the analysis *Triticum monococcum* and *T. boeoticum, Secale cereale* and *S. vavilovii, Pseudoroegneria tauri* and *Ps. libanotica, Taeniatherum caput-medusae* and *Tae. crinitum, Agropyron cristatum* and *Agr. cimmericum* were each subsumed under a single species name (Additional file 1: Table S1). (JPEG 1082 kb)

Acknowledgements
We like to thank E-M Willing and K Schneeberger for designing the *ndh*F baits, R Brandt for performing the Illumina sequencing, and C Koch and B Wedemeier for technical assistance. We are grateful for seeds obtained from ICARDA, IPK, USDA, the Czech Crop Research Institute, and the Kyoto University Laboratory of Plant Genetics. We also thank JM Saarela and two anonymous reviewers for helpful comments on earlier versions of the manuscript.

Funding
This work was supported by the German Research Foundation (DFG) [BL462/10].

Authors' contributions
NB, FRB, BK designed the study. BK provided data or materials. NB performed the experiments. NB and JB analysed the data. NB and FRB wrote the initial manuscript. All authors contributed to and approved the final version.

Competing interests
The authors declare that they have no competing interests.

Author details
[1]Leibniz Institute of Plant Genetics and Crop Plant Research (IPK), Gatersleben, Germany. [2]Present address: Crop Trust, Bonn, Germany. [3]German Centre for Integrative Biodiversity Research (iDiv) Halle-Jena-Leipzig, Leipzig, Germany.

References
1. Barkworth ME, R von B. Scientific names in the Triticeae. In: Muehlbauer GJ, Feuillet C, editors. Genetics and Genomics of the Triticeae. US: Springer; 2009. p. 3–30.
2. Feldman M, Levy AA. Origin and evolution of wheat and related Triticeae species. In: Molnár-Láng M, Ceoloni C, Doležel J, editors. Alien Introgression in Wheat: Springer; 2015. p. 21–76.
3. Hsiao C, Chatterton NJ, Asay KH, Jensen KB. Phylogenetic relationships of the monogenomic species of the wheat tribe, Triticeae (Poaceae), inferred from nuclear rDNA (internal transcribed spacer) sequences. Genome. 1995;38:211–23.
4. Kellogg EA, Appels R, Mason-Gamer RJ. When genes tell different stories: The diploid genera of Triticeae (Gramineae). Syst Bot. 1996;21:321–47.
5. de Bustos A, Jouve N. Phylogenetic relationships of the genus *Secale* based on the characterisation of rDNA ITS sequences. Plant Syst Evol. 2002;235:147–54.
6. Mason-Gamer RJ, Orme NL, Anderson CM. Phylogenetic analysis of North American *Elymus* and the monogenomic Triticeae (Poaceae) using three chloroplast DNA data sets. Genome. 2002;45:991–1002.
7. Petersen G, Seberg O, Yde M, Berthelsen K. Phylogenetic relationships of *Triticum* and *Aegilops* and evidence for the origin of the A, B, and D genomes of common wheat (*Triticum aestivum*). Mol Phylogenet Evol. 2006;39:70–82.
8. Mason-Gamer RJ. Allohexaploidy, introgression, and the complex phylogenetic history of *Elymus repens* (Poaceae). Mol Phylogenet Evol. 2008;47:598–611.
9. Bordbar F, Rahiminejad MR, Saeidi H, Blattner FR. Phylogeny and genetic diversity of D-genome species of *Aegilops* and *Triticum* (Triticeae, Poaceae) from Iran based on microsatellites, ITS, and *trn*L-F. Plant Syst Evol. 2011;291:117–31.

10. Escobar JS, Scornavacca C, Cenci A, Guilhaumon C, Santoni S, Douzery EJ, et al. Multigenic phylogeny and analysis of tree incongruences in Triticeae (Poaceae). BMC Evol Biol. 2011;11:181.

11. Yan C, Sun G. Multiple origins of allopolyploid wheatgrass *Elymus caninus* revealed by *RPB2*, *PepC* and *Trn* D/T genes. Mol Phylogenet Evol. 2012;64:441–51.

12. Bernhardt N. Taxonomic treatments of Triticeae and the wheat genus *Triticum*. In: Molnár-Láng M, Ceoloni C, Doležel J, editors. Alien Introgression in Wheat: Springer; 2015. p. 1–19.

13. Kellogg EA. Flowering Plants. Monocots. Poaceae. In: Kubitzki K, editor. The Families and Genera of Vascular Plants, XIII: Springer; 2015.

14. Soltis DE, Soltis PS. Polyploidy: Recurrent formation and genome evolution. Trends Ecol Evol. 1999;14:348–52.

15. Mason-Gamer RJ. Reticulate evolution, introgression, and intertribal gene capture in an allohexaploid grass. Syst Biol. 2004;53:25–37.

16. Jakob SS, Blattner FR. Two extinct diploid progenitors were involved in allopolyploid formation in the *Hordeum murinum* (Poaceae: Triticeae) taxon complex. Mol Phylogenet Evol. 2010;55:650–9.

17. Brassac J, Jakob SS, Blattner FR. Progenitor-derivative relationships of *Hordeum* polyploids (Poaceae, Triticeae) inferred from sequences of *TOPO6*, a nuclear low-copy gene region. PLoS One. 2012;7:e33808.

18. Brassac J, Blattner FR. Species-level phylogeny and polyploid relationships in *Hordeum* (Poaceae) inferred by next-generation sequencing and *in silico* cloning of multiple nuclear loci. Syst Biol. 2015;64:792–808.

19. Kellogg EA. Has the connection between polyploidy and diversification actually been tested? Curr Opin Plant Biol. 2016;30:25–32.

20. Marcussen T, Sandve SR, Heier L, Spannagl M, Pfeifer M, Jakobsen KS, et al. Ancient hybridizations among the ancestral genomes of bread wheat. Science. 2014;345:1250092.

21. Seberg O, Petersen G. A critical review of concepts and methods used in classical genome analysis. Bot Rev. 1998;64:372–417.

22. Kellogg EA. Comments on genomic genera in the Triticeae (Poaceae). Am J Bot. 1989;796–805.

23. Seberg O, Frederiksen S. A phylogenetic analysis of the monogenomic Triticeae (Poaceae) based on morphology. Bot J Linn Soc. 2001;136:75–97.

24. Sang T. Utility of low-copy nuclear gene sequences in plant phylogenetics. Crit Rev Biochem Mol Biol. 2002;37:121–47.

25. Bänfer G, Moog U, Fiala B, Mohamed M, Weising K, Blattner FR. A chloroplast genealogy of myrmecophytic *Macaranga* species (Euphorbiaceae) in Southeast Asia reveals hybridization, vicariance and long-distance dispersals. Mol Ecol. 2006;15:4409–24.

26. Petersen G, Seberg O. Phylogenetic analysis of the Triticeae (Poaceae) based on *rpoA* sequence data. Mol Phylogenet Evol. 1997;7:217–30.

27. Golovnina KA, Glushkov SA, Blinov AG, Mayorov VI, Adkison LR, Goncharov NP. Molecular phylogeny of the genus *Triticum* L. Plant Syst Evol. 2007;264:195–216.

28. Seberg O, Petersen G. Phylogeny of Triticeae (Poaceae) based on three organelle genes, two single-copy nuclear genes, and morphology. Aliso: J Syst Evol Bot. 2007;23:362–71.

29. Ogihara Y, Isono K, Kojima T, Endo A, Hanaoka M, Shiina T, et al. Chinese spring wheat (*Triticum aestivum* L.) chloroplast genome: Complete sequence and contig clones. Plant Mol Biol Report. 2000;18:243–53.

30. Gornicki P, Zhu H, Wang J, Challa GS, Zhang Z, Gill BS, et al. The chloroplast view of the evolution of polyploid wheat. New Phytol. 2014;204:704–14.

31. Middleton CP, Senerchia N, Stein N, Akhunov ED, Keller B, Wicker T, et al. Sequencing of chloroplast genomes from wheat, barley, rye and their relatives provides a detailed insight into the evolution of the Triticeae tribe. PLoS One. 2014;9:e85761.

32. Saarela JM, Wysocki WP, Barrett CF, Soreng RJ, Davis JI, Clark LG, et al. Plastid phylogenomics of the cool-season grass subfamily: Clarification of relationships among early-diverging tribes. AoB Plants. 2015;7:plv046.

33. Zeng QX, Yuan JH, Wang LY, Xu JQ, Nyima T. The complete chloroplast genome of Tibetan hulless barley. Mitochondrial DNA 2015;0:1–2.

34. Gogniashvili M, Jinjikhadze T, Maisaia I, Akhalkatsi M, Kotorashvili A, Kotaria N, et al. Complete chloroplast genomes of *Aegilops tauschii* Coss. and *Ae. cylindrica* Host sheds light on plasmon D evolution. Curr Genet. 2016:1–8.

35. Catalán P, Kellogg EA, Olmstead RG. Phylogeny of Poaceae subfamily Pooideae based on chloroplast *ndh*F gene sequences. Mol Phylogenet Evol. 1997;8:150–66.

36. Weitemier K, Straub SCK, Cronn RC, Fishbein M, Schmickl R, McDonnell A, et al. Hyb-Seq: Combining target enrichment and genome skimming for plant phylogenomics. Appl Plant Sci. 2014;2:1400042.

37. Doležel J, Greilhuber J, Suda J. Estimation of nuclear DNA content in plants using flow cytometry. Nat Protoc. 2007;2:2233–44.

38. Doyle J, Doyle JL. Genomic plant DNA preparation from fresh tissue-CTAB method. Phytochem Bull. 1987;19:11–5.

39. Meyer M, Kircher M. Illumina sequencing library preparation for highly multiplexed target capture and sequencing. Cold Spring Harb Protoc. 2010;2010:t5448.

40. Himmelbach A, Knauft M, Stein N. Plant sequence capture optimised for Illumina sequencing. Bio-Protoc. 2014;4:e1166.

41. Esling P, Lejzerowicz F, Pawlowski J. Accurate multiplexing and filtering for high-throughput amplicon-sequencing. Nucleic Acids Res. 2015:gkv107.

42. Katoh K, Standley DM. MAFFT multiple sequence alignment software version 7: Improvements in performance and usability. Mol Biol Evol. 2013;30:772–80.

43. Bahieldin A, Al-Kordy MA, Shokry AM, Gadalla NO, Al-Hejin AMM, Sabir JSM, et al. Corrected sequence of the wheat plastid genome. C R Biol. 2014;337:499–502.

44. Swofford DL. PAUP*. Phylogenetic analysis using parsimony (* and other methods). Version 4.b10. Sunderland Massachusetts: Sinauer Associates; 2003.

45. Clement M, Posada D, Crandall KA. TCS: A computer program to estimate gene genealogies. Mol Ecol. 2000;9:1657–9.

46. Darriba D, Taboada GL, Doallo R, Posada D. jModelTest 2: More models, new heuristics and parallel computing. Nat Methods. 2012;9:772.

47. Schwarz G. Estimating the dimension of a model. Ann Stat. 1978;6:461–4.

48. Posada D, Crandall KA. Selecting the best-fit model of nucleotide substitution. Syst Biol. 2001;50:580–601.

49. Ronquist F, Teslenko M, van der Mark P, Ayres DL, Darling A, Höhna S, et al. MrBayes 3.2: efficient Bayesian phylogenetic inference and model choice across a large model space. Syst Biol. 2012;61:539–42.

50. Miller MA, Pfeiffer W, Schwartz T. Creating the CIPRES Science Gateway for inference of large phylogenetic trees. Gatew Comput Environ Workshop GCE. 2010;2010:1–8.

51. Huelsenbeck JP, Larget B, Alfaro ME. Bayesian phylogenetic model selection using reversible jump Markov chain Monte Carlo. Mol Biol Evol. 2004;21:1123–33.

52. Raftery AE, Newton MA, Satagopan JM, Krivitsky PN. Estimating the integrated likelihood via posterior simulation using the harmonic mean identity. Bayesian Stat. 2007;

53. Baele G, Lemey P, Bedford T, Rambaut A, Suchard MA, Alekseyenko AV. Improving the accuracy of demographic and molecular clock model comparison while accommodating phylogenetic uncertainty. Mol Biol Evol. 2012;29:2157–67.

54. Rambaut A, Suchard M, Xie W, Drummond A. Tracer v. 1.6. Institute of Evolutionary Biology, University of Edinburgh. 2014.

55. Kass RE, Raftery AE. Bayes factors. J Am Stat Assoc. 1995;90:773–95.

56. Xie W, Lewis PO, Fan Y, Kuo L, Chen M-H. Improving marginal likelihood estimation for Bayesian phylogenetic model selection. Syst Biol. 2010;60:150–60.

57. Brown JM, Lemmon AR. The importance of data partitioning and the utility of Bayes factors in Bayesian phylogenetics. Syst Biol. 2007;56:643–55.

58. Pelser PB, Nordenstam B, Kadereit JW, Watson LE. An ITS phylogeny of tribe Senecioneae (Asteraceae) and a new delimitation of *Senecio* L. Taxon. 2007;56:1077–104.

59. Guindon S, Gascuel O. A simple, fast, and accurate algorithm to estimate large phylogenies by maximum likelihood. Syst Biol. 2003;52:696–704.

60. Lanfear R, Calcott B, Ho SYW, Guindon S. PartitionFinder: Combined selection of partitioning schemes and substitution models for phylogenetic analyses. Mol Biol Evol. 2012;29:1695–701.

61. Bouckaert R, Heled J, Kühnert D, Vaughan T, Wu C-H, Xie D, et al. BEAST 2: A software platform for Bayesian evolutionary analysis. PLoS Comput Biol. 2014;10:e1003537.

62. eMonocot: An online resource for monocot plants. Available from: http://www.emonocot.org/. Accessed 20 Oct 2016.

63. Huelsenbeck JP, Rannala B. Phylogenetic methods come of age: testing hypotheses in an evolutionary context. Science. 1997;276:227–32.

64. Drummond AJ, Ho SY, Phillips MJ, Rambaut A. Relaxed phylogenetics and dating with confidence. PLoS Biol. 2006;4:699.

65. Ayres DL, Darling A, Zwickl DJ, Beerli P, Holder MT, Lewis PO, et al. BEAGLE: An application programming interface and high-performance computing library for statistical phylogenetics. Syst Biol. 2012;61:170–3.

66. Rambaut A. FIGTREE v. 1.4.2. Available from: http://tree.bio.ed.ac.uk/software/figtree. Accessed 20 Mar 2015.

67. Jakob SS, Meister A, Blattner FR. The considerable genome size variation of Hordeum species (Poaceae) is linked to phylogeny, life form, ecology, and speciation rates. Mol Biol Evol. 2004;21:860–9.

68. Tavaré S. Some probabilistic and statistical problems in the analysis of DNA sequences. Lect Math Life Sci. 1986;17:57–86.

69. Yang Z. Maximum likelihood phylogenetic estimation from DNA sequences with variable rates over sites: Approximate methods. J Mol Evol. 1994;39:306–14.

70. Asay KH, Jensen KB, Hsiao C, Dewey DR. Probable origin of standard crested wheatgrass, Agropyron desertorum Fisch. ex Link, Schultes. Can J Plant Sci. 1992;72:763–72.

71. Jauhar PP. Chromosome pairing in hybrids between hexaploid bread wheat and tetraploid crested wheatgrass (Agropyron cristatum). Hereditas. 1992;116:107–9.

72. Yu H, Fan X, Zhang C, Ding C, Wang X, Zhou Y. Phylogenetic relationships of species in Pseudoroegneria (Poaceae: Triticeae) and related genera inferred from nuclear rDNA ITS (internal transcribed spacer) sequences. Biologia. 2008;63:498–505.

73. Cabi E, Doğan M. Taxonomic study on the genus Eremopyrum (Ledeb.) Jaub. et Spach (Poaceae) in Turkey. Plant Syst Evol. 2010;287:129–40.

74. Kilian B, Mammen K, Millet E, Sharma R, Graner A, Salamini F, et al. Aegilops. In: Kole C, editor. Wild Crop Relatives: Genomic and Breeding Resources: Springer; 2011. p. 1–76.

75. Frederiksen S, Petersen G. A taxonomic revision of Secale (Triticeae, Poaceae). Nord J Bot. 1998;18:399–20.

76. Jakob SS, Rödder D, Engler JO, Shaaf S, Özkan H, Blattner FR, et al. Evolutionary history of wild barley (Hordeum vulgare subsp. spontaneum) analyzed using multilocus sequence data and paleodistribution modeling. Genome Biol Evol. 2014;6:685–702.

77. Eilam T, Anikster Y, Millet E, Manisterski J, Feldman M. Genome size in diploids, allopolyploids, and autopolyploids of Mediterranean Triticeae. J Bot. 2010;2010:1–12.

78. Özkan H, Tuna M, Kilian B, Mori N, Ohta S. Genome size variation in diploid and tetraploid wild wheats. AoB Plants. 2010;plq015.

79. Guo Y, Long J, He J, Li C-I, Cai Q, Shu X-O, et al. Exome sequencing generates high quality data in non-target regions. BMC Genomics. 2012;13:194.

80. Samuels DC, Han L, Li J, Quanghu S, Clark TA, Shyr Y, et al. Finding the lost treasures in exome sequencing data. Trends Genet. 2013;29:593–9.

81. Mascher M, Muehlbauer GJ, Rokhsar DS, Chapman J, Schmutz J, Barry K, et al. Anchoring and ordering NGS contig assemblies by population sequencing (POPSEQ). Plant J. 2013;76:718–27.

82. Kuilman T, Velds A, Kemper K, Ranzani M, Bombardelli L, Hoogstraat M, et al. CopywriteR: DNA copy number detection from off-target sequence data. Genome Biol. 2015;16:1–15.

83. Mason-Gamer RJ. Phylogeny of a genomically diverse group of Elymus (Poaceae) allopolyploids reveals multiple levels of reticulation. PLoS One. 2013;8:e78449.

84. Li L-F, Liu B, Olsen KM, Wendel JF. A re-evaluation of the homoploid hybrid origin of Aegilops tauschii, the donor of the wheat D-subgenome. New Phytol. 2015;208:4–8.

85. Watson L, Clifford HT, Dallwitz MJ. The classification of Poaceae: Subfamilies and supertribes. Aust J Bot. 1985;33:433–84.

86. Soreng RJ, Davis JI, Doyle JJ. A phylogenetic analysis of chloroplast DNA restriction site variation in Poaceae subfamily Pooideae. Plant Syst Evol. 1990;172:83–97.

87. Bouchenak-Khelladi Y, Salamin N, Savolainen V, Forest F, Bank M van der, Chase MW, et al. Large multi-gene phylogenetic trees of the grasses (Poaceae): Progress towards complete tribal and generic level sampling. Mol Phylogenet Evol 2008;47:488–505.

88. Kellogg EA. Tools for studying the chloroplast genome in the Triticeae (Gramineae): an EcoRI Map, a diagnostic deletion, and support for Bromus as an outgroup. Am J Bot. 1992;79:186–97.

89. Schneider J, Döring E, Hilu KW, Röser M. Phylogenetic structure of the grass subfamily Pooideae based on comparison of plastid matK gene-3′ trnK exon and nuclear ITS sequences. Taxon. 2009:405–24.

90. Hochbach A, Schneider J, Röser M. A multi-locus analysis of phylogenetic relationships within grass subfamily Pooideae (Poaceae) inferred from sequences of nuclear single copy gene regions compared with plastid DNA. Mol Phylogenet Evol. 2015;87:14–27.

91. Barkworth ME, Jacobs SW. The Triticeae (Gramineae) in Australasia. Telopea. 2011;13:37–56.

92. Mason-Gamer RJ, Kellogg EA. Phylogenetic analysis of the Triticeae using the starch synthase gene, and a preliminary analysis of some North American Elymus species. In: Jacobs S, Everett J, editors. Grasses. Systematics and Evolution. Collingwood, Victoria, Australia: CSIRO Publishing; 2000. p. 102–9.

93. Mason-Gamer RJ. The β-amylase genes of grasses and a phylogenetic analysis of the Triticeae (Poaceae). Am J Bot. 2005;92:1045–58.

94. Petersen G, Seberg O. Molecular evolution and phylogenetic application of DMC1. Mol Phylogenet Evol. 2002;22:43–50.

95. Petersen G, Seberg O, Salomon B. The origin of the H, St, W, and Y genomes in allotetraploid species of Elymus L. and Stenostachys Turcz. (Poaceae: Triticeae). Plant Syst Evol. 2011;291:197–210.

96. Mahelka V, Kopecký D, Paštová L. On the genome constitution and evolution of intermediate wheatgrass (Thinopyrum intermedium: Poaceae, Triticeae). BMC Evol Biol. 2011;11:127.

97. van Slageren MW. Wild wheats: A monograph of Aegilops L. and Amblyopyrum (Jaub. & Spach) Eig (Poaceae). Wageningen Agricultural University Papers; 1994.

98. Li L-F, Liu B, Olsen KM, Wendel JF. Multiple rounds of ancient and recent hybridizations have occurred within the Aegilops–Triticum complex. New Phytol. 2015;208:11–2.

99. Sandve SR, Marcussen T, Mayer K, Jakobsen KS, Heier L, Steuernagel B, et al. Chloroplast phylogeny of Triticum/Aegilops species is not incongruent with an ancient homoploid hybrid origin of the ancestor of the bread wheat D-genome. New Phytol. 2015;208:9–10.

100. Wang G-Z, Miyashita NT, Tsunewaki K. Plasmon analyses of Triticum (wheat) and Aegilops: PCR–single-strand conformational polymorphism (PCR-SSCP) analyses of organellar DNAs. Proc Natl Acad Sci USA. 1997;94:14570–7.

101. Haider N. The origin of the B-genome of bread wheat (Triticum aestivum L.). Russ J Genet. 2013;49:263–74.

102. Dorofeev VF, Filatenko AA, Migushova EF, Udaczin RA, Jakubziner MM. Wheat. In: Dorofeev VF, Korovina ON, editors. Flora of Cultivated Plants vol I. Leningrad (St. Petersburg): Kolos; 1979.

103. Smith DR. Mutation rates in plastid genomes: They are lower than you might think. Genome Biol Evol. 2015;7:1227–34.

104. Bendich AJ. Why do chloroplasts and mitochondria contain so many copies of their genome? BioEssays. 1987;6:279–82.

105. Khakhlova O, Bock R. Elimination of deleterious mutations in plastid genomes by gene conversion. Plant J. 2006;46:85–94.

106. Jakob SS, Blattner FR. A chloroplast genealogy of Hordeum (Poaceae): Long-term persisting haplotypes, incomplete lineage sorting, regional extinction, and the consequences for phylogenetic inference. Mol Biol Evol. 2006;23:1602–12.

107. Blattner FR. Progress in phylogenetic analysis and a new infrageneric classification of the barley genus Hordeum (Poaceae: Triticeae). Breed Sci. 2009;59:471–80.

5

Phylogenetic conservatism in skulls and evolutionary lability in limbs – morphological evolution across an ancient frog radiation is shaped by diet, locomotion and burrowing

Marta Vidal-García* [iD] and J. Scott Keogh

Abstract

Background: Quantifying morphological diversity across taxa can provide valuable insight into evolutionary processes, yet its complexities can make it difficult to identify appropriate units for evaluation. One of the challenges in this field is identifying the processes that drive morphological evolution, especially when accounting for shape diversification across multiple structures. Differential levels of co-varying phenotypic diversification can conceal selective pressures on traits due to morphological integration or modular shape evolution of different structures, where morphological evolution of different modules is explained either by co-variation between them or by independent evolution, respectively.

Methods: Here we used a 3D geometric morphometric approach with x-ray micro CT scan data of the skull and bones of forelimbs and hindlimbs of representative species from all 21 genera of the ancient Australo-Papuan myobatrachid frogs and analysed their shape both as a set of distinct modules and as a multi-modular integrative structure. We then tested three main questions: (i) are evolutionary patterns and the amount and direction of morphological changes similar in different structures and subfamilies?, (ii) do skulls and limbs show different levels of integration?, and (iii) is morphological diversity of skulls and limbs shaped by diet, locomotion, burrowing behavior, and ecology?.

Results: Our results in both skulls and limbs support a complex evolutionary pattern typical of an adaptive radiation with an early burst of phenotypic variation followed by slower rates of morphological change. Skull shape diversity was phylogenetically conserved and correlated with diet whereas limb shape was more labile and associated with diet, locomotion, and burrowing behaviour. Morphological changes between different limb bones were highly correlated, depicting high morphological integration. In contrast, overall limb and skull shape displayed semi-independence in morphological evolution, indicating modularity.

Conclusions: Our results illustrate how morphological diversification in animal clades can follow complex processes, entailing selective pressures from the environment as well as multiple trait covariance with varying degrees of independence across different structures. We suggest that accurately quantifying shape diversity across multiple structures is crucial in order to understand complex evolutionary processes.

Keywords: Morphology, Modularity, Morphological integration, 3D morphology, Geometric morphometrics, Phylomorphospace

* Correspondence: marta.vidalga@gmail.com
Research School of Biology, The Australian National University, Canberra,
Australia

Background

Understanding morphological evolution, and the underlying mechanisms that generate the enormous phenotypic diversity we see, is a central aim in evolutionary biology [1–4]. Phenotypic diversity often is correlated with ecology and behaviour, especially in traits for which form and function are tightly associated due to evolutionary and ecological pressures [5–8]. However, while some clades display extensive ecological and morphological variation that is correlated with lifestyle, others retain ancestral environmental niches and conserved body shape patterns that are better explained by phylogenetic conservatism on a shared ancestral lifestyle [9, 10]. These differing patterns of diversification are best illustrated in related groups of species where one group might display more phenotypic diversification than another due to different selective pressures [4, 11]. There are many examples of this in the species-rich radiations of characiform fishes [11], gobies and cardinal fishes [12], passerine birds [13], archosaurs [14], and many others.

While diverse evolutionary processes can generate phenotypic change, morphological evolution is typically inferred from integration or co-variation among multiple traits [15]. Body shape patterns can usually be broken down into 'modules', which are characterized by more internal integration within them, than externally among them [15]. Therefore, each module displays a certain amount of independence from other modules and can differ developmentally, genetically, and in the way they respond to selection [15, 16]. While many phenotypic changes across a radiation are modular in this way [17], shape diversification can follow a more complex pattern of integrative co-variation between modules and show correlated morphological variation among them [15, 18]. The degree of shape-co-variation between modules is due to the interplay between morphological integration and modularity, where morphological modules evolve in concert with others and in which morphology evolves independently among different structures, respectively [15]. High morphological diversity could be correlated with modularity, as autonomy among different structural units might promote higher independent morphological changes due to the evolutionary lability necessary for adaptive shifts [19–21]. Conversely, morphological integration could be one of the causes leading to convergence among unrelated clades [22, 23]. Integration and co-variation among modules should also shape the morphological evolution of individual organisms, as some modules might be subject to strong selective pressures from the environment, whereas others might be phylogenetically constrained. Therefore, identifying the patterns of variation in each module, while accounting for integration among them, is crucial in order to study morphological evolution and the processes that might have driven it [23].

Due to the close relationship between form and function, some morphological traits are likely to be more closely linked to the ecology of an organism than others [24]. For example, Zaaf & Van Damme [25] proposed the idea of evaluating morphological differences between and within distinct modules in limbs, in relation to functional traits like locomotion, and tested it in climbing and ground-dwelling geckos. Limb shape might provide the most insight into the ecotype a species occupies, as it is closely correlated with its performance, and thus, locomotion through the environment [26, 27]. Similarly, Cornette et al. [28] looked at both the skull and mandible in shrews in order to disentangle the relationship between diet, ecological factors, and head shape evolution. On the other hand, some modules might be correlated with life history traits or not be under selection as functional traits [29]. Moreover, inferring adaptive processes by looking at the 'wrong' structure might be uninformative, and in some cases even misleading. Assessing morphological evolution in a group of organisms provides more valuable information when looking at a wide range of phenotypic traits, but may also increase the difficulty of data interpretation, due to complex co-variation processes between different structures.

Anuran amphibians are an ideal model group in which to investigate morphological evolutionary patterns: they display a highly derived morphology compared to other terrestrial vertebrates [30], yet their body plan has been relatively conserved since the early Jurassic [31, 32]. Despite phylogenetic constraints on their appendicular skeleton as an adaptation to saltatory locomotion [33], substantially different body shape patterns have evolved independently across several clades [34]. Frogs and toads have adapted to a wide array of extreme environments through a combination of behavioural, physiological, and morphological mechanisms. Extreme morphological shifts are usually associated with unique locomotor types, such as gliding in "flying" frogs [35], or with specialised locomotion, such as the improved swimming ability in frogs like pipids [36]. Similarly, strong shape changes are observed in burrowing frogs and toads that have adapted to desiccating conditions in arid and semi-arid environments [26, 37]. Morphological convergence in burrowing frogs has been documented across numerous clades, in both forward (head and forelimbs first), and backward (hindlimbs first) burrowing species, with backward burrowing being the most common digging type in frogs and toads (~95%), yet unique among vertebrates [38]. These diverse morphological adaptations make frogs an ideal system in which to study modularity and integration, as they relate to ecology.

The family Myobatrachidae is an old Gondwanan lineage endemic to Australia and New Guinea with its closest relatives in South America [39]. The family currently comprises 133 described species across 21 genera, accounting for 57% of the Australian frog diversity [40]. Australia's large landmass is characterised by a wide range of biomes and has a complex history of isolation, aridification and broad climatic changes that have had a strong impact on the evolutionary processes in its biota [41]. Myobatrachid frogs are extremely diverse in ecology (from tropical rainforest dwellers to exclusive alpine species or desert-specialists; [42]), locomotion (including excellent swimmers, jumpers, hoppers, and walkers), reproductive systems (egg deposition, calls, parental care modes, etc.; [43, 44]), and also body shape patterns [45]. Thus, they stand out as a model system to examine morphological diversification patterns on a diverse and species rich radiation across a whole continent.

We sought to address three broad questions: (i) is morphological evolution similar in different body parts, (ii) do skulls and limbs show different levels of integration?, and (iii) is morphological diversity of skulls and limbs shaped by diet, locomotion, burrowing behavior, and ecology? To do this we used 3D imaging across all genera of myobatrachids, combined with geometric morphometric analyses, to discriminate the morphological integration hypothesis and the modularity hypothesis in different structures. We used 3D data from the skull and several limb bones of the appendicular skeleton (radioulna, humerus, tibiofibula and femur), and studied their shape both as a set of distinct modules and jointly as a multi-modular integrative structure. First, we sought to quantify skull and limb shape differences across representatives of all 21 genera of myobatrachid frogs by using 3D microCT scans and geometric morphometric (GM) techniques. We then addressed three major aims. First, we tested the hypothesis that evolutionary patterns and morphological disparity are similar in the two major clades of myobatrachids across different structures. We predicted that both skull and limbs followed an evolutionary pattern typical of an adaptive radiation, and that dispersion across morphospace would be correlated with species richness, with this trend being consistent across most modules. We then determined whether there were differences in dispersion and direction of shape diversification in skulls and limbs, and whether morphological evolution acts independently in each module, or if there was some integration across different structures. We predicted a high degree of morphological integration, especially among limb modules, due to selective pressures derived from environmental correlates and associated adaptations such as burrowing behavior and locomotion. Finally, we tested for relationships between morphology and burrowing, locomotion, and environment. We predicted that form would be correlated with function, i.e. ecology, locomotion, and burrowing behavior would have been key drivers in shaping morphological evolution on the limbs, whereas head shape would be more phylogenetically conserved due to a lower functional pressure imposed by the environment.

Methods

Study samples and morphological data

This study is based on 41 ethanol-preserved specimens from 21 species of the Australo-Papuan myobatrachid frog radiation. Sampling covered all genera from this family, and with the exception of the monotypic *Spicospina flammocaerulea* where only one specimen was available, we used two representative specimens of the same species per genus as a previous study across all myobatrachid species showed high morphological conservatism within genera [45]. Species and voucher number details are presented in Additional file 1: Appendix S1. Since sexual dimorphism is known in some myobatrachid species (e.g. *Adelotus brevis*), we only sampled adult females in order to avoid morphological differences due to sexual dimorphism. We gathered data for burrowing behavior from several sources [42, 46] and classified each species into three categories based on the type of burrowing: (a) forward burrowers which use their forelimbs, (b) backwards burrowers which use their hindlimbs, and (c) non-burrowers. Locomotion information was gathered from Anstis (2013) and Cogger (2014), and locomotor mode categories were defined according to basic characteristics of their stride: (a) walkers are species that are strictly walkers or crawlers, (b) hoppers are species that can only hop, or hop and walk, and not jump (an average jumping distance that is less than five times their body length), and (c) jumpers/swimmers are species that can jump and/or swim (whose average jumping distance is greater than five times their body length and are proficient swimmers). Even though some genera display multiple states for burrowing and locomotor modes, the analyses were performed using the state present on the selected species. Data for habitat type or ecoregions was gathered taking into account each species' distribution and the seven main ecoregions in Australia [42, 47]: (a) tropical and subtropical moist broadleaf forests, (b) temperate broadleaf and mixed forests, (c) tropical and subtropical grassland, savannas and shrublands, (d) temperate grasslands, savannas and shrublands, (e) montane grasslands and shrublands, (f) mediterranean forests, woodlands and shrubs, and (g) deserts and xeric shrublands. Dietary information [48–58] was gathered for all species in this study (except for the little-known species *Spicospina flammocaerulea*, for which we inferred diet from its close relatives and based on similarities in other life-history traits), which was classified into two categories: (a) generalists have multiple taxa represented in their diet, regardless of their size) and (b) specialists only feed on certain taxa (mostly termites and ants). Data on burrowing behavior, locomotion, diet, and ecoregions is summarized on Additional file 1: Table S1. All morphological data was gathered using three different X-ray micro-CT scanners, depending on the size of the individual frog: Skyscan 1174 (Bruker micro-CT) for small frogs, MicroXCT-400 (Xradia system) for intermediate sized frogs, and a custom-made double-helical x-ray micro CT scanner from the Australian National University for the larger specimens. The settings for each CT scanner were as follows: Skyscan 1174–40 kV source voltage, 800 µA source current, voxel size of 32.47 µm, 0.7° rotational step, 1.6 s exposure time, and 360° rotational angle scanning. The acquired images (angular projections) were reconstructed into a virtual stack of 2D cross-section slices using the NRecon (Skyscan) software interface. Xradia MicroXCT-400 - 50 kV, 360° rotational angle scanning, 2 s exposure, and voxel size of 49.13 µm. Acquired images were reconstructed in the MicroXCT and exported to a virtual stack of 2D cross-section slices (8-bit BMP format) using Avizo software system (version 8.0, Mercury Computer Systems, Inc., Germany). Custom-made double-helical x-ray micro CT - 80 kV, 100 µA, voxel size of 43 µm, using a 0.3 mm Al filter, 3.4 s exposure, and 0.143° rotational step, resulting in 2520 angular projections. This RAW data was also then reconstructed into 2D cross-section slices (NC format). Each stack of reconstructed images was then converted into 3D data, using the volume-rendering software *Drishti* [59].

Shape analyses

Skull and limbs' bone shape differences were identified using geometric morphometric (GM) methods. We used rendering software *Drishti* [59] in order to digitise 3D landmarks of the skull and limb bones, and also sliding semilandmarks on limb bones (Additional files 2 and 3). We then averaged each dataset of morphometric data by species with *geomorph* [60], in order to allow analyses in a phylogenetic context and focus on morphological variation among genera and clades. We also performed GM analyses with all raw data sets before taking species means to ensure that interspecific variation was greater than intraspecific variation. Each data set was subjected to a generalised Procrustes sumperimposition fit with the package *geomorph* [60–62]. We performed a Principal Component Analysis (PCA) on the projected Procrustes coordinates into the tangent space for each set of morphological data. Each data set of GM data was analysed separately, but also joined, considering each long bone as a distinct module. To do so, we translated and rigidly rotated all landmarks and semi-landmarks from each data set using a newly developed Rigid Rotation equation, with the R package *ShapeRotator* [63]. This allowed us to set up all the different modules in the same position, angle and torsion and thus allow us to analyse different mobile structures as a whole (as modules would

be in the same position relative to each other). We then analysed shape and size differences across all genera in each module and also in each different group of modules: (a) forelimbs (H + RU), (b) hindlimbs (F + TF) and limbs (H + RU + F + TF). In order to test our modularity and morphological integration hypotheses we also analysed morphological co-variation between: (a) skull and the four modules in the limbs (H + RU + F + TF), (b) co-variation between forelimbs (H + RU) and hindlimbs (F + TF), (c) whithin each limb, so between radioulna and humerus in forelimbs and in femur and tibiofibula in hindlimbs, and among different modules within the skull (Additional file 1: Table S2).

Statistical analyses

In order to investigate patterns of morphological evolution across the myobatrachid frog family we used a phylogeny for the group based on two mtDNA genes (ND2 and 12S) and two nDNA loci (Rag1 and Rhodopsin). This is the same phylogeny we used in a previous study of shape evolution in these frogs ([45]; toplogy available on dryad: https://datadryad.org/resource/doi:10.5061/dryad.1vb63) We used the R package *ape* [64] to prune this tree to only include the species used in this study, and to produce an ultrametric tree with branch lengths approximating proportions of their total age. The resulting phylogeny was projected onto morphospace (previously obtained through PCA of the Procrustes coordinates) with *geomorph* [60] to visualise shape differences in a phylogenetic context for each of the data sets [11, 65, 66]. We also used thin-plate spline (TPS) deformation grids to visualise shape changes in the skull in the three dimension (TPS grids for x, y and x, z) using *geomorph* [60]. To test for the strength of phylogenetic signal in our shape data we calculated the K-statistic's generalization for multivariate data (K_{mult}; [67]) with *geomorph* [60] on the Procrustes-aligned coordinates for each GM data set. We considered a strong phylogenetic signal (K_{mult} presenting values grater 1) as the null hypothesis which means that closely-related taxa would occupy similar regions in morphospace [68]. We tested which evolutionary model of phenotypic evolution best fits our data, for both the skull and the limbs (all four limbs bones) shape data sets, using the R packages *geiger* [69] and *ouch* [70] in the first five Principal Components. Since the results were not congruent among each PC, we decided to take a multi-variate approach using the R package *mvMORPH* [71], which allows complex model fitting in multivariate data. We tested the best fit for multiple models of morphological evolution in the first ten PCs of both the skull and the limbs data sets, and selected diet as a shift since it was found to be correlated to shape differences in both skulls and limbs. The models tested were: BM (Brownian Motion), BM two rates (based on diet), OU (Ornstein–Uhlenbeck), OU with two adaptive optima,

EB (Early Burst), and twelve different models with a shift from two different processes at a given point in time in which some had independent rates on each time slice (Table 1).

We tested for evolutionary allometry by performing a regression of shape variation on size variation among different species in a phylogenetic context [72]. In order to test whether shape variation was correlated to burrowing behavior, locomotor mode, or ecoregion, we performed phylogenetic ANOVAs using the function procD.pgls() in *geomorph* [60] on Procrustes-aligned co-ordinates from each GM data set for diet, locomotor mode, burrowing behaviour, and ecology (bioregions). We also performed a phylogenetic ANOVA with all the factors, and factorial phylogentic ANOVAs with pairs of factors and their interactions. We performed a Mantel test using the R package *vegan* [73] to test whether there was an association between the species distribution in the skull and the limbs shape data sets, using a Spearman's Rank correlation coefficient.

Table 1 Summary statistics for the fit of models of phenotypic evolution in the multivariate shape datasets of Skull and limbs (all four limb bones' analysed together): maximum likelihood estimate (ln L), sample-size corrected Akaike's Information Criterion (AICc), and Delta AICc (ΔAICc, difference between a model and the model with the lowest AICc)

Variable	SKULL			LIMBS		
	ln L	AICc	ΔAICc	ln L	AICc	ΔAICc
BM1	395.321	−851.309	100.238	586.680	−1234.026	117.504
BMM	437.403	−925.206	26.341	645.680	−1341.761	9.769
OU1	414.072	−878.544	73.003	611.689	−1273.778	77.752
OUM	427.652	−904.941	46.606	629.454	−1308.544	42.986
EB	392.314	−844.888	106.659	583.708	−1227.677	123.853
BM_EB	401.074	−862.409	89.138	586.723	−1233.707	117.823
EB_BM	400.259	−860.780	90.767	586.626	−1233.512	118.018
BM_EBi	447.326	−944.969	6.578	649.317	−1348.950	2.580
EB_BMi	**450.615**	**−951.547**	**0**	**650.606**	**−1351.530**	**0**
BM_OU	445.226	−940.852	10.695	620.337	−1291.075	60.455
OU_BM	425.734	−901.867	49.680	618.049	−1286.499	65.031
BM_OUi	448.231	−943.881	7.666	645.977	−1339.372	12.158
OU_BMi	330.765	−708.949	242.598	404.802	−857.024	494.506
EB_OU	408.027	−866.371	85.177	598.161	−1246.638	104.892
OU_EB	430.634	−911.585	39.962	617.650	−1285.623	65.907
EB_OUi	446.735	−940.855	10.692	649.460	−1346.304	5.226
OU_EBi	332.339	−712.063	239.484	407.160	−861.704	489.826

We tested the fit of the following evolutionary models: BM1 = Brownian Motion (unique rate), BMM = Brownian Motion (multiple rates), EB = Early Burst, and 12 evolutionary models with shifts from one model to another (e.g. BM_EB = shift of a BM to EB process, EB_BM = shift of EB to BM, BM_EBi = BM_EB with independent rates, EB_BMi = EB_BM with independent rates, etc.). Analyses were performed in R using the functions *mvBM(), mvOU(), mvEB()* and *mvSHIFT()* from the R package *mvMORPH* (Clavel et al., [71])

Finally, we also tested for morphological disparity among the main four clades and subfamilies in the myobatrachid frog radiation in each GM data set, in relation to the number of genera per clade and the age of each clade. We used Procrustes variance (mean squared Procrustes distance of each genera from the mean shape of their clade) as a measure of morphological disparity which was calculated using *geomorph* [60]. Finally, we used two-block partial least squares (PLS) analysis in order to quantify shape co-variation between different structures, using *geomorph* [60]. We performed two-block PLS analyses between: a) skull and the overall limb shape (RU + H + TF + F), b) forelimbs (RU + H) and hindlimbs (TF + F), c) radioulna and humerus, and d) tibiofibula and femur. All two-block PLS analyses were performed on the Procrustes-aligned coordinates from each GM data set. We also assessed phylogenetic morphological integration between all these modules using the function phylo.integration() in *geomorph* [60].

Results
Size and shape variation

Evolutionary allometry did not account for a significant amount of variance on skull shape: the multivariate regression of Procrustes-aligned coordinates (shape) on log-transformed centroid size (size) demonstrate that only 6.77% of the total shape variation is correlated to size variation ($p = 0.23$). Similarly, evolutionary allometry of limb bones was also low: only 4.29% of the total variance in total limb shape (RU + H + TF + F; $p = 0.15$) was correlated to size changes, 3.76% for forelimb shape (RU + H; $p = 0.19$), and it was slightly higher for hindlimb shape, with size correlates explaining 10.7% of the variance in shape (TF + F; $p = 0.03$). Given the small impact of size on shape variation we performed further analyses on the raw morphometric data sets without removing the allometric effects.

We depict skull shape variation and shape diversity across the four limb bones in Figs. 1 and 2, respectively. In the skull shape data set, the first five principal components (PC) accounted for 82.23% of the total variance (Additional file 1: Table S3), with PC_{skull} 1 and PC_{skull} 2 explaining 41.58% and 19.72% of morphological variation, respectively. The primary axis of variation (PC_{skull} 1) corresponded to width and height of the cranium, and separated burrowing species (both forward burrowers and backward burrowers) and non-burrowing species (Fig. 3). The second axis of variation (PC_{skull} 2) mainly corresponded to variation in the shape of the snout (from pointy to very rounded snouts), and clearly grouped the main two clades in different regions of the morphospace (Fig. 3). Cranium variation is also depicted in Fig. 1 through TPS grids of individuals that present

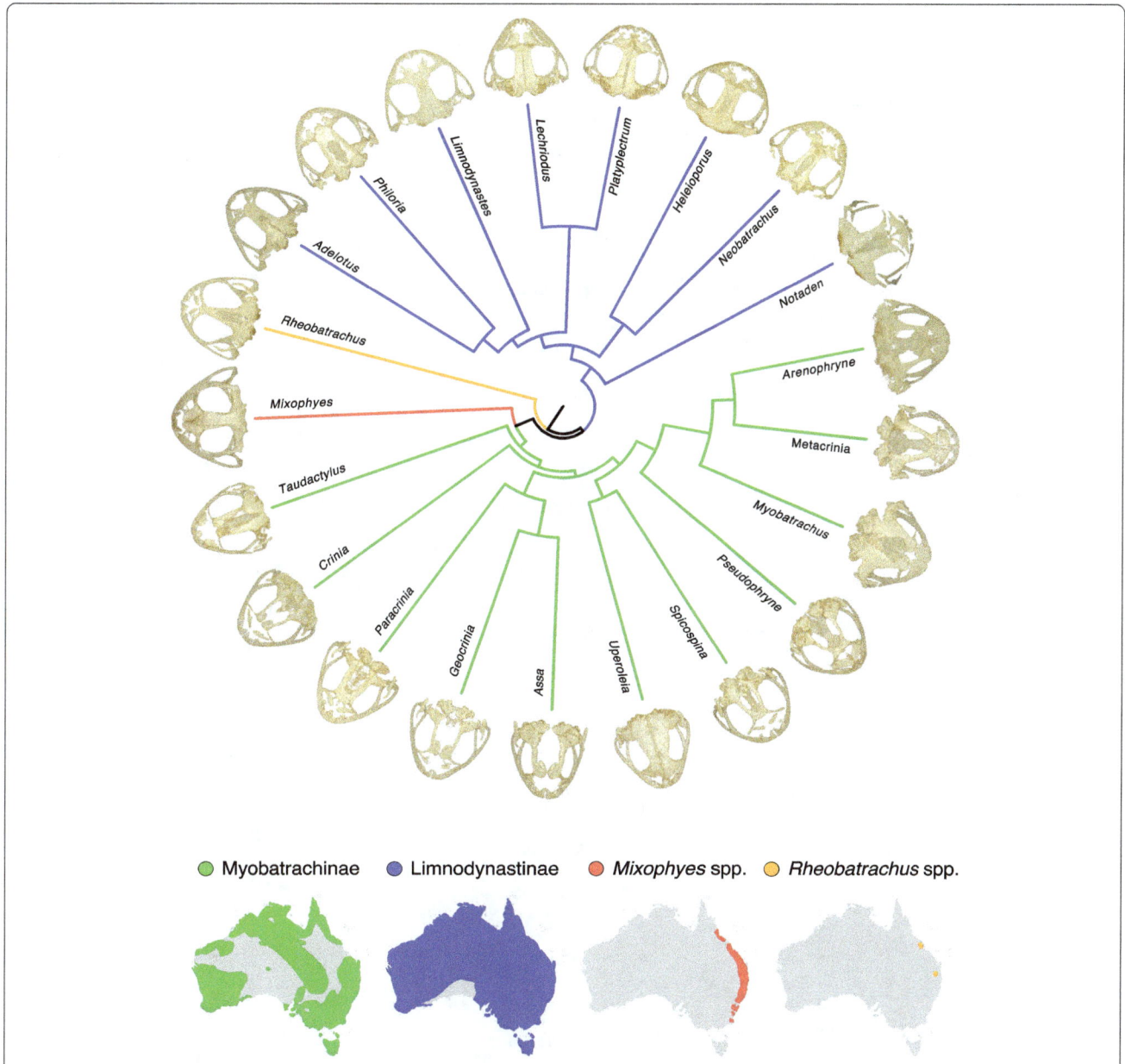

Fig. 1 Dorsal view of skull diversity across all genera of myobatrachid frogs. The four maps display the distribution across Australian of each of the four main clades within the myobatrachids

the most extreme morphological variation from the consensus cranium shape.

For radioulna (RU) shape variation, the first five PCs explained 78.06% of the variance (Additional file 1: Table S3), with PC_{RU} 1 representing 57.99%, and being mostly correlated with arching on the diaphysis of the radioulna (ranging from extremely curved and constricted radioulnas in the medial part of the diaphysis to an almost straight radioulnas). PC_{RU} 2 only added an additional 7.23% (Additional file 4: Figure S1a), and was correlated with the shape of the epiphysis. The first five PCs of the humerus (H) data set accounted for 76.79% of the

overall variance (Additional file 1: Table S3), with PC_H 1 representing 39.38%, and PC_H 2 23.14%, mostly accounting for relative size of the deltoid tuberosity and robustness of the whole humerus (Additional file 4: Figure S1c). On the joined data set of RU and H, the first five PCs explained 82.6% of the total shape variability (Additional file 1: Table S3), with PC_{RU+H} 1 accounting for 47.22% of the variance and PC_{RU+H} 2 another 12.24%, and most of the morphological variability represented robustness of both humerus and radioulna, and the length of the radioulna relative to the humerus (Additional file 4: Figure S1e). On the hindlimb bones data sets, shape

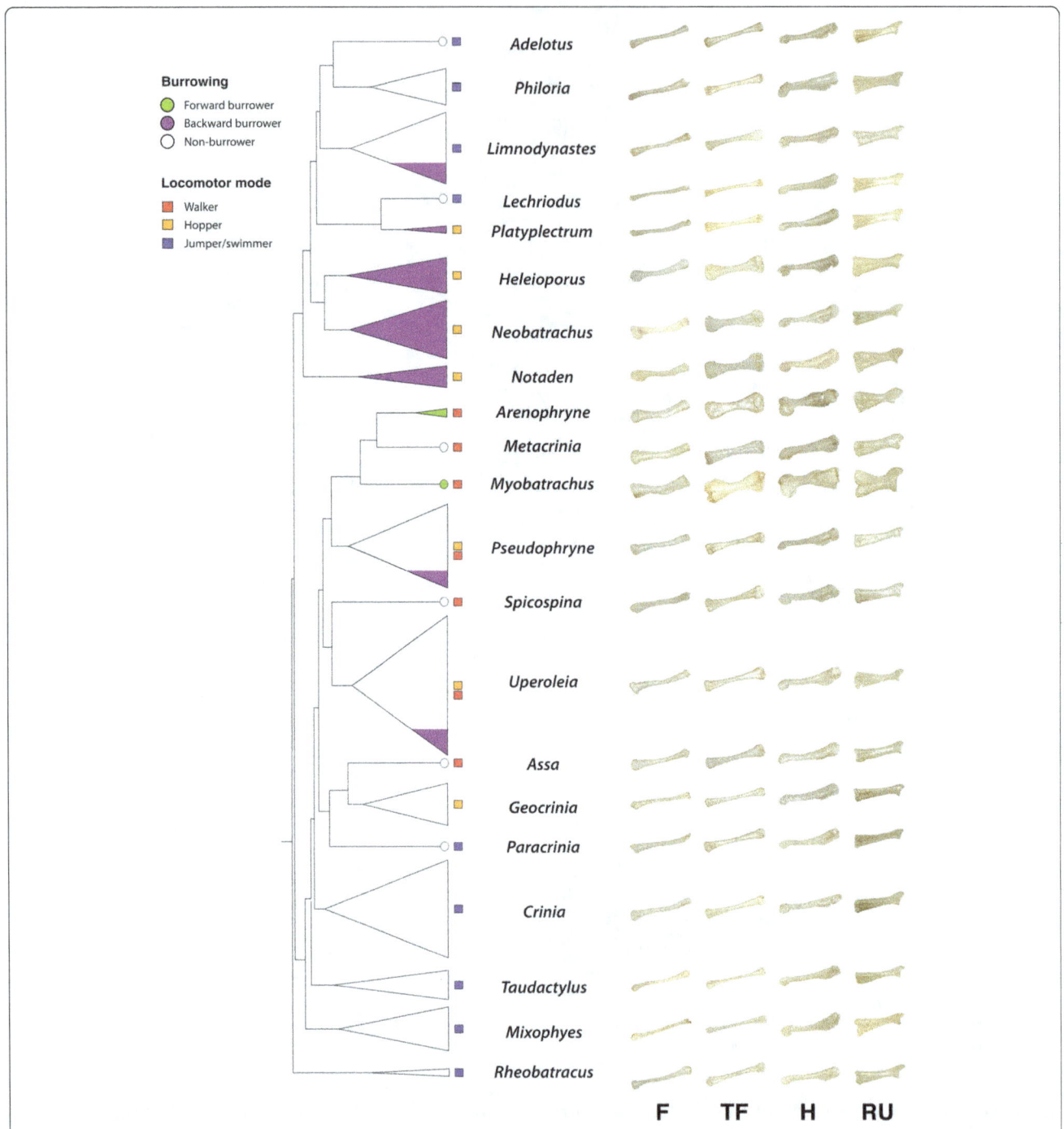

Fig. 2 Shape diversity of limb bones in each genera of myobatrachid frogs: femur (F), tibiofibular (TF), humerus (H), and radioulna (RU). Branches on each genera have been collapsed while retaining information on the species richness of each genus. The legend depicts the three burrowing modes (forward, backward, and non-burrower) and locomotor modes (walker, hopper, and jumper/swimmer). Clades with only few species adapted to fossoriality have been indicated in the figure (*Limnodynastes* spp., *Pseudophryne* spp., and *Uperoleia* spp.)

variation was mostly accounted within the first five PCs, with 94.82% of the total variance in tibiofibula (TF) and 97.45% in femur (F; Additional file 1: Table S3). The first axis of variation in the TF data set (PC$_{TF}$ 1) explained most of the morphological variation as it represented 62.01% of the overall variance (Additional file 1: Table S3)

and was highly correlated with the robustness of the tibiofibula and the degree of constriction in the medial par of the diaphysis. PC$_{TF}$ 2 only added an additional 13.58% (Additional file 4: Figure S1b). On the F data set, PC$_F$ 1 explained 81.23% of the total morphological variance, while PC$_F$ 2 only added an additional 9.88% (Additional file 4:

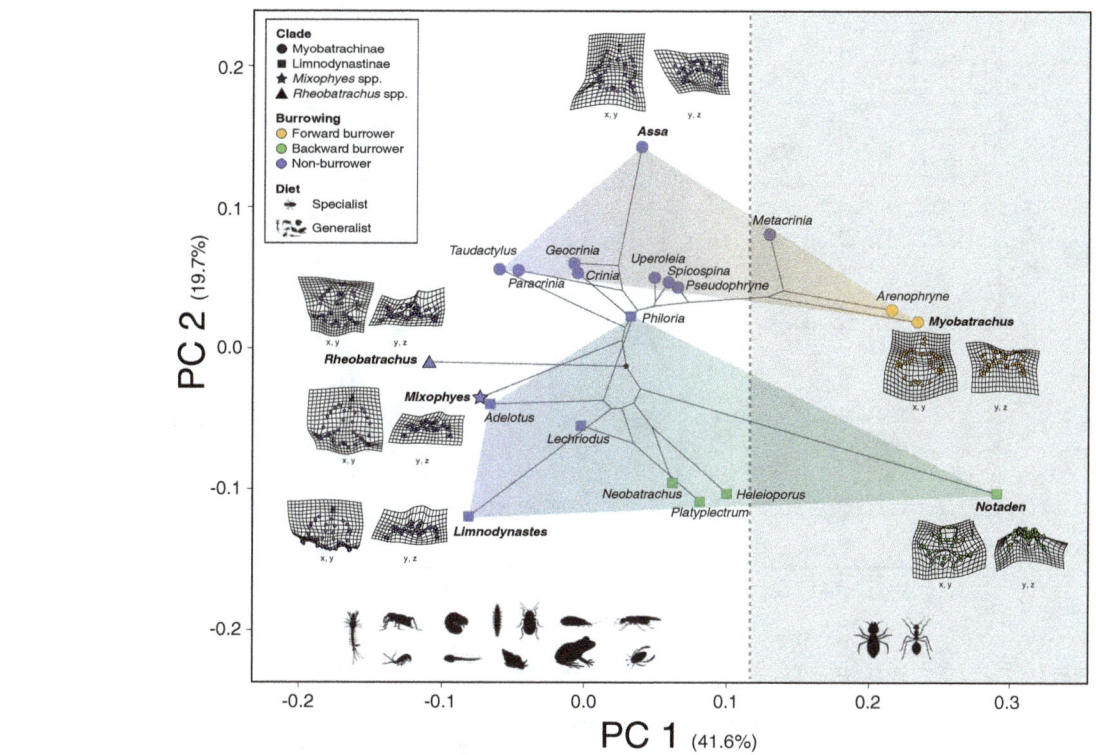

Fig. 3 Phylomorphospace of PCA values on skull shape variation based on species means, using the R package *geomorph*. Each clade is depicted with a different shape, while burrowing behavior is indicated by different colouring. The two main different diet types (specialist and generalist) are also indicated by a schematic of each type and background colouring. Thin-plate spline (TPS) deformation grids are displayed to indicate extreme variation on skull shape among different species (names in bold), and only species from the outer limits of the morphospace are depicted

Figure S1d), and most of the morphological variance was correlated with the degree of arching in the medial part of the diaphysis. On the joined data set of TF and F, the first five PCs accounted for 94.72% of the variance (Additional file 1: Table S3), with PC_{TF+F} 1 representing for 75.34% and PC_{TF+F} 2 an additional 9.77% (Additional file 4: Figure S1f). In contrast with the TF and F data sets, the morphospace hindlimb shape axes (PC_{TF+F} 1 and PC_{TF+F} 2) were mostly correlated with the robustness of both the femur and tibiofibula, the amount of arching observed in the femur, the degree of constriction in the medial part of the diaphysis, and the length of the tibiofibula relative to the femur. Finally, in the overall limb bones shape data set (radioulna + humerus, + tibiofibula + femur), the first five PCs accounted for 91.19% of the morphological variation, with PC_{limbs} 1 representing 45.46% of the variance and PC_{limbs} 2 an additional 37.47% (Fig. 4). Most of the shape changes in the first two axes were associated with general robustness of all four bones, and were correlated with locomotor mode: walker species displayed the most negative values in both PC_{limbs} 1 and PC_{limbs} 2 and occupied distinct regions in the morphospace, while hoppers and jumper/swimmer species overlapped and usually displayed neutral or positive values in both axes.

Patterns of morphological evolution in heads and limbs

We found strong phylogenetic signal on the skull Procrustes-aligned coordinates with K_{mult} values equivalent or greater than 1, and this was significant for the skull, femur, tibiofibular, and limbs (RU + H + TF + F; Additional file 1: Table S4). This means that more closely related species to resemble each other under a Brownian Motion process. The fitting of evolutionary models to univariate data (first five PCs) in both skull and the limb (all four limbs bones) datasets supported different models for each PC (Additional file 1: Table S5). Tests for the best fitting model of phenotypic evolution in multivariate data (first ten PCs) showed support for the same complex process in both skulls and limbs: a model of Early Burst followed by a Brownian Motion process with two different rates based on diet (Table 1). Morphological disparity of skull shape was quite similar in the two most species-rich clades, with Procrustes variance ($Proc_{var}$) of 0.022 in Myobatrachinae, and $Proc_{var}$ = 0.030 in Limnodynastinae. In the limbs (RU + H + TF + F), morphological disparity was higher in Limnodynastinae ($Proc_{var}$ = 0.006) than Myobatrachinae ($Proc_{var}$ = 0.003). In forelimbs disparity was higher in Myobatrachinae ($Proc_{var}$ = 0.007) than Limnodynastinae ($Proc_{var}$ = 0.003). In each forelimb module separately, morphological

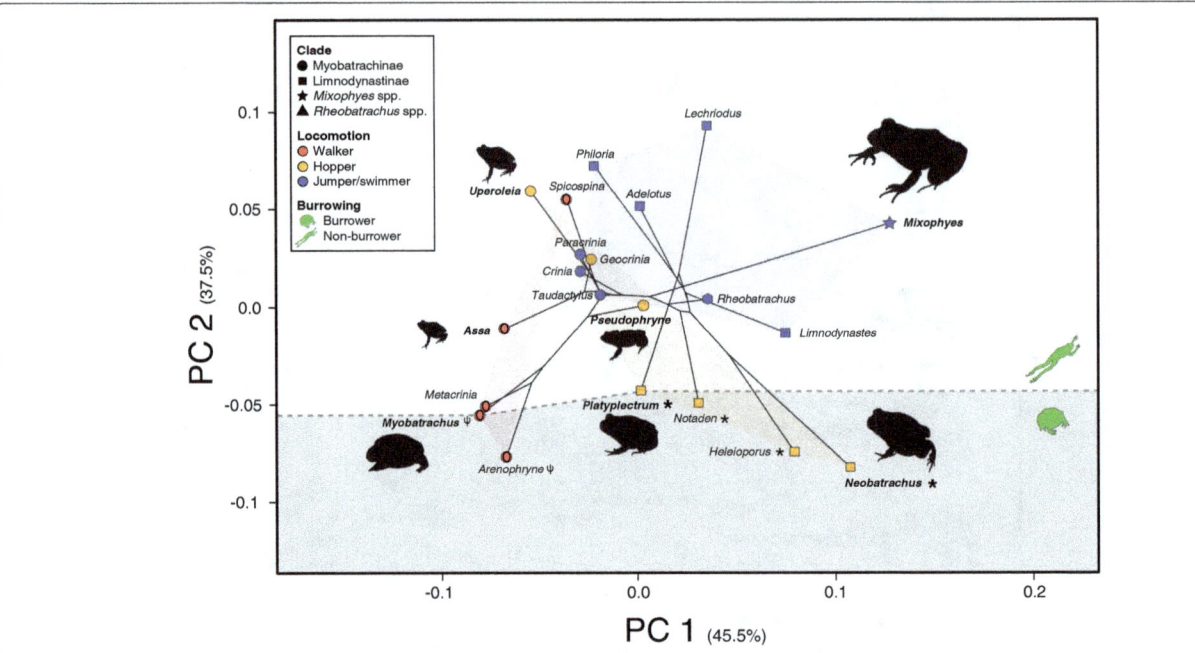

Fig. 4 Phylomorphospace of PCA values on overall limb shape variation based on total shape variation of radioulna, humerus, and femur, and generated with geomorph. Different shapes depict each of the four clades, while colour represents locomotor mode: Walker, Hopper, and Jumper/Swimmer. Burrowing behavior is also indicated in the figure by a schematic of each type (burrower and non-burrower) and background colouring, and the signs ψ and * indicate whether the burrower is forward-burrowing (head and forelimbs first) or backward-burrowing (hindlimbs first), respectively. Outlines of overall body shape are displayed in species with the most extreme limb shape variation

disparity of the radioulna was higher in Myobatrachinae ($Proc_{var}$ = 0.007; $Proc_{var}$ = 0.002 in Limnodynastinae), and also in the humerus ($Proc_{var}$ = 0.005 in Myobatrachinae; $Proc_{var}$ = 0.002 in Limnodynastinae). Procrustes distances in both clades were equal in hindlimbs (TF + F; $Proc_{var}$ = 0.004), higher in the femurs of Limnodynastinae ($Proc_{var}$ = 0.004 in Myobatrachinae; $Proc_{var}$ = 0.006 in Limnodynastinae), and higher in the tibiofibula of Myobatrachinae ($Proc_{var}$ = 0.013 in Myobatrachinae; $Proc_{var}$ = 0.009 in Limnodynastinae). The Mantel test performed on the dissimilarity matrices extracted from the PC components of skull and limbs shape data sets was not significant (r = −0.048, p = 0.678), supporting the null hypothesis that there is no association between the species distribution in the skull and the limb morphospace.

Testing morphological integration and modularity hypotheses

The two-block partial least squares (PLS) analysis between skull and overall limb shape (RU + H + TF + F) indicates that there was slight morphological integration between head and all four limbs (r-PLS = 0.685, p = 0.011; r-PLS = 0.694, p = 0.013 after phylogenetic correction), suggesting semi-independent morphological evolution. However, this result does not hold when we look at the relationship between the head and the fore

and hindlimbs separately: morphological co-variation was much higher when assessed independently on only head and forelimb (r-PLS = 0.923, p < 0.001; r-PLS = 0.909, p = 0.001 after phylogenetic correction), and even higher on head and hindlimb (r = 0.983, p = 0.002; r-PLS = 0.946, p = 0.018 after phylogenetic correction). Morphological integration between forelimbs (RU + H) and hindlimbs (TF + F) was moderate (r-PLS = 0.767, p < 0.001), but it was much higher after correcting for phylogenetic effects (r-PLS = 0.897, p = 0.001) Shape co-variation between the two modules in hindlimbs (F + TF) was extremely high (r-PLS = 0.968, p < 0.001), even after considering phylogenetic correlates (r-PLS = 0.976, p = 0.001). Similarly, the two-block PLS on the forelimbs was also high, supporting strong morphological integration between humerus and radioulna (r-PLS = 0.925, p < 0.001), even after phylogenetic correction (r-PLS = 0.932, p < 0.001). Thus, these results suggest that selective pressures acted on the two modules of hindlimbs (F + TF) and forelimbs as if it was a single integrative structure, but there was certain degree of independence between fore and hindlimbs. All the two-block PLS analyses among different modular partitions within the skull (both raw and taking phylogenetic relationships into account) displayed high levels of integration (Additional file 1: Table S5 and Additional file 5: Figure S2), pointing out that morphological features in the different substructures within the skull have evolved in concert.

Ecology, locomotion and burrowing behaviour

Phylogenetic ANOVAs performed on Procrustes-aligned coordinates of the skull data set were statistically significant for diet ($F_{20,1}$ = 6.058, p = 0.001). Similarly, they were also significant for burrowing ($F_{20,2}$ = 2.806, p = 0.021), and locomotor modes ($F_{20,2}$ = 3.208, p = 0.001). Conversely, they were not significant for the broad eco-regions based on Australian biomes ($F_{20,4}$ = 1.233, p = 0.235). In the phylogenetic ANOVA with the three significant factors (diet + burrowing + locomotion), only diet was significant ($F_{20,1}$ = 3.184, p = 0.005). In the factorial phylogenetic ANOVA between burrowing and locomotion, neither the factors ($F_{20,2}$ = 3.111, p = 0.157 and $F_{20,2}$ = 1.952, p = 0.158, respectively) nor the interaction ($F_{20,1}$ = 1.053, p = 0.316) were significant. In the factorial phylogenetic ANOVA between burrowing and diet, both diet ($F_{20,1}$ = 3.368, p = 0.003) and the interaction ($F_{20,1}$ = 2.373, p = 0.006) were significant, whereas burrowing was not ($F_{20,2}$ = 1.551, p = 0.074). Finally, in the factorial phylogenetic ANOVA between diet and locomotion, both factors ($F_{20,1}$ = 7.629, p = 0.001 and $F_{20,2}$ = 2.855, p = 0.018, respectively) and the interaction ($F_{20,1}$ = 2.304, p = 0.003) were significant.

On the combined limb GM data set (RU + H + TF + F) phylogenetic ANOVAs, burrowing ($F_{20,2}$ = 3.113, p = 0.028), locomotion ($F_{20,2}$ = 2.848, p = 0.012), and diet ($F_{20,2}$ = 3.219, p = 0.006) had a significant effect on overall limb shape, whereas ecorregions ($F_{20,4}$ = 1.086, p = 0.404) did not. In the phylogenetic ANOVA with combined factors of burrowing + locomotion + diet on the combined limb data set, none of the factors were significant ($F_{20,2}$ = 3.265, p = 0.167; $F_{20,2}$ = 1.502, p = 0.314; $F_{20,1}$ = 0.877, p = 0.388; respectively). Similarly, in the factorial phylogenetic ANOVA between burrowing and locomotion, neither the factors ($F_{20,2}$ = 3.161, p = 0.186 and $F_{20,2}$ = 1.454, p = 0.344, respectively) nor the interaction ($F_{20,1}$ = 0.371, p = 0.818) were significant. The factorial ANOVA between burrowing and diet, and the factorial phylogenetic ANOVA between diet and locomotion were also not significant. These results were slightly different when looking at forelimbs and hindlimbs data sets separately. On the forelimbs GM data set (RU + H), burrowing ($F_{20,2}$ = 3.8343, p = 0.003) and diet ($F_{20,1}$ = 4.383, p = 0.002) were significant, whereas locomotor mode ($F_{20,2}$ = 1.310, p = 0.196) and biome were not significant ($F_{20,4}$ = 0.6608, p = 0.271). In the phylogenetic ANOVA with the three factors (diet + burrowing + locomotion), only burrowing was significant ($F_{20,2}$ = 5.543, p = 0.012). In the factorial ANOVA between burrowing and locomotion, only burrowing ($F_{20,2}$ = 4.453, p = 0.021) was significant. In the factorial ANOVA between diet and burrowing, both factors were significant ($F_{20,1}$ = 5.534, p = 0.004 and $F_{20,2}$ = 3.448,

p = 0.018, respectively) but the interaction was not ($F_{20,1}$ = 1.095, p = 0.292). Finally, in the factorial ANOVA between diet and locomotion, only diet ($F_{20,1}$ = 4.202, p = 0.012) was significant. Finally, on the hindlimbs GM data set, burrowing ($F_{20,2}$ = 5.177, p = 0.013) was also significant, whereas locomotor mode ($F_{20,2}$ = 1.316, p = 0.251), diet ($F_{20,1}$ = 0.881, p = 0.252), and biome ($F_{20,4}$ = 0.708, p = 0.687) were not. In the phylogenetic ANOVA with the three factors (burrowing + locomotion + diet) on the hindlimb GM data set, only burrowing was significant ($F_{20,2}$ = 6.361, p = 0.038). In the factorial ANOVA between burrowing and locomotion, only burrowing was significant ($F_{20,2}$ = 7.084, p = 0.027). None of the factors nor the interactions were significant in the factorial ANOVAs between burrowing and diet, and diet and locomotion.

Discussion

We evaluated morphological differences in skulls and limb bones on representative species from all 21 genera of Australian myobatrachid frogs, using a 3D geometric morphometric approach on multiple structures. With this method we were able to focus on the tempo and mode of morphological evolution in this old Gondwanan radiation by asking three main questions: (1) whether morphological evolutionary patterns are similar for different structures, (2) if the amount and direction of morphological change differs for each structure and clade, and (3) if morphological evolution is correlated to functional traits such as locomotion, burrowing, or diet. We found that both head and limbs followed a complex evolutionary pattern typical of adaptive radiation, followed by a Brownian Motion process. Nevertheless, there was a low level of morphological integration between the skull and the limbs and there were significant differences in the mode of morphological evolution between the head and limbs. Skull morphology was phylogenetically conserved and correlated to diet, whereas limb morphology was more labile within clades and appeared to be shaped by diet, burrowing behavior and locomotion. Morphological differences among different limb modules suggest co-variation and strong morphological integration due to selection and functional constraints imposed by burrowing and locomotion. Our results illustrate how morphological diversification in animal clades can follow complex processes, entailing selective pressures from the environment as well as multiple trait covariance with varying degrees of independence across different structures. We discuss each of these topics in turn, and suggest that accurately quantifying shape diversity across multiple structures is crucial in order to understand complex evolutionary processes.

We showed that different phylogenetic clades were separated in skull morphospace, suggesting an early diversification of head shape in myobatrachid frogs, which was supported by an Early Burst model of phenotypic evolution followed by a Brownian Motion process. The majority of skull differences were correlated with fenestration: the subfamily Limnodynastinae displayed bigger and rounder orbits, and robust sphenethmoids and parasphenoids, while species from the Myobatrachinae subfamily generally showed more elongate orbits and larger antorbital fenestrae. The two other major clades, *Rheobatrachus* (comprising the two extinct gastric-brooding frog species) and *Mixophyes* spp. (8 extant species of barred frogs), displayed skull shapes that were intermediate to Limnodynastinae and Myobatrachinae. This pattern of early morphological diversification suggests that occupancy of new morphospace regions by ancestral lineages of myobatrachids could have been associated with major ecological niche filling processes that are typical of diversifying lineages [11, 74, 75]. Analogous with skull shape diversification, morphological evolution in the limbs was best explained by a complex model of an Early Burst process, followed by Brownian Motion. This was unexpected, as myobatrachid frogs are an old Gondwanan adaptive radiation, displaying an exceptionally high degree of ecological, behavioural and morphological diversity across the whole Australian continent. However, phylogenetic non-independence of highly dimensional data, such as 3D GM data, involves some potential pitfalls when inferring complex evolutionary processes, as exposed by Uyeda et al. [76]. For example, Early Burst processes could arise as an artefact from discretising highly dimensional data sets and examining a relatively small sample from multivariate patterns. Thus, caution should be used when inferring evolutionary processes on high effective dimensionality, and our initial results on myobatrachid clades would most likely benefit from a more extensive sampling within genera.

Myobatrachid frogs have experienced several major geological and climatic processes that would have affected diversification and ecological and morphological evolution [41]. Our PCA analyses of the humerus, radioulna, and whole forelimb (H + RU), distributed forward burrowers and most walkers in one broad region of the morphospace, with jumpers/swimmers, backward and non-burrowers, and one walker grouped together on the opposite side of the morphospace. The most extreme forelimb shape was exhibited by forward burrowers that displayed a stronger and more robust humerus, with larger lateral epicondyles, extremely robust radioulnas with large olecranons and a conspicuous longitudinal groove between the radius and ulna. In contrast, good jumpers or swimmer species from wet environments, such as *Lechriodus* sp. and the extinct *Rheobatrachus* spp., generally displayed slender forelimb bones with less pronounced

arching, and a faint longitudinal groove in the radioulna. For the hindlimbs (F + TF), shape diversity was mostly strongly correlated with burrowing behavior (both forward burrowers and clades in which all species are backward burrowers). Both the femur and tibiofibula were shorter and thicker in burrowing species, and displayed pronounced arching and tuberosities (such as the third trochanter) to facilitate muscle attachments.

We found that diet, burrowing, and locomotion played an important role in shaping morphological diversification of myobatrachid frogs. Skull morphology was associated with diet, with ant and termite specialist feeders displaying shorter snouts than generalist species. Several other taxa, including lizards [77], crocodiles [78], mammals [22, 23], and turtles [79] also show clear associations between diet and skull shape. We also found a strong correlation between skull shape and functional traits, such as burrowing and locomotion. That is not an unexpected result, while ecotype, habitat, and other environmental and climate has been found to not have an impact in skull shape diversification in some clades [80], it can also greatly influence head shape in some clades [79, 81, 82], it can also have no impact in others.

Shape diversification of limb bones was not as strongly correlated with phylogenetic history, and instead, diet, locomotor type and burrowing behavior seemed to be important contributors to the morphological variation observed among species. Even though each limb bone displayed slight differences in their shape diversification and its correlation with different ecological variables, they did not differ substantially overall, probably due to their high morphological integration. There was, however, certain degree of independence between fore and hindlimbs, mostly due to functional requirements. Both fore and hindlimbs were correlated to burrowing behavior, but only forelimbs were associated with dietary requirements. The fact that locomotion was strongly correlated with the overall limb shape but not each module or fore and hind limbs independently is not surprising, as fore-to-hind-limb ratios have been proved to be important in explaining locomotor abilities in different frog clades [37]. While most frogs and toads explosive jumping energy is produced by the hind-limbs, they land on their adducted forelimbs, which play a critical role in locomotion by determining the landing and stabilizing actions that enable the next jumping phase [83–85]. Thus, our results suggest that limb shape might have evolved as a response to locomotion constraints imposed by different structural habitats, which would constrain the locomotor modes. This concurs with results found in other amphibian clades, where variation in habitat use and locomotor behaviour seem to correlate with particular ratios between fore and hindlimb lengths [26, 27, 37]. The same trend also is noticeable in other vertebrate groups, such as phrynosomatid lizards [86], anoles [87], sauropods [88], and carnivorous mammals [89, 90], in which ecotype or locomotor type is correlated with limb morphology or distinct proportions between fore and hindlimbs.

The study of locomotion is fundamental to understanding animal biology, as it links morphology with the use of different environments through navigation, feeding, and escape from predators [91]. In addition, factors such as ecology and some less-conspicuous behavioural aspects could also contribute to morphological evolution, making inferences about evolutionary history difficult [24]. Because multiple variables can create selective pressures in different directions on phenotypic traits, their interactions could potentially lead to trade-offs. For example, morphological optimisation for burrowing creates opposing pressures from optimisation for jumping, due to discordances in functional morphological requirements for each behaviour [38]. Although myobatrachid frogs generally display phylogenetic conservatism in morphology, burrowing behavior and other ecological correlates still appear to have a strong effect on limb shape. Morphological adaptations in forward burrowers are primarily associated with forelimb bones in both amphibians and other fossorial vertebrates [92]. Similarly, despite not being found in any other vertebrate, backward burrowing represents 95% of all burrowing types in anurans [34]. The evolution of both forward and backward burrowing likely led to reduced length and increased robustness of fore and hindlimbs respectively, which would almost certainly have resulted in reduced locomotor abilities [30, 38]. This trend towards shorter and more robust limbs in burrowing anurans is likely also a beneficial adaptation to arid environments. Amphibians have adapted to a wide range of extreme climatic conditions, despite experiencing more constraints than any other terrestrial vertebrate due to rapid evaporative water loss through their permeable skin [34]. By reducing limb length, total surface area of the body can also be reduced and with it, evaporative water loss.

Despite high overall morphological disparity among different myobatrachid genera [45], some structures (e.g. limbs) displayed morphological integration and co-variation leading to convergent phenotypes, while other structures (e.g. skulls) followed semi-independent evolutionary processes. Despite the low integration between skull and all post-cranial modules, hind- and forelimbs were more tightly correlated to the skull when assessed independently, especially the hindlimbs. These results could be due to a certain degree of integration between head and postcranial modules, which could follow different directions in the morphospace for each limbs module, resulting in semi-independent pattern of morphological evolution of the head versus the rest of the body. Our

results, therefore, support the modularity or semi-independent hypothesis when looking at morphological evolution between skulls and limbs, but favours the morphological integration hypothesis for shape diversification within different limb modules, or the different substructures within the skull. Thus, while high evolutionary lability experienced by limbs is a result of selective pressures from the environment, skulls instead display relatively high phylogenetic conservatism. This suggests that morphological diversification might have occurred rapidly quite early in the myobatrachid frog radiation, followed by a decrease in shape disparity, which is conspicuous through the different areas of skull morphospace. Head shape in anurans appears to have undergone extreme morphological change very early in the evolutionary history of modern amphibians, which is especially conspicuous through a substantial widening of the skull and orbits, and enlargement of fenestrae [76, 93]. Moreover, strong phylogenetic structure on skull shape is not unusual among other amphibian groups older than 50 MY (e.g. caecilians [75]), in contrast to younger vertebrate radiations that typically display greater morphological disparity, with weaker phylogenetic signal [94].

Phylogenetic conservatism and morphological diversification in functional traits can provide insight into evolutionary processes [24], but the interplay between different potential drivers of adaptation can blur the link between form and function. For example, limb morphology might appear strongly correlated with locomotion type, but habitat use or burrowing behaviour might be equally important correlates. In this way adaptive traits often cannot (and should not) be explained by just one adaptive process. Morphological integration or modularity also can affect the accuracy of evolutionary inferences on adaptation to certain ecological, locomotor or behavioural factors [95]. Furthermore, rates of phenotypic evolution can be correlated with species diversification rates within clades, as morphological traits typically have slower evolutionary rates than other traits such as behaviour [67, 96]. Moreover, closely related clades might display unequal magnitudes of morphological change, thus hindering or boosting apparent morphological diversification, especially early in their evolutionary history [11].

Conclusions

Our study is the first to accurately identify evolutionary processes that drive morphological diversity in the context of modularity and morphological integration of several structures in an old adaptive radiation and across a whole continent. Our results highlight how form is usually tightly linked to function, and that different structures can evolve semi-independently, while in other modules morphological evolution is tightly coupled. There was strong morphological co-variation among different modules in the limbs due to strong selective pressures from the environment and functional trade-offs (e.g. burrowing and locomotion), whereas skull shape was correlated to diet, and yield a pattern of very early morphological diversification followed by strong phylogenetic conservatism. Our results also show that even when different structures evolve following the same evolutionary models, patterns of morphological diversification can be drastically different. The complex interplay between selective pressures and different levels of co-varying morphological evolution makes it harder to accurately identify processes that drive clade diversification and infer their evolutionary history. Thus, we highlight the importance of accurately assessing morphological evolution in multiple structures in order to properly understand complex evolutionary processes that generate the phenotypic diversity we see today.

Additional files

Additional file 1: Table S1. Summary of several ecological and behavioural traits of the myobatrachid frogs studied here, used in posterior analyses: burrowing behaviour, locomotor mode, habitat type or ecoregion, and diet type. **Table S2.** Summary of the different landmarks used for each module (m1, m2 or m3) in all five models of modular partitions (bimodular or trimodular) within the skull that correspond to the models displayed on Additional file 5: Figure S2. **Table S3.** Principal Component Analyses of shape variation for different sets of Procrustes-aligned species means, using *geomorph*. **Table S4.** Summary of phylogenetic signal tests, using *geomorph* (Adams & Otarola-Castillo, 2013). K 95% confidence interval for values expected under a Brownian Motion model of trait evolution = [0.799, 1.318]. **Table S5.** Summary statistics for the fit of models of phenotypic evolution in the first five principal components of the Skull shape dataset and the limbs shape dataset (all four limb bones together): maximum likelihood estimate (ln L), sample-size corrected Akaike's Information Criterion (AICc), and Delta AICc (ΔAICc, difference between a model and the model with the lowest AICc). We tested the fit of the following evolutionary models: BM = Brownian Motion, EB = Early Burst, white = nonphylogenetic, OU = Ornstein-Uhlenbeck, OU2_diet = Ornstein-Uhlenbeck with two optima based on diet, OU3_loc = OU with three optima based on locomotion, and OU3_burr = OU with three optima based on burrowing behaviour. Analyses were performed in R using *geiger* [68] and *ouch* [69]. **Table S6.** Results from the *integration.test* function in geomorph (Adams & Otrola-Castillo, 2013) in order to quantify the degree of modularity between the two or three modules in each modular configuration (a-e), using the landmark coordinate data. **Appendix S1.** Species and specimen codes for all the individuals used in this study, by museums. (PDF 171 kb)

Additional file 2: Video displaying the 42 landmarks used in the GM analyses of the skull. (PDF 25415 kb)

Additional file 3: Video displaying both the landmarks and semi-landmarks used in the GM analyses of the four limb bones: radioulna, humerus, tibiofibular and femur. (PDF 6874 kb)

Additional file 4: Figure S1. (a) Phylomorphospace of PCA values on shape variation of radioulna (RU); (b) Phylomorphospace of PCA values on shape variation of tibiofibula (TF); (c) Phylomorphospace of PCA values on shape variation of humerus (H); (d) Phylomorphospace of PCA values on shape variation of femur (F); (e) Phylomorphospace of PCA values on fore-limb shape variation (RU + H); (f) Phylomorphospace of PCA values on hind-limb shape variation (TF + F). (PDF 332 kb)

Additional file 5: Figure S2. Modular configurations modeled for the skull with two or three different partitions, based on different

evolutionary hypothesis based on biological relevant regions. The different colours depict different modules. (a) The first module (green) includes the tip of the snout and the olfactory area (premaxilla, maxilla, and nasal), as it captures a lot of morphological variation among frog species, whereas the second module (blue) includes the rest of the skull; (b) this configuration captures skull depth – the first module includes the dorsal region of the skull, and the second module captures morphological information from the ventral region; (c) this tripartite model splits the skull in three modules: snout (green), squamosal (orange, which is part of the suspensory apparatus), and the rest of the skull (blue); (d) The first module depicts the snout (green), the second includes the medial part of the skull (orange), and the third module includes the most posterior region of the skull (blue); (e) this tripartite configuration includes a first module (green) with the snout morphology, a second module (orange) that encompasses the brain region (from the sphenethmoid to the exoccipital and foramen magnum, including the frontoparietal), and a third module (blue) for the rest of the skull. (PDF 275 kb)

Acknowledgements

We are grateful to all museum curators for specimens' loans: R. Sadlier, M. Hutchinson, P. Doughty. We thank L. Joseph, R. Palmer and M. Cawsey (ANWC); M. Foley (ACMM); and M. Turner and T. Senden (ANU) for facilitating access to CT scan facilities. We also thank M. Mahony, S. Clulow and D. Roberts for their personal communications on frog diet. We are indebted to A. Pyron, M. Pepper, I. Brennan, T. Semple, and an anonymous reviewer for useful comments on this manuscript. Thank you to E. Walsh for her drawings for the burrowing legends in Fig. 4.

Funding

All the data was gathered as part of MVG's PhD at the Australian National University. MVG was supported by the Peter Rankin Trust for Herpetology and the SSAR. JSK was funded by the ARC.

Authors' contributions

MVG and SK conceived the study. MVG collected, processed, and analysed the data, and drafted the initial version of the manuscript. Both authors read, edited, and approved the final version of the manuscript.

Competing interests

The authors declare that they have no competing interests.

References

1.   Russell ES. Form and function. A contribution to the history of animal morphology. 1917;
2.   La Barbera M. Analysing body size as a factor in ecology and evolution. Annu Rev Ecol Syst. 1989;20:97–117.
3.   Blackburn TM, Gaston KJ. Animal body size distributions: patterns, mechanisms and implications. Trends Ecol Evol. 1994;9:471–4.
4.   Collar DC, Near TJ, Wainwright PC. Comparative analysis of morphological diversity: does disparity accumulate at the same rate in two lineages of centrarchid fishes? Evolution. 2005;59:1783–94.
5.   Ricklefs RE. Community diversity: relative roles of local and regional processes. Science. 1987;235:167–71.
6.   Wainwright PC. Ecomorphology: experimental functional anatomy for ecological problems. Am Zool. 1991;31:680–93.
7.   Wainwright PC, Reilly SM. Ecological morphology: integrative organismal biology [internet]: University of Chicago Press; 1994.
8.   Losos JB. The evolution of form and function: morphology and locomotor performance in West Indian Anolis lizards. Evolution. 1990;44:1189–203.
9.   Foote M. The evolution of morphological diversity. Annu. Rev. Ecol. Syst. 1997;28:129–52.
10. Crisp MD, Arroyo MTK, Cook LG, Gandolfo MA, Jordan GJ, McGlone MS, et al. Phylogenetic biome conservatism on a global scale. Nature. 2009;458:754–6.
11. Sidlauskas B. Continuous and arrested morphological diversification in sister clades of characiform fishes: a phylomorphospace approach. Evolution. 2008;62:3135–56.

12. Thacker CE. Species and shape diversification are inversely correlated among gobies and cardinalfishes (Teleostei: Gobiiformes). Org Divers Evol. 2014;14:419–36.

13. Claramunt S. Discovering exceptional diversifications at continental scales: the case of the endemic families of neotropical suboscine passerines. Evolution. 2010;64:2004–19.

14. Brusatte SL, Nesbitt SJ, Irmis RB, Butler RJ, Benton MJ, Norell MA. The origin and early radiation of dinosaurs. Earth-Science Rev. 2010;101:68–100.

15. Klingenberg CP. Morphological integration and developmental modularity. Annu Rev Ecol Evol Syst. 2008;39:115–32.

16. Mitteroecker P, Bookstein F. The conceptual and statistical relationship between modularity and morphological integration. Syst Biol. 2007;56:818–36.

17. Gatesy SM, Dial KP. Locomotor modules and the evolution of flight. Evolution. 1996;50:331–40.

18. Feilich KL. Correlated evolution of body and fin morphology in the cichlid fishes. Evolution. 2016;70:2247–67.

19. Hansen TF. Is modularity necessary for evolvability? Biosystems. 2003;69:83–94.

20. Raff RA. The shape of life: genes, development, and the evolution of animal form: University of Chicago Press; 2012.

21. Fruciano C, Franchini P, Meyer A. Resampling-based approaches to study variation in morphological modularity. PLoS One. 2013;8:e69376.

22. Goswami A, Polly PD. The influence of modularity on cranial morphological disparity in Carnivora and primates (Mammalia). PLoS One. 2010;5:e9517.

23. Goswami A. Notes and comments cranial modularity shifts during mammalian. Evolution. 2012;168:270–80.

24. Ricklefs RE, Miles DB. Ecological and evolutionary inferences from morphology: an ecological perspective. in Ecological morphology: integrative organismal biology. University of Chicago Press; 1994. p. 13–41.

25. Zaaf A, Van Damme R. Limb proportions in climbing and ground-dwelling geckos (Lepidosauria, Gekkonidae): a phylogenetically informed analysis. Zoomorphology. 2001;121:45–53.

26. Moen DS, Irschick DJ, Wiens JJ. Evolutionary conservatism and convergence both lead to striking similarity in ecology, morphology and performance across continents in frogs. Proc Biol Sci. 2013;280:2013–156.

27. Enriquez-Urzelai U, Montori A, Llorente GA, Kaliontzopoulou A. Locomotor mode and the evolution of the Hindlimb in western Mediterranean anurans. Evol Biol. 2015;42:199–209.

28. Cornette R, Baylac M, Souter T, Herrel A. Does shape co-variation between the skull and the mandible have functional consequences? A 3D approach for a 3D problem. J Anat. 2013;223:329–36.

29. Wagner GP, Pavlicev M, Cheverud JM. The road to modularity. Nat Rev Genet. 2007;8:921–31.

30. Hall BK. Fins into limbs: evolution, development, and transformation: University of Chicago Press; 2008.

31. Shubin NH, Jenkins FA. An early Jurassic jumping frog. Nature. 1995;377:49–52.

32. Jenkins F A., Shubin NH. Prosalirus bitis and the anuran caudopelvic mechanism. J Vertebr Paleontol 1998;18:495–510.

33. Reilly S, Essner R, Wren S, Easton L, Bishop PJ. Movement patterns in leiopelmatid frogs: insights into the locomotor repertoire of basal anurans. Behav Process. 2015;121:43–53.

34. Wells KD. The ecology and behavior of amphibians: University of Chicago Press; 2010.

35. Emerson SB, Koehl MAR. The interaction of behavioral and morphological change in the evolution of a novel locomotor type: "flying" frogs. Evolution. 1990;44:1931–46.

36. Wilson RS, James RS, Van Damme R. Trade-offs between speed and endurance in the frog Xenopus laevis: a multi-level approach. J Exp Biol. 2002;205:1145–52.

37. Vidal-García M, Keogh JS. Convergent evolution across the Australian continent: ecotype diversification drives morphological convergence in two distantly related clades of Australian frogs. J Evol Biol. 2015;28:2136–51.

38. Emerson SB. Burrowing in frogs. J Morphol. 1976;149:437–58.

39. Pyron RA. Biogeographic analysis reveals ancient continental Vicariance and recent oceanic dispersal in amphibians. Syst Biol. 2014;63:779–97.

40. Frost DR. Amphibian Species of the World: an Online Reference. Version 6.0. [cited. Oct 15. 2016; Available from: http://research.amnh.org/herpetology/amphibia/index.html

41. Byrne M, Yeates DK, Joseph L, Kearney M, Bowler J, Williams MAJ, et al. Birth of a biome: insights into the assembly and maintenance of the Australian arid zone biota. Mol Ecol. 2008;17:4398–417.

42. Anstis M. Tadpoles and frogs of Australia. Sydney, NSW: New Holland Publishing Pty Ltd; 2013.

43. Roberts J, Standish R, Byrne P, Doughty P. Synchronous polyandry and multiple paternity in the frog Crinia georgiana (Anura: Myobatrachidae). Anim Behav. 1999;57:721–6.

44. Byrne PG, Roberts JD, Simmons LW. Sperm competition selects for increased testes mass in Australian frogs. J Evol Biol. 2002;15:347–55.

45. Vidal-García M, Byrne PG, Roberts JD, Keogh JS. The role of phylogeny and ecology in shaping morphology in 21 genera and 127 species of Australo-Papuan myobatrachid frogs. J Evol Biol. 2014;27:181–92.

46. Cogger H. Reptiles and amphibians of Australia. 7th ed. Collingwood, VIC: Csiro Publishing; 2014.

47. National Reserve System. National Reserve System (NRS). 2016. Available from: https://www.environment.gov.au/land/nrs/about-nrs

48. Harrison L. Notes on some western Australian frogs, with descriptions of new species. Rec Aust Museum. 1927;15:277–87.

49. Calaby JH. The food habits of the frog, Myobatrachus gouldii (Gray). West Aust Nat. 1956;5:93–4.

50. Calaby JH. A note on the food of Australian desert frogs. West. Aust. Nat. 1960;7:79–80.

51. Lee AK. Studies in Australian amphibia II..Taxonomy, ecology and evolution of the genus Heleioporus Gray (Anura : Leptodactylidae). Aust. J. Zool 1967; 15:367–439.

52. Tyler MJ, Roberts JD, Davies M. Field observations on Arenophryne rotunda Tyler, a Leptodactylid frog of coastal Sandhills. Aust Wildl Res. 1980;7:295–304.

53. Mac Nally RC. Trophic relationships of two sympatric species of Ranidella (Anura). Herpetologica. 1983:130–40.

54. Winter J, McDonald R. Eungella, the land of cloud. Aust Nat Hist. 1986;22:39–43.

55. Cappo M. Frogs as predators of organisms of aquatic origin in the Magela Creek system. Northern Territory. Thesis: University of Adelaide, Department of Zoology; 1987.

56. Katsikaros K, Shine R. Sexual dimorphism in the tusked frog, Adelotus brevis (Anura:Myobatrachidae): the roles of natural and sexual selection. Biol J Linn Soc. 1997;60:39–51.

57. Lima AP, Magnusson WE, Williams DG. Differences in diet among frogs and lizards coexisting in subtropical forests of Australia. J Herpetol. 2000;34:40–6.

58. Mahony M, Clulow S, Roberts JD. Personal communication. 2016.

59. Limaye A. Drishti: a volume exploration and presentation tool. SPIE 8506, Dev. X-Ray Tomogr. 2012;8506:85060X.

60. Adams DC, Otárola-Castillo E. geomorph : an R package for the collection and analysis of geometric morphometric shape data. 2013;63:685–697.

61. Klingenberg CP, Gidaszewski N. Testing and quantifying phylogenetic signals and homoplasy in morphometric data. Syst Biol. 2010;59:245–61.

62. Rohlf F, Slice D. Extensions of the Procrustes method for the optimal superimposition of landmarks. Syst Biol. 1990;39:40–59.

63. Vidal-Garcia M, Bandara L, Keogh JS. ShapeRotator: an R package for standardised rigid rotations of articulated Three-Dimensional structures with application for geometric morphometrics. bioRxiv. 2017.

64. Paradis E, Claude J, Strimmer K. APE: analyses of Phylogenetics and evolution in R language. Bioinformatics. 2004;20:289–90.

65. Maddison WP. Squared-change parsimony reconstructions of ancestral states for continuous-valued characters on a phylogenetic tree. 2013;40:304–314.

66. Rohlf FJ. Geometric morphometrics and phylogeny. In: MacLeod N, Forey PL, editors. Morphol. shape phylogeny. 2002. p. 175–193.

67. Adams DC. A generalized K statistic for estimating phylogenetic signal from shape and other high-dimensional multivariate data. Syst Biol. 2014;63:685–97.

68. Blomberg SP, Garland T, Ives AR. Testing for phylogenetic signal in comparative data: behavioral traits are more labile. Evolution. 2003;57:717–45.

69. Harmon LJ, Weir JT, Brock CD, Glor RE, Challenger W. GEIGER: investigating evolutionary radiations. Bioinformatics. 2007;24(1):129–31.

70. King AA, Butler MA. ouch: Ornstein-Uhlenbeck models for phylogenetic comparative hypotheses (R package). 2009. Available from: http://ouch.r-forge.r-project.org.

71. Clavel J, Escarguel G, Merceron G. mvMORPH: an R package for fitting multivariate evolutionary models to morphometric data. Methods Ecol. Evolution. 2015;6:1311–9.

72. Klingenberg CP, Ekau W. A combined morphometric and phylogenetic analysis of an ecomorphological trend: pelagization in Antarctic fishes (Perciformes: Nototheniidae). Biol J Linn Soc. 1996;59:143–77.

73. Dixon P. Vegan, a package of R functions for community ecology. J Veg Sci. 2003;14:927–30.

74. Schluter D. Ecological character displacement in adaptive radiation. Am Nat. 2000;156:S4–S16.

75. Sherratt E, Gower DJ, Peter C, Mark K. Evolution of cranial shape in caecilians (Amphibia : Gymnophiona). Evol Biol. 2014;4:528–45.

76. Uyeda JC, Caetano DS, Pennell MW. Comparative analysis of principal components can be misleading. Syst Biol. 2015;64(4):677–89.

77. Stayton CT. Morphological Evolution of the Lizard Skull : A Geometric Morphometrics Survey. 2005;59:47–59.

78. Pierce SE, Angielczyk KD, Rayfield EJ. Morphospace occupation in thalattosuchian crocodylomorphs : skull shape variation, species delineation and temporal patterns. Palaeontology. 2009;52:1057–97.

79. Claude J, Pritchard P, Tong H, Paradis E, Auffray J-C. Ecological correlates and evolutionary divergence in the skull of turtles: a geometric morphometric assessment. Syst Biol. 2004;53:933–48.

80. Kohlsdorf T, Grizante MB, Navas CA, Herrel A. Head shape evolution in Tropidurinae lizards: does locomotion constrain diet? J Evol Biol. 2008;21:781–90.

81. Kaliontzopoulou A, Carretero MA, Llorente GA. Intraspecific ecomorphological variation : linear and geometric morphometrics reveal habitat-related patterns within Podarcis bocagei wall lizards. J Evol Biol. 2010;23:1234–44.

82. Kaliontzopoulou A, Adams DC, van der Meijden A, Perera A. Carretero M a. Relationships between head morphology, bite performance and ecology in two species of Podarcis wall lizards. Evol. Ecol. 2012;26:825–45.

83. Nauwelaerts S, Stamhuis E, Aerts P. Swimming and jumping in a semi-aquatic frog. Anim Biol. 2005;55:3–15.

84. Essner RL, Suffian DJ, Bishop PJ, Reilly SM. Landing in basal frogs: evidence of saltational patterns in the evolution of anuran locomotion. Naturwissenschaften. 2010;97:935–9.

85. Gillis GB, Akella T, Gunaratne R. Do toads have a jump on how far they hop? Pre-landing activity timing and intensity in forelimb muscles of hopping Bufo marinus. Biol Lett. 2010;6:486–9.

86. Herrel A, Meyers JJ, Vanhooydonck B. Relations between microhabitat use and limb shape in phrynosomatid lizards. Biol J Linn Soc. 2002;77:149–63.

87. Losos JB. Ecomorphology, performance capability, and scaling of West Indian Anolis lizards: an evolutionary analysis. Ecol Monogr. 1990;60:369–88.

88. Bonnan MF. Morphometric analysis of humerus and femur shape in Morrison sauropods: implications for functional morphology and paleobiology. Paleobiology. 2004;30:444–70.

89. Ercoli MD, Prevosti FJ, Álvarez A. Form and function within a phylogenetic framework : locomotory habits of extant predators and some Miocene Sparassodonta (Metatheria). Zool J Linnean Soc. 2012;165:224–51.

90. Fabre A-C, Cornette R, Goswami A, Peigné S. Do constraints associated with the locomotor habitat drive the evolution of forelimb shape? A case study in musteloid carnivorans. J Anat. 2015;226:596–610.

91. Gray J. How animals move. Harmondsworth: Penguin Books; 1959.

92. Piras P, Sansalone G, Teresi L, Kotsakis T, Colangelo P. Loy a. Testing convergent and parallel adaptations in talpids humeral mechanical performance by means of geometric morphometrics and finite element analysis. J. Morphology. 2012;273:696–711.

93. Sigurdsen T, Bolt JR. The lower Permian amphibamid Doleserpeton (Temnospondyli: Dissorophoidea), the interrelationships of amphibamids, and the origin of modern amphibians. J Vertebr Paleontol. 2010;30:1360–77.

94. Losos JB. Adaptive radiation, ecological opportunity, and evolutionary determinism. Am Nat. 2010;175:623–39.

95. Young N. Modularity and integration in the hominoid scapula. J Exp Zool B Mol Dev Evol. 2004;302:226–40.

96. Harmon LJ, Losos JB, Davies TJ, Gillespie RG, Gittleman JL. Jennings WB, et al. Early bursts of body size and shape evolution are rare in comparative data. 2010:2385–96.

6

Molecular phylogeny reveals food plasticity in the evolution of true ladybird beetles (Coleoptera: Coccinellidae: Coccinellini)

Hermes E. Escalona[1,2], Andreas Zwick[2], Hao-Sen Li[3], Jiahui Li[4], Xingmin Wang[5], Hong Pang[3], Diana Hartley[2], Lars S. Jermiin[6], Oldřich Nedvěd[7,8], Bernhard Misof[1], Oliver Niehuis[9], Adam Ślipiński[2] and Wioletta Tomaszewska[10]*

Abstract

Background: The tribe Coccinellini is a group of relatively large ladybird beetles that exhibits remarkable morphological and biological diversity. Many species are aphidophagous, feeding as larvae and adults on aphids, but some species also feed on other hemipterous insects (i.e., heteropterans, psyllids, whiteflies), beetle and moth larvae, pollen, fungal spores, and even plant tissue. Several species are biological control agents or widespread invasive species (e.g., *Harmonia axyridis* (Pallas)). Despite the ecological importance of this tribe, relatively little is known about the phylogenetic relationships within it. The generic concepts within the tribe Coccinellini are unstable and do not reflect a natural classification, being largely based on regional revisions. This impedes the phylogenetic study of important traits of Coccinellidae at a global scale (e.g. the evolution of food preferences and biogeography).

Results: We present the most comprehensive phylogenetic analysis of Coccinellini to date, based on three nuclear and one mitochondrial gene sequences of 38 taxa, which represent all major Coccinellini lineages. The phylogenetic reconstruction supports the monophyly of Coccinellini and its sister group relationship to Chilocorini. Within Coccinellini, three major clades were recovered that do not correspond to any previously recognised divisions, questioning the traditional differentiation between Halyziini, Discotomini, Tytthaspidini, and Singhikaliini. Ancestral state reconstructions of food preferences and morphological characters support the idea of aphidophagy being the ancestral state in Coccinellini. This indicates a transition from putative obligate scale feeders, as seen in the closely related Chilocorini, to more agile general predators.

Conclusions: Our results suggest that the classification of Coccinellini has been misled by convergence in morphological traits. The evolutionary history of Coccinellini has been very dynamic in respect to changes in host preferences, involving multiple independent host switches from different insect orders to fungal spores and plants tissues. General predation on ephemeral aphids might have created an opportunity to easily adapt to mixed or specialised diets (e.g. obligate mycophagy, herbivory, predation on various hemipteroids or larvae of leaf beetles (Chrysomelidae)). The generally long-lived adults of Coccinellini can consume pollen and floral nectars, thereby surviving periods of low prey frequency. This capacity might have played a central role in the diversification history of Coccinellini.

Keywords: Coccinelloidea, Ladybugs, Diet shifts, Evolution, Feeding strategies, Food preferences, Taxonomy

* Correspondence: wiolkat@miiz.waw.pl
[10]Museum and Institute of Zoology, Polish Academy of Sciences, Wilcza 64, 00-679 Warszawa, Poland
Full list of author information is available at the end of the article

Background

Ladybirds (Coccinellidae) are a well-defined monophyletic group of small to medium sized beetles of the superfamily Coccinelloidea, the superfamily formerly known as the Cerylonid Series within the superfamily Cucujoidea [1–3]. The relationships between the currently recognized 15 families of Coccinelloidea are not well understood, but comprehensive molecular phylogenetic analyses of Coccinelloidea [2] suggested that Eupsilobiidae, a mycophagous group of small brown beetles, previously included as a subfamily of Endomychidae [4, 5], are the sister group of Coccinellidae. Coccinellidae, which comprises 360 genera and about 6000 species world-wide, is by far the largest family of coccinelloid beetles and, with the notable exception of the parasitic Bothrideridae, the only predominantly predatory lineage of Coccinelloidea. The ancestor of Coccinellidae presumably lived in the Jurassic (~ 150 Mya [6]) and even a Permian-Triassic origin of Coccinelloidea has been suggested [7]. The development of a predatory life style in the ancestor of Coccinellidae, was possibly a relevant event for the evolutionary history of this beetle lineage, with herbivory, sporophagy and pollenophagy being derived from this predatory mode of life.

Most of the traditional classifications of Coccinellidae [8–10] recognize six or seven subfamilies (i.e., Chilocorinae, Coccidulinae, Coccinellinae, Epilachninae, Scymninae, Sticholotidinae, and sometimes Ortaliinae, each with numerous tribes). The foundation of this system was developed by Sasaji [11, 12] based on comparative morphological analyses of adults and larvae from species of the Palaearctic Region, mostly Japan. Kovář [9] presented a major modification of Sasaji's classification on a global scale, recognizing seven subfamilies and 38 tribes. The classifications proposed by Sasaji [11] and Kovář [9] were found to be largely artificial and phylogenetically unacceptable by Ślipiński [13], who argued for a basal split of Coccinellidae into two subfamilies, Microweiseinae and Coccinellinae, with the latter containing most of the tribes, including Coccinellini. Six subsequent papers on the molecular phylogeny of the family Coccinellidae [14–17] and Cucujoidea [2, 3] corroborated the monophyly of the family and of the two subfamilies recognized by Ślipiński [13]. They also provided strong evidence for the monophyly of Coccinellini. Based on results of phylogenetic analyses of molecular data and a combination of molecular and morphological data from Coccinellidae, Ślipiński and Tomaszewska [18] and Seago et al. [17] formalized the taxonomic status of Coccinellini as a tribe within the broadly defined Coccinellinae.

Coccinellini, commonly referred to as 'true ladybirds', comprises 90 genera and over 1000 species world-wide. The tribe includes many charismatic and easily recognised beetles that are often seen on aphid-infested trees and shrubs in the natural and urban landscapes. It is also one of the most frequently studied groups of beetles, the subject of thousands of peer-reviewed scientific papers on biology, genetics, colour polymorphism, physiology and biological control, summarized in various influential books [19–22].

Coccinellini are generally viewed as predators of aphids, but their diet is much more diverse and often includes other hemipterous insects (i.e., heteropterans, psyllids), beetle and moth larvae, pollen, fungal spores, and even plant tissues. Coccinellini display extraordinary morphological diversity in all life stages and are among the most conspicuously and attractively coloured beetles, often bearing strikingly red or yellow elytra, with contrasting black spots, stripes, or fasciae (Figs. 1 and 2). These vivid colours are aposematic, warning predators that these beetles are distasteful and produce noxious or poisonous alkaloids [23] excreted as droplets of fluid during a 'reflex bleeding' behaviour. Many species of Coccinellini are also of great economic importance as biological control agents or unwanted invaders on a scale of entire

Fig. 1 Representative spectrum of Coccinellini morphologies and feeding habits: **a** *Coccinella septempunctata*, adult feeding on aphids; **b** *Coelophora variegata*, adult feeding on aphids; **c** *Heteroneda reticulata*, pupa being parasitized by a phorid fly; **d** *Cleobora mellyi*, larva feeding on larva of *Paropsis charybdis* (Chrysomelidae); **e** *Halyzia sedecimguttata*, larva feeding on mildew; **f** *Harmonia conformis*, adult feeding on psyllids; **g, h** *Bulaea lichatschovi*, larva and adult, feeding on leaves and buds of *Bassia prostrata*. Photographs credits: **a, b** Paul Zborowski; **c** Melvyn Yeo; **d** Andrew Bonnitcha; **e** Gilles San Martin; **f** Nick Monaghan; **g, h** Maxim Gulyaev

Fig. 2 Representative spectrum of Coccinellidae morphologies and feeding habits. **a** *Psyllobora vigintiduopunctata*, adult feeding on mildew; **b** *Hippodamia variegata*, adults feeding on pollen; **c** *Scymnus* sp., larva with dense waxy covering; **d** *Harmonia axyridis*, larva showing droplets of haemolymph at abdominal segments; **e** *Harmonia axyridis*, pupa with nymph of parasitic mite; **f** *Anatis ocellata*, adult with excreted droplets of haemolymph; **g** *Halyzia sedecimguttata*, adult with excreted haemolymph droplets on legs; **h** *Illeis galbula*, adult, showing strongly expanded terminal maxillary palpomere; **i** *Phrynocaria astrolabiana*, female terminalia showing glands (indicated by *arrows*); **j** *Archegleis kingi*, pupa lateral showing gin traps between abdominal tergites (indicated by *arrows*). Photographs credits: **a** Jelle Devalez; **b** Nick Monaghan; **c** Paul Zborowski; **d** Gilles San Martin; **e** Bruce Marlin; **f** Remy Ware; **g** John Jeffery; **h** Steve Axford

continents (e.g., multicoloured Asian ladybird beetle, *Harmonia axyridis* [24]).

Surprisingly, relatively little is known about the phylogenetic relationships and the evolutionary history of this ecologically important and species-rich beetle lineage. The phylogeny of the tribe has not been studied in detail and its subordinated taxonomic classification is largely regional and non-phylogenetic, impeding comparative analyses of important features of coccinellid evolution, such as host preferences, on a global scale. So far, published research on the evolutionary history of Coccinellidae has focussed on the phylogeny of the entire family and included only a very limited set of Coccinellini. The study by Magro et al. [15] included more species and genera of Coccinellini (i.e., 32 species, 15 genera) than any other investigation, but the authors' taxon sampling was heavily focused on European

species. Their data set differed from a smaller set of Asian taxa (24 species, 15 genera) analysed by Aruggoda et al. [16] and a similar sized but more global data set (23 species, 16 genera) by Robertson et al. [2]. In addition to the taxonomically different data sets, the molecular hypotheses put forth in the cited papers had very weak support especially at deeper nodes within Coccinellini, and each study recovered incongruent relationships among the genera. More comprehensive morphological and molecular research is required to improve the global classification of Coccinellini and to establish a reliable generic classification for this tribe.

Here we present molecular phylogenetic analyses based on a world-wide and taxonomically broad sampling of Coccinellini, representing all major lineages and analysing the phylogenetic signal of four genes (one mitochondrial and three nuclear) using Bayesian and Maximum-Likelihood (ML) phylogenetic approaches. The aims of our study are to: (1) assess the monophyly of Coccinellini; (2) generate the first comprehensive phylogenetic hypothesis about generic relationships within the tribe Coccinellini; (3) test if some formerly recognised tribes of Coccinellini (i.e., Discotomini, Halyziini, Singhikaliini, and Tytthaspidini) merit recognition as subtribes; and (4) reconstruct the evolution of selected morphological characters and of food preferences within Coccinellini.

Methods

Taxon sampling and morphology

We analysed 38 species of Coccinellini belonging to 32 of 90 genera. They represent all previously proposed tribes currently included in Coccinellini (i.e., Coccinellini – 23 of 67 genera, Discotomini – 1 of 5 genera, Halyziini – 3 of 8 genera, Singhikaliini – 1 genus (monotypic tribe), and Tytthaspidini (=Bulaeini) – 4 of 9 genera) and 14 outgroup species, representing a variety of coccinellid subfamilies and tribes, and two species of Corylophidae. Our taxon sampling was not designed to assess relationships within the family Coccinellidae, but was aimed at inferring the relationships within the tribe Coccinellini and tracing the evolution of morphological traits and food preferences. We selected species with known biology and food preferences, if tissue samples containing DNA were available to us. The biology of *Seladia beltiana* Gorham (former Discotomini) and *Singhikalia duodecimguttata* Xiao (former Singhikaliini) is unknown, but the examination of gut contents of two specimens of *Seladia* sp. revealed abundant fungal spores, suggesting that this species may be fungivorous (A. Ślipiński, personal observation). Gut contents of *Singhikalia duodecimguttata* Xiao from China and *S. latemarginata* (Bielawski) from Papua New Guinea showed a mixture of unrecognizable cuticular pieces and fungal conidia (A. Ślipiński, personal observation), which indicates a mixed or fungal diet.

We compiled a data matrix with essential information on food preference and the state of six morphological characters of adults and immatures for each species in our study (Additional file 1: Table S3). Morphological characters selected (adult pubescence, female colleterial glands [13], larval dorsal gland openings, larval wax secretions, and pupal gin traps) have been used as diagnostic characters for Coccinellini [8, 10, 12, 13] or (mandible type) used in discussions about the food preferred by adult beetles [25, 26] but none of these have been phylogenetically tested. Morphological characters were obtained from voucher specimens at the Australian National Insect Collection (CSIRO) and the literature [8, 9, 11]. The primary food preference (essential food source) of each species was established from the dissected guts of several representatives of each species and from the literature [8, 14, 21, 27–32].

DNA sequencing of target genes

DNA was extracted from ethanol preserved specimens following the standard protocol for animal tissues of the Qiagen DNeasy Blood and Tissue kit. Generally, one specimen per species was used for the extraction. Four nuclear and one mitochondrial gene fragments were amplified by PCR (i.e., two sections of carbamoylphosphate synthetase / aspartate transcarbamylase / dihydroorotase (CAD: CADMC and CADXM), topoisomerase I (TOPO), wingless (WGL), and cytochrome oxidase I (COI). These genes contrary to the widely used ribosomal genes (e.g. 18S, 28S) can be aligned with more accuracy. The amplification strategy [33], using degenerate primers with M13 (–21) / M13REV tails attached to the 5′ ends of the forward and the reverse primer, respectively. The primers had either been published previously [34] or were developed by us in context of the present study (CADXM2; Additional file 2: Table S1). Depending on the PCR yield, PCR products were either sequenced directly or re-amplified in a second and/or third PCR with hemi-nested and / or M13 primers. Initial PCRs were performed in 50-µL reaction volumes (32.8 µL of water, 5 µL of 10× buffer, 4 µL of 25 mM MgCl$_2$, 2 µL of 10 mM dNTP mix, 2 µL of each 10 mM forward and reverse primer, 0.2 µL of 5 U/µL KAPA taq polymerase, 2 µL of template DNA) and a touch-down temperature profile that stepped from 55 °C down to 45 °C for conveniently amplifying with all primer pairs, irrespective of their specific binding temperature, 25 cycles with 94 °C for 30 s., 55 °C [–0.4 °C each cycle] for 30 s., and 72 °C, for 60 s. [+ 2 s. each cycle], followed by 13 cycles with 94 °C for 30 s., 45 °C for 30 s, 72 °C for 120 s [+ 3 s. each cycle], followed by 72 °C for 600 s. [35]. Reamplifications also used 50-µl PCR reactions, but a simplified three-step hot-start temperature profile (22 cycles with 94 °C for 30 s., 50 °C for 30 s, and 72 °C for 60 s. [+ 2 s. each cycle], followed by 72 °C for 600 s.). All PCR products were bidirectionally sequenced using Sanger sequencing technology provided by LGC Genomics (Berlin, Germany). All raw reads were assembled with Geneious (v9.1.5; Biomatters, New Zealand [36]) and manually checked for sequencing errors, ambiguities and if necessary, manually edited.

Phylogenetic analyses

The coding DNA sequence of each gene was translated to the corresponding amino-acid sequence with the software Virtual Ribosome (version 2.0; [37]). The amino-acid sequences of each gene (CADMC, CADXM, TOPO, WGL, COI) were aligned using MAFFT (version 7.164b; [38]) and the original nucleotide sequences were mapped onto the alignments of amino-acid with a Perl script to generate a codon-based alignment of the nucleotide sequences (available upon request to AZ). The nucleotide and amino-acid multiple sequence alignments (MSAs) were visually inspected, and ambiguously aligned or gapped areas were excluded from downstream analyses (i.e., 194 of 3485 sites in the MSAs). All nucleotide sequences were queried against GenBank (NCBI [39]) using the software BLAST+ [40] to check for potential contaminations (e.g., gut content, fungi). We also inferred a neighbour-joining tree (PAUP*4.0b10, Linux, Sinauer Associates, MA, USA; [41]), from the nucleotide sequence of each gene fragment to check for potential cross-contaminations and sample swapping (the results are not shown because these were carried out on a more inclusive data set (Tomaszewska et al., in preparation)).

The five MSAs of nucleotides (CADMC: 693 bp, CADXM: 735 bp, TOPO: 678 bp, WGL: 420 bp, COI: 765 bp) were concatenated to form a supermatrix (five-fragment MSAs, Additional file 3: Supermatrix S1, 52 sequences, 3291 columns and 1571 informative sites) with a custom Perl script, that also generates character sets corresponding to the concatenated gene boundaries (available on request from AZ). To explore potential conflicting phylogenetic signal between the individual gene fragments in the concatenation, each one was excluded in turn from the MSAs and the resulting four-fragment MSAs were analysed using ML as implemented in RAxML (v8.0.26; [42]) (Additional file 4: Fig. S1a–e). The best ML topology and support values from 100 rapid bootstrap pseudo-replicates were compared to the analysis results of the five-fragment MSAs, not showing conflict among well-supported nodes (bootstrap values >85%) between topologies.

We inferred the optimal substitution models and partitioning scheme with PartitionFinder (version 1.1.1; [43]) using data blocks by gene fragment (CADMC, CADXM, TOPO, WGL, COI) and codon position as input, applying a greedy search approach with branch lengths linked across partitions and the Bayesian Information Criterion (BIC). The best partitioning scheme with corresponding substitution models (Additional file 5: Table S2) was then used to infer phylogenetic trees under the ML optimality criterion, as implemented in GARLI (version 2.01; [44]) (Additional file 6: Fig. S4). A total of 1080 heuristic tree searches were carried out on CSIRO compute cluster system, Pearcey (Dell PowerEdge M630), and the tree with the highest likelihood score selected. Bootstrap support values were obtained from 500 non-parametric bootstrap replicates with 10 heuristic tree search replicates each. Bootstrap values were mapped onto the ML tree using SumTrees (DendroPy version 3.12.2; [45]) and visualised with FigTree (version 1.4.2; https://github.com/rambaut/figtree, accessed May 8, 2015). The data was also analysed using a Bayesian method, as implemented in MrBayes (version 3.2.6; [46]) and the BEAGLE library (version 2.1.2; [47]). All model parameters, except branch lengths, remained unlinked, and two independent phylogenetic analyses were run with four chains each, sampling for 10 million generations every 1000th generation. The standard deviation of split frequencies was found to be <0.01, and convergence of the two runs was assessed using Tracer (version 1.6.0; [48]). The first 25% of the sampled trees were discarded as burn-in and the remaining sampled trees from the two runs were pooled. A 50% majority rule consensus tree with clade frequencies (posterior probabilities) was calculated with SumTrees and printed with FigTree (Additional file 7: Fig. S2).

To check for potentially detrimental influence of synonymous substitutions, the five-fragment MSA was fully degenerated with their respective genetic codes, using the software Degeneracy Coding (version 1.4; [49]). The resulting degenerated MSAs was analysed using RAxML with the same setting as for the four-fragment MSAs data set (which refers to the five gene fragment MSA less one gene fragment) (Additional file 8: Fig. S3).

Ancestral character state reconstruction
The ancestral character states of six discrete morphological and of one behavioural character (Additional file 1: Table S3) were inferred using the maximum parsimony (MP) and ML methods as implemented in Mesquite (version 3.1; [50]) and using the ML tree (Fig. 3) as backbone. The Mk1 model, also implemented in Mesquite, was used to calculate the ML probabilities of the ancestral states.

Results
Phylogenetic analyses
The ML and Bayesian phylogenetic analyses of the five-fragment MSA resulted in identical topologies (Fig. 3, ML topology, log likelihood –55,518.302879 and Additional file 7: Fig. S2, the result from the Bayesian analysis). In both cases, the topology is mostly well supported, with 30 of 49 edges having a bootstrap support value of at least 75% and 35 of 49 edges having a

posterior probability of at least 0.95 (both here subjectively regarded as "at least moderately supported").

The ML analysis of the degeneracy-coded five-fragment MSA resulted in a similar topology with 21 of 49 edges at least moderately supported (Additional file 8: Fig. S3). These 21 edges were also all present in the above topology generated from the non-degenerated data (Fig. 3). Except for the sister-group relationship between Coccinellini and Chilocorini (bootstrap values of 76% and 60% with the degenerated and non-degenerated data, respectively), support values from the degenerated data are not much higher than those from analysing the non-degenerated data set. The higher bootstraps support values of the analysis with non-degenerated data and the topological congruence between results based on non-degenerated (Fig. 3) and degenerated data (Additional file 8: Fig. S3), for at least 21 edges with moderate to very strong support, are both indicative of the utility of the synonymous changes for the estimation of the Coccinellini phylogeny. The subsequent discussion, therefore, focuses on analyses of the non-degenerated data set (Fig. 3).

Coccinellini – Monophyly and sister relationship
To assess the support for the monophyly of Coccinellini, we used a comprehensive taxon sampling that represents all previously recognized tribes of Coccinellinae: Coccinellini, Discotomini, Halyziini, Singhikaliini, and Tytthaspidini (incl. Bulaeini). The outgroup includes twelve species of ladybirds classified as Microweiseinae (two species) and the Coccinellinae tribes Chilocorini (two species), Epilachnini (two species), Aspidimerini (two species), Noviini (one species), Sticholotidini (one species) and Coccidulini (two species). In addition, we included two species of fungivorous Corylophidae as more distant outgroup taxa within the superfamily Coccinelloidea [2] (Additional file 9: Table S4). The monophyly of Coccinellini sensu *lato* [13] was strongly supported with a bootstrap value of 100% and a posterior probability of 1.0.

Despite recent attempts to establish phylogenetic relationships within Coccinellidae, there is still no satisfactory resolution within the broadly defined subfamily Coccinellinae, that would lead to a stable tribal classification [17]. In previous studies, Chilocorini and Coccinellini were repeatedly recovered as monophyletic groups and as sister taxa of each other [2, 15, 17]. Our results from the ML and Bayesian analyses are consistent with these findings, but the support is weak (Bootstrap Support (BS) 60%, Posterior Probability (PP) 0.57; Fig. 3). Only moderate support was obtained when analysing the degenerated data set (BS 76%; Additional file 8: Fig. S3). Other previously-suggested phylogenetic positions of Coccinellini [14, 16] were not supported in our analyses.

Major clades within the tribe
Our analyses recovered three strongly supported clades within Coccinellini (Fig. 3), but relationships between these

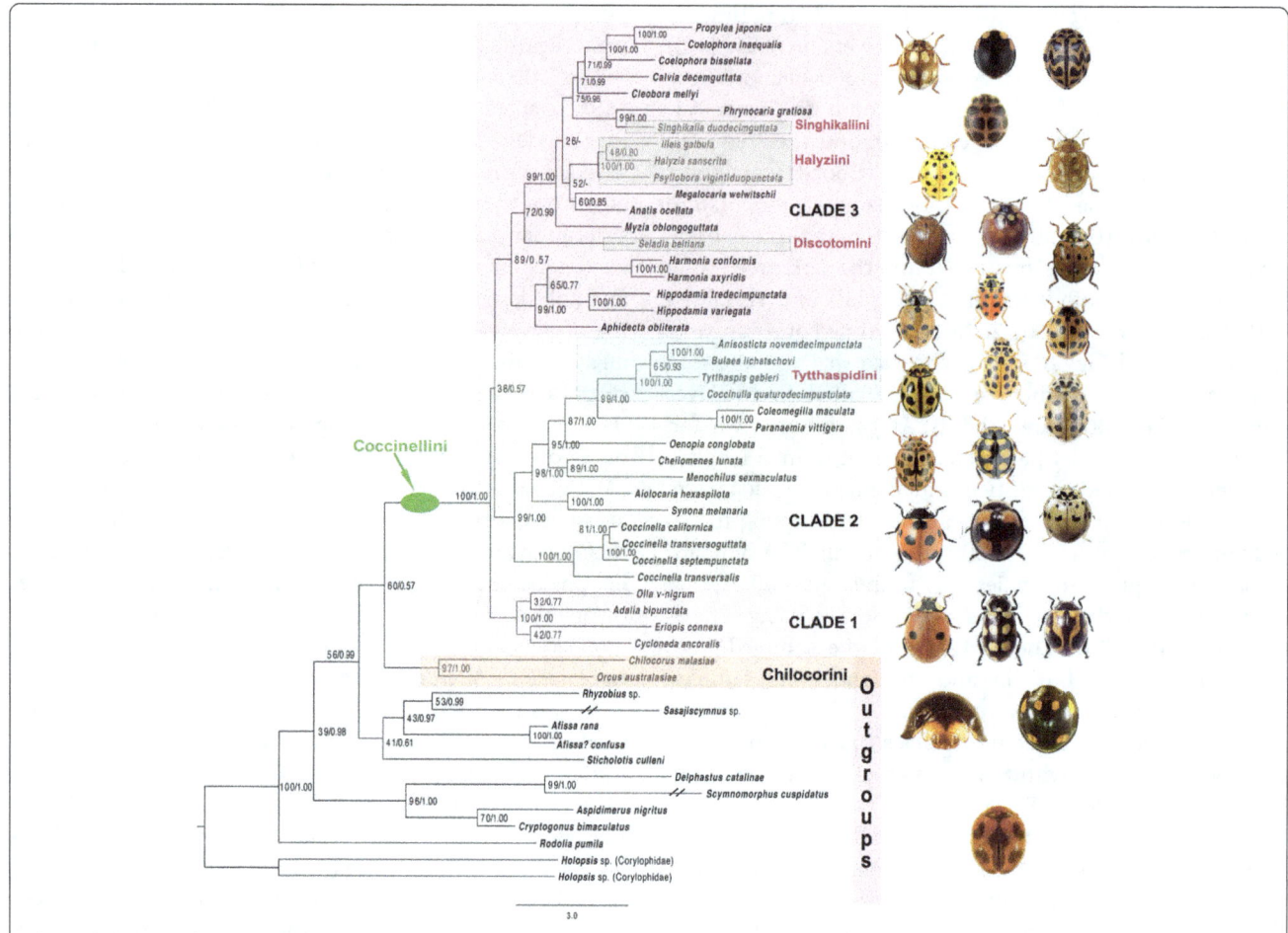

Fig. 3 Phylogeny of Coccinellini based on ML best topology; number above branches show bootstrap support and posterior probability value above 0.50. Clades 1–3 of Coccinellini are discussed in the text. Taxa formerly classified in tribes Halyziini, Singhikaliini, Discotomini and Tytthaspidini are showed in different colour

clades remain unresolved, as they are connected by short edges with low support values (i.e., BS < = 38 and PP < = 0.57). Clade 1 consists of species of the widespread genus *Adalia* and of the three New World genera *Olla*, *Cycloneda*, and *Eriopis* that are speciose in Central and South America [51]. Clades 2 and 3 comprise large radiations of primarily Old World species. Clade 2 is composed of species of the Holarctic genus *Coccinella*, of species of several genera formerly included in Tytthaspidini of species of the Holarctic genera *Coleomegilla* and *Paranaemia* (sometimes classified as Hippodamiini), and of species of the genera *Oenopia*, *Cheilomenes*, *Aiolocaria* and *Synona*. Within Clade 2, the genus *Coccinella* forms the sister group to the other species included in this clade. Clade 3 includes many genera. The Holarctic genera *Aphidecta* and *Hippodamia* and diverse Old World genus *Harmonia* constitute a well-supported group (BS 99%, PP 1.0). The Neotropical genus *Seladia* (formerly Discotomini) forms a moderately supported (BS 72%, PP 0.99) sister group to a phylogenetically unresolved complex of genera (BS 99%, PP 1.0) that

includes the Old World *Cleobora*, *Coelophora*, *Propylea*, all genera of the former Halyziini, the Chinese species *Singhikalia duodecimguttata* (former Singhikaliini) (BS 99%, PP 1.0), and the widely-distributed Indo-Australian species *Phrynocaria gratiosa*. Interestingly, very large species of Coccinellini feeding on Hemiptera, *Anatis ocellata* (aphids) and *Megalocaria* (heteropterans), form a sister, albeit weakly supported (BS 52%, PP < = 0.50) group to powdery mildew fungi feeding taxa of the former Halyziini (*Halyzia*, *Illeis* and *Psyllobora*).

Ancestral state reconstruction

The results from the ancestral state reconstruction of adult pubescence, mandible type, female colleterial glands, larval dorsal gland openings, larval wax secretions, and pupal gin traps are presented on Additional files 10, 11, 12, 13, 14 and 15: Figs. S5–S10. Both the ML and MP approaches to ancestral state reconstruction are congruent and revealed that the female colleterial glands (Additional file 10: Fig. S5) and pupal gin traps (Additional file 11: Fig. S6) were most likely present in the most recent common ancestor of Coccinellini, strongly supporting the monophyletic origin of this clade. The the common ancestor of Chilocorini and Coccinellini lacked adult dorsal pubescence (Additional file 14: Fig. S9), but it was regained in *Singhikalia*, the only known genus of Coccinellini with dorsal pubescence. The highly agile larvae of Coccinellini lack both defensive gland openings and protective waxes (Additional files 12 and 13: Figs. S7 and S8), and the ancestral state reconstruction analyses indicate that these features were lost in the most recent common ancestor of Chilocorini and Coccinellini. In our data set, larval and pupal waxes are present in only a few genera of Coccidulini (*Rodolia, Sasajiscymnus, Rhyzobius*) and appear to have evolved convergently (Additional file 13: Fig. S8).

With respect to the food preferences and associated structural modifications of the adult mouth parts (Additional file 15: Fig. S10), the ancestral state reconstruction analyses suggest that preying upon aphids is the ancestral state of Coccinellini, and that feeding on other Hemiptera, beetle larvae, mildew, spores, pollen and plant tissue has occurred multiple times independently.

Discussion

Phylogenetic analyses

In agreement with previously published molecular phylogenetic studies [2, 14–17] the monophyly of Coccinellini was resolved with high confidence in our analyses. Our studies are also consistent with the research based on nuclear and mitochondrial markers [2, 15, 17] recovering Chilocorini as the sister taxon of Coccinellini. The traditional idea of Coccinellini and Epilachnini being sister groups [9, 11], derived from studying morphological characters, remained unsupported by our analyses, as they were in other molecular analyses, which recovered Epilachnini at the base of the tree of Coccinellidae [15], within the taxa classified in Coccidulini (incl. Scymnini; [14, 16, 17]) or as sister to Coccidulini [52]. Our results (Fig. 4) suggest that the relatively large and aposematically coloured adults of aphid-feeding Coccinellini and herbivorous Epilachnini, both living on exposed surfaces and capable of strong reflex bleeding, are independently derived from smaller scale feeding ancestors. Epilachnini, which nest in our inferences within the "Coccidulinae" clade, retained densely pubescent bodies, while the last common ancestor of Coccinellini and Chilocorini lost this character (Additional file 14: Fig. S9), with the exception of the genus *Singhikalia* (former Singhikaliini), the only known pubescent Coccinellini. The genus *Singhikalia* is deeply nested within the tribe Coccinellini and represents an interesting case of convergence, possibly because it is mimicking local members of Epilachnini. *Singhikalia ornata* Kapur (India, Vietnam, Taiwan) and *S. duodecimguttata* Xiao (China) are reddish with black colour markings, while *S. latemarginata* (Bielawski) (Papua

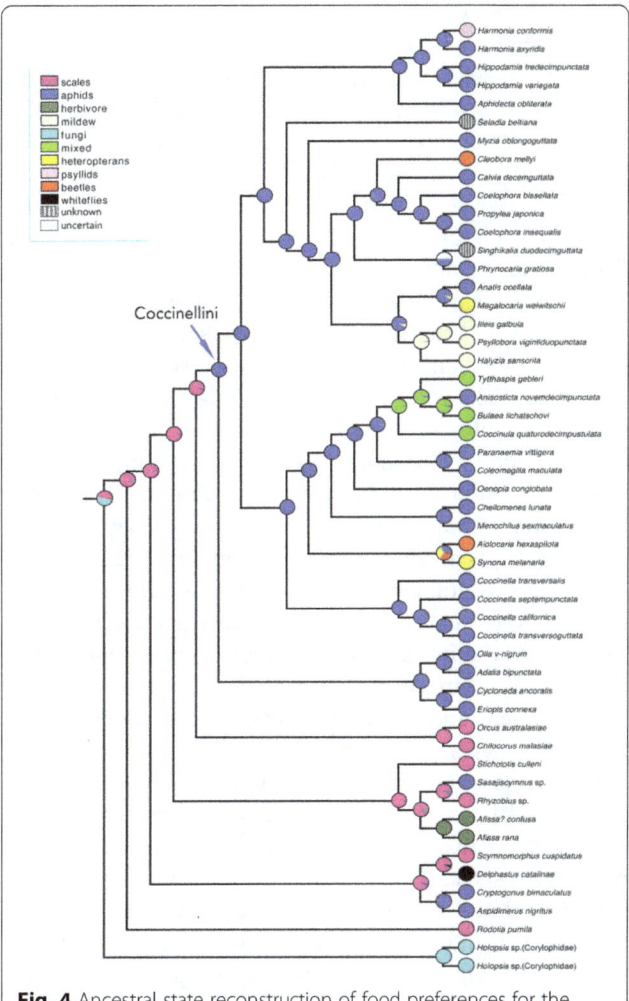

Fig. 4 Ancestral state reconstruction of food preferences for the Coccinellini based on maximum likelihood method in Mesquite

New Guinea) is almost entirely black. In this respect, all *Singhikalia* species match local members of Epilachnini very closely, to the extent that they are often found in the same series in museum collections, suggesting that they may co-occur in the same area and host plants.

The former tribe Halyziini forms strongly supported monophyletic group placed within the Clade 3 and comprises of speciose but very poorly defined Old World genera *Coelophora, Calvia, Phrynocaria,* and *Propylea,* the Asian *Singhikalia,* Old World *Megalocaria,* and species poor Holarctic genera *Myzia* and *Anatis.* In spite of taxon sampling differences the relationships between some of these taxa are in agreement with previous studies [2, 15, 16]. The second branch of the Clade 3 consists of Holarctic *Aphidecta* as a sister taxon to *Harmonia* and *Hippodamia.* The relationship between the last two genera has been recovered before [2, 16] but the placement of *Aphidecta* in this clade is a new position.

The exclusively Meso- and South American former tribe Discotomini is a poorly known group of 5 genera diagnosed by their strongly serrate or pectinate antennae. Their placement within traditional Coccinellinae has been uncertain and the molecular studies [2, 14] published so far placed Discotomini as a sister group to the remaining Coccinellini. The combined molecular and morphological analysis of Seago et al. [17] recovered *Seladia* as a sister group to the former Halyziini. We have recovered *Seladia* deeply embedded within Clade 3 at the base of large primarily Old World taxa, including former Halyziini.

The placement of the small Old World genera *Tytthaspis* and *Bulaea* in Coccinellini varied considerably in the past but their close affinity has been recognised by Iablokoff-Khnzorian [53], who pointed out similarities in male and female genitalia of several genera, later recognized as Tytthaspidini (= Bulaeini) by Kovář [9]. Most of the genera of former Tytthaspidini form strongly supported monophyletic group within the Clade 2 with *Oenopia* as the sister group, which is in agreement with Magro et al. [15].

The close relationships between *Olla*, *Adalia* and *Cycloneda* recovered in the Clade 1 has been suggested before [2, 15] but not the inclusion of the Neotropical *Eriopis* in this clade. Such arrangement suggests that the endemic New World genera and almost cosmopolitan *Adalia* have had long and independent evolutionary history from the much more diverse and speciose Coccinellini of the Old World.

As none of the previously recognised tribes, Discotomini, Halyziini, Singhikaliini, and Tytthaspidini, correspond with major clades within Coccinellini re-granting any of them subtribal status would render Coccinellini paraphyletic. Our results indicate that the newly-discovered clades, Clades 1–3 (Fig. 3), should receive recognition as formal taxonomic units, but it requires corroboration by analysis of a larger data set (Tomaszewska et al., in preparation).

Ancestral state reconstruction
Morphological characters
Coccinellini are generally recognised by relatively large adults having glabrous and convex dorsal surfaces, often with aposematic colouration, rather long and feebly clavate antennae inserted in front of large eyes, and strongly expanded "securiform" terminal maxillary palpomeres and 'handle and blade' female coxal plates [9, 11]. Most adults of Coccinellini can be distinguished by a combination of the above listed characters, but these are known to occur in taxa classified in other coccinellid groups. Ślipiński [13] expanded the list of diagnostic characters of Coccinellini, arguing that the presence of large paired reservoirs, associated with female terminalia called "colleterial glands" (Fig. 2i) and the development of the "gin traps" between abdominal tergites of the pupa (Fig. 2j) were unique developments within Coccinellidae and may constitute autapomorphies of Coccinellini. The functions of both these structures are not well understood but the secretion of the colleterial glands have been linked to mating behaviour or

egg deposition in batches [54] while the "gin traps" significantly contribute to the pupal defence [55] facilitating a quick body flicking and by creating sharp edges between segments to pinch legs or entire bodies of predatory and parasitic invertebrates to discourage oviposition or predation (Figs. 1c, 2e).

To test the hypotheses by Ślipiński [13], we traced the evolution of pupal gin traps and female colleterial glands along our main ML tree with Mesquite. Using MP and ML methods applied to character evolution, we found that the above-mentioned characters originated in the common ancestor of Coccinellini (Additional files 10 and 11: Figs. S5 and S6) and consequently regard these characters as autapomorphies of the tribe.

In addition to the above traits, we investigated the development of larval dorsal abdominal glands (Additional file 12: Fig. S7) and protective larval waxes (Additional file 13: Fig. S8), present in many groups of ladybirds, but absent in Coccinellini. The function and homology of the dorsal glands in larvae of Coccinelloidea has not been thoroughly investigated, but paired openings on abdominal tergites are present in larvae of most Corylophidae, some Endomychidae and Coccinellidae [13]. They are absent in several ladybird groups, including Coccinellini. Adults of Coccinellidae are known to reflex-bleed by excreting droplets of alkaloid loaded hemolymph to deter or tangle apparent predators. This process is less studied in ladybird larvae, but the "bleeding" from the dorsal glands has been observed in *Hyperaspis maindroni* Sicard (J. Poorani, ICAR-NRCB, India, personal information) and *Orcus bilunnulatus* (Chilocorini, A. Ślipiński, personal observation). Larvae of several species of Coccinellini have not been observed to excrete hemolymph when disturbed (A. Ślipiński, personal observation). However, this process has been documented in larvae of *Harmonia axyridis* [56, 57], with droplets originating from intersegmental membranes on most abdominal segments, and more recently in larvae of *Hippodamia variegata* (O. Nedvěd, personal observation). It is unclear whether this is a species-specific behaviour or whether it has been overlooked in other Coccinellini.

The generalised carnivorous type of the adult mandible with bifid apex and molar part bearing two unequal teeth [9], that has originated in the ancestor of Coccinellidae has been carried over with very little modification in all predatory lineages of ladybirds with several independent origins of a single sharp apical incisor in specialized scale predators (Chilocorini, Microweiseinae). All known Coccinellini have apically bifid mandibles, used to pierce their prey, suck body fluids or to masticate the entire prey [26]. The mildew or microphagous feeding taxa (former Halyziini and Tytthaspidini) have the same type of mandible with additional serration along the incisor edge (Halyziini) or relatively stiff and comb like prostheca used to scoop the spores and pollen (Tytthaspidini). Interestingly, the mandible of sometimes phytophagous *Bulaea* does not differ from the microphagous type found in *Tytthaspis*, but markedly differs from strongly modified mandibles in phytophagous Epilachnini.

Food preferences

The evolution of food preferences in Coccinellidae is a very complex issue that has received much attention due to the importance of ladybirds as biological control agents [14, 58] and, more recently, due to the recognition of the environmental impact of introduced or invading ladybirds [59] on populations of native species. Some groups of ladybirds (e.g., Noviini, Stethorini, Telsimiini and most Chilocorini) show remarkably stable food preferences, feeding mostly on taxonomically narrow groups of invertebrates [22]. Coccidophagy, preying upon scale insects, which are gregarious organisms of limited mobility, has been evolved as an ancestral food preference in Coccinellidae [14, 17]. Coccidophagous coccinellids are often morphologically and physiologically adapted to a given prey [27, 60]. But even very specialized ladybirds feed and develop occasionally on a very different host (e.g., Stethorini, which are specialised on spider mites (Tetranychidae) can develop on whiteflies [61]).

Most species of Coccinellini are "general predators", feeding in principle on aphids. Character-state reconstruction indicates that the transition from feeding on coccids to aphidophagy was acquired in the ancestor of Coccinellini (Fig. 4), but this feeding preference has independently arisen a few times in Coccinellinae (e.g., in Aspidimerini), in some genera of Coccidulini (e.g., *Coccidula, Sasajiscymnus*), Scymnini (*Scymnus*), and Platynaspidini (*Platynaspis*).

Coccinellini have diverse food preferences. While being primarily aphidophagous, they consume a broad spectrum of food that also includes other invertebrates, pollen, nectar, and often spores [14]. These opportunistic predators regularly cannibalize eggs and larvae of ladybirds, including those of their own species, and change diet depending on season and availability of prey. The gut contents of many species of Coccinellini examined during this study often consisted of predominantly sternorrhynchan Hemiptera mixed with pollen, and sometimes, with fungal spores.

Within Coccinellini, our results revealed repeated and phylogenetically independent food preference transitions from aphidophagy to other food sources (Fig. 4 and Additional file 16: Fig. S11): (a) to specialized and obligate mycophagy in the taxa classified in the former tribes Halyziini (*Halyzia, Illeis, Psyllobora*), feeding on hymenium and conidia of powdery mildew fungi (Erysiphales); (b) to a mixed diet in *Bulaea, Coccinula* and *Tytthaspis* (Tytthaspidini), with their known diet including spores of various Ascomycete fungi [32, 62], but also plant tissue (*Bulaea*), pollen (mainly of Asteraceae in *Coccinula*), acari and Thysanoptera (*Tytthaspis*); (c) to specialised predation on

nymphs of the plataspid bugs in various phylogenetically independent lineages of some *Megalocaria* [63] and of *Synona* [64]; (d) to predation on larvae of Chrysomelidae in at least Asian *Aiolocaria* [65], Australian *Cleobora mellyi* [66] and the New World *Neoharmonia* sp. [67] (N. Vandenberg, USDA-Smithsonian, USA, personal communication; not included in our data), and (e) to psyllids (Psylloidea) as the essential food of *Harmonia conformis* at least in some geographic areas [68].

Conclusions

This study represents the first molecular phylogenetic analysis of the tribe Coccinellini with a world-wide taxonomic sampling. Our phylogenetic analyses revealed strong support for Coccinellini sensu *lato* [13] being monophyletic and a sister group to Chilocorini. Three major clades were identified within Coccinellini, suggesting that Old and New World taxa, especially South American Coccinellini, have probably evolved separately. None of the major clades correspond to the previously recognised tribes Discotomini, Halyziini, Singhikaliini, or Tytthaspidini. Consequently, we suggest that these taxonomic units should no longer be used. Further testing with more taxa, especially from South America, is required to corroborate the constitution of and relationships between the three major clades of Coccinellini proposed in this study. Our study also provides an understanding of the diversification of Coccinellini and character evolution within this tribe, particularly the evolution of food preferences. The switch from obligate coccidophagy to aphidophagy in ancestral Coccinellini was accompanied by larval changes (losing dorsal defensive glands and strong dorsal ornamentation) for increased agility, and the pupae shedding larval skins completely and exposing dorsal gin traps.

Additional files

Additional file 1: Table S3. Character states of six discrete morphological and of one behavioural character. (DOCX 29 kb)

Additional file 2: Table S1. Primers used for PCR amplification of the genes. (DOCX 17 kb)

Additional file 3: Supermatrix S1. The nucleotide multiple sequence alignment (MSA), used for the phylogenetic analysis, its partitions, nucleotide composition information and percentage of gaps and ambiguities per taxa. It contains 52 taxa identified by the species name, their voucher number and the gene fragments: WGL, TOPO, COI, CADXM and CADMC. (NEX 171 kb)

Additional file 4: Figs. S1a–e. The best RAxML ML trees with bootstrap values from 100 rapid bootstrap pseudo-replicates for the four fragment MSA that lack WGL (Fig. S1a), TOPO (Fig. S1b), COI (Fig. S1c), CADXM (Fig. S1d) and CADMC (Fig. S1e). (PDF 5145 kb)

Additional file 5: Table S2. The best (BIC) multiple sequence alignment partitioning scheme with corresponding substitution models generated with PartitionFinder (version 1.1.1; [41]). (TXT 2 kb)

Additional file 6: Fig. S4. The best GARLI ML tree for the non-degenerated five-fragment MSA set with bootstrap support values (equivalent to Fig. 3 without Posterior Probabilities). (PDF 127 kb)

Additional file 7: Fig. S2. The resulting tree of the MrBayes Bayesian analysis with posterior probabilities. (PDF 132 kb)

Additional file 8: Fig. S3. The best RAxML ML tree with bootstrap values for the fully degenerated five-fragment MSA. (PDF 131 kb)

Additional file 9: Table S4. Specimens used in this study with voucher identification (Australian National Insect Collection), data, and corresponding GenBank accession numbers. (XLS 35 kb)

Additional file 10: Fig. S5. Ancestral state reconstruction based on parsimony (A) and maximum likelihood (B) for female colleterial glands in Coccinellidae. The ancestral states are present (blue) and absent (green). The topology is derived from the ML tree in Fig. 3. (TIFF 39987 kb)

Additional file 11: Fig. S6. Ancestral state reconstruction based on parsimony (A) and maximum likelihood (B) for pupal gin traps in Coccinellidae. The ancestral states are present (green), absent (blue), unknown (shadowed) and uncertain (grey). The topology is derived from the ML tree in Fig. 3. (TIFF 53439 kb)

Additional file 12: Fig. S7. Ancestral state reconstruction based on parsimony (A) and maximum likelihood (B) for larval dorsal glands in Coccinellidae. The ancestral states are present (blue), absent (green), unknown (shadowed) and uncertain (grey). The topology is derived from the ML tree in Fig. 3. (TIFF 60213 kb)

Additional file 13: Fig. S8. Ancestral state reconstruction based on parsimony (A) and maximum likelihood (B) for larval waxes in Coccinellidae. The ancestral states are present (blue), absent (green), unknown (shadowed) and uncertain (grey). The topology is derived from the ML tree in Fig. 3. (TIFF 41014 kb)

Additional file 14: Fig. S9. Ancestral state reconstruction based on parsimony (A) and maximum likelihood (B) for presence of dorsal pubescence in Coccinellidae. The ancestral states are present (blue) and absent (green). The topology is derived from the ML tree in Fig. 3. (TIFF 59853 kb)

Additional file 15: Fig. S10. Ancestral state reconstruction based on parsimony (A) and maximum likelihood (B) for mandible type in Coccinellidae. The ancestral states are separated on fungivorous (light blue), mildew (purple), carnivorous1 (red), microphagous (yellow), phytophagous (green), carnivorous2 (black). The topology is derived from the ML tree in Fig. 3. (TIFF 60492 kb)

Additional file 16: Fig. S11. Ancestral state reconstruction based on parsimony (A) and maximum likelihood (B) for food preferences in Coccinellidae. The ancestral states are separated on scales (purple), aphids (navy blue), herbivore (blue), mildew (light green), fungi (dark green), mixed (light blue), heteropterans (yellow), psyllids (orange), beetles (red), whiteflies (black), unknown (shadowed), uncertain (grey). The topology is derived from the ML tree in Fig. 3. (TIFF 58415 kb)

Acknowledgments
We are grateful to A. Vogler (Imperial College London and Natural History Museum, London) and two anonymous reviewers for their input. Thanks to C. Lemann (ANIC-CSIRO), L. Ashman (Australian National University), S. Pinzon-Navarro, O. Hlinka (CSIRO) and D. McKenna (University of Memphis) for technical help; L. Bocak, M. Bocakova, J. McHugh, K. Miller, M.A. Ivie, Y. Lee, N. Lord, J. Robertson, H. Yoshitomi and M. Whiting for providing specimens. Thanks to L. Borowiec, G. Gonzales, K. Makarov, B. Katayev, P. Zborowski, M. Yeo, J. Devalez, A. Bonnitcha, G. San Martin, N. Monaghan, M. Gulyaev, B. Marlin, J. Jeffery, R. Ware and S. Axford for permission to use their photographs or illustrations.

Funding
Funding for this study was provided by a grant from the Polish National Science Center (Narodowe Centrum Nauki), No. 2012/07/B/NZ8/02815 to W. Tomaszewska and A. Ślipiński; grant GA JU 152/2016/P by University of South Bohemia to O. Nedvěd; Alexander von Humboldt-Foundation Postdoctoral Fellowship to H. Escalona; National Natural Science Foundation of China, Grants No. 31572052 to H. Pang and No. 31501884 to X. Wang and National Basic Research Program of China (973 Program), Grant No. 2013CB127605 to H. Li and H. Pang.

Authors' contributions
AŚ, WT, HP, JL and HE conceived and designed the project. AŚ, JL, XW, DH and HE carried out field work or laboratory work or both: sorting and identifying samples, DNA extraction, PCR, sample preparation for sequencing, PCR protocols, generated DNA consensus sequence, multiple segment alignment and phylogenetic analysis. O. Nedvěd contributed with biological information. AZ and HSL conducted phylogenetic analyses. AŚ, WT and HE designed and carried out the morphological character reconstruction. AŚ and HE designed the main plates. AŚ, WT, HE and AZ wrote the manuscript with contributions from LSJ, O. Niehuis and BM. All authors approved its final submission.

Competing interests
The authors declare that they have no competing interests.

Author details
[1]Centre for Molecular Biodiversity Research (ZMB), Museum Alexander Koenig, Adenauerallee, 53113 Bonn, Germany. [2]Australian National Insect Collection, CSIRO, GPO Box 1700, Canberra ACT 2601, Australia. [3]State Key Laboratory of Biocontrol, Key Laboratory of Biodiversity Dynamics and Conservation of Guangdong Higher Education Institute, College of Ecology and Evolution, Sun Yat-Sen University, Guangzhou 510275, China. [4]College of Environment and Plant Protection, Hainan University, No. 58 Renmin Avenue, Haikou 570228, China. [5]Key Laboratory of Bio-Pesticide Innovation and Application, Guangdong Province, Guangzhou, China. [6]Centre for Biodiversity Analysis, Australian National University, ACT, Acton 2601, Australia. [7]Institute of Entomology, Biology Centre, Branišovská 31, -37005 České Budějovice, CZ, Czech Republic. [8]University of South Bohemia, Branišovská, 31 České Budějovice, Czech Republic. [9]Department of Evolutionary Biology and Ecology, Institute of Biology I (Zoology) Albert Ludwig University of Freiburg, Hauptstr. 1, 79104 Freiburg, Germany. [10]Museum and Institute of Zoology, Polish Academy of Sciences, Wilcza 64, 00-679 Warszawa, Poland.

References
1. Crowson RA. The natural classification of the families of Coleoptera. London: Lloyd; 1955.
2. Robertson JA, Ślipiński A, Moulton M, Shockley FW, Giorgi A, Lord NP, et al. Phylogeny and classification of Cucujoidea and the recognition of a new superfamily Coccinelloidea (Coleoptera: Cucujiformia). Syst Entomol. 2015; 40:745–78.
3. Robertson JA, Whiting MF, McHugh JV. Searching for natural lineages within the Cerylonid series (Coleoptera: Cucujoidea). Mol Phylogenet Evol. 2008;46: 193–205.
4. Sasaji H. In: Paper on Entomology presented to Prof. Takehiko Nakane in Commemoration of his Retirement. In: Systematic position of the genus *Eidoreus* sharp (Coleoptera: Clavicornia). Tokyo: Jpn Soc Coleopterol; 1986. p. 229–35.
5. Sasaji H. On the higher classification of the Endomychidae and their relative families (Coleoptera). Entomol J Fukui. 1987;1:44–51.
6. Mckenna DD, Wild AL, Kanda K, Bellamy CL, Beutel RG, Caterino MS, et al. The beetle tree of life reveals that Coleoptera survived end of Permian mass extinction to diversify during the cretaceous terrestrial revolution. Syst Entomol. 2015;40(4):835–80.
7. Toussaint EF, Seidel M, Arriaga-Varela E, Hájek J, Král D, Sekerka L, et al. The peril of dating beetles. Syst Entomol 2017;42(1):1–10.
8. Gordon RD. The Coccinellidae (Coleoptera) of America north of Mexico. J N Y Entomol Soc. 1985;93:654–78.
9. Kovář I. Phylogeny. In: Hodek I, Honěk A, editors. Ecology of Coccinellidae. Dordrecht: Kluwer Academic Pub. 1996. p. 19–31.
10. Vandenberg NJ. Family 93. Coccinellidae Latreille 1807. In: Arnett RH, Jr., Thomas MC, Skelley PE, Frank JH, editors. American Beetles. Volume 2. Polyphaga: Scarabaeoidea through Curculionoidea. CRC Press LLC, Boca Raton, FL; 2002. p. 19.
11. Sasaji H. Phylogeny of the family Coccinellidae (Coleoptera). Etizenia. 1968;35:1–37.
12. Sasaji H. Fauna Japonica. Coccinellidae (Insecta: Coleoptera). Academic Press of Japan, Keigaku Publishing Tokyo; 1971.
13. Ślipiński A. Australian ladybird beetles (Coleoptera: Coccinellidae): their biology and classification. Canberra: Australian Biological Resources Study; 2007.

14. Giorgi JA, Vandenberg NJ, McHugh JV, Forrester JA, Ślipiński SA, Miller KB, et al. The evolution of food preferences in Coccinellidae. Biol Control. 2009;51:215–31.

15. Magro A, Lecompte E, Magne F, Hemptinne J-L, Crouau-Roy B. Phylogeny of ladybirds (Coleoptera: Coccinellidae): are the subfamilies monophyletic. Mol Phylogenet Evol. 2010;54:833–48.

16. Aruggoda AGB, Shunxiang R, Baoli Q. Molecular phylogeny of ladybird beetles (Coccinellidae: Coleoptera) inferred from mitochondria 16S rDNA sequences. Trop Agric Res. 2010;21(2):209–17.

17. Seago AE, Giorgi JA, Li J, Ślipiński A. Phylogeny, classification and evolution of ladybird beetles (Coleoptera: Coccinellidae) based on simultaneous analysis of molecular and morphological data. Mol Phylogenet Evol. 2011;60(1):137–51.

18. Ślipiński A, Coccinellidae Latreille TW. Handbook of Zoology, Vol. 2, Coleoptera. In: RAB L, Beutel RG, Lawrence JF, editors. , vol. 2010. Berlin/New York: Walter de Gruyter GmbH & co. KG; 1802. p. 454–72.

19. Hodek I. Biology of Coccinellidae. Prague, The Hague: Academia Publishing and W. Junk; 1973.

20. Majerus M. Ladybirds. The New Naturalist Library. London: Harper Collins Publishers. 357 pp; 1994.

21. Hodek I, Honěk A. Ecology of Coccinellidae. Series Entomologica, 54. Dordrecht: Kluwer Academic Pub. 1996.

22. Hodek I, van Emden HF, Honěk A. Ecology and behaviour of the ladybird beetles (Coccinellidae). Chichester: Wiley-Blackwell. 604 pp; 2012.

23. Pettersson J. Coccinellids and semiochemicals. In: Ecology and behaviour of the ladybird beetles (Coccinellidae). John Wiley & Sons, Ltd. 2012; p. 444–64.

24. Brown PMJ, Thomas CE, Lombaert E, Jeffries DL, Estoup A, Lawson Handley L-J. The global spread of Harmonia axyridis (Coleoptera: Coccinellidae): distribution, dispersal and routes of invasion. BioControl. 2011;56:623–41.

25. Kovář I. Morphology and anatomy. In: Hodek I, Honěk A, editors. Ecology of Coccinellidae. Dordrecht: Kluwer Academic Publishers; 1996. p. 1–18.

26. Samways MJ, Osborn R, Saunders TL. Mandible form relative to the main food type in ladybirds (Coleoptera: Coccinellidae). Biocontrol Sci Tech. 1997;7:275–86.

27. Hodek I, Honěk A. Scale insects, mealybugs, whiteflies and psyllids (Hemiptera, Sternorrhyncha) as prey of ladybirds. Biol Control. 2009;51(2):232–43.

28. Kuznetsov VN. Lady beetles of the Russian far east. Center for Systematic Entomology Memoir. 1997;1:1–248.

29. Evans EW. Lady beetles as predators of insects other than Hemiptera. Biol Control. 2009;51(2):255–67.

30. Savoiskaya GI. Coccinellids of the Alma-Ata reserve. Trudy Alma Atinskogo Gosudarstvennogo Zapovednika. 1970;9:163–87.

31. Savoiskaya GI. Larvae of coccinellids (Coleoptera, Coccinellidae) of the fauna of the USSR. Leningrad: Zoologicheski Institut; 1983.

32. Sutherland AM, Parrella MP. Mycophagy in Coccinellidae: review and synthesis. Biol Control. 2009;51(2):284–93.

33. Regier JC, Shi D. Increased yield of PCR product from degenerate primers with nondegenerate, nonhomologous 5′ tails. BioTechniques. 2005;38:34–8.

34. Wild AL, Maddison DR. Evaluating nuclear protein-coding genes for phylogenetic utility in beetles. Mol Phylogenet Evol. 2008;30;48(3):877–91.

35. Regier JC, Shultz JW, Ganley ARD, Hussey A, Shi D, Ball B, et al. Resolving arthropod phylogeny: exploring phylogenetic signal within 41 kb of protein-coding nuclear gene sequence. Syst Biol. 2008;57(6):920–38.

36. Kearse M, Moir R, Wilson A, Stones-Havas S, Cheung M, Sturrock S, et al. Geneious Basic: an integrated and extendable desktop software platform for the organization and analysis of sequence data. Bioinformatics. 2012;28(12):1647–9.

37. Wernersson R. Virtual ribosome – a comprehensive DNA translation tool with support for integration of sequence feature annotation. Nucleic Acids Res. 2006;34:385–8.

38. Katoh K, Standley DM. MAFFT multiple sequence alignment software version 7: improvements in performance and usability. Mol Biol Evol. 2013;30(4):772–80.

39. Sayers EW, Barrett T, Benson DA, Bolton E, Bryant SH, Canese K, et al. Database resources of the National Center for biotechnology information. Nucleic Acids Res. 2011;39(1):D38–51.

40. Altschul SF, Gish W, Miller W, Myers EW, Lipman DJ. Basic local alignment search tool. J Mol Biol. 1990;215(3):403–10.

41. Swofford DL. PAUP*. Phylogenetic analysis using parsimony (* and other methods). Version 4. Sinauer Associates, Sunderland, Massachusetts; 2003.

42. Stamatakis A. RAxML version 8: a tool for phylogenetic analysis and post-analysis of large phylogenies. Bioinformatics. 2014;30(9):1312–3.

43. Lanfear R, Calcott B, Ho SYW, Guindon S. PartitionFinder: combined selection of partitioning schemes and substitution models for phylogenetic analyses. Mol Biol Evol. 2012;29:1695–701.

44. Zwickl DJ. Genetic algorithm approaches for the phylogenetic analysis of large biological sequence data sets under the maximum likelihood criterion. PhD thesis, The University of Texas; 2006. https://repositories.lib.utexas.edu/handle/2152/2666. Accessed 10 Jan 2014.

45. Sukumaran J, Holder MT. DendroPy: a python library for phylogenetic computing. Bioinformatics. 2010;26(12):1569–71.

46. Huelsenbeck JP, Ronquist F. MRBAYES: Bayesian inference of phylogeny. Bioinformatics. 2001;17:754–5.

47. Ayres DL, Darling A, Zwickl DJ, Beerli P, Holder MT, Lewis PO, et al. BEAGLE: an application programming interface and high-performance computing library for statistical phylogenetics. Syst Biol. 2012;61:170–3.

48. Rambaut A, Suchard MA, Xie D, Drummond AJ. Tracer v1.6. Available from http://beast.bio.ed.ac.uk/Tracer; 2014.

49. Zwick A, Regier JC, Zwickl DJ. Resolving discrepancy between nucleotides and amino-acids in deep-level arthropod phylogenomics: differentiating serine codons in 21-amino-acid models. PLoS One. 2012;7(11):e47450.

50. Maddison WP, Maddison DR. Mesquite: A modular system for evolutionary analysis. Version 2.7. 2008. http://mesquiteproject.org Accessed 5 Jun 2015.

51. Vandenberg NJ. Revision of the new world lady beetles of the genus Olla and description of a new allied genus (Coleoptera: Coccinellidae). Ann Entomol Soc Am. 1992;85:370–92.

52. Szawaryn K, Bocak L, Ślipiński A, Escalona HE, Tomaszewska W. Phylogeny and evolution of phytophagous ladybird beetles (Coleoptera: Coccinellidae: Epilachnini), with recognition of new genera. Syst Entomol. 2015;40(3):547–69.

53. Iablokoff-Khnzorian SM. Les Coccinelles, Coleopteres-Coccinellidae, tribu Coccinellini des Regions Palearctique et Orientale. Paris: Societe Nouvelle des Editions Boubee; 1982.

54. Hemptinne JJ, Leclercq-Smekens M, Naisse J. Structure and function of the exocrine glands of the genitalia of females of the two-spot ladybird Adalia bipunctata (Linnaeus, 1758) (Coleoptera: Coccinellidae). Belg J Zool. 1991;121:27–37.

55. Eisner T, Eisner M. Operation and defensive role of "gin traps" in a Coccinellid pupa (Cycloneda sanguinea). Psyche. 1992;99:225–73.

56. Sato S, Kushibuchi K, Yasuda H. Effect of reflex bleeding of a predatory ladybird beetle, Harmonia axyridis (Pallas) (Coleoptera: Coccinellidae), as a means of avoiding intraguild predation and its cost. Appl Entomol Zool. 2009;44(2):203–6.

57. Nedvěd O. Slunéčko východní (Harmonia axyridis) – pomocník v biologické ochraně nebo ohrožení biodiverzity? České Budějovice: Jihočeská univerzita; 2014.

58. Weber DC, Lundgren JG. Assessing the trophic ecology of the Coccinellidae: their roles as predators and as prey. Biol Control. 2009;51(2):199–214.

59. Roy HE, Adriaens T, Isaac N, Kenis M, Onkelinx T, San Martin G, et al. Invasive alien predator causes rapid declines of native European ladybirds. Divers Dist. 2012;18(7):717–25.

60. Hodek I. Food relationships. In: Hodek I, Honěk A, editors. Ecology of Coccinellidae. Dordrecht Boston London: Kluwer Academic Pub. 1996.

61. Biddinger DJ, Weber DC, Hull LA. Coccinellidae as predators of mites: Stethorini in biological control. Biol Control. 2009;51(2):268–83.

62. Ricci C. Sulla constituzione e funzione delle mandibole delle larve di Tytthaspis sedecimpunctata (L.) e Tytthaspis trilineata (Weise). Frustula Entomol. 1982;3:205–12.

63. Dejean A, Orivel J, Gibernau M. Specialized predation on plataspid heteropterans in a coccinellid beetle: adaptive behavior and responses of prey attended or not by ants. Behav Ecol. 2002;13:154–9.

64. Poorani J, Ślipiński A, Booth R. A revision of the genus Synona pope (Coleoptera: Coccinellidae, Coccinellini). Ann Zool. 2008;58(3):350–62.

65. Iwata K. On the biology of two large lady-birds in Japan. Trans Kansai Entomol Soc. 1932;3:13–26.

66. Bashford R. Predation by ladybird beetles (coccinellids) on immature stages of the eucalyptus leaf beetle Chrysoparta bimaculata (Olivier). Tasforest. 1999;11:77–86.

67. Whitehead DR, Duffield RM. An unusual specialized predator prey association (Coleoptera: Coccinellidae, Chrysomelidae): failure of a chemical defense and possible practical application. Coleopt Bull. 1982;36:96–7.

68. Pope RD. A revision of the Australian Coccinellidae (Coleoptera). Part 1. Subfamily Coccinellinae. Invertebr Syst. 1989;2[1988]:633–735.

7

The influence of molecular markers and methods on inferring the phylogenetic relationships between the representatives of the *Arini* (parrots, Psittaciformes), determined on the basis of their complete mitochondrial genomes

Adam Dawid Urantowka[1*†], Aleksandra Kroczak[2] and Paweł Mackiewicz[2*†] (iD)

Abstract

Background: Conures are a morphologically diverse group of Neotropical parrots classified as members of the tribe *Arini*, which has recently been subjected to a taxonomic revision. The previously broadly defined *Aratinga* genus of this tribe has been split into the 'true' *Aratinga* and three additional genera, *Eupsittula*, *Psittacara* and *Thectocercus*. Popular markers used in the reconstruction of the parrots' phylogenies derive from mitochondrial DNA. However, current phylogenetic analyses seem to indicate conflicting relationships between *Aratinga* and other conures, and also among other *Arini* members. Therefore, it is not clear if the mtDNA phylogenies can reliably define the species tree. The inconsistencies may result from the variable evolution rate of the markers used or their weak phylogenetic signal. To resolve these controversies and to assess to what extent the phylogenetic relationships in the tribe *Arini* can be inferred from mitochondrial genomes, we compared representative *Arini* mitogenomes as well as examined the usefulness of the individual mitochondrial markers and the efficiency of various phylogenetic methods.

Results: Single molecular markers produced inconsistent tree topologies, while different methods offered various topologies even for the same marker. A significant disagreement in these tree topologies occurred for *cytb*, *nd2* and *nd6* genes, which are commonly used in parrot phylogenies. The strongest phylogenetic signal was found in the control region and RNA genes. However, these markers cannot be used alone in inferring *Arini* phylogenies because they do not provide fully resolved trees. The most reliable phylogeny of the parrots under study is obtained only on the concatenated set of all mitochondrial markers. The analyses established significantly resolved relationships within the former *Aratinga* representatives and the main genera of the tribe *Arini*. Such mtDNA phylogeny can be in agreement with the species tree, owing to its match with synapomorphic features in plumage colouration.

(Continued on next page)

* Correspondence: adam.urantowka@up.wroc.pl;
pamac@smorfland.uni.wroc.pl
†Equal contributors
[1]Department of Genetics, Wroclaw University of Environmental and Life
Sciences, ul. Kożuchowska7, 51-631, Wroclaw, Poland
[2]Department of Genomics, Faculty of Biotechnology, University of Wrocław,
ul. Fryderyka Joliot-Curie 14a, 50-383 Wrocław, Poland

(Continued from previous page)

Conclusions: Phylogenetic relationships inferred from single mitochondrial markers can be incorrect and contradictory. Therefore, such phylogenies should be considered with caution. Reliable results can be produced by concatenated sets of all or at least the majority of mitochondrial genes and the control region. The results advance a new view on the relationships among the main genera of *Arini* and resolve the inconsistencies between the taxa that were previously classified as the broadly defined genus *Aratinga*. Although gene and species trees do not always have to be consistent, the mtDNA phylogenies for *Arini* can reflect the species tree.

Keywords: *Arini*, Molecular markers, Parrots, Phylogenetic methods, Phylogenetic tree, Psittaciformes

Background

The most species-rich group of the New World parrots is the subfamily *Arinae* [1, 2]. In the present taxonomy, it includes 164 species classified into 35 extant genera and one extinct genus *Conuropsis* [3]. A remarkable feature of this group is the extraordinarily high diversity in morpho-behavioural characters across its genera and even species [4]. Understandably, the exact number of genera and species belonging to *Arinae* still remains questionable and controversial. The current formal separation of some *Arinae* taxa is often based on quite intuitive and subjective morphological differences. Molecular analyses do not provide conclusive results and the phylogenetic position of many *Arinae* taxa varies depending on the published molecular datasets [5, 6]. At present, the subfamily *Arinae* is divided into the following tribes: [3] *Amoropsittacini*, *Androglossini*, *Arini* and *Forpini*. Among them, the *Arini* is represented by the largest number of genera. Of the 19 currently recognized genera, nine (*Aratinga*, *Enicognathus*, *Eupsittula*, *Guaruba*, *Leptosittaca*, *Ognorhynchus*, *Psittacara*, *Pyrrhura* and *Thectocercus*) form a morphologically diverse group called conures [7].

From the taxonomical point of view, the genus *Aratinga* is arguably the most controversial albeit noteworthy. Until now, the genus has been defined in different ways. One of the researchers identified 23 species within the genus [2]. Notwithstanding, Ribas and Miyaki [8]as well as Tavares et al. [9] called into question the monophyly of the genus. Silveira et al. [10] provided also some molecular details showing that the genus consists of three separate lineages, which were confirmed by further studies based on much broader taxon sampling [6, 11, 12]. Finally, Urantowka et al. [6] supplied the first molecular data for *Aratinga acuticaudata* and found that the species is more closely related to the representatives of three monotypic genera (*Diopsittaca*, *Guaruba* and *Leptosittaca*) than to any other member of the genus *Aratinga*. *Aratinga acuticaudata* was formerly included by Silveira et al. [10] in one of three separated clades of *Aratinga*, drawing on only some morphological features. The delimitation of the clade consisting of *Aratinga acuticaudata*, two other conures (*Guaruba*, *Leptosittaca*) and the smallest macaw (*Diopsittaca*

nobilis) resulted in reviving the monotypic genus *Thectocercus* for the former *Aratinga acuticaudata*. By virtue of the molecular revision of the broadly defined genus *Aratinga*, two additional genera, *Eupsittula* and *Psittacara*, were elevated for two of three previously recognized *Aratinga* lineages, whereas the 'true' *Aratinga* genus was established as the third lineage [7].

The currently recognized sensu stricto genus *Aratinga* comprises six South American species: *auricapillus*, *jandaya*, *maculata*, *nenday*, *solstitialis* and *weddellii*. *Aratinga solstitialis* and *Aratinga nenday* are two species which are predominantly used as representatives of this genus in parrot phylogenies. Although the taxonomic status of the genus *Aratinga* within *Arini* tribe was clarified in molecular studies, the relationships of the former *Aratinga* members still remain debatable and controversial. The most comprehensive and species-rich phylogenetic analysis of the *Arinae* subfamily, which has been performed so far, classifies the species *Aratinga* as sister to *Eupsittula* clade [4]. However, the node consisting of these two genera is supported only by a moderate Bayesian posterior probability. Other analyses showed that the species *Aratinga* are sister to the extinct Carolina parakeet [11], the species of *Eupsittula* [8] or macaws genera, such as *Ara*, *Cyanopsitta*, *Orthopsittaca* and *Primolius* [9, 12]. These discrepancies depend on the molecular markers used, among which the most popular are those derived from mitochondrial genomes. Thus, it is important to check whether mtDNA can reliably infer the species tree or phylogenies are affected by short length of sequences under study [11] and/ or incompleteness of combined alignments [4].

Summarizing, the phylogenetic position of the present species *Aratinga* is still disputable and more molecular data are required to reconstruct their precise and well-resolved phylogeny. It was demonstrated that the complete mitochondrial genomes can provide useful information for the evolutionary studies of many taxa [13]. Hence, to obtain a better resolved phylogeny of *Arini* and solve the controversies about their relationships, we carried out comprehensive phylogenetic analyses based on the representative and complete mitochondrial genomes, applying various methodological approaches. We determined the usefulness and applicability of different phylogenetic methods and various molecular markers in the phylogeny of parrots at the tribe level. Even though various molecular markers (mostly *cox1*, *cytb* and *nd2*) have been used so far in inferring the phylogenetic relationships in parrots, including *Arini*, a variety of tree topologies have been produced [4, 8, 9, 11, 12]. The phylogenetic inconsistencies may result from various evolutionary rates of individual molecular markers and the quality of phylogenetic signal. Conserved markers can contain too low variation to provide a sufficient resolution of trees, whereas the phylogenetic signal in non-conserved sequences can be blurred because of multiple substitutions and homoplasy. The biased signal in individual mitochondrial markers was shown for metazoans [14], vertebrates [15–17], amniotes [18], insects [19], teleosts [20] as well as selected groups of amphibians [21, 22] and mammals [23–25]. Yet, such systematic studies were not undertaken for birds, although reconstruction of their phylogenetic relationships relied upon one or several, often arbitrarily selected, mitochondrial markers. This is why it is necessary to assess systematically the performance and usefulness of individual markers in phylogenetic studies of this group, and to verify if mtDNA phylogenies can credibly infer its species trees. To address these problems, we considered the case of the parrot tribe *Arini* and checked the consistency between the markers as well as how, on the strength of these markers, the phylogenies correspond with morphology.

Methods

Sequence data

The molecular markers under study were derived from ten complete mitochondrial genomes of the representative genera *Arini* (Table 1, see Additional file 1). We created two sets of taxa to ascertain also the influence of taxon sampling on the reconstructed phylogenetic relationships. The 7-taxa

Table 1 Complete mitochondrial genomes of parrots analysed in the study

Species	Abbreviation	Accession	Length [bp]	Reference
Amazona barbadensis	Ab	JX524615	18,983	[84]
Ara glaucogularis	Ag	JQ782215	16,983	[85]
Aratinga solstitialis	As	JX441869	16,984	[62]
Eupsittula pertinax	Ep	HM640208	16,980	[50]
Guaruba guarouba	Gg	JQ782217	17,008	[86]
Orthopsittaca manilata	Om	KJ579139	16,985	[87]
Primolius couloni	Pc	KF836419	16,995	[88]
Psittacara mitratus	Pm	JX215256	16,984	[89]
Pyrrhura rupicola	Pr	KF751801	16,994	[90]
Rhynchopsitta terrisi	Rt	KF010318	17,027	[91]
Thectocercus acuticaudatus	Ta	JQ782214	16,998	[6]

set included *Ara glaucogularis* and members of the former broadly defined genus *Aratinga*, namely *Aratinga solstitialis*, *Eupsittula pertinax*, *Phyrrhura rupicola*, *Psittacara mitratus* and *Thectocercus acuticaudatus*. The 11-taxa set contained in addition *Guaruba guarouba*, *Orthopsittaca manilata*, *Primolius couloni* and *Rhynchopsitta terrisi*. *Amazona barbadensis* from the closely related tribe *Androglossini* was used as an outgroup. Its inclusion allowed us to determine the branching order and the basal taxon within the ingroup *Arini*.

We analysed all mitochondrial markers, both separately and in concatenated sets: 13 protein-coding genes, 12S and 16S rRNA, 22 tRNA sequences and the control region (CR) (Additional file 2). Drawing on the correspondence analysis (CA) performed in Statistica [26], tRNA gene sequences were clustered into two sets according to their nucleotide composition, which was related with their location on heavy and light DNA strands (Fig. 1a). Finally, we investigated 18 single markers (or their groups in the case of tRNAs) as well as the concatenated set of all nucleotide markers (called ALL set) and the set of all markers, excluding the control region (called ALL-CR) – Additional file 2.

The homologous sequences were aligned in MAFFT v7.215 using slow and accurate algorithm L-INS-i with 1000 cycles of iterative refinement [27]. The alignments were then edited manually in JalView [28]. Sites suitable for phylogenetic studies from the alignments of rRNAs and CR were selected in GBlocks 0.91b [29]. To maximize the number of these sites, we applied less stringent criteria, i.e. smaller final blocks, gap positions within the final blocks and less strict flanking positions, as it is illustrated on the web site http://molevol.cmima.csic.es/castresana/Gblocks_server.html.

Phylogenetic and statistical analyses

We applied eight phylogenetic approaches for nucleotide data sets: two Bayesian (BA) analyses in MrBayes [30] and PhyloBayes [31], two maximum likelihood (ML) methods in TreeFinder [32] and PAUP [33], as well as maximum parsimony (MP) and distance methods: neighbour joining (NJ), minimum evolution (ME) and weighted least squares (WLS), as implemented in PAUP [33].

The best fitting models were applied for particular alignments (Additional file 2). For each individual protein-coding gene, we used separate nucleotide substitution models for each codon position in MrBayes and TreeFinder. In the concatenated sets ALL and ALL-CR, all the first, second and third codon positions from 12 protein-coding genes (PCGs) were examined as three separate partitions. The three codon positions of *nd6* were considered as three additional separate partitions because the sequence of this gene showed a biased nucleotide composition in comparison to other PCGs (Fig. 1b), which was related to the location of these genes on the heavy

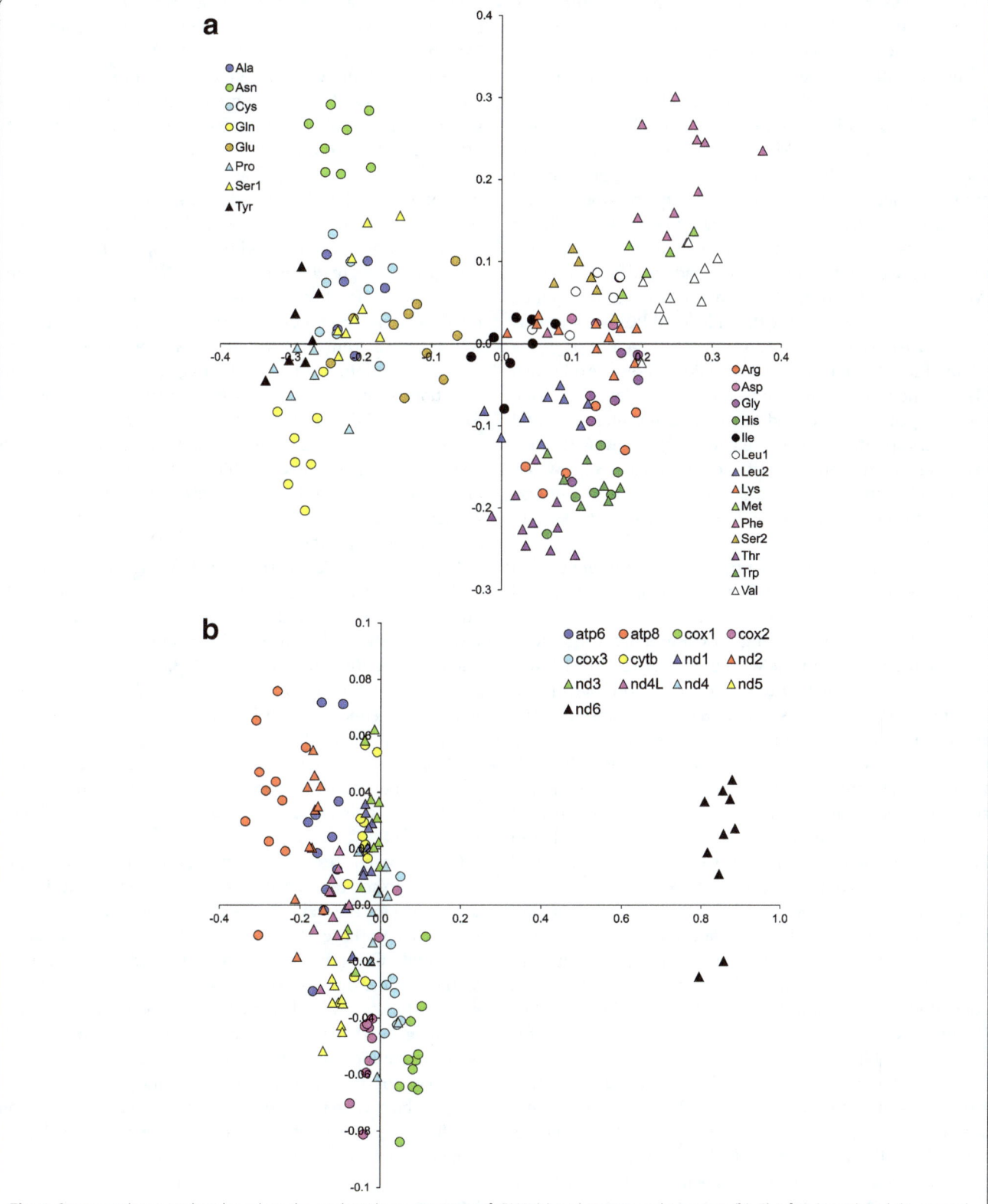

Fig. 1 Correspondence analysis based on the nucleotide composition of tRNA (**a**) and protein-coding genes (**b**). The first (X axis) and the second (Y axis) dimensions explain 67.0% and 26.9% variance for **a** panel, and 97.1% and 1.9% variance for **b** panel, respectively

and light DNA strands, respectively. The remaining five markers (two rRNAs, two tRNAs sets, CR) were also described by their own substitution models.

The best fitting nucleotide substitution models in PAUP analyses, were selected from 1624 models in jModelTest 2.1 [34], whereas in TreeFinder, the models were chosen according to Propose Model module in this program (Additional file 2). In MrBayes analyses, we applied mixed models rather than fixed ones to specify appropriate substitution models across the large parameter space [35], but the models describing heterogeneity rate across sites (invariant and discrete gamma) were adopted according to jModelTest (Additional file 2). In PhyloBayes, we applied CAT + GTR + Γ model for the individual PCGs, as well as ALL and ALL-CR data sets, whereas GTR + Γ model was adopted for the remaining nucleotide markers (Additional file 2). The CAT is an infinite mixture model accounting for site-specific amino-acid or nucleotide preferences and then appropriate for concatenated sequences with sites evolving in different substitution patterns [31]. When gamma-distributed rate variation across the sites was implemented, we approximated it by five discrete rate categories.

To reduce a potential compositional heterogeneity in sequences related with AT or GC bias, we recoded respective nucleotides for purines (R) and pyrimidines (Y) [36–38]. The RY-coding was applied for the ALL data set and the alignments were then analysed in MrBayes and TreeFinder, assuming the selection of both partitioned and not-partitioned sets. In MrBayes, we adopted the assumptions as described above, whereas in TreeFinder, we selected the two-state model for nucleotides (GTR2) with the best fitting assumptions for rate variation across sites, i.e. gamma and invariant models. We also performed analyses in PhyloBayes assuming CAT + Poisson + Γ and Poisson + Γ models. Moreover, we analysed the recoded alignments in PAUP using ML, WLS, ME, NJ and MP methods with Cavender's and Felsenstein's models [39] combined with gamma and invariant models (CF + Γ + I), because such variant appeared best fitting according to AIC (Akaike Information Criterion) [40], AICc (the corrected Akaike Information Criterion) [41] and HQC (Hannan–Quinn Information Criterion) [42] criteria for 7-taxa and 11-taxa sets.

In MrBayes analyses, two independent runs starting from random trees, each using four Markov chains, were applied. The trees were sampled every 100 generations for 10,000,000 generations. In the final analysis, we selected the last 22,360 to 78,090 trees, depending on the data set, that reached the stationary phase and convergence (i.e. when the standard deviation of split frequencies stabilized and was lower than the proposed threshold of 0.01). In PhyloBayes, the number of components, weights and profiles of the applied models were inferred from the data.

Two independent Markov chains were run for 50,000 generations with one tree sampled for each generation. Depending on the data set, the last 5000 to 45,000 trees from each chain were collected to compute posterior consensus trees after reaching convergence, when the largest discrepancy observed across all bipartitions (maxdiff) was below the recommended threshold 0.1.

All possible tree topologies, corresponding to 945 unrooted and 10,395 rooted trees, were evaluated for the 7-taxa set in TreeFinder and the following PAUP approaches: maximum likelihood (ML), minimum evolution (ME), weighted least squares (WLS) and maximum parsimony (MP). For the 11-taxa set, we set the search depth at 2 in TreeFinder, whereas in PAUP, the final tree was searched from ten starting trees constructed by stepwise and random sequence addition, followed by the tree-bisection-reconnection (TBR) branch-swapping algorithm for ML, ME, WLS and MP methods. To assess the significance of the individual branches, non-parametric bootstrap analyses were performed on 1000 replicates in TreeFinder and PAUP. In the bootstrap analysis, we applied the branch-and-bound algorithm for the 7-taxa set and the TBR algorithm for the 11-taxa set in PAUP.

Topology tests were carried out in Consel [43] assuming 1000,000 replicates based on site-wise log-likelihoods calculated in TreeFinder under the best fitting substitution models for the ALL and ALL-CR data sets. We also compared competitive topologies in MrBayes using Bayes factor based on the stepping-stone method to estimate the marginal likelihood. In this approach, we assumed four independent runs, 50 steps of the sampling algorithm and 10,000,000 generations of the MCMC simulation. The distances between trees were calculated as symmetric difference of Robinson and Foulds in Treedist from Phylip package [44].

Codon-based tests of positive selection (Z-test and Fischer's Exact test) as well as the average evolutionary divergence for the individual *Arini* markers were computed in MEGA package [45, 46]. The distance was based on Maximum Composite Likelihood method, and expressed by the number of nucleotide substitutions per site. Standard error estimate was calculated by a bootstrap procedure assuming 1000 replicates. Pearson correlation coefficients were determined in Statistica [26] between the mean tree distances for the individual markers and their lengths as well as the divergence. Additionally, we calculated Pearson coefficients correlating the number of clades with the assumed support thresholds and the two latter parameters.

Results

Evolutionary diversity of mitochondrial markers

To compare evolutionary divergence among all mitochondrial markers of ten *Arini* representatives, we computed the mean number of nucleotide substitutions per site

(Fig. 2). Generally, the majority of tRNA genes were characterized by a very low number of substitutions per site among the sequences under study (from 0.015 to 0.072). The most conserved were tRNA-Ser2 and tRNA-Pro. However, tRNA-Glu and tRNA-Phe, with the divergence of 0.13 and 0.14, clearly deviated from the other tRNAs. Two rRNA genes had the divergence of 0.073 and 0.079, close to the upper limit of the conserved tRNAs. Protein coding genes evolved with a rate generally greater than RNAs. Among PCGs, genes encoding three subunits of cytochrome c oxidase (*cox1*, *cox2*, *cox3*) showed the lowest divergence of about 0.11. On the other hand, the most divergent were genes coding for NADH dehydrogenase subunits (*nd4* and *nd6*) and ATP synthase F0 subunit 8 (*atp8*). These genes showed 0.14 and 0.16 substitutions per site, respectively. The control region (CR) was characterized by the largest divergence, i.e. 0.19.

Comparison of phylogenies for individual mitochondrial markers

We obtained 144 phylogenetic trees for 18 markers using eight methodological approaches for each of the two data sets (Additional files 3 and 4). The trees with 7 taxa represented 72 different relationships and 39 topologies were inferred only once. Only two topologies were retrieved

several times: one was obtained in five approaches for the control region as well as in one method for *atp6* and *nd2* (Additional file 3), the other was found by five approaches for *cox2* and two methods for *nd3*. For the 11-taxa set, 105 different topologies were produced and 81 of them were obtained only once (Additional file 4). In this case, only one topology dominated and was retrieved by seven approaches for the control region.

To compile the results from various tree-building approaches for individual markers, we calculated for them the 50%-consensus from individual trees (Figs. 3 and 4). There was a rather weak consistency between the tree topologies inferred by various phylogenetic methods for the same molecular marker. For the 7-taxa set, the weakest agreement among these methods was for *cytb* and *nd2* genes, whose consensus trees were fully polytomous for *Arini* taxa (Fig. 3). In the case of *atp6* and *nd3*, only one clade was inferred by more than half of the methods. The fully resolved consensus trees were found for six markers. The number of clades that were predicted by all the methods was the highest for the CR and tRNA1 trees, which had three such clades out of four possible. In the case of the 11-taxa set, the greatest disagreement among the methods was for *nd6*, whose consensus tree had only three resolved clades out of

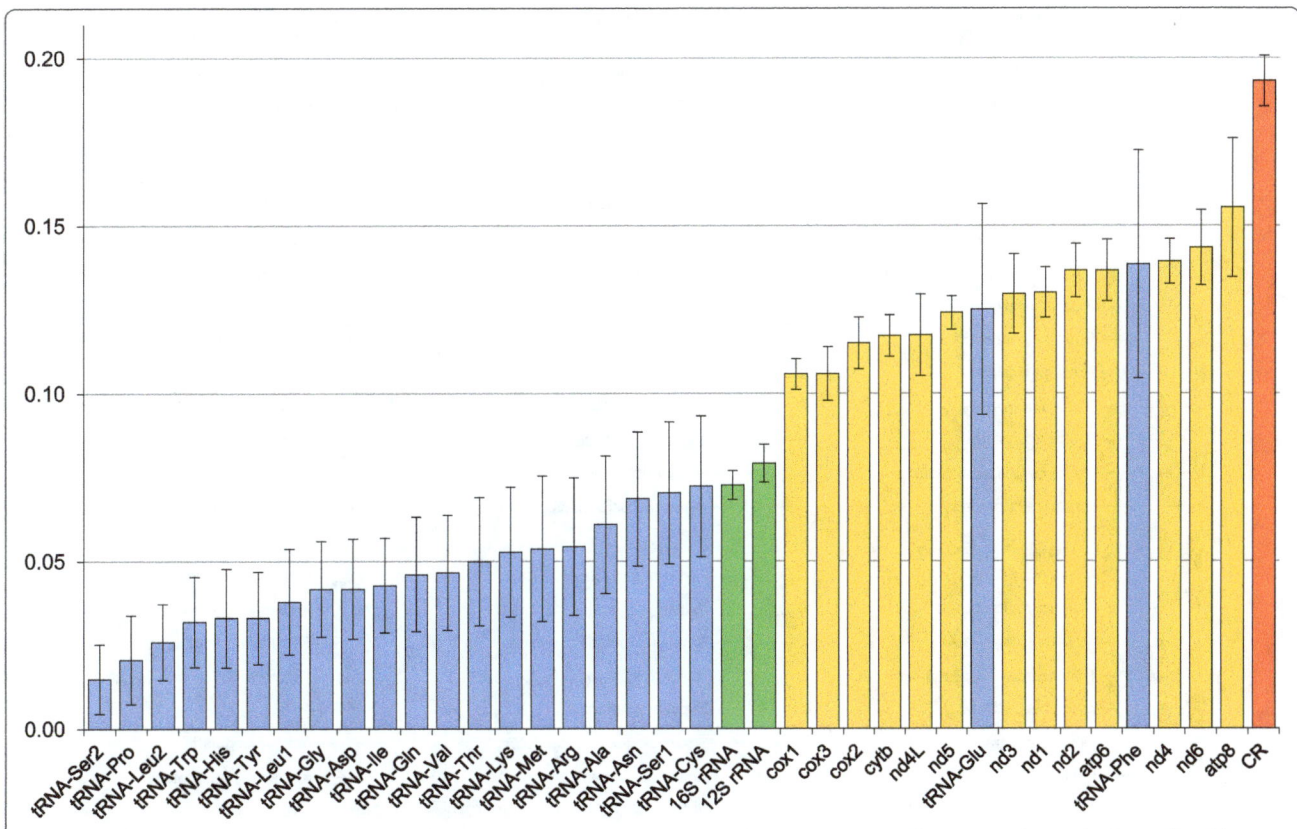

Fig. 2 The average evolutionary divergence, i.e. the mean number of nucleotide substitutions per site, for individual sets of *Arini* markers. The whiskers indicate standard error calculated using the bootstrap approach

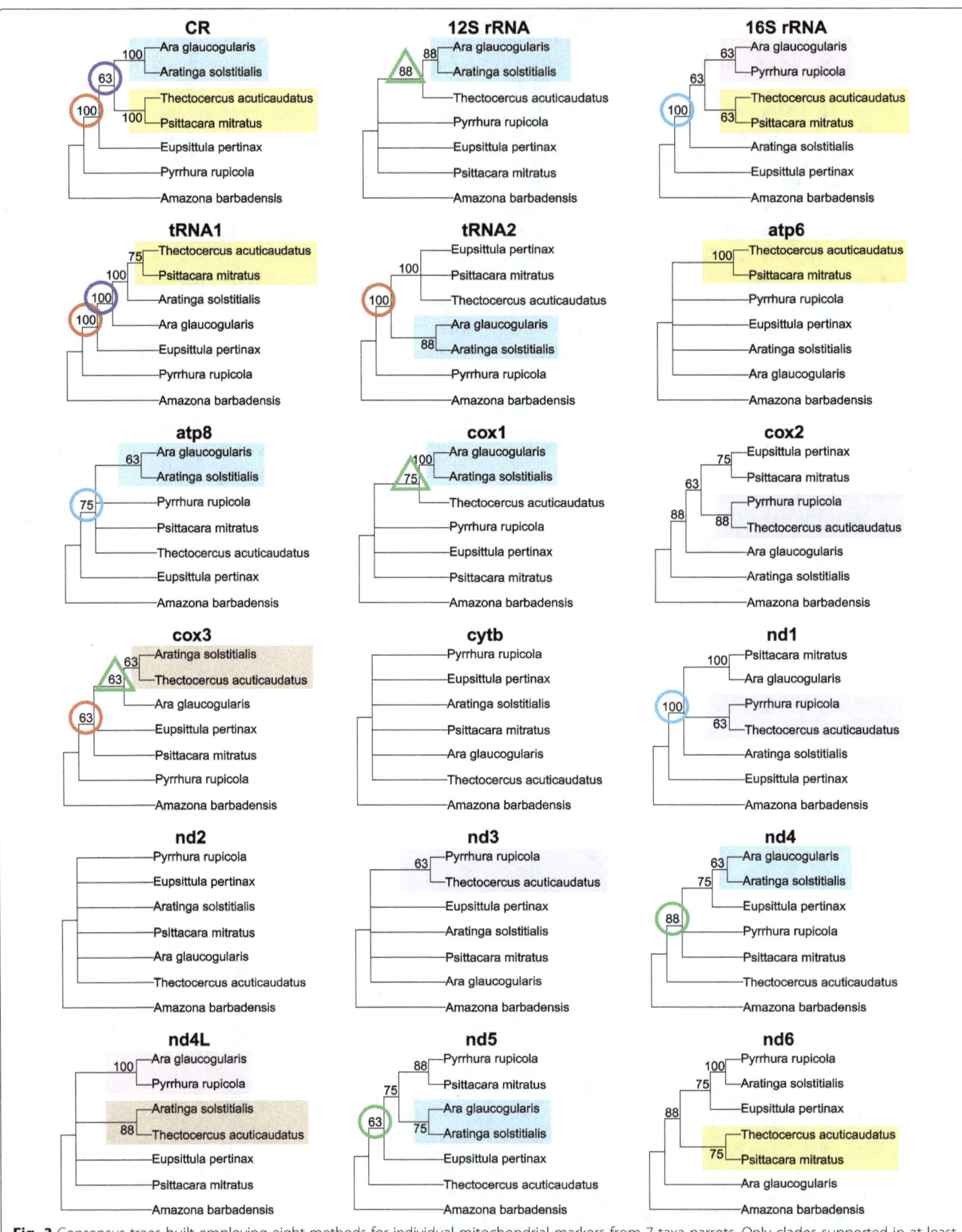

Fig. 3 Consensus trees built employing eight methods for individual mitochondrial markers from 7 taxa parrots. Only clades supported in at least 50% of trees were shown. The same groupings of taxa are marked by the same colours or graphical symbols

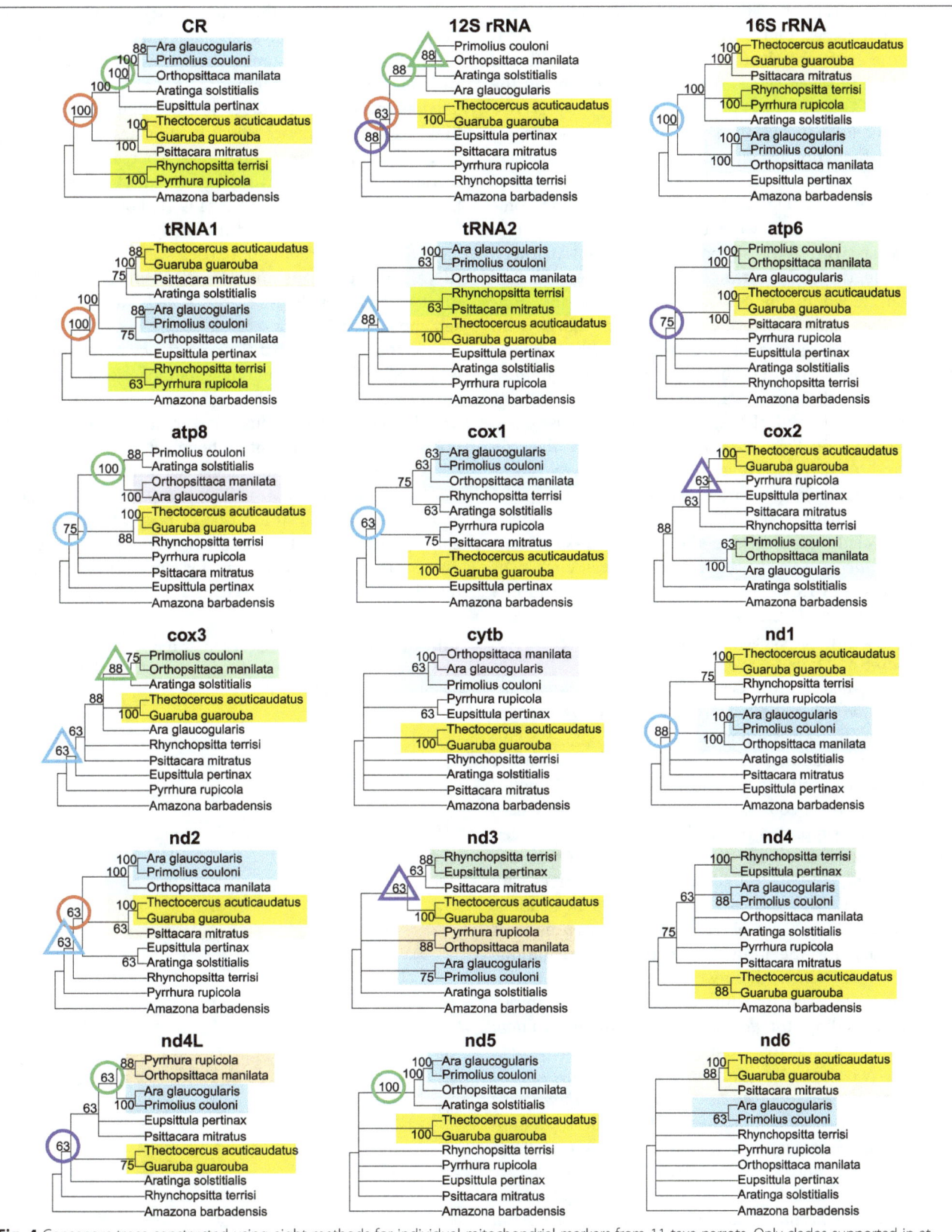

Fig. 4 Consensus trees constructed using eight methods for individual mitochondrial markers from 11 taxa parrots. Only clades supported in at least 50% were shown. The same groupings of taxa are marked by the same colours or graphical symbols

possible eight (Fig. 4). The trees for *cytb* and *nd5* had half of such clades. Only two trees, for CR and tRNA1, had all the clades resolved. The trees based on CR and 16S rRNA had seven clades that were inferred by all the methods.

A lack of full concordance was also visible when we compared the consensus trees for individual markers (Figs. 3 and 4). All of the six fully resolved trees with 7 taxa were different (Fig. 3). It was possible to find competitive groupings that were obtained by all the eight methods, i.e. with 100% support, for different markers: for example, the clade *Aratinga solstitialis* + *Ara glaucogularis* occurred in the CR and *cox1* trees, and conversely, the clade clustering *Ara glaucogularis* with *Psittacara mitratus* was in *nd1* tree or with *Pyrrhura rupicola* in *nd4L* tree. The basal placement of *Pyrrhura rupicola* with other *Arini* was supported with 100% certainty in the CR and tRNAs trees, and likewise such basal location of *Eupsittula pertinax* was in the 16S rRNA and *nd1* trees. Nevertheless, there were some groupings that were present in several consensus trees for the individual markers (Fig. 3). The most frequently occurring clade grouped together *Aratinga solstitialis* and *Ara glaucogularis*. It was present in seven consensus trees for CR, 12S rRNA, tRNA2, *atp8*, *cox1*, *nd4* and *nd5* markers. The clade *Thectocercus acuticaudatus* + *Psittacara mitratus* was in the consensus trees for CR, 16S rRNA, tRNA1, *atp6* and *nd6*.

For the 11-taxa set, the clade *Thectocercus acuticaudatus* + *Guaruba guarouba* was present in all the 18 consensus trees, with 100%-support in 15 cases (Fig. 4). This clade, with the addition of *Psittacara mitratus*, occurred in six trees. The grouping of *Ara glaucogularis* + *Primolius couloni* was also frequently recovered in 12 consensus trees and was supported by all the methods in six trees. The clade containing these two taxa, as well as *Orthopsittaca manilata*, could be found in 11 consensus trees with the full support in six trees. However, deep bipartitions were much less reproducible and all the consensus trees with 11 taxa for the individual markers were unique (Fig. 4). Consequently, there were many mutually exclusive clades with the maximum support. For example, the clade including *Ara glaucogularis* + *Primolius couloni* was in the trees for 16S rRNA, tRNA1, *nd1*, *nd2*, *nd4L* and *nd5*, whereas *Ara glaucogularis* + *Orthopsittaca manilata* in trees for *atp8* and *cytb*, and *Orthopsittaca manilata* + *Primolius couloni* in *atp6* tree. The 100%-supported clade *Rhynchopsitta terrisi* + *Pyrrhura rupicola* occurred in CR and 16S rRNA trees and its alternative clade, *Rhynchopsitta terrisi* + *Eupsittula pertinax* in *nd4* tree. *Eupsittula pertinax* was at the base to other *Arini* taxa in the tree for 16S rRNA, whereas such position was occupied by *Rhynchopsitta terrisi* + *Pyrrhura rupicola* in CR and tRNA1 trees.

We also expressed the self-consistency of the applied phylogenetic approaches for individual markers by the mean of pairwise distances between the trees built by various phylogenetic methods. The distance was calculated as the symmetric difference of Robinson and Foulds (Table 2). The distance equal 0 indicates the same topology of compared individual trees, whereas the maximum possible value for the topology in question, with seven taxa, is eight and for eleven taxa is 16. For both the 7- and 11-taxa sets, the smallest distance was for the control region. Quite congruent topologies were also inferred for tRNA2 in 7-taxa sets and 16S rRNA in 11-taxa set. The most variable topologies were for *cytb* and *nd2* in 7-taxa set, and *nd6* in 11-taxa set.

One could assume that the consistency of tree topologies is related to the length and the evolutionary rate of molecular markers. However, the correlation between the mean tree distances for individual markers and their lengths was not statistically significant, although it was not unexpectedly negative, namely, the longer the sequence, the smaller the stochastic error: $r = -0.270$, $p = 0.279$ (for 7-taxa set) and $r = -0.364$, $p = 0.137$ (for 11-taxa set). Similarly, the correlation between the mean tree distances and the mean number of substitution per site for alignments of individual markers was low and not statistically significant: $r = 0.165$,

Table 2 Mean distances calculated as the symmetric difference of Robinson and Foulds between trees constructed using different methods for particular markers

Marker	Data set	
	7 taxa	11 taxa
12 s rRNA	4.00	7.96
16 s rRNA	3.21	1.89
atp6	3.62	6.39
atp8	5.24	6.42
cox1	3.49	8.06
cox2	3.83	5.80
cox3	4.62	6.39
CR	1.07	0.50
cytb	5.86	8.25
nd1	2.96	6.07
nd2	5.76	6.57
nd3	5.29	8.32
nd4	3.79	6.47
nd4L	3.66	7.73
nd5	4.31	6.04
nd6	3.88	8.89
tRNA1	2.97	3.72
tRNA2	1.56	8.07

The minimum distance = 0 and maximum distance = 8 for 7-taxa set and 16 for 11-taxa set

$p = 0.512$ for 7-taxa set and $r = -0.150$, $p = 0.553$ for 11-taxa set.

To check if the different phylogenetic approaches can generate similar topologies, we calculated 50% consensus trees constructed by the individual methods for all markers. However, the consensus trees with 7 taxa were fully polytomous for almost every method used. Only the maximum parsimony method resolved two clades, but with the marginal support of ten markers out of 18. The consensus trees based on the 11-taxa set were also poorly resolved. Only three clades of closely related taxa were produced. *Thectocercus acuticaudatus + Guaruba guarouba* clade was recovered by all the methods with the support of 17 or 18 markers. The second most frequent clade of *Ara glaucogularis + Primolius couloni* was retrieved for 10 to 12 markers by all the methods, with the exception of maximum parsimony. The clade containing the above-mentioned two taxa and *Orthopsittaca manilata* did not appear in the consensus tree of minimum evolution and neighbour joining trees, and in others was supported by 10 to 13 markers. The other clades were not resolved. It seems to indicate that the phylogenetic signal in individual markers is extremely variable and that the impact of selecting a given tree-building method rather than another is negligible.

Significance of phylogenies for individual mitochondrial markers

We also counted the number of clades in the trees for the individual markers, which were supported by posterior probability (PP) or bootstrap percentage (BP) for three levels of significance: PP > 0.70 or BP > 50%, PP > 0.95 or BP > 70% as well as PP > 0.99 or BP > 90%. Among all the 568 identified nodes for all the methods, all the markers and the 7-taxa set, there were 194 cases (34%) that satisfied the first level of significance. For the other two levels, the number of the supported nodes drastically decreased to 59 (10%) and 12 (2%), respectively. The inclusion of 11 taxa did not improve the support because for all 1152 possible nodes there were only 294 (26%), 128 (11%) and 69 (6%) cases for the three corresponding significance levels. The mean numbers of clades in trees based upon individual markers were presented in Table 3. For the topology in question with 7-taxa, there are four possible clades that can receive any support. The largest number of clades, i.e. 3.4 on average, with the first level of significance occurred for the trees based on the control region. The worst supported trees were reconstructed for *nd2* and *nd5* genes. More significantly supported clades (with PP > 0.95 or BP > 70%) occurred only for eight markers. The CR trees again had three such clades on average. However, highly supported clades with the third level of significance were found only in CR and tRNA2 trees. Similarly, the trees with 11 taxa based on only the control region contained a substantial number of significant nodes with PP > 0.70 or BP > 50%, i.e. 7.3 on average, out of eight possible nodes (Table 3). The worst markers for the first level of significance were *nd3*, *nd4L* and *nd6*. Nodes with PP > 0.95 or BP > 70% were absent from the trees based on *nd3*, *nd4L* and tRNA1. The highest level of significance was obtained usually by single nodes for the majority of markers, except for CR, whose trees had 3.6 such nodes on average.

For the trees with 11 taxa, we found a significant correlation between the length of the markers and the number of the clades with the nodes supported by three levels of significance: PP > 0.70 or BP > 50% ($r = 0.668$, $p = 0.002$), PP > 0.95 or BP > 70% ($r = 0.568$, $p = 0.014$) and PP > 0.99 or BP > 90% ($r = 0.610$, $p = 0.007$). The correlation was also significant between the sequence divergence and the second ($r = 0.510$, $p = 0.031$) and the third level of significance ($r = 0.481$, $p = 0.043$). However, we did not observe such significant relationships for data with 7 taxa.

Phylogenetic analyses based on concatenated sets of all markers in comparison with single markers

Since the trees for individual markers were not fully resolved and showed contradictory relationships among the taxa under study, we carried out phylogenetic analyses using concatenated alignments of all markers to enhance the phylogenetic signal and reduce the stochastic error. The eight approaches produced only two tree topologies for 7 taxa, assigned as A and B in Fig. 5. The first one was favoured by two Bayesian methods in MrBayes and Phylo-Bayes as well as two maximum likelihood methods in Tree-Finder and PAUP, whereas the second tree was produced by maximum parsimony and three distance methods, i.e. neighbour joining, minimum evolution and weighted least squares, in PAUP. In both trees, *Pyrrhura rupicola* had a basal position to the other *Arini* representatives, *Aratinga solstitialis* was grouped together with *Ara glaucogularis*, whereas *Psittacara mitratus* with *Thectocercus acuticaudatus*. The only difference between these topologies existed in the case of the placement of *Eupsittula pertinax*. In the topology A, it was sister to the clade of *Psittacara mitratus + Thectocercus acuticaudatus*, whereas in the topology B, it diverged before the split of *Aratinga solstitialis*, *Ara glaucogularis*, *Psittacara mitratus* and *Thectocercus acuticaudatus*.

For the concatenated alignment of all markers with 11 taxa, we arrived at three topologies called F, G and H (Figs. 6 and 7). Similarly to the 7-taxa set, topology F was obtained by Bayesian and maximum likelihood methods, whereas topology G by distance methods and topology H only by maximum parsimony. All the trees contained the clade including four genera: *Ara glaucogularis*, *Primolius couloni*, *Orthopsittaca manilata* and *Aratinga solstitialis* (APOA clade). The clade grouping together *Thectocercus acuticaudatus*, *Guaruba guarouba*

Table 3 Mean number of clades with a given support, posterior probability (PP) or bootstrap percentage (BP) obtained by different methods for particular markers

Marker	7 taxa			11 taxa		
	PP > 0.70 or BP > 50	PP > 0.95 or BP > 70	PP > 0.99 or BP > 90	PP > 0.70 or BP > 50	PP > 0.95 or BP > 70	PP > 0.99 or BP > 90
12 s rRNA	1.88	0.00	0.00	3.38	1.38	1.00
16 s rRNA	0.88	0.38	0.00	4.25	1.25	1.00
atp6	0.63	0.00	0.00	2.50	1.25	0.38
atp8	1.13	0.00	0.00	2.88	1.13	0.13
cox1	1.25	0.34	0.00	2.50	1.13	1.00
cox2	1.25	0.00	0.00	1.75	0.75	0.38
cox3	1.25	0.00	0.00	2.13	1.13	0.63
CR	3.38	2.88	1.38	7.25	4.88	3.63
cytb	1.13	0.00	0.00	3.00	1.38	1.00
nd1	1.88	1.00	0.00	3.00	2.75	1.13
nd2	0.13	0.00	0.00	2.88	2.50	1.25
nd3	0.63	0.00	0.00	1.50	0.00	0.00
nd4	2.25	0.50	0.00	3.38	1.13	0.50
nd4L	1.13	0.50	0.00	1.00	0.00	0.00
nd5	0.25	0.00	0.00	4.25	2.38	1.25
nd6	0.63	0.00	0.00	1.63	1.00	0.88
tRNA1	2.63	1.13	0.00	2.75	0.00	0.00
tRNA2	2.00	0.63	0.13	3.63	1.38	0.38

The maximum number of possible clades that can receive any support is 4 for 7-taxa set and 8 for 11-taxa set

and *Psittacara mitratus* (TGP clade) was present in F and G trees, but H tree exhibited only *Thectocercus acuticaudatus* + *Guaruba guarouba* grouping. All the trees differed in the location of *Eupsittula pertinax*. In tree F, it was grouped together with APOA clade, in tree G it was at the base to APOA + TGP clades and in tree H it clustered with *Psittacara mitratus*. Tree F showed also common origin of *Rhynchopsitta terrisi* and *Pyrrhura rupicola* lineages, which branched off successively at the base of other trees. Comparing trees from 7- and 11-taxa sets, it can be safely concluded that only tree B and G show corresponding topologies.

Having the topologies based on the most character-rich and presumably reliable set, it was interesting to check which individual markers and methods were capable of producing such topologies. Hence, we calculated the distances between these topologies and every tree constructed by the various methods for the individual markers. Unexpectedly, among all the 144 possible combinations of markers and methods, only nine trees with 7 taxa had the same topologies as the trees based on the concatenated set (Additional file 5). The trees built by three distance methods for the control region had the same topology as the tree A, whereas Bayesian, maximum likelihood and maximum parsimony topologies based on this marker were identical with the topology B. Interestingly, only MP

tree for *atp6* showed the topology B. It is noteworthy that two tRNA sets for the many methods were capable of yielding trees quite similar, with the distance 2, to the topologies derived from all markers. The ML trees of *nd2* were also very close to the final topologies. Nonetheless, the most different topologies, with the distance 8, were very often generated because within the total of 18 markers 12 had at least two such trees. We did not notice any preference of methods in favour of such disparate topologies. Generally, the deviated trees were most often produced for *nd1*, *nd4L*, *nd3* and *cox2* sequences. Averaged for all the methods, their distance to the topologies based on the full-marker set was greater than 7.

The weak correspondence between the trees for individual markers and concatenated alignments appeared also for the trees with 11 taxa (Additional file 6). The trees exclusively based on the control region and produced by all the methods except for maximum parsimony were identical with tree F. Moreover, only tRNA1 tree inferred by weighted least squares method had the same topology as G tree. Quite small distances (2 and 4) to the F, G or H topologies were noticed in other tRNA1 trees and several trees for tRNA2, *nd2*, *nd5* and *atp6*. Still, the vast majority of the other trees based on individual markers were considerably different from those based on the concatenated alignment. Many topologies deviating from the expected topology were

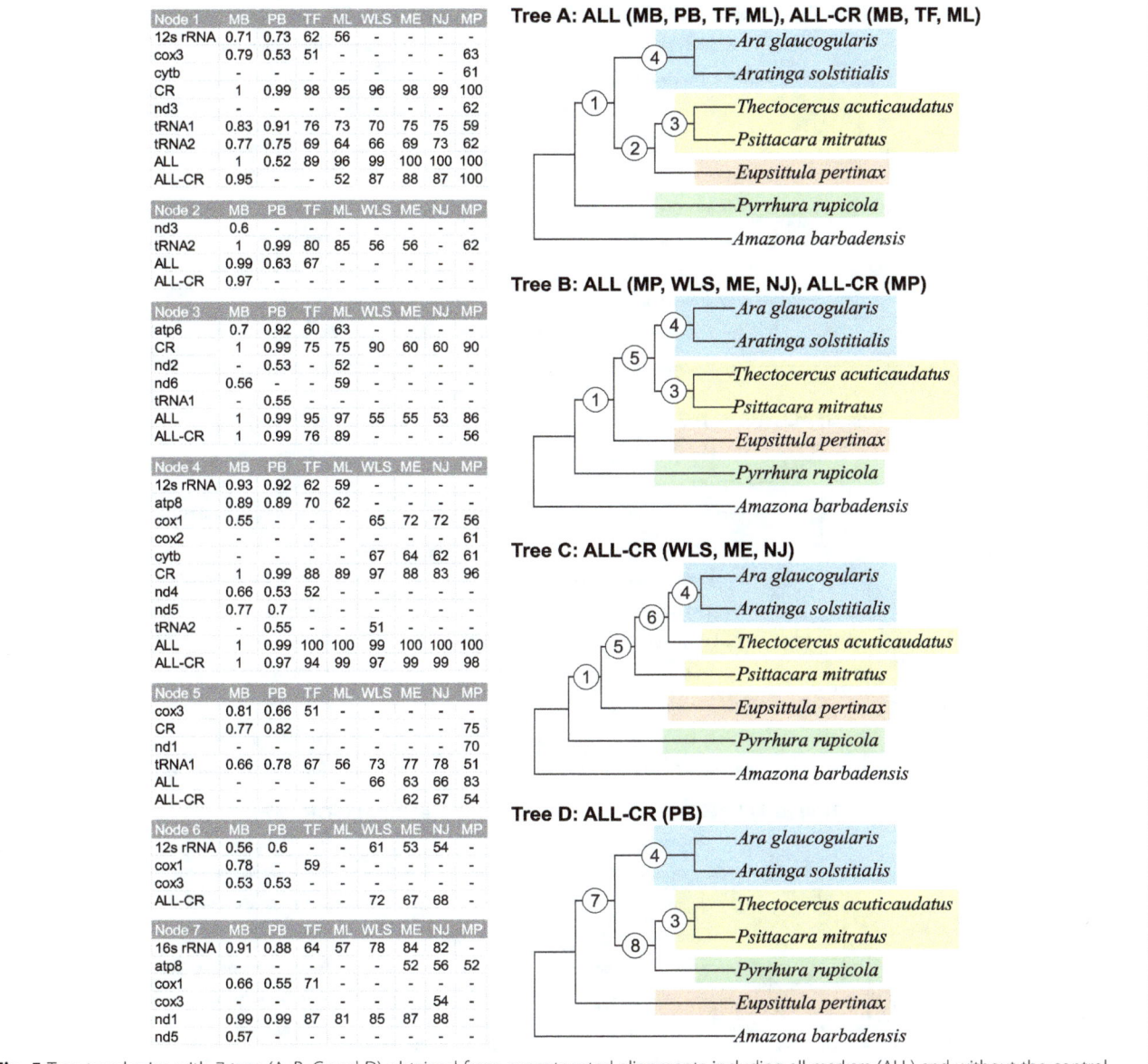

Fig. 5 Tree topologies with 7 taxa (A, B, C and D) obtained from concatenated alignments including all markers (ALL) and without the control region (ALL-CR) for the following phylogenetic approaches: Bayesian in MrBayes (MB) and PhyloBayes (PB), maximum likelihood with partitioned data in TreeFinder (TF) and not-partitioned in PAUP (ML) as well as neighbour joining (NJ), minimum evolution (ME), weighted least squares (WLS) and maximum parsimony (MP) in PAUP. The abbreviations of the data sets and methods that produced the given topology are shown above the corresponding tree. Support values at individual nodes for various combinations of alignment sets and methods are presented in accompanying tables. The posterior probabilities ≤0.5 and bootstrap percentages ≤50 were indicated by a dash "-". The node 8 received no support under these thresholds

reconstructed by virtue of *atp8*, *cox1*, *cytb*, *nd3*, *nd4* and *nd4L* sequences. Such topologies were usually generated by distance methods.

Phylogenetic analyses based on concatenated sets of markers without the control region

Since the control region alone was instrumental in recovering the tree topologies based on the concatenated sequences of all markers and in providing the best supported and consistent trees, we removed it from the concatenated alignment to find which phylogenetic signal lasted in the remaining markers. Surprisingly, we also obtained the same tree topologies as for the whole set. In the case of the 7-taxa set, the topology A was again supported by MrBayes and two maximum likelihood methods, whereas the topology B by maximum parsimony (Fig. 5). In addition, two other topologies appeared, assigned as C, supported by three distance methods, and D, produced by PhyloBayes. Both C and D indicated a close relationship between *Aratinga solstitialis* and *Ara*

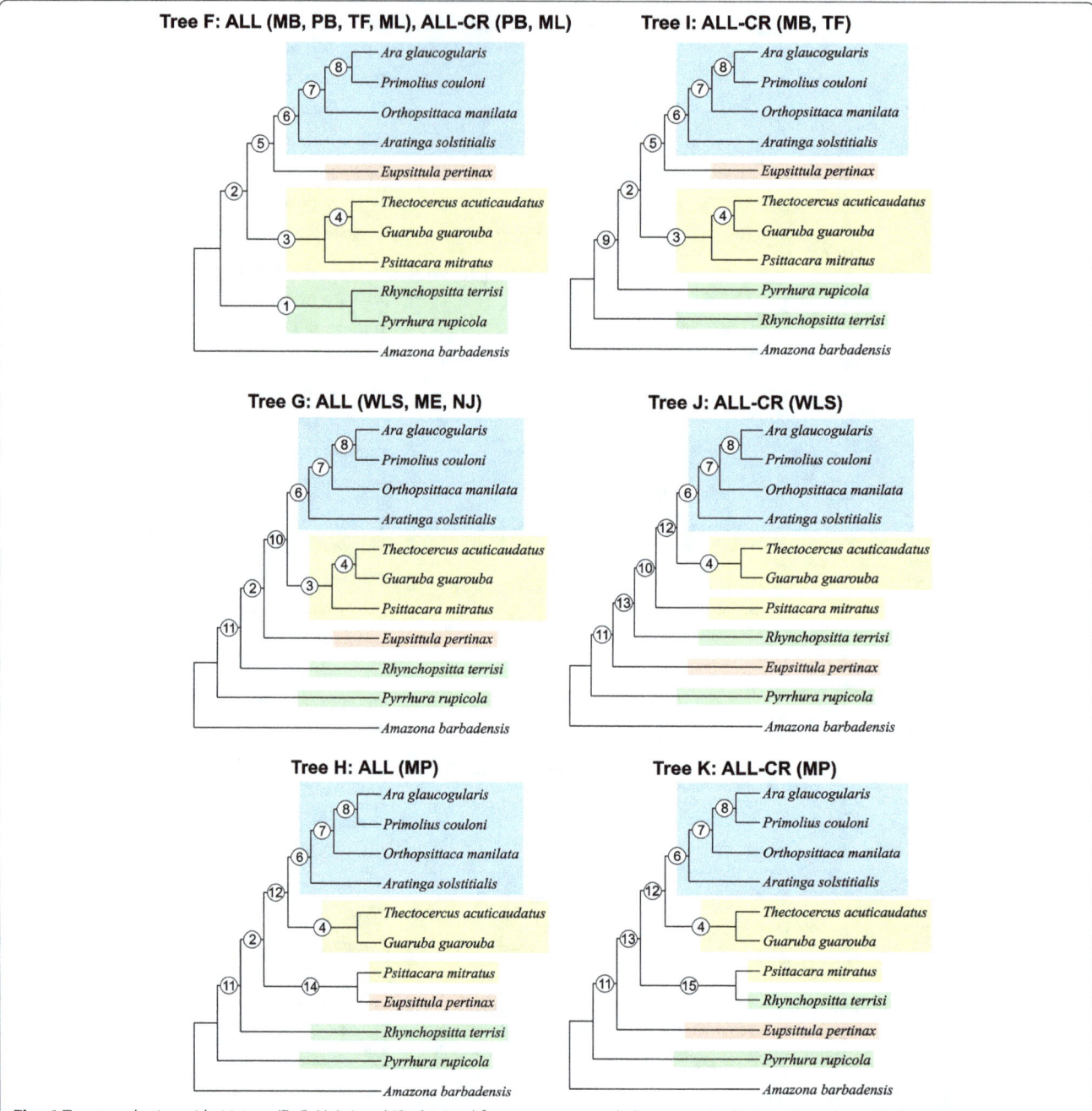

Fig. 6 Tree topologies with 11 taxa (F, G, H, I, J and K) obtained from concatenated alignments including all markers (ALL) and without the control region (ALL-CR) for the following phylogenetic approaches: Bayesian in MrBayes (MB) and PhyloBayes (PB), maximum likelihood with partitioned data in TreeFinder (TF) and not-partitioned in PAUP (ML) as well as neighbour joining (NJ), minimum evolution (ME), weighted least squares (WLS) and maximum parsimony (MP) in PAUP. The abbreviations of the data sets and methods that produced the given topology are shown above the corresponding tree. Support values at individual nodes were presented in Fig. 7

glaucogularis. The D topology contained the clade *Psittacara mitratus* + *Thectocercus acuticaudatus*, like the A and B topologies, but it was next joined to *Pyrrhura rupicola* placing *Eupsittula pertinax* basal with other *Arini*. The tree C maintained the early evolving *Pyrrhura rupicola* lineage, like A and B, but other taxa (*Eupsittula pertinax*, *Psittacara mitratus* and *Thectocercus acuticaudatus*)

branched off successively and did not create clusters like in the other topologies.

Phylogenetic analyses of concatenated markers from 11 taxa, excluding the control region, produced four topologies (Fig. 6). One of them obtained by PhyloBayes and maximum likelihood method in PAUP was the same as the topology F for the full set of markers. The other

Node 1	MB	PB	TF	ML	WLS	ME	NJ	MP
16s rRNA	0.97	0.94	61	62	69	70	69	74
CR	0.66	0.66	55	54	66	60	58	51
nd1	-	0.52	-	-	-	-	-	-
nd6	0.69	-	-	51	-	-	-	-
tRNA1	0.57	0.53	55	55	-	-	-	-
ALL	0.91	0.87	67	71	-	-	-	-
ALL-CR	-	0.62	-	52	-	-	-	-

Node 2	MB	PB	TF	ML	WLS	ME	NJ	MP
12s rRNA	0.96	0.95	58	56	-	-	-	-
atp6	0.57	-	-	-	-	-	-	-
CR	1	1	100	99	100	98	100	100
tRNA1	0.95	0.95	61	60	-	-	-	52
ALL	1	0.99	98	99	99	94	96	94
ALL-CR	1	0.65	67	60	-	-	-	-

Node 3	MB	PB	TF	ML	WLS	ME	NJ	MP
16s rRNA	0.89	0.85	-	-	-	-	-	-
atp6	0.96	0.95	70	72	56	52	54	-
CR	1	0.99	90	94	99	94	91	75
nd6	0.58	-	-	52	-	-	-	-
tRNA1	0.85	0.92	-	-	-	-	-	-
ALL	1	1	99	99	72	73	63	-
ALL-CR	1	0.99	81	93	-	-	-	-

Node 4	MB	PB	TF	ML	WLS	ME	NJ	MP
12s rRNA	1	1	100	100	100	100	100	99
16s rRNA	1	1	100	100	100	99	99	100
atp6	1	0.99	92	92	84	72	74	87
atp8	0.99	0.98	89	89	91	86	90	89
cox1	1	0.99	100	100	99	99	99	100
cox2	1	0.99	87	92	81	58	64	91
cox3	1	0.99	98	99	94	88	90	91
CR	1	1	100	100	100	100	100	100
cytb	1	1	100	100	99	96	97	100
nd1	1	0,99	100	100	100	100	100	100
nd2	1	0,99	100	100	98	98	96	100
nd3	0.55	0.83	-	68	-	-	-	53
nd4	1	0.99	97	96	71	63	59	100
nd5	1	0.94	88	89	85	73	79	99
nd6	1	0.99	91	96	98	95	95	99
tRNA1	0.67	0.68	52	51	58	-	-	64
tRNA2	0.96	0.94	60	60	68	66	60	57
ALL	1	1	100	100	100	100	100	100
ALL-CR	1	1	100	100	100	100	100	100

Node 5	MB	PB	TF	ML	WLS	ME	NJ	MP
CR	0.99	0.99	67	66	56	58	52	52
ALL	1	0.99	72	79	-	-	-	-
ALL-CR	0.7	-	-	-	-	-	-	-

Node 6	MB	PB	TF	ML	WLS	ME	NJ	MP
12s rRNA	0.98	0.97	53	60	-	-	-	-
atp8	0.93	0.95	59	62	-	-	-	-
cox3	0.97	0.94	-	64	-	-	-	-
CR	0.98	0.98	78	77	96	67	56	64
nd5	0.92	0.82	57	56	73	63	65	-
ALL	1	1	100	100	100	100	100	95
ALL-CR	1	0.99	99	100	99	98	99	88

Node 15	MB	PB	TF	ML	WLS	ME	NJ	MP
tRNA2	0.65	0.63	-	-	-	-	-	-

Node 7	MB	PB	TF	ML	WLS	ME	NJ	MP
16s rRNA	0.95	0.94	52	61	-	-	-	-
atp6	-	-	-	-	-	-	-	60
cox1	1	0.56	55	56	-	-	-	-
cox2	0.95	0.86	55	-	-	-	-	54
CR	1	1	97	94	97	97	97	89
cytb	0.51	-	-	-	-	-	-	56
nd1	1	0.97	72	74	-	-	-	-
nd2	1	0.99	95	93	89	82	83	70
nd5	1	0.99	98	98	97	94	93	89
tRNA1	0.95	0.94	54	55	-	-	-	-
tRNA2	-	-	53	-	66	57	58	70
ALL	1	1	100	100	100	100	100	100
ALL-CR	1	1	100	100	100	100	100	100

Node 8	MB	PB	TF	ML	WLS	ME	NJ	MP
16s rRNA	0.78	0.78	53	53	-	52	-	-
cox1	0.66	-	-	-	-	-	-	-
CR	0.56	0.59	51	-	66	66	65	-
nd1	0.64	0.77	75	75	-	-	-	-
nd2	0.94	0.53	86	76	78	70	71	81
nd3	-	0.55	-	51	-	-	-	-
nd4	0.8	0.8	60	75	-	-	-	56
nd4L	0.81	-	-	-	59	-	-	67
nd5	0.97	0.84	92	91	66	57	58	62
nd6	0.86	0.95	-	58	-	-	-	-
tRNA1	0.63	0.61	-	52	-	-	-	-
tRNA2	1	0.98	91	94	89	89	88	89
ALL	1	1	100	100	97	97	94	78
ALL-CR	1	1	100	100	98	96	94	91

Node 9	MB	PB	TF	ML	WLS	ME	NJ	MP
12s rRNA	0.83	0.84	68	63	-	51	51	-
atp6	-	-	-	51	-	-	-	-
cytb	-	-	53	-	-	-	-	-
nd4L	0.78	-	-	-	-	-	-	58

Node 10	MB	PB	TF	ML	WLS	ME	NJ	MP
cox3	0.75	-	-	-	-	-	-	-
nd5	0.82	-	-	-	-	-	-	-
tRNA1	0.66	0.8	-	-	-	-	-	-
ALL	-	-	-	-	65	70	64	-

Node 11	MB	PB	TF	ML	WLS	ME	NJ	MP
cox3	-	-	-	-	-	-	-	51
cytb	-	-	-	-	-	-	-	58
nd3	-	-	-	-	-	-	-	57
tRNA2	0.92	0.92	68	75	60	58	59	-
ALL	-	-	-	-	58	72	60	76
ALL-CR	-	-	-	-	72	76	76	94

Node 12	MB	PB	TF	ML	WLS	ME	NJ	MP
12s rRNA	-	-	-	-	-	53	-	-
cox3	0.61	0.61	-	-	-	-	-	-
ALL-CR	-	-	-	-	74	69	84	60

Node 13	MB	PB	TF	ML	WLS	ME	NJ	MP
nd1	1	-	-	-	-	-	-	72
ALL-CR	-	-	-	-	-	54	52	-

Node 14	MB	PB	TF	ML	WLS	ME	NJ	MP
12s rRNA	0.82	0.83	55	55	-	-	-	-
cox2	0.57	-	-	-	-	-	-	-
ALL	-	-	-	-	-	-	-	65

Fig. 7 Support values at individual nodes for various combinations of alignment sets and methods for topologies with 11 taxa presented in Fig. 6. The posterior probabilities ≤0.5 and bootstrap percentages ≤50 were indicated by a dash "-"

topology named I, retrieved by MrBayes and TreeFinder, differed from F only in the separation of *Rhynchopsitta terrisi* and *Pyrrhura rupicola*. However, the other two topologies obtained by weighted least squares and maximum parsimony methods were markedly different. Not only did they separate *Psittacara mitratus* from *Thectocercus acuticaudatus* and *Guaruba guarouba*, but also placed *Eupsittula pertinax* between *Rhynchopsitta terrisi* and *Pyrrhura rupicola*. The trees for neighbour joining and minimum evolution methods were not fully resolved.

Significance of phylogenies based on concatenated sets of markers

In Figs. 5, 6 and 7, we had additionally compiled support values for individual nodes of all ten topologies; these were based on concatenated sets of markers. Only posterior probabilities >0.5 and bootstrap percentages >50 were shown. The most firmly supported node 4, present in all four topologies with 7 taxa, joined *Aratinga solstitialis* and *Ara glaucogularis* (Fig. 5). This node corresponds with node 6 in topologies with 11 taxa (Figs. 6 and 7). These two nodes were supported with maximal or almost maximal posterior probabilities and bootstrap values for trees by all methods in both sets of the concatenated markers. Among individual markers, the control region supported this relationship quite significantly using a number of methods, too. Likewise, trees based on eight other markers from 7 taxa and four markers from 11 taxa corroborated this grouping for many phylogenetic approaches. The relationships within the clade converging to the node 6 in topologies with 11 taxa were very well resolved (Figs. 6 and 7). The same relationships were recovered in all topologies with maximal and very large support values not only in the trees based on concatenated sets but also on several individual markers. Interestingly, node 8 was only weakly supported by trees based on the control region in contrast to other markers, e.g. tRNA2.

The viability of node 3 (*Psittacara mitratus* + *Thectocercus acuticaudatus*) occurring in topologies A, B and D with 7-taxa (Fig. 5) and the corresponding node 3 in topologies F, G and I with 11-taxa (Figs. 6 and 7) was confirmed by all methods for the ALL set; only Bayesian and maximum likelihood methods produced very significant support. Again, trees for all the methods using the CR alignment contained this bipartition with moderate-to-high support. Curiously, three other markers including these nodes were the same in two sets with different number of taxa. In trees with 11 taxa, the clade supported by the node 3 included also *Guaruba guarouba*, which was clustered with *Thectocercus acuticaudatus* (Figs. 6 and 7). This grouping showed the largest support across almost all individual markers.

The node 1 separating *Pyrrhura rupicola* and other *Arini* was also present in three topologies with 7 taxa: A, B and C (Fig. 5). It was very well supported by all methods

(except PhyloBayes tree) for the ALL set. All methods for the control region and two tRNA alignments also favoured this clade, the former very significantly and the latter moderately. This node corresponds to node 2 in topologies F, G, H and I with 11 taxa, which also enjoyed a very high support from the set based on all markers and the control region (Figs. 6 and 7). In topology F, *Pyrrhura rupicola* was grouped together with *Rhynchopsitta terrisi*, which was moderately and weakly supported by Bayesian and maximum likelihood methods for the ALL set. Such a cluster appeared in trees based on five individual markers, too. Five other topologies with 11 taxa had these two species separated. In topologies G, H, J and K, *Pyrrhura rupicola* lineage diverged earlier than *Rhynchopsitta terrisi* and in topology I the reverse was true. Such arrangements, though, claimed weaker support than the clade grouping these species.

The position of *Eupsittula pertinax* was the least stable. In topology A with 7 taxa, it grouped together with *Psittacara mitratus* + *Thectocercus acuticaudatus* in node 2, which was supported by posterior probability from 0.97 to 1 in the MrBayes trees for the ALL and ALL-CR sets as well as the MrBayes and PhyloBayes trees for tRNA2 and some other methods (Fig. 5). However, such placement of *Eupsittula pertinax* was not accomplished in any trees with 11 taxa. In trees F and I based on Bayesian and maximum likelihood methods, this species was most significantly clustered with the clade of *Ara-Primolius-Orthopsittaca-Aratinga*: node 5 in Figs. 6 and 7. This node had very high posterior probabilities and moderate-to-weak bootstrap values in trees based on the control region and ALL set. In tree B and C (node 5), *Eupsittula pertinax* was sister to *Aratinga solstitialis*, *Ara glaucogularis*, *Psittacara mitratus* and *Thectocercus acuticaudatus*, making the last four taxa monophyletic. Such position was supported by, for instance, maximum parsimony and distance methods for the concatenated sets and all trees based on tRNA1. Their monophyly was also present in topologies G and J but was poorly supported. Other nodes were very weakly supported or did not gain significant posterior probability or bootstrap percentages and can be disregarded.

Topological test of *Arini* phylogenies

We carried out six topological tests for the ALL and ALL-CR data sets to assess the statistical significance of the differences between the best found tree and competitive topologies with 7 and 11 taxa (Fig. 8, Table 4). The topology A was the best for the ALL and ALL-CR sets and was significantly better (with *p*-value <0.05) in four cases than topologies B, D and E. Topology C fared poorly, compared with A in ten cases. Similarly, Bayes factor test also demonstrated substantial differences between the topology A and other topologies (Table 5). A log

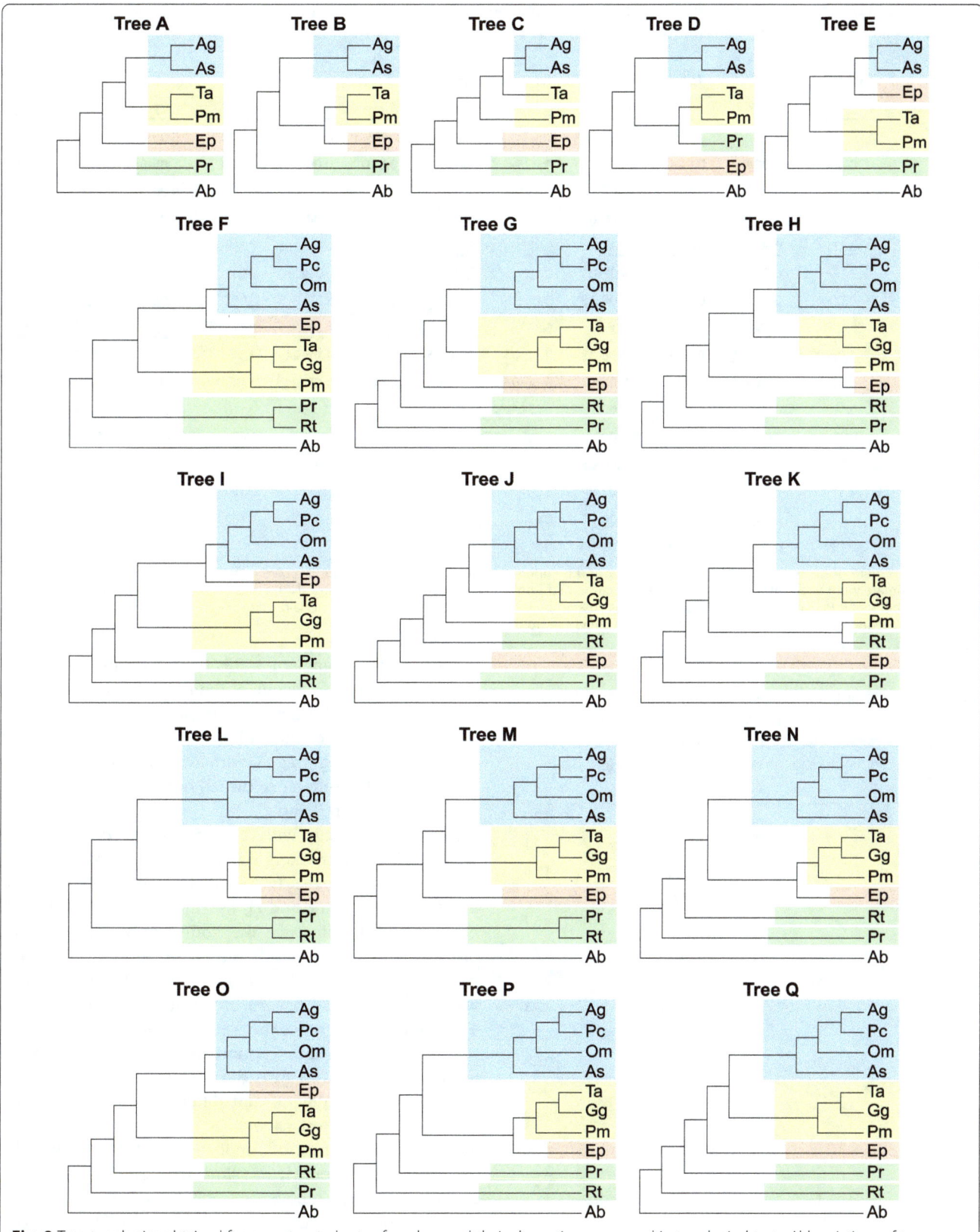

Fig. 8 Tree topologies obtained for concatenated sets of markers and their alternatives compared in topological tests. Abbreviations of taxa names are explained in Table 1

Table 4 Results of tests (*p*-values) comparing the best tree topology with alternatives for data sets with 7 and 11 taxa including all mitochondrial markers (ALL) and excluding the control region (ALL-CR)

Data set	Tree	ALL						ALL-CR					
		AU	NP	BP	PP	SH	wSH	AU	NP	BP	PP	SH	wSH
7 taxa	A	0.892	0.756	0.756	0.995	0.942	0.965	0.841	0.680	0.680	0.986	0.930	0.917
	B	0.257	0.105	0.105	**0.003**	0.525	0.486	0.183	**0.037**	**0.037**	**0.003**	0.434	0.414
	C	**0.008**	**0.004**	**0.004**	**6.0E-12**	**0.025**	**0.029**	**0.020**	**0.009**	**0.009**	**1.0E-08**	**0.050**	0.059
	D	0.075	**0.039**	**0.039**	**2.0E-06**	0.184	0.161	0.303	0.229	0.229	**0.006**	0.450	0.457
	E	0.225	0.096	0.096	**2.0E-03**	0.501	0.447	0.171	**0.045**	**0.045**	**0.004**	0.449	0.421
11 taxa	F	0.833	0.500	0.495	0.900	0.976	0.989	0.552	0.160	0.158	0.194	0.926	0.921
	G	0.133	**0.014**	**0.014**	**5.0E-05**	0.525	0.501	0.386	0.056	0.055	**0.044**	0.883	0.882
	H	**0.004**	**0.001**	**0.001**	**7.0E-15**	**0.031**	**0.022**	0.108	0.056	0.055	**6.0E-06**	0.310	0.303
	I	0.428	0.150	0.150	0.062	0.820	0.725	0.603	0.188	0.185	0.231	0.942	0.946
	J	**1.0E-07**	**4.0E-08**	**0.000**	**8.0E-38**	**2.0E-05**	**3.0E-05**	**0.001**	**1.0E-04**	**1.0E-04**	**1.0E-15**	**0.008**	**0.009**
	K	**5.0E-07**	**5.0E-08**	**3.0E-07**	**9.0E-42**	**6.0E-06**	**3.0E-05**	**0.001**	**1.0E-04**	**1.0E-04**	**4.0E-16**	**0.007**	**0.016**
	L	0.323	0.101	0.101	**0.002**	0.671	0.616	0.498	0.113	0.110	0.095	0.894	0.900
	M	0.326	0.095	0.094	**0.002**	0.702	0.643	0.436	0.080	0.078	0.066	0.889	0.885
	N	0.113	**0.009**	**0.009**	**4.0E-05**	0.497	0.474	0.398	0.054	0.053	0.058	0.886	0.896
	O	0.319	0.085	0.084	**0.034**	0.809	0.660	0.536	0.132	0.129	0.163	0.935	0.924
	P	0.172	**0.025**	**0.024**	**9.0E-05**	0.530	0.516	0.479	0.104	0.102	0.089	0.888	0.882
	Q	0.180	**0.027**	**0.026**	**1.0E-04**	0.553	0.536	0.411	0.076	0.075	0.061	0.880	0.859

The topologies were presented in Fig. 8. Underlined letters and numbers refer to the best topology for the given data set. The following tests were used: approximately unbiased (AU), bootstrap probability calculated from all sets of the scaled replicates (NP), bootstrap probability calculated from one set of replicates (BP), Bayesian posterior probability calculated by the BIC approximation (PP), Shimodaira-Hasegawa (SH) and weighted Shimodaira-Hasegawa (wSH). Values smaller than 0.05 are shown in **bold**

Table 5 Results of Bayes factor test (expressed as the difference in log likelihood units) comparing the best tree topology with alternatives for data sets with 7 and 11 taxa including all mitochondrial markers (ALL) and excluding the control region (ALL-CR)

Data set	Tree	ALL	ALL-CR
7 taxa	A	0.0	0.0
	B	**5.6**	**3.1**
	C	**24.9**	**15.8**
	D	**13.1**	2.5
	E	**5.8**	2.9
11 taxa	F	0.0	0.2
	G	**9.4**	2.2
	H	**33.6**	**12.3**
	I	1.1	0.0
	J	**86.0**	**35.8**
	K	**94.6**	**35.6**
	L	**5.8**	**3.04**
	M	**5.7**	2.7
	N	**10.7**	2.2
	O	**3.3**	1.7
	P	**9.9**	2.3
	Q	**9.0**	2.8

The topologies were presented in Fig. 8. Zero refers to the best topology. Values larger than 3 are shown in **bold**

difference in the range of 3–5 is typically considered to be a strong evidence in favour of the better hypothesis and a log difference larger than 5 is regarded as a very strong significance [47]. For the ALL set, the difference exceeded 5 when comparing tree A with others, whereas for the ALL-CR set, the difference turned out to be above 3 when comparing A with B and C.

The best topology with 11 taxa for the concatenated alignments of all markers was F. It appeared significantly better than topologies L, M and O in one test, G, N, P and Q in three tests, and H, J, K in all six tests (Table 4). No significant difference was found only between topology F and I, which differed in position of *Pyrrhura rupicola* and *Rhynchopsitta terrisi* (Fig. 8). The results are compatible with Bayes factor test, which also showed a significant difference between topology F and other topologies, with the exception of topology I (Table 5). This topology was in turn the best one for the ALL-CR set and revealed a significant difference from topologies G and H in one test as well as J and K in all six tests. The common feature of topologies H, J and K was the disruption of monophyly of *Psittacara-Guaruba-Thectocercus* (Fig. 8). These topologies were also substantially worse than topology I by virtue Bayes factor test (Table 5). Topology L was marginally worse than I with Bayes Factor 3.04.

The best topologies F and I for the 11-taxa set corresponded roughly to the alternative topology E for the 7-

taxa set, whereas the best topology A with 7 taxa was represented by the alternative topologies G, M or Q in the 11-taxa set. The differences between these best topologies and the alternatives representing the best tree from the data set with different number of taxa were statistically significant at least for the concatenated alignment of all markers.

Phylogenies based on recoded nucleotides sequences

Since a compositional bias related with A + T content may contribute to artificial relationships [36–38], we recoded respective nucleotides for purines and pyrimidines in the alignment, including all mitochondrial markers to eliminate the potential compositional heterogeneity. The applied eleven approaches produced three trees with 7 taxa and five trees with 11 taxa. However, only one tree for each of these data sets received significant support (Fig. 9). The tree with 7 taxa (Fig. 9a) was produced by

four approaches (partitioned and non-partitioned in MrBayes, non-partitioned in TreeFinder and maximum parsimony) and is identical with the topology A, which was found for the not-recoded sequences. The tree with 11 taxa (Fig. 9b) was reconstructed by five approaches (non-partitioned in MrBayes and TreeFinder, maximum likelihood, weighted least squares and minimum evolution in PAUP) and is identical with the topology F, which was also found by the largest number of methods for not-recoded sequences. Other trees based on the recoded data differed in the location of *Eupsittula pertinax* and *Pyrrhura rupicola* and/or broke the monophyly of *Pyrrhura rupicola* and *Rhynchopsitta terrisi*. However, the conflicting clades did not get posterior probabilities >0.5 or bootstrap percentages >50 and any supported clades in these trees were also present in the highly supported trees. Only a moderately supported bipartition with 67% bootstrap value was obtained

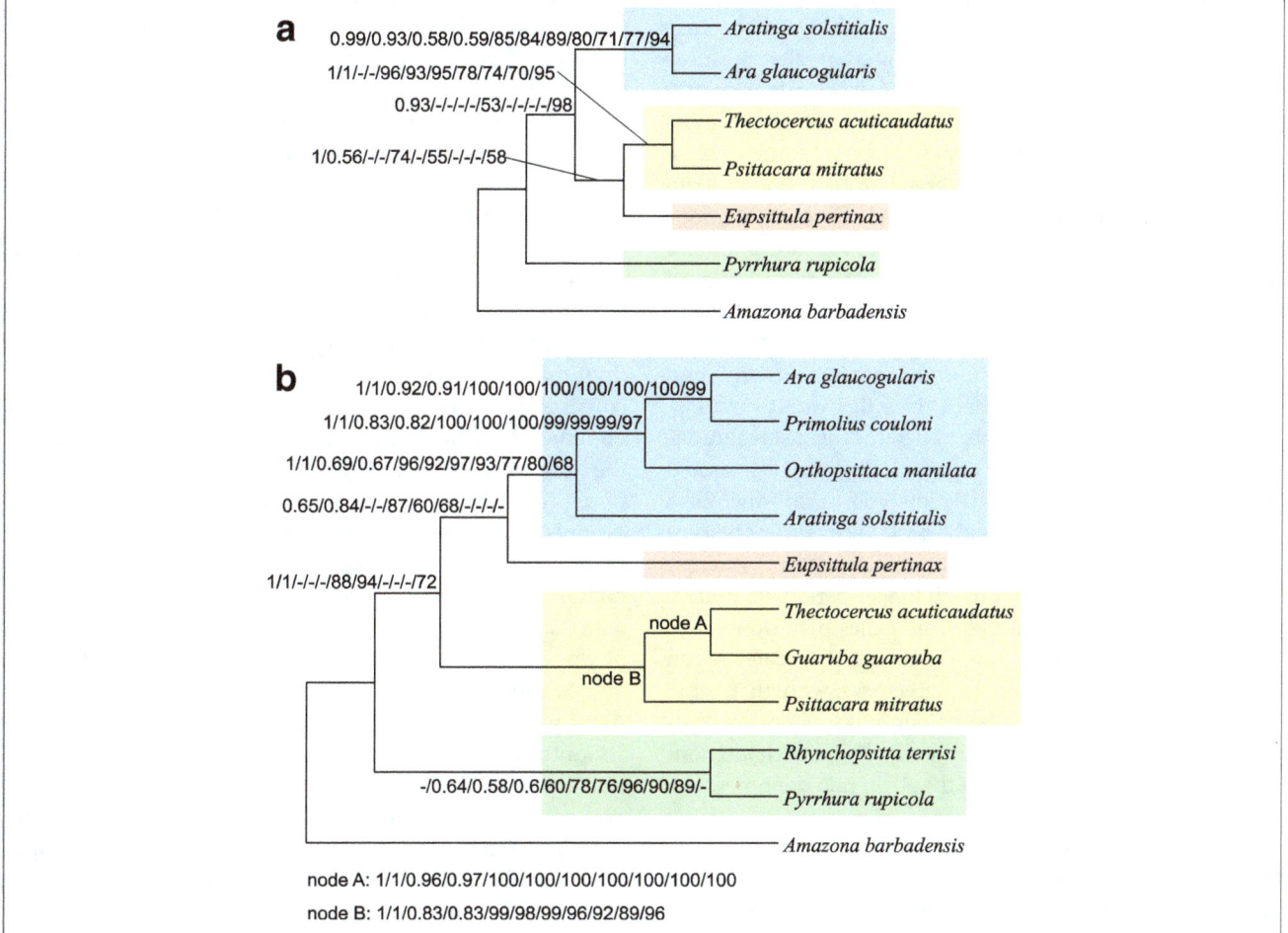

Fig. 9 Tree topologies obtained for the ALL data set subjected to RY-coding for 7- (**a**) and 11-taxa set (**b**). Numbers at nodes correspond to support values calculated using the following methods: MrBayes under partitioned and not-partitioned model, PhyloBayes assuming CAT and Poisson model, TreeFinder under partitioned and not-partitioned model, maximum likelihood, weighted least squares, minimum evolution, neighbour joining and maximum parsimony in PAUP. The posterior probabilities ≤0.5 and bootstrap percentages ≤50 were indicated by a dash "-"

for the 11-taxa set by maximum parsimony and separated *Amazona barbadensis* and *Rhynchopsitta terrisi* from other studied taxa.

Discussion

Sequence variation of mitochondrial markers

Our study showed that tRNAs are on average the most slowly evolving genes, which may be attributed to the strong constraints on their structure and function. Sequences of rRNA genes changed on average 1.7 times faster, protein-coding genes 3.1 times and the control region showed the greatest rate, which was 4.4 times higher than tRNAs. Yet, there were two outstanding genes, tRNA-Glu and tRNA-Phe, that showed divergence 3.1 times greater than other tRNAs in *Arini* mitogenomes. Two variable tRNA genes, albeit for other amino acids (Val and Ser), were found in the genomes from Epinephelidae fishes [48]. The order of the genes according to their evolutionary divergence is similar to the results calculated for mitochondrial genes in mammals [49] and birds, including parrots [50]. However, in the last group, there are some genes (*cox1* and *cox3*) that evolve more slowly than rRNAs. In our analyses of *Arini* members, all PCGs showed greater substitution rates than rRNAs. The studies of avian mitogenomes [50] seem to be in agreement with our findings: the most conserved PCGs are the two genes for cytochrome oxidase subunits I and III. The next conserved protein gene in *Arini* mitochondrial genomes was *cox2*, which, however, demonstrated much greater substitution rate in the global studies of bird and parrot mitogenomes [50]. The most diverged gene in *Arini* mitochondrial genomes was *atp8* and *nd6* was next line. In the global analyses of Aves and Psittaciformes [50], the most rapidly evolving gene was *nd2* and next *atp6*, which in *Arini* did not show the widest variation. The discrepancy is attributable to heterogeneous evolutionary rate of genes in time and various lineages, which should be included in global phylogenetic studies using members of different groups. Our study took into account short evolutionary distances (about 14 million years), whereas Pacheco et al. [50] studied a much longer period: 85 million years. The substitution rate of some genes may slow down or accelerate in the tribe *Arini*, but on a global scale we can observe an opposite effect. In agreement with that, *atp8* showed the least clock-like behaviour and was not the fastest evolving gene in the avian study [50] but it demonstrated the greatest variation in the *Arini* mitogenomes.

Performance of individual mitochondrial markers in inferring phylogenies

Using data for the complete mitochondrial genomes in the representatives of the main evolutionary lineages of *Arini*, we tried to infer their phylogenetic relationships using individual mitochondrial molecular markers as well as concatenated sets. The results for some markers

were, however, not satisfactory. The trees based on the same marker and obtained by various methods most often showed disparate evolutionary histories. Similarly, the same method applied to the various markers also produced different phylogenetic relationships. It should be emphasized that it was possible to find conflicting groupings of the same taxa that were supported by all methods for various markers. It implies that inferring phylogenetic relationships based on individual mitochondrial markers can lead to wrong and contradictory conclusions. The individual trees were usually characterized by poor resolution. Generally, only one tenth of all identified nodes for all the methods and all the markers had posterior probability >0.95 or bootstrap percentage > 70%, both for 7- and 11-taxa sets.

The sharpest disagreement of methods was registered for *cytb*, *nd2* and *nd6* genes. The trees based on *nd2*, *nd3*, *nd4L*, *nd5* and *nd6* belonged also to the worst resolved. It is remarkable that *cytb* and *nd2* markers are very often used in phylogenetic analyses, including parrots [4, 9, 11, 12]. Therefore, their results should be treated with caution. The most similar trees produced by different methods were for CR and next for tRNA and 16S rRNA genes. The trees constructed on the control region contained the largest number of well resolved nodes, too. The bias of individual markers in inferring phylogenetic relationships may be the effect of systematic and stochastic errors, too poor phylogenetic signal, unequal and differential substitution rate as well as the convergent changes in unrelated lineages resulting from similar mutational or selection pressures [21, 51–58].

The variable phylogenetic signal in various mitochondrial markers demonstrated in our study explains the inconsistent or unresolved phylogenies of *Arini* members obtained previously for various markers. For example, the tree based on *cox1* and *nd2* [11] and enhanced by two nuclear introns (*trop* and *tgfb2*) and coded gaps [12] grouped *Aratinga solstitialis* and *Ara glaucogularis* together in one clade, whereas in the tree based on *cox1*, *cytb*, *nd2* and *rag-1*, *Aratinga solstitialis* was clustered with *Eupsittula pertinax* [4]. The weak or conflicting phylogenetic signal in the used markers produced poorly or not significantly resolved relationships between other *Arini* members in the trees [4, 6, 9, 11, 12].

Contradictory tree topologies obtained from various mitochondrial markers were reported also for various groups of animals [14–25]. It could be hypothesised that the consistency of tree topologies is related to the length of molecular markers. Longer sequences with stronger phylogenetic signal could cause smaller stochastic error and trees based on them can show similar topologies regardless of the method used. Similarly, the level of agreement among tree topologies for individual markers could also depend on their evolutionary divergence. One could expect that the trees based on more conserved sequences are more stable and concordant, whereas highly variable sequences may lead to phylogenetic artefacts, which may be emphasized by some methods (e.g. long-branch attraction by maximum parsimony). On the other hand, less variable sequences may be affected by a poor phylogenetic signal and generate stochastic error. Nevertheless, our analyses demonstrated that the inconsistency and efficiency in recovering the correct nodes did not depend on the length and evolutionary rate of the markers [17, 18, 22]. Our results also revealed that the consistency of the trees computed using various methods did not significantly correlate with the length of the individual markers; on the other hand, we found a significant positive correlation between the resolution of trees, measured by number of supported nodes, and the divergence of markers measured by the mean number of substitution per site. This implies that the resolution of trees at the parrot tribe level may be improved by using fast evolving markers with sufficient number of variable and informative sites such as the control region.

The usage of single markers was also criticized because their phylogenies did not match those based on many markers or entire mitochondrial genomes [24, 59]. For example, the commonly used control region did not allow well resolved trees and failed to identify unique mitochondrial lineages in contrast to complete mitogenomes [59]. The marker selection and their number have also an influence on the molecular time estimation as it was proved for birds [60]. Generally, *cox1*, *nd1*, *nd2* and the concatenated RNAs can provide molecular time estimates most similar to those based on the complete mitogenomes. The most informative markers or group of markers needed to reconstruct reliable phylogenies vary with groups and should be selected individually not only for big lineages [14] but even for a given family or genus [23].

On the other hand, Seixas et al. [61] showed that full mitogenomes cannot provide better resolved phylogenies than single genes, which contrasts with other studies [24, 59]. The usefulness of single or numerous concatenated markers depends probably on the group under study and its level of evolutionary divergence as well as on the differences in the evolutionary rates of individual markers. For larger evolutionary distances, analysed by e.g. Seixas et al. [61], there is a larger probability that individual genes will display differentiated evolutionary patterns. Therefore, the combined contradictory signals can produce a noise rather than enhance the phylogenetic signal. However, for closely related taxa, such as those dealt with in this study, the phylogenetic signal is not so variable and the inclusion of many markers decreases the stochastic error, yielding more phylogenetic information. To alleviate the effect of the differentiated evolutionary rate of concatenated markers, it is important to apply partitioned models assuming separate substitution

patterns and rates for individual markers and codon positions in protein-coding genes. In agreement with that, we found that more sophisticated methods, i.e. Bayesian and maximum likelihood assuming partitioned models produced trees with large support values for the concatenated alignments more often than simpler methods.

Nevertheless, it is difficult to indicate an individual marker that could reliably recover the phylogenetic relationship within *Arini* although the strongest phylogenetic signal was included in the control region and tRNA genes. The trees based on these markers were identical or very similar to the final topologies inferred from all markers for some methods. Interestingly, these two types of markers are characterized by opposite evolutionary rates in mitochondrial genomes. The former is rapidly evolving, whereas the latter is conserved on average. The usefulness of tRNA genes was usually underestimated. They were not commonly used in the reconstruction of phylogenetic relationships although their usefulness was also proved for bony fishes [20] and amphibians [22]. Nevertheless, neither CR nor tRNAs may be used alone because not all the nodes based on these markers were significantly supported and the topologies depended on the method used.

Our results indicate that the most reliable phylogenetic relationship within *Arini* may be obtained on the data sets consisting of concatenated alignments of at best all mitochondrial markers. The exclusion of the control region, which provided alone the strongest phylogenetic signal, did not remove vital information about the phylogenetic relationships. The most robust methods, Bayesian and maximum likelihood, were able to recover the same tree topology as for the complete data set. It is suggestive of sufficient information also present in other mitochondrial markers.

Which phylogenetic relationships within *Arini* are reliable? To infer phylogenetic relationships within *Arini* we applied eight methods on two data sets with 7 and 11 taxa to verify also the influence of the taxa number. The resulting topologies showed consistently the presence of two highly supported clades: (1) *Aratinga solstitialis* + *Ara glaucogularis* with the addition of *Primolius couloni* and *Orthopsittaca manilata* in the case of 11-taxa set, and (2) *Psittacara mitratus* + *Thectocercus acuticaudatus* with the addition of *Guaruba guarouba* in the 11-taxa set. The clades and the relationships within them were strongly supported and occurred also in many consensus topologies derived from individual trees constructed utilising various methods for individual markers.

Such groupings are supported by some common features present in mitochondrial genomes of the *Arini* [62] (see Additional file 1). Three species in the first clade share the same stop codons in *nd5* gene, whereas all members of the second clade have the same starts and stops in all genes. Two members of the first clade have one-nucleotide spacer between tRNA-Val and 16S rRNA genes, whereas two members of the second clade share exclusively a two-nucleotide spacer between tRNA-Ala and tRNA-Asn genes. All species in the second clade have the longest 12S rRNA gene (more than 975 bp) of all studied *Arini*.

The trees based on all markers also indicated an early divergence of *Pyrrhura rupicola*. In the case of 11-taxa set, this species clustered with *Rhynchopsitta terrisi* in all Bayesian and maximum likelihood trees based on all markers and almost all trees based on RY-recoded alignment. This grouping obtained higher statistical support than the separation of these species. Therefore, the monophyly of these genera seems to be the most probable. They also share the longer C-box (a conserved motif in the control region), whereas other *Arini* have a deletion (see Additional file 1). In support of the uniqueness of this lineage and the close relationship of other *Arini* members, *Pyrrhura* and *Rhynchopsitta* genomes are characterized by the lowest and highest A + T content, as well as the most and the least biased DNA symmetry (skew) among the *Arini* under study (see Additional file 1).

Among the inferred relationships, the position of *Eupsittula pertinax* is controversial. In the most probable tree for 7-taxa set (topology A), this taxon grouped together with the clade *Psittacara mitratus* + *Thectocercus acuticaudatus*. This tree was inferred on the set of all markers by powerful methods such as Bayesian and maximum likelihood, in contrast to maximum parsimony and distance methods, which favoured the alternative topology B. The topology A, which showed also greater support values for the conflicting clade, was significantly favoured by topology tests and was recovered on the alignment that was subjected to RY-recoding to reduce compositional bias. On the other hand, in the tree F, based on 11-taxa set and also supported by the powerful methods on the set of all markers and the RY-recoding alignment, *Eupsittula* was sister to the clade including *Ara glaucogularis*, *Primolius couloni*, *Orthopsittaca manilata* and *Aratinga solstitialis*. This clustering was also statistically more significant in support values and topological tests than alternative placements of *Eupsittula*. Tree A and F, including these two groupings, were much more probable than other competing topologies. These results illustrate that taxon sampling may influence tree topology. It is difficult to point to the most probable relationship on the basis of these data because support values for them are comparable. The topology F seems to be more probable because it is based on the set containing a larger number of taxa. The difficulties in determining phylogenetic relationships within *Arini* may result from a rapid divergence and differentiation of its members, as evidenced by short internal branches in phylogenetic trees.

As far as mitochondrial phylogenies are considered, we should be aware that mtDNA might not reflect the true species trees. The disagreement between the mtDNA and species tree may stem from introgression and incomplete lineage sorting following recent speciation [63]. However, these processes do not have to occur in birds. Following Haldane's rule [64], the introgression of maternally inherited mtDNA is restricted between heterogametic avian species because female hybrids are characterized by a reduced viability [65–70]. Nevertheless, according to Rheindt and Edwards [71], it is true only for loci that are not subjected to natural selection. If the premise of neutrality is violated, the positive selection can lead to mtDNA introgression. Therefore, we tested the hypothesis about the positive selection for each of 13 protein-coding genes from the 11-taxa set, using Z-test and Fischer's Exact test with all available models in the MEGA package [46]. In each case, we found that the null hypothesis of neutrality cannot be rejected and the alternative hypothesis of the positive selection should not be accepted with p-value = 1. It suggests that a sufficient number of non-synonymous substitutions did not accumulate during the evolution of the tribe *Arini* to cause the selection-dependent introgression.

Since mitochondrial markers are single-copy genes, they can be less susceptible to the incomplete lineage sorting than the nuclear ones. The nuclear genes more often duplicate and their different copies can disappear in various lineages, which can lead to hidden paralogy and disagreement between gene and species trees [54, 72–79]. However, the incomplete sorting may influence mtDNA if multiple population divergences or speciation events were closely spaced in time [80], which cannot be definitely excluded for the *Arini* taxa studied here.

The application of nuclear markers could provide additional information about phylogenetic relationships among parrots. However, the resolution of phylogenetic relationships based on available nuclear markers is much worse than mitochondrial ones [9, 81, 82]. For example, an alignment of nuclear RAG1 gene sequences from 36 *Arini* representatives has only 3.6% of variable sites (98/2544 bp), whereas an alignment based on mitochondrial *nd2* sequences from the same taxa has 49.9% (446/894 bp) of variable sites.

Comparison of phylogenetic relationships with morphological data

Even a well-resolved phylogenetic tree does not have to reflect true relationships between the taxa under study, i.e. the species tree. To check to what extent the mtDNA phylogenies may correspond to the real *Arini* relationships, we confronted the phylogenetic results with the morphological data. Curiously, drawing on the molecular markers the proposed most probable relationships do not contradict the classification based on the colouration of the wings and tail, which were believed by Remsen et al. [5] to discriminate the new genera of this tribe.

Greater primary coverts of *Aratinga* are violet blue and its tail is olive-green tipped violet [5]. The sister taxon to *Aratinga* is *Ara*, which shows greater differentiation. Apart from four species that are predominantly green, two others are mostly blue and yellow [2]. Two additional species are characterized by a domination of red colouring [2]. However, greater primary coverts of all the eight *Ara* species are blue and the tail is also blue-tipped, regardless of interspecies differences in rectrices colouration [2]. *Primolius* have also blue colouration of greater primary coverts and tail tip [83]. Although *Orthopsittaca* is mostly green, its greater primary coverts are also bluish [83]. It seems that the blue or violet-blue colouration of greater primary coverts is a common feature of the clade including *Aratinga* and three macaws, and could be present in its common ancestor. *Eupsittula*, which is sister to this clade in the trees with 11 taxa, also has blue greater wing coverts, like *Aratinga*. However, the blue is paler, reduced to a faint tinge in cactorum and does not extend to the outer primary coverts [5].

In contrast to *Aratinga*, *Eupsittula* and three macaw genera, the greater primary coverts, as well as primaries and secondaries of *Psittacara*, *Thectocercus* and *Guaruba* are green [83]. However, other body parts, including the tail, are mainly yellow in adult *Guaruba* (Fig. 10b) unlike *Psittacara* and *Thectocercus*, which are predominantly green. *Thectocercus*'s tail is additionally red in its inner webs. Nevertheless, it is quite probable that *Guaruba*'s yellow colouration evolved from an ancestral green-coloured state because *Guaruba* chicks and juvenile individuals have mixed yellow and green plumage (Fig. 10a). Moreover, some double-coloured feathers can be found even in individuals much older than juveniles (Fig. 10c). Therefore, it seems probable that the green colouration is an ancestral feature of the clade *Psittacara-Thectocercus-Guaruba*.

Guaruba and *Thectocercus*, which are closely related in phylogenetic trees, share many similar morphological features, such as: relatively large size compared to other members of previously broadly defined *Aratinga*; fusiform body shape with strongly build chest (Fig. 11a); bill shape from the lateral view (Fig. 11b); very wide bill, whose lower mandible is more sizeable in width than in depth (Fig. 11c); slender, acute and ridged tip of the upper mandible (Fig. 11b); large and whitish eye ring (Fig. 11d); orange eye iris (Fig. 11d); underside (Fig. 12a) and upperside (Fig. 12b) colouration of remiges. These genera are characterized also by some interesting developmental features. A grey colour of the legs in young individuals is replaced by a flesh-pink colour in mature individuals (Fig. 11e). *Guaruba*'s upper mandible horn-coloured with grey-coloured tip (Fig. 11b, c), resembles that of all the five recognized *Thectocercus* subspecies. The lower mandible of *Guaruba* is generally also horn-coloured, but a small grey area close to the edge of its central part is often noticeable (Fig. 11b, c). Only two *Thectocercus* subspecies share this horn-grey pattern of the lower mandible, whereas three other subspecies have the lower mandible dark grey (Fig. 11a, b). However, their young chicks also display the horn-coloured state of the lower mandible [83], which is subsequently replaced by the grey colouration with the intermediate horn-grey pattern observed in *Guaruba* (Fig. 11b, c).

Since the basal taxa *Pyrrhura* and *Rhynchopsitta* have mainly a green plumage, an ancestral state of the *Arini* was most probably of green colouration. This feature survived in related *Psittacara* and *Thectocercus*; in *Guaruba*, *Eupsittula* and the lineage *Aratinga* + macaws, however, more differentiated colouring evolved, including blue, violet, yellow, black, brown and red.

Conclusions

Using the complete mitochondrial genomes of the main representatives of the tribe *Arini*, we validated the usefulness

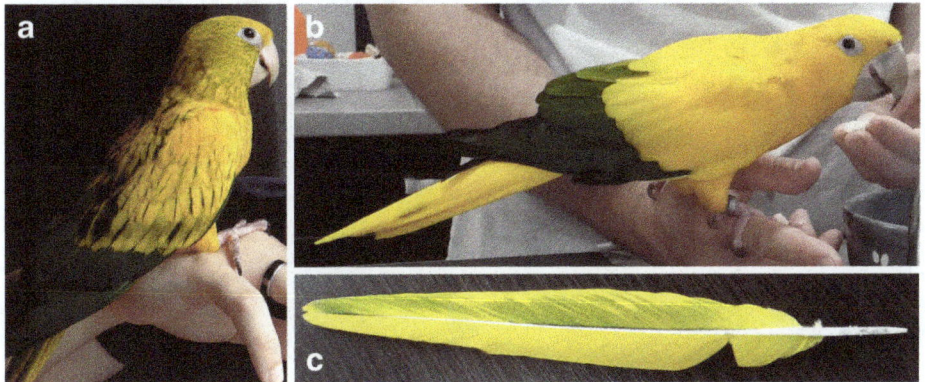

Fig. 10 *Guaruba guarouba* colouration. **a**: Juvenile individual with mixed yellow and green plumage. **b** Adult with typical mainly yellow plumage. **c** Double coloured feather from an adult

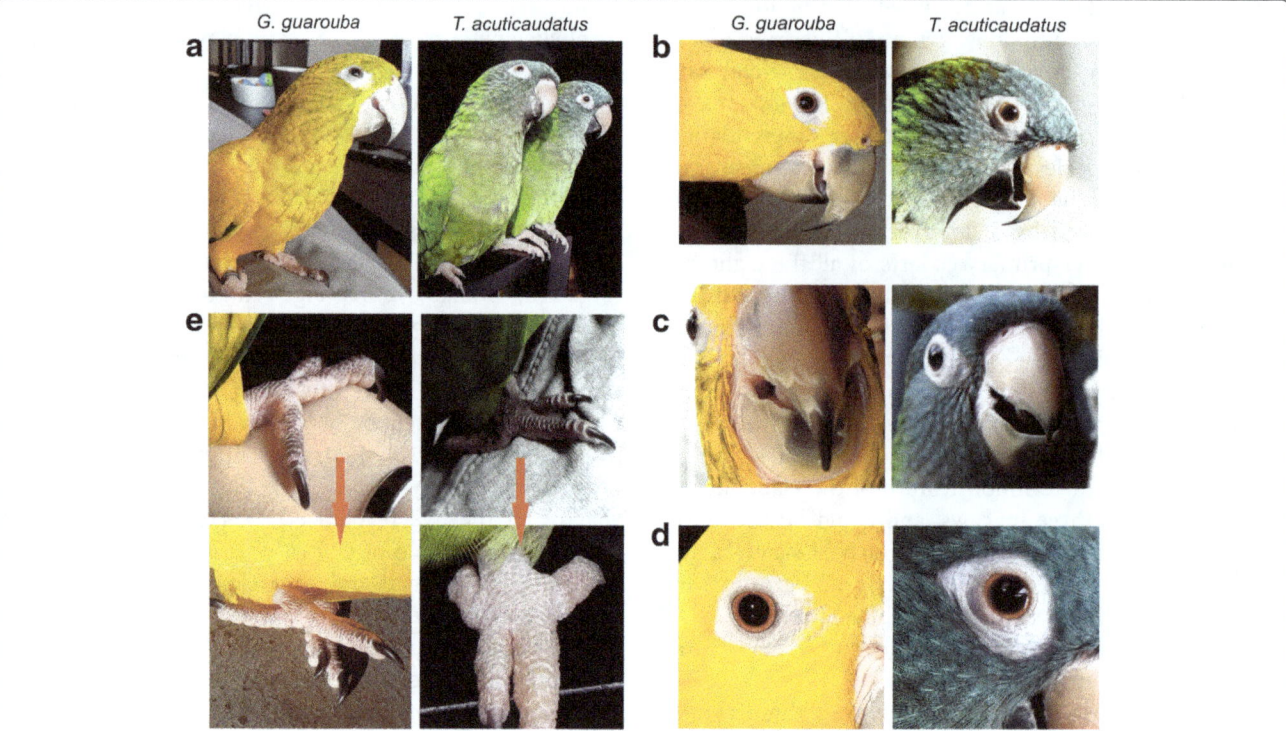

Fig. 11 Morphological comparison of *Guaruba guarouba* and *Thectocercus acuticaudatus acuticaudatus*. **a** Body shape. **b** Head with bill from the lateral view. **c** Bill from the front view. **d** Eye. **e** Change of leg colour from grey in juvenile to flesh-pink in adult

Fig. 12 Colouration of remiges in *Guaruba guarouba*-underside (**a**) and upperside (**b**)

and applicability of different phylogenetic methods and various molecular markers in the phylogeny of parrots. The individual markers provided conflicting tree topologies, and different tree-building methods produced various topologies even for the same marker. Therefore, the results of phylogenetic analyses relying on single approaches and markers need to be interpreted with caution. It applies in particular to mitochondrial *cytb*, *nd2* and *nd6* genes, commonly used in parrot phylogenies, because the tree topologies based on them showed a substantial disagreement across the methods used and a sharp difference from the tree based on all molecular markers. Among individual markers, the strongest phylogenetic signal was recorded in the control region as well as in tRNA and 16S rRNA genes, but the trees based on them were not fully resolved either. Uncertainty in the inferred phylogenies may arise from taxon sampling, rapid evolution of the taxa under study and a short time interval between the divergences of particular lineages. The best resolved phylogenetic relationships of these parrots can be obtained only on the concatenated set of all mitochondrial markers, especially with the application of robust methods such as Bayesian and maximum likelihood. The molecular phylogenies are supported by several features in the mitochondrial genomes and are consistent with plumage colouration, which may reflect an agreement between the mtDNA phylogenies and the species tree. However, it seems advisable to incorporate additional molecular markers to verify the phylogenetic relationships within *Arini* presented in this paper.

Additional files

Additional file 1: Comparison of mitochondrial genomes from ten *Arini* representatives. (PDF 5796 kb)

Additional file 2: Sequence data sets and nucleotide substitution models used in the present study. (PDF 53 kb)

Additional file 3: Tree topologies obtained by eight methods for individual mitochondrial markers from 7 parrot taxa. When some methods yielded several equally viable topologies, the majority-rule consensus tree was presented. The following phylogenetic approaches were applied: Bayesian analyses in MrBayes (MB) and PhyloBayes (PB), maximum likelihood analyses with partitioned data in TreeFinder (TF) and not partitioned in PAUP (ML) as well as neighbour joining (NJ), minimum evolution (ME), weighted least squares (WLS) and maximum parsimony (MP) in PAUP. The same tree topologies have the same background colour. (PDF 32 kb)

Additional file 4: Tree topologies built employing eight methods for individual mitochondrial markers from 11 parrot taxa. When some methods yielded several equally probable topologies, the majority-rule consensus tree was presented. The following phylogenetic approaches were applied: Bayesian analyses in MrBayes (MB) and PhyloBayes (PB), maximum likelihood analyses with partitioned data in TreeFinder (TF) and not partitioned in PAUP (ML) as well as neighbour joining (NJ), minimum evolution (ME), weighted least squares (WLS) and maximum parsimony (MP) in PAUP. The same tree topologies have the same background colour. (PDF 647 kb)

Additional file 5: Data set with 7-taxa: distance of individual marker trees from trees A and B based on the concatenated alignment of all markers. The maximum distance = 8. When some methods yielded several equally probable topologies, we averaged the calculated distances for these

trees. The following phylogenetic approaches were applied: Bayesian analyses in MrBayes (MB) and PhyloBayes (PB), maximum likelihood analyses with partitioned data in TreeFinder (TF) and not partitioned in PAUP (ML) as well as neighbour joining (NJ), minimum evolution (ME), weighted least squares (WLS) and maximum parsimony (MP) in PAUP. (PDF 49 kb)

Additional file 6: Data set with 11 taxa: distance of individual marker trees from trees F, G and H based on the concatenated alignment of all markers. The maximum distance = 16. When some methods yielded several equally probable topologies, we averaged the calculated distances for these trees. The following phylogenetic approaches were applied: Bayesian analyses in MrBayes (MB) and PhyloBayes (PB), maximum likelihood analyses with partitioned data in TreeFinder (TF) and not partitioned in PAUP (ML) as well as neighbour joining (NJ), minimum evolution (ME), weighted least squares (WLS) and maximum parsimony (MP) in PAUP. (PDF 53 kb)

Abbreviations
ALL set: the concatenated set of all nucleotide markers; ALL-CR: the set of all markers excluding the control region; BP: bootstrap percentage; CA: correspondence analysis; CR: control region; ME: minimum evolution; MP: maximum parsimony; NJ: neighbour joining; PCGs: protein-coding genes; PP: posterior probability; WLS: weighted least squares

Acknowledgments
We are very grateful to the Reviewers for their valuable comments and insightful remarks that have significantly improved the paper. We would like to express our warmest thanks to Krzysztof Grabowski, MSC, for his technical assistance in the laboratory.

Authors' contribution
Planned and designed the research: PM, AU. Performed bioinformatics analyses: PM, AK. Performed laboratory (experimental) analyses: AU, AK. Analysed the data: PM, AU, AK. Wrote the manuscript: PM, AU. Managed the manuscript submission, corrected the manuscript and responded to Editor and Reviewers: PM. All authors participated in the improvement of the manuscript and approved the final version.

Funding
The study was supported by National Science Centre Poland (Narodowe Centrum Nauki, Polska) grant no. 2015/17/B/NZ8/02402. The publication was supported financially by Wroclaw Centre for Biotechnology programme "The Leading National Research Centre (KNOW) for years 2014-2018". None of the authors has a conflict of interests.

Competing interests
The authors declare that they have no competing interests.

References
1. Joseph L, Toon A, Schirtzinger EE, Wright TF, Schodde R. A revised nomenclature and classification for family-group taxa of parrots (Psittaciformes). Zootaxa. 2012;3205:26–40.
2. Forshaw JM. Parrots of the world. London: A & C Black Publishers Ltd; 2010.
3. Schodde R, Remsen JV, Schirtzinger EE, Joseph L, Wright TF. Higher classification of new world parrots (Psittaciformes; Arinae), with diagnoses of tribes. Zootaxa. 2013;3691(5):591–6.
4. Schweizer M, Hertwig ST, Seehausen O, Ebach M. Diversity versus disparity and the role of ecological opportunity in a continental bird radiation. J Biogeogr. 2014;41(7):1301–12.
5. Remsen JV, Schirtzinger EE, Ferraroni A, Silveira LF, Wright TF. DNA-sequence data require revision of the parrot genus Aratinga (Aves: Psittacidae). Zootaxa. 2013;3641(3):296–300.
6. Urantowka AD, Grabowski KA, Strzala T. Complete mitochondrial genome of blue-crowned parakeet (Aratinga Acuticaudata)–phylogenetic position of the species among parrots group called Conures. Mitochondrial DNA. 2013; 24(4):336–8.

7.   Remsen JVJ, Areta JI, Cadena CD, Jaramillo A, Nores M, Pacheco JF, Pérez-Emán J, Robbins MB, Stiles FG, Stotz DF et al. A classification of the bird species of South America. American Ornithologists' Union. 2016. http://www.museum.lsu.edu/~Remsen/SACCBaseline.html. Accessed 22 June 2016.

8.   Ribas CC, Miyaki CY. Molecular systematics in Aratinga parakeets: species limits and historical biogeography in the 'solstitialis' group, and the systematic position of Nandayus Nenday. Mol Phylogenet Evol. 2004;30(3):663–75.

9.   Tavares ES, Baker AJ, Pereira SL, Miyaki CY. Phylogenetic relationships and historical biogeography of Neotropical parrots (Psittaciformes : Psittacidae: Arini) inferred from mitochondrial and nuclear DNA sequences. Syst Biol. 2006;55(3):454–70.

10.   Silveira LF, De Lima FCT, Hofling E. A new species of Aratinga parakeet (Psittaciformes: Psittacidae) from Brazil, with taxonomic remarks on the Aratinga Solstitialis Complex. Auk. 2005;122(1):293–305.

11.   Kirchman JJ, Schirtzinger EE, Wright TF. Phylogenetic relationships of the extinct Carolina parakeet (Conuropsis Carolinensis) inferred from DNA sequence data. Auk. 2012;129(2):197–204.

12.   Schirtzinger EE, Tavares ES, Gonzales LA, Eberhard JR, Miyaki CY, Sanchez JJ, Hernandez A, Mueller H, Graves GR, Fleischer RC, et al. Multiple independent origins of mitochondrial control region duplications in the order Psittaciformes. Mol Phylogenet Evol. 2012;64(2):342–56.

13.   Nabholz B, Uwimana N, Lartillot N. Reconstructing the phylogenetic history of long-term effective population size and life-history traits using patterns of amino acid replacement in mitochondrial genomes of mammals and birds. Genome Biol Evol. 2013;5(7):1273–90.

14.   Havird JC, Santos SR. Performance of single and concatenated sets of mitochondrial genes at inferring metazoan relationships relative to full Mitogenome data. PLoS One. 2014;9(1):e84080.

15.   Cummings MP, Otto SP, Wakeley J. Sampling properties of DNA-sequence data in phylogenetic analysis. Mol Biol Evol. 1995;12(5):814–22.

16.   Russo CAM, Takezaki N, Nei M. Efficiencies of different genes and different tree-building methods in recovering a known vertebrate phylogeny. Mol Biol Evol. 1996;13(3):525–36.

17.   Zardoya R, Meyer A. Phylogenetic performance of mitochondrial protein-coding genes in resolving relationships among vertebrates. Mol Biol Evol. 1996;13(7):933–42.

18.   Lambret-Frotte J, Perini FA, Russo CAD. Efficiency of nuclear and mitochondrial markers recovering and supporting known Amniote groups. Evol Bioinforma. 2012;8:463–73.

19.   Talavera G, Vila R. What is the phylogenetic signal limit from mitogenomes? The reconciliation between mitochondrial and nuclear data in the Insecta class phylogeny. BMC Evol Biol. 2011;11:315.

20.   Miya M, Nishida M. Use of mitogenomic information in teleostean molecular phylogenetics: a tree-based exploration under the maximum-parsimony optimality criterion. Mol Phylogenet Evol. 2000;17(3):437–55.

21.   Mueller RL. Evolutionary rates, divergence dates, and the performance of mitochondrial genes in Bayesian phylogenetic analysis. Syst Biol. 2006;55(2):289–300.

22.   San Mauro D, Gower DJ, Massingham T, Wilkinson M, Zardoya R, Cotton JA. Experimental Design in Caecilian Systematics: phylogenetic information of mitochondrial genomes and nuclear rag1. Syst Biol. 2009;58(4):425–38.

23.   Duchene S, Archer FI, Vilstrup J, Caballero S, Morin PA. Mitogenome phylogenetics: the impact of using single regions and partitioning schemes on topology, substitution rate and divergence time estimation. PLoS One. 2011;6(11):e27138.

24.   Rohland N, Malaspinas AS, Pollack JL, Slatkin M, Matheus P, Hofreiter M. Proboscidean mitogenomics: chronology and mode of elephant evolution using mastodon as outgroup. PLoS Biol. 2007;5(8):1663–71.

25.   Willerslev E, Gilbert MT, Binladen J, Ho SY, Campos PF, Ratan A, Tomsho LP, da Fonseca RR, Sher A, Kuznetsova TV, et al. Analysis of complete mitochondrial genomes from extinct and extant rhinoceroses reveals lack of phylogenetic resolution. BMC Evol Biol. 2009;9:95.

26.   StatSoft Inc. STATISTICA (data analysis software system), version 10. 2011. www.statsoft.com.

27.   Katoh K, Standley DM. MAFFT multiple sequence alignment software version 7: improvements in performance and usability. Mol Biol Evol. 2013; 30(4):772–80.

28.   Waterhouse AM, Procter JB, Martin DM, Clamp M, Barton GJ. Jalview version 2–a multiple sequence alignment editor and analysis workbench. Bioinformatics. 2009;25(9):1189–91.

29.   Talavera G, Castresana J. Improvement of phylogenies after removing divergent and ambiguously aligned blocks from protein sequence alignments. Syst Biol. 2007;56(4):564–77.

30.   Ronquist F, Teslenko M, van der Mark P, Ayres DL, Darling A, Hohna S, Larget B, Liu L, Suchard MA, Huelsenbeck JP. MrBayes 3.2: efficient Bayesian phylogenetic inference and model choice across a large model space. Syst Biol. 2012;61(3):539–42.

31.   Lartillot N, Philippe H. A Bayesian mixture model for across-site heterogeneities in the amino-acid replacement process. Mol Biol Evol. 2004;21(6):1095–109.

32.   Jobb G, von Haeseler A, Strimmer K. RETRACTED ARTICLE: TREEFINDER: a powerful graphical analysis environment for molecular phylogenetics. BMC Evol Biol. 2004;4:18.

33.   Swofford DL. Phylogenetic analysis using parsimony (*and other methods). Version 4. In. Sunderland MA: Sinauer Associates; 1998.

34.   Darriba D, Taboada GL, Doallo R, Posada D. jModelTest 2: more models, new heuristics and parallel computing. Nat Methods. 2012;9(8):772.

35.   Huelsenbeck JP, Larget B, Alfaro ME. Bayesian phylogenetic model selection using reversible jump Markov chain Monte Carlo. Mol Biol Evol. 2004;21(6):1123–33.

36.   Harrison GL, McLenachan PA, Phillips MJ, Slack KE, Cooper A, Penny D. Four new avian mitochondrial genomes help get to basic evolutionary questions in the late cretaceous. Mol Biol Evol. 2004;21(6):974–83.

37.   Delsuc F, Phillips MJ, Penny D. Comment on "hexapod origins: monophyletic or paraphyletic?". Science. 2003;301(5639):1482. author reply 1482

38.   Phillips MJ, Penny D. The root of the mammalian tree inferred from whole mitochondrial genomes. Mol Phylogenet Evol. 2003;28(2):171–85.

39.   Cavender JA, Felsenstein J. Invariants of phylogenies in a simple case with discrete states. J Classif. 1987;4:57–71.

40.   Akaike H. A new look at the statistical model identification. IEEE Trans Autom Control. 1974;19:716–23.

41.   Sugiura N. Further analysis of the data by akaike's information criterion and the finite corrections. Commun Stat Theory Methods. 1978;A7:13–26.

42.   Hannan EJ, Quinn BG. Determination of the order of an autoregression. J Roy Stat Soc B Met. 1979;41(2):190–5.

43.   Shimodaira H, Hasegawa M. CONSEL: for assessing the confidence of phylogenetic tree selection. Bioinformatics. 2001;17(12):1246–7.

44.   Felsenstein J. PHYLIP (Phylogeny Inference Package) version 3.6. Distributed by the author. Department of Genome Sciences, University of Washington, Seattle. 2005.

45.   Tamura K, Stecher G, Peterson D, Filipski A, Kumar S. MEGA6: molecular evolutionary genetics analysis version 6.0. Mol Biol Evol. 2013;30(12):2725–9.

46.   Kumar S, Stecher G, Tamura K. MEGA7: molecular evolutionary genetics analysis version 7.0 for bigger datasets. Mol Biol Evol. 2016;33(7):1870–4.

47.   Kass RE, Raftery AE. Bayes factors. J Am Stat Assoc. 1995;90:773–95.

48.   Zhuang X, Qu M, Zhang X, Ding S. A comprehensive description and evolutionary analysis of 22 grouper (perciformes, epinephelidae) mitochondrial genomes with emphasis on two novel genome organizations. PLoS One. 2013;8(8):e73561.

49.   Pesole G, Gissi C, De Chirico A, Saccone C. Nucleotide substitution rate of mammalian mitochondrial genomes. J Mol Evol. 1999;48(4):427–34.

50.   Pacheco MA, Battistuzzi FU, Lentino M, Aguilar RF, Kumar S, Escalante AA. Evolution of modern birds revealed by Mitogenomics: timing the radiation and origin of major orders. Mol Biol Evol. 2011;28(6):1927–42.

51.   Felsenstein J. Cases in which parsimony or compatibility methods will be positively misleading. Syst Zool. 1978;27(4):401–10.

52.   Susko E, Spencer M, Roger AJ. Biases in phylogenetic estimation can be caused by random sequence segments. J Mol Evol. 2005;61(3):351–9.

53.   Delsuc F, Brinkmann H, Philippe H. Phylogenomics and the reconstruction of the tree of life. Nat Rev Genet. 2005;6(5):361–75.

54.   Gribaldo S, Philippe H. Ancient phylogenetic relationships. Theor Popul Biol. 2002;61(4):391–408.

55.   Lopez P, Casane D, Philippe H. Heterotachy, an important process of protein evolution. Mol Biol Evol. 2002;19(1):1–7.

56.   Jeffroy O, Brinkmann H, Delsuc F, Philippe H. Phylogenomics: the beginning of incongruence? Trends Genet. 2006;22(4):225–31.

57.   Phillips MJ, Delsuc F, Penny D. Genome-scale phylogeny and the detection of systematic biases. Mol Biol Evol. 2004;21(7):1455–8.

58.   Naylor GJP, Brown WM. Amphioxus mitochondrial DNA, chordate phylogeny, and the limits of inference based on comparisons of sequences. Syst Biol. 1998;47(1):61–76.

59. Knaus BJ, Cronn R, Liston A, Pilgrim K, Schwartz MK. Mitochondrial genome sequences illuminate maternal lineages of conservation concern in a rare carnivore. BMC Ecol. 2011;11:10.

60. Pereira SL, Baker AJ. A mitogenomic timescale for birds detects variable phylogenetic rates of molecular evolution and refutes the standard molecular clock. Mol Biol Evol. 2006;23(9):1731–40.

61. Seixas VC, Paiva PC, CAdM R. Complete mitochondrial genomes are not necessarily more informative than individual mitochondrial genes to recover a well-established annelid phylogeny. Gene Reports. 2016;5:10–7.

62. Urantowka AD, Strzala T, Kroczak A, Mackiewicz P. The first complete genome of "true" Aratinga genus in comparison to mitogenomes of other parrots from Arini tribe. Mitochondrial DNA Part B: Resources. 2016;1(1):853–5.

63. Galtier N, Nabholz B, Glemin S, Hurst GD. Mitochondrial DNA as a marker of molecular diversity: a reappraisal. Mol Ecol. 2009;18(22):4541–50.

64. Haldane JBS. Sex ratio and unisexual sterility in hybrid animals. J Genet Camb. 1922;12:101–9.

65. Tegelstrom H, Gelter HP. Haldane rule and sex biased gene flow between 2 hybridizing flycatcher species (Ficedula-Albicollis and F-Hypoleuca, Aves, Muscicapidae). Evol; Int J Org Evol. 1990;44(8):2012–21.

66. Carling MD, Brumfield RT. Haldane's rule in an avian system: using cline theory and divergence population genetics to test for differential introgression of mitochondrial, autosomal, and sex-linked loci across the Passerina bunting hybrid zone. Evol Int J Org Evol. 2008;62(10):2600–15.

67. Saetre GP, Borge T, Lindroos K, Haavie J, Sheldon BC, Primmer C, Syvanen AC. Sex chromosome evolution and speciation in Ficedula flycatchers. P Roy Soc B-Biol Sci. 2003;270(1510):53–9.

68. Saetre GP, Borge T, Lindell J, Moum T, Primmer CR, Sheldon BC, Haavie J, Johnsen A, Ellegren H. Speciation, introgressive hybridization and nonlinear rate of molecular evolution in flycatchers. Mol Ecol. 2001;10(3):737–49.

69. Turelli M, Orr HA. The dominance theory of Haldane's rule. Genetics. 1995; 140(1):389–402.

70. Brumfield RT, Jernigan RW, McDonald DB, Braun MJ. Evolutionary implications of divergent clines in an avian (Manacus : Aves) hybrid zone. Evol Int J Org Evol. 2001;55(10):2070–87.

71. Rheindt FE, Edwards SV. Genetic introgression: an integral but neglected component of speciation in birds. Auk. 2011;128(4):620–32.

72. Kuraku S. Palaeophylogenomics of the vertebrate ancestor-impact of hidden Paralogy on hagfish and lamprey gene phylogeny. Integr Comp Biol. 2010; 50(1):124–9.

73. Kuraku S, Takio Y, Tamura K, Aono H, Meyer A, Kuratani S. Noncanonical role of Hox14 revealed by its expression patterns in lamprey and shark. Proc Natl Acad Sci U S A. 2008;105(18):6679–83.

74. Feiner N, Begemann G, Renz AJ, Meyer A, Kuraku S. The origin of bmp16, a novel Bmp2/4 relative, retained in teleost fish genomes. BMC Evol Biol. 2009;9

75. Kuraku S. Impact of asymmetric gene repertoire between cyclostomes and gnathostomes. Semin Cell Dev Biol. 2013;24(2):119–27.

76. Maddison WP. Gene trees in species trees. Syst Biol. 1997;46(3):523–36.

77. Moore WS. Inferring phylogenies from Mtdna variation - mitochondrial-gene trees versus nuclear-gene trees. Evol Int J Org Evol. 1995;49(4):718–26.

78. Page RDM. Extracting species trees from complex gene trees: reconciled trees and vertebrate phylogeny. Mol Phylogenet Evol. 2000;14(1):89–106.

79. Martin AP, Burg TM. Perils of paralogy: using HSP70 genes for inferring organismal phylogenies. Syst Biol. 2002;51(4):570–87.

80. Zink RM, Barrowclough GF. Mitochondrial DNA under siege in avian phylogeography. Mol Ecol. 2008;17(9):2107–21.

81. de Kloet RS, de Kloet SR. The evolution of the spindlin gene in birds: sequence analysis of an intron of the spindlin W and Z gene reveals four major divisions of the Psittaciformes. Mol Phylogenet Evol. 2005;36(3):706–21.

82. Mayr G. Parrot interrelationships – morphology and the new molecular phylogenies. Emu. 2010;110:348–57.

83. Juniper T, Parr M. Parrots: a guide to the parrots of the world. London: Yale University Press; 1998.

84. Urantowka AD, Hajduk K, Kosowska B. Complete mitochondrial genome of endangered yellow-shouldered Amazon (Amazona Barbadensis): two control region copies in parrot species of the Amazona genus. Mitochondrial DNA. 2013;24(4):411–3.

85. Urantowka AD. Complete mitochondrial genome of critically endangered blue- throated macaw (Ara Glaucogularis): its comparison with partial mitogenome of scarlet macaw (Ara Macao). Mitochondrial DNA. 2016; 27(1):422–4.

86. Urantowka AD, Strzala T, Mackiewicz P. Complete mitochondrial genome of golden conure (Guaruba Guarouba). Mitochondrial DNA Part B: Resour. 2017;2(1):33–4.

87. Urantowka AD. Complete mitochondrial genome of red-bellied macaw (Orthopsittaca Manilata): its comparison with mitogenome of blue-throated macaw (Ara Glaucogularis). Mitochondrial DNA A. 2016;27(3):2110–1.

88. Urantowka AD. Complete mitochondrial genome of blue-headed macaw (Primolius Couloni): its comparison with mitogenome of blue-throated macaw (Ara Glaucogularis). Mitochondrial DNA A. 2016;27(3):2106–7.

89. Urantowka AD, Mackiewicz P, Strzala T. Complete mitochondrial genome of Mitred Conure (Psittacara mitratus): its comparison with mitogenome of Socorro Conure (Psittacara brevipes). Mitochondrial DNA A DNA MappSeq Anal. 2016;27(5):3363–4.

90. Urantowka AD, Strzala T, Grabowski KA. The first complete mitochondrial genome of Pyrrhura sp. - question about conspecificity in the light of hybridization between Pyrrhura Molinae and Pyrrhura Rupicola species. Mitochondrial DNA. 2016;27(1):471–3.

91. Urantowka AD, Strzala T, Grabowski KA. Complete mitochondrial genome of endangered maroon-fronted parrot (Rhynchopsitta Terrisi) - conspecific relation of the species with thick-billed parrot (Rhynchopsitta Pachyrhyncha). Mitochondrial DNA. 2014;25(6):424–6.

8

Inferring Methionine Sulfoxidation and serine Phosphorylation crosstalk from Phylogenetic analyses

Juan Carlos Aledo (ID)

Abstract

Background: The sulfoxidation of methionine residues within the phosphorylation motif of protein kinase substrates, may provide a mechanism to couple oxidative signals to changes in protein phosphorylation. Herein, we hypothesize that if the residues within a pair of phosphorylatable-sulfoxidable sites are functionally linked, then they might have been coevolving. To test this hypothesis a number of site pairs previously detected on human stress-related proteins has been subjected to analysis using eukaryote ortholog sequences and a phylogenetic approach.

Results: Overall, the results support the conclusion that in the eIF2α protein, serine phosphorylation at position 218 and methionine oxidation at position 222, belong to the same functional network. First, the observed data were much better fitted by Markovian models that assumed coevolution of both sites, with respect to their counterparts assuming independent evolution (p-value = 0.003). Second, this conclusion was robust with respect to the methods used to reconstruct the phylogenetic relationship between the 233 eukaryotic species analyzed. Third, the co-distribution of phosphorylatable and sulfoxidable residues at these positions showed multiple origins throughout the evolution of eukaryotes, which further supports the view of an adaptive value for this co-occurrence. Fourth, the possibility that the coevolution of these two sites might be due to structure-driven compensatory mutations was evaluated. The results suggested that factors other than those merely structural were behind the observed coevolution. Finally, the relationship detected between other modifiable site pairs from ataxin-2 (S814-M815), ataxin-2-like (S211-M215) and Pumilio homolog 1 (S124-M125), reinforce the view of a role for phosphorylation-sulfoxidation crosstalk.

Conclusions: For the four stress-related proteins analyzed herein, their respective pairs of PTM sites (phosphorylatable serine and sulfoxidable methionine) were found to be evolving in a correlated fashion, which suggests a relevant role for methionine sulfoxidation and serine phosphorylation crosstalk in the control of protein translation under stress conditions.

Keywords: Coevolution, Sulfoxide, eIF2α, Oxidative stress, Cytoplasmic stress granules, Post-translational modification

Background

Under stress conditions, cells respond with a global reduction of protein synthesis. Although all the phases of translation are susceptible of being affected, initiation of translation is considered to be the most important regulatory step of the translation cycle [1]. In eukaryotes, translation initiation is a complex and highly regulated process that requires the action of at least a dozen of protein factors. One of these factors, eIF2, is a stable heterotrimeric (subunits α, β and γ) GTPase that binds

Correspondence: caledo@uma.es
Departamento de Biología Molecular y Bioquímica, Facultad de Ciencias, Universidad de Málaga, 29071 Málaga, Spain

the Met-RNA$_i^{Met}$ and delivers it to the small ribosomal subunit. Pairing between the anticodon of the Met-RNA$_i^{Met}$ and the AUG start codon from the mRNA, triggers hydrolysis of GTP by eIF2 and eIF2·GDP is released. Since eIF2·GDP cannot bind Met-RNA$_i^{Met}$, eIF2·GDP should be converted to eIF2·GTP by the heteropentameric exchange factor eIF2B before it can start a new initiation cycle. In response to stress, phosphorylation of eIF2α on Ser-51 stabilizes the complex eIF2·GDP·eIF2B and inhibits the GDP-GTP exchange, which prevents the liberation of an active eIF2·GTP, thereby reducing initiation of translation [2].

Despite the evidence supporting this description, recent results suggest that the picture is incomplete. In many cases, this mechanism has been presumed to be responsible of the observed inhibition of protein synthesis after stress, simply because eIF2α was found phosphorylated at Ser-51. Indeed, the importance of this mechanism of translation attenuation may have been overestimated as convincingly argued by Knutsen and coworkers [3]. These authors found that when the fission yeast *Schizosaccharomyces pombe* was exposed to hydrogen peroxide, protein synthesis was drastically reduced and eIF2α was phosphorylated on Ser-52 (homologous to Ser-51 in mammalian cells). However, when the mutant eIF2α-S52A was employed in these experiments, the translation was reduced to the same extent as in wild-type cells, despite the inability of this mutant to phosphorylate eIF2α at Ser-52 [3]. Results obtained using other model organisms such as the budding yeast *Saccharomyces cerevisiae* and mammalian cells, also point to the existence of a hitherto unrecognized mechanism contributing to the regulation of translation after stress, independent of the phosphorylation at Ser-51 on eIF2α [3–5].

One possibility, that should not be ruled out, is that residues from eIF2α others than Ser-51 may be post-translationally modified and contribute to the regulation of the translation initiation. In this sense, during a high-throughput study aimed at identifying substrates of kinases related with the cell cycle, eIF2α was found phosphorylated at Ser-218 [6]. However, whether or not this phosphorylation has a functional effect remains yet to be investigated. On the other hand, Met-222 has also been reported to suffer extensive oxidation to methionine sulfoxide (MetO) after treating the cells with H_2O_2 [7]. Although oxidation of protein-bound methionine has been traditionally perceived as an inevitable damage derived from aerobic metabolism, it is now emerging as another post-translational modification (PTM) able to regulate protein activity during stress conditions [8]. To this respect, we have recently shown that oxidation of methionine harbored within phosphorylation motifs is a process highly selective among stress-related proteins, including eIF2α and other proteins belonging to stress granules (SGs) [9]. In the current study we have addressed the working hypothesis that both post-translational modifications, sulfoxidation of Met-222 and phosphorylation of Ser-218, may be functionally relevant and interrelated. To this end, we have followed an evolutionary approach.

Molecular coevolution between two positions of a protein occurs when amino acid substitutions at one of these positions affect the rates of substitution at the other position [10]. The forces leading to coevolution derive from functional and/or structural selective pressures acting to maintain specific combinations of residues at the coevolving positions. During the last two decades, much effort has been devoted to investigate

molecular coevolution and a plethora of methods aimed at detecting coevolving positions have been described (reviewed in [10, 11]). These can be broadly divided into those methods that attempt to model coevolution in a phylogenetic context and, on the other hand, those methods based on analyzing covariation in multiple sequence alignments. Thus, popular approaches to search for coevolving sites involve substitution correlations [12], mutual information of amino acid frequencies [13] or a global statistical model of the multiple sequence alignment, as is the case of direct coupling analysis [14, 15] and protein sparse inverse covariance [16].

Covariation-based methods, which are simpler and much more popular than those based on phylogenetic grounds, have been shown useful to predict residue contacts [15], and have even proved to be valuable tools for ab initio protein structure predictions. However, as it has recently been brought to our attention by Talavera and coworkers, covariation is a poor measure of molecular coevolution [17]. Therefore, we have followed a phylogenetic approach to study the evolutionary interrelationship between PTM sites. Evidence suggesting a tight relationship between sulfoxidation and phosphorylation among stress-related proteins involved in translation regulation, will be presented and discussed herein.

Methods

Phylogenetic trees and multiple sequence alignment data sets

Phylogenetic trees and multiple sequences alignments (MSAs) were initially obtained from eggNOG 4.5, a public resource (http://eggnogdb.embl.de) that provides ortholog groups with integrated functional annotations [18], using the human eIF2α (P05198) as query protein, and the Eukaryota as the target taxon. In this way, 272 sequences belonging to 233 species were retrieved. When multiple paralogs from one species were available, only the one with the smallest editing distance to the human homolog was included. Similarly, the original tree was pruned to reflect the phylogenetic relationship between the 233 ortholog proteins. In addition to the pre-computed tree from eggNOG, the phylogenetic relationship between these 233 sequences was also reconstructed using the neighbor-joining (NJ) [19]. Trees based on maximum parsimony (MP) employing the Fitch algorithm and the nearest neighbor interchange rearrangement strategy, and trees based on maximum likelihood analysis using a general time-reversible model with four discrete gamma rate categories, were computed with the assistance of two R packages: *phangorn* 2.0.4 [20] and *ape* 3.5 [21]. All these trees, as well as the MSA and raw data related to eIF2α, can be downloaded from https://github.com/jcaledo/PTM_sites_coevolution. Trees and MSAs for other SG related proteins such as ataxin-2 (Q99700), ataxin-2-like (Q8WWM7) and Pumi-lio homolog 1 (Q14671) were obtained from eggNOG 4.5. When the MSAs were not intended to reconstruct phylogenetic relationships, but to compute the state (the amino acid found) at different positions in the ortholog proteins, the MSA was further modified to remove those columns corresponding with gaps in the human protein used as reference.

A four-state continuous-time Markov chain model of evolution

To model the evolution of the residues found at positions 218 and 222, we used the theoretical framework of continuous time Markov processes. Since proteins are built up from a pool of twenty proteinogenic amino acids, the state space of these markovian models should, supposedly, be composed by 400 elements. However, most sites in ortholog proteins generally exist in a limited number of residue states. Furthermore, because our interest was focused on detecting coevolving PTM sites, the cardinality of the state space can be drastically reduced. Thus, for the character residue at position 218 (random variable X) two states are possible. When the residue found is a phosphorylatable one we set X = 1, and X = 0 otherwise. Similarly, for the character residue at position 222 (random variable Y) the accessible states are also two: a sulfoxidable methionine is found at that position (Y = 1) or any other non sulfoxidable amino acid is observed at that position (Y = 0). Four combinations of states are possible when these two binary variables are simultaneously considered. Each of these four states can either stay the same over the length of a branch of the phylogeny, or change to one of the three other states. Fig. 1 links the four combinations of states by arrows with parameters that describe the evolutionary rates of transitions between two states of one character, holding constant the state of the other. This Markov process is defined by the instantaneous rate matrix:

$$Q = \begin{pmatrix} -\sum q_{1j} & q_{12} & q_{13} & 0 \\ q_{21} & -\sum q_{2j} & 0 & q_{24} \\ q_{31} & 0 & -\sum q_{3j} & q_{34} \\ 0 & q_{42} & q_{43} & -\sum q_{4j} \end{pmatrix}$$

(1)

The values of the rate parameters describing instantaneous double changes, such as $(0,0) \rightarrow (1,1)$, $(0,1) \rightarrow (1,0)$, $(1,0) \rightarrow (0,1)$ and $(1,1) \rightarrow (0,0)$ are set to zero because the probability of both traits changing in the same instant (dt) is negligibly small and can be ignored. The model, however, allows both traits to change over a longer time

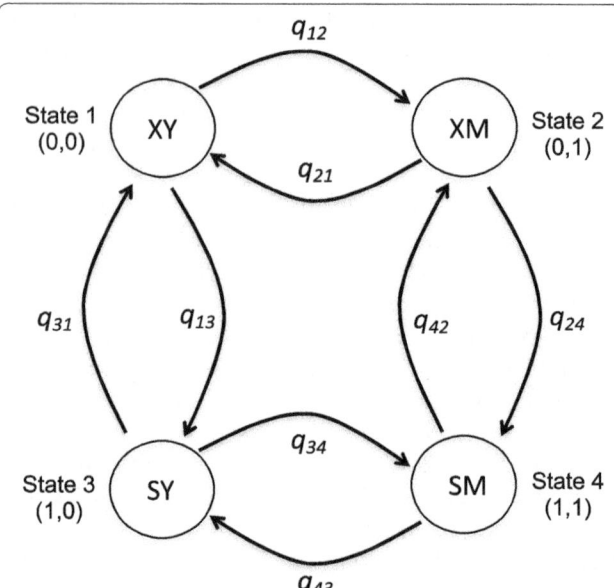

Fig. 1 Diagram of the Markovian model of evolution for a pair of PTM sites. For two binary traits, four combined states are possible. Each state is defined by an ordered pair where the first entry is either 0 or 1 depending on the absence or presence, respectively, of a phophorylatable residue at the considered position (trait X). The second entry of the pair again will be 0 or 1 depending now on the absence or presence of a methionyl residue at the analyzed position (trait Y). The allowed transitions between states are indicated by the *arrows* and their associated rate parameters, q_{ij}

period, t, but they must have transitioned through intermediate states. This approach is particularly suitable to detect evolutionary relationship between two traits, in which the state of one trait affects the probability of a change in the other [22].

The constraints that each row sums to zero, and that dual transitions have a probability of zero over time dt, means that the rate matrix, Q, defining the Markov process, is fully specified by eight parameters. However, if the two traits have evolved independently of one another, then the rate of change between the two states of one character will not depend on the background state of the other. For instance, if the rate of gaining an oxidable methionine at position 222 does not depend on the amino acid found at position 218, then $q_{12} = q_{34}$. More generally, the model of independent evolution can be defined by setting $q_{12} = q_{34}$, $q_{13} = q_{24}$, $q_{21} = q_{43}$ and $q_{31} = q_{42}$. Thus, the model of independent evolution uses a maximum of four parameters, while the model of dependent evolution does not place any restrictions on the parameters, allowing some kinds of transitions to depend on the background state of the other trait. When that happens, pairs of states will tend to be associated with each other in the species data more often than expected by chance, and the dependent model will provide a better description of the data.

Likelihood function optimization and likelihood ratio test

Given the assumption of either independence (Q_I) or dependence (Q_D) between X and Y, the corresponding substitution probability matrices can be calculated using $P_I(t) = e_I^{tQ}$ and $P_D(t) = e_D^{tQ}$, respectively. Eigenvalues and eigenvectors necessary for determining these expressions were calculated using standard numerical methods. To find the likelihood of our data given a chosen model, we have to consider all possible assignments of character states at the interior nodes. The likelihood of each of these single realizations is given by the product over all of the branches of the tree of the appropriate probabilities, as derived from the previously computed substitution probability matrices using the appropriate branch lengths. The likelihood of the data will be the sum over all possible realizations, which will be expressed as a function of the model parameters. In the Appendix A, from Additional file 1, we have considered a simple hypothetical phylogeny and data set to illustrate these calculations. The maximum likelihood estimates of parameters are the parameter values that maximize the likelihood function, and were found numerically using iterative optimization algorithms implemented in R [23]. Once the likelihoods for the independence $L(I)$ and dependence model $L(D)$ were computed, the value of the likelihood ratio test statistic was obtained according to the following equation:

$$LRT = -2\ln\frac{L(I)}{L(D)} \qquad (2)$$

Although we routinely use the R function *ace* from the package *ape* [21] to fit the so-called M*k* models (in our case M4 models) other functions such as *fitDiscrete* (from the package *GEIGER*) [23] and *fitMk* from the package *phytools* [24] were also used yielding similar results. Different assumptions regarding the stationary frequencies of each character state, only involved slight differences in parameter estimates and likelihoods. Thus, regardless of the frequency vector used ('equal frequencies', 'estimated frequencies from the stationary distribution of Q' or 'observed frequencies') the conclusion was always the same: the dependent models of evolution explained much better the data than the independent models (LRT = 15.9, 15.4 and 15.6, respectively).

Pairwise comparison analyses

Using the multiple protein sequence alignment of eIF2α, pairs of species contrasting in both binary characters (absence/presence of phosphorylatable residue at 218 and absence/presence of sulfoxidable residues at 222) were searched and analyzed as described by Maddison [25] and implemented in Mesquite 3.10.

Stochastic character mapping

We used stochastic character mapping [26] to infer 10 possible evolutionary histories of residues at positions 218 and 222 on the phylogenetic tree of eukaryotic IF2α protein. To this purpose we employed the function *make.simmap* in the *phytools* package (v. 0.5–38) for R [24]. For the parametrization of *make.simmap*, we used a model of unequal rates. To check the robustness of the obtained results, we tested two methods, Q = "empirical", which first fits a continuous-time Markov model for the evolution of our combined characters, and then simulates stochastic character histories using that model and the data (the tip states on the tree). Alternatively, Q = "mcmc", first samples Q 10 times from the posterior probability distribution of Q using Markov chain Monte Carlo, and then it simulates 10 stochastic maps conditioned on each sample value of Q. Apart from slightly higher variances in the number of transitions between states through the simulated histories, both methods provided similar results.

Protein stability of double-mutants

The thermodynamic stability changes ($\Delta\Delta G$) of single and double-mutants at different positions of eIF2α were computed using the protein design tool FoldX version 4.0 [27]. FoldX uses a full atomic description of the protein structure to provide a quantitative estimation of the importance of the interactions contributing to the stability of the protein. For this purpose, the different energy terms taken into account, which have been described in detail somewhere else [28], have been weighted using empirical data obtained from protein engineering experiments. The 3D structure of eIF2α (1Q8K) was subjected to an optimization procedure using the RepairPDB command from FoldX. Afterwards, 400 double-mutant models were built for each pair of positions. For instance, for the study of the pair S218X-M222Y, 400 models were built and analyzed (with X and Y belonging to the set of twenty proteinogenic amino acids).

Results

Residues at positions 218 and 222 from eIF2α have been coevolving

Two positions of a protein are said to be coevolving if they mutually influence their evolutionary rates. As it has already been pointed in the Introduction, covariation is a poor measure of molecular coevolution [17]. Therefore, we rather resorted to a maximum likelihood methodology for detecting correlated evolution on phylogenies. The method, which is an adaptation of that originally proposed by Mark Pagel in a seminal work published in 1994 [22], relies on a model based on Markov chain in continuous time and the optimization of the associated likelihood function to obtain the model's parameters that best fit the data. Our data set encompasses a phylogenetic tree of the eIF2α protein from 233 eukaryotic species, including branch lengths, as well as information about the residues found at positions 218 (character or variable X) and 222 (character or variable Y) in these ortholog proteins (Additional file 1: Figure S1).

To test for correlated evolution between variables X and Y, we used a likelihood ratio test (LRT) to compare the performance of two models, one assuming independent evolution of the characters and one assuming coevolution (see Methods). The likelihood ratio test significantly supported the dependent model (LRT = 15.9, df = 4, p = 0.003). Although the χ^2 distribution is the asymptotic distribution for the LRT statistic when the compared models are nested, it may not always apply directly. Whether the LRT statistic is distributed as a χ^2 may depend upon the values of the rate parameters and the amount of data [22], as well as the tree structure and the equilibrium state frequencies [29]. Therefore, to conclude with confidence that the coevolutionary model is better model for our data, we turned to the Monte Carlo parametric bootstrapping technique. Briefly, we started by finding the maximum likelihood estimates (MLE) of the four parameters of the model of independent evolution that best fitted to our data. These MLE parameters were then used to evolve the two characters along the established phylogeny in 1000 simulations. Using the data from these simulations, the dependent and independent models were fitted and their likelihoods computed. The likelihood ratios, of the two models for each simulation, were used to form the null distribution that is shown in Fig. 2 as a bar histogram plot. As it can be observed, the empirical Monte Carlo distribution of the statistics LRT matched quite closely the chi-squared distribution with four degree of freedom. Therefore, the model of independent evolution can be rejected with high confidence in favor of the coevolution model.

The evolutionary codependence between positions 218 and 222 is linked to PTM

The existence of correlated evolution between the sites 218 and 222 has been examined above by linking to each position a biological character. Concretely, the presence/absence of phosphorylatable residues at 218 and the presence/absence of a sulfoxidable amino acid at 222. Thus, the concluded codependence may be related to the PTMs studied. However, residue-residue interactions unrelated to the phosphorylation-sulfoxidation interplay may lead to the observed coevolutionary signal when collapsing residues in a reduced dictionary. To investigate this possibility, we have explored the codependence between these sites using different sets of residues to define the model states. For instance, one of the different

alternative that were assessed was: X = 1 if at position 218 is found a residue sensitive to deamidation (asparagine or glutamine) and Y = 1 if the non-polar valine is present at position 222. In this way, the evolutionary codependence of these traits, X and Y, was studied for 16 alternative definitions of states. As it can be observed in Table 1, none of these alternative definitions led to a significant coevolutionary signal. This lack of codependence, in these other reduced dictionaries, support a link between the coevolution at these positions and the PTMs undergone for the residues present at these sites.

Robustness of the coevolution model with respect to the phylogeny

A fundamental assumption of the maximum likelihood approach is a correct topology of the used tree, as well as an accurate estimation of their branch lengths [22]. To check the robustness of our conclusion regarding the evolutionary relationship between Ser-218 and Met-222, we carried out the coevolutionary analysis described above, but employing trees that were reconstructed using different methods. The results of such analyses are summarized in Table 2. Although the trees obtained using methods based on genetic distances, maximum parsimony and maximum likelihood were slightly different between them, in all the cases we concluded that the model of coevolution between both PTM sites fit the observed data significantly better than the model of independent evolution, regardless of the method used to reconstruct the phylogeny.

Reliability of the maximum likelihood approach to detect functional PTMs

Residues that are close in the three-dimensional structure of the protein are expected to mutually influence their evolution more often than residues that are far away from each other, merely due to structural reasons [29]. Since both PTM sites (Ser-218 and Met-222) are four residues away from each other, we wanted to explore whether causes other than structural may underline the observed coevolution between them. To this end, we carried out the following control analyses.

Ser-218 and Met-222 are located in a loop between helix α6 and strand β7 from the C-terminal domain of eIF2α. Thus, we selected all the pair sites found outside helices and strands that included a modifiable residue (either Ser, Thr, Tyr or Met) together with the amino acid found four positions downstream from it, but still remaining within the same loop (Fig. 3). In this way, five control site pairs were selected, including a positive control site pair such as that formed by Ser-51 and the hydrophobic residue Ile-55, which is expected to show a high degree of coevolution due to the well known functional relevance of these sites [30]. On the other hand,

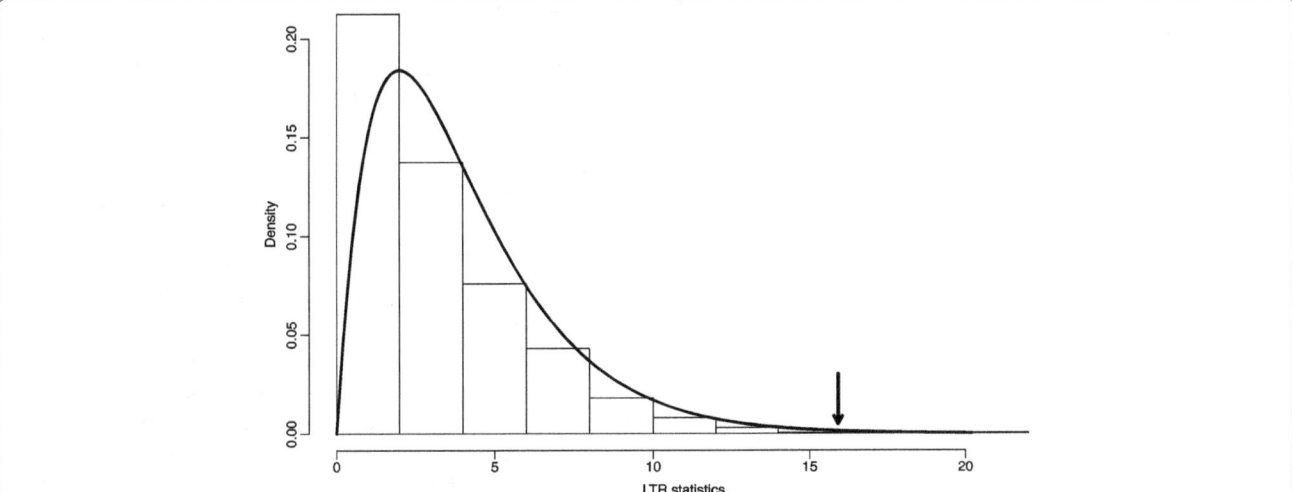

Fig. 2 Comparing models of evolution for the S218-M222 PTM site pair on eIF2α. The compared models were one assuming independent evolution of both PTM sites (null model), and an other assuming coevolution (alternative model). The likelihood value under a given model measures the fit of that model to data. Hence, the two models can be compared by comparing their respective likelihood values. The computed LRT statistic value for the independent versus dependent evolution was 15.9 (*arrow*). Since in our case both models were nested (see Methods), the probability distribution of the LRT statistic, assuming that the null model is true, can be approximated by a χ^2 distribution with four degree of freedom (*black continuous curve*). In this way, the null model (independent evolution of both sites) could be rejected with a *p*-value of 0.003. The same conclusion was reached when the distribution of the LRT statistic was derived by Monte Carlo simulation (histogram bars)

the remaining four site pairs are referred to as negative control because they include non-regulatory sites [31]. That is, residues that may be modified only as conse-, quence of off-target interactions and therefore are thought to be of little, if any, functional significance [32]. All these control site pairs were subjected to analysis using the same maximum likelihood methodology used to analyze the pair Ser-218/Met-222 under study. The results of such coevolutionary analyses are summarized in Table 3. As it can be observed, data related to non-functional site pairs are best explained by assuming independent evolution, in spite of the proximity between the sites. In contrast, the functional pair Ser-51/Ile-55, used as positive control, yielded a high LRT indicative of the good performance of the current procedure to detect functionally important PTMs.

To further explore the possibility that the three-dimensional distance between these pairs of residues may influence the observed coevolutionary signal, we extended the set of control pairs to include all the pairs formed by a target residue (either Ser-218 or Met-222) and a second residue in the vicinity within the α6/β7 loop (Fig. 4). Using all these pairs, we failed to detect any relationship between the distance (in ångströms) separating the pair members and their evolutionary codependence, as estimated by their LRT statistic values (Table 3 and Additional file 1: Figure S2). In addition, we also addressed whether the structural importance of the pair of sites might be a key determinant of their evolutionary codependence. To this

end, for each pair of sites the mean thermodynamic stability change ($\Delta\Delta G$) for the 400 possible double-mutants was computed and plotted against its LRT. Again we failed to observe a relationship between these variables (Additional file 1: Figure S3).

Vigilance against over-interpretation
In a recent work Maddison and Fitzjohn warn against the risks of over-interpreting the results of phylogenetic correlation tests based on Pagel's method. Although these well-respected methods are commonly used in many biological fields, they do not eliminate pseudo-replications derived from a single evolutionary event [33]. They argue that, in some circumstances, the co-distribution between two characters along the phylogeny may be the result of sharing a pair of synapomorphic characters, each emerged by its own independent causes in a common ancestor, and then maintained each by its own causes in the descendants. Such a situation is what these authors call Darwin's scenario. Therefore, if we want to support the conclusion that serine phosphoryl-ation and methionine oxidation form part of the same functional network, Darwin's scenario should be ruled out. In Fig. 5a the same phylogeny of eIF2α is mirrored to show the state of character X (absence/presence of phosphorylatable residue) at left, and Y (absence/pres-ence of methionine) at right. On the other hand, Fig. 5b shows a hypothetical co-distribution of both characters made up to illustrate Darwin's scenario. As it can be deduced from this figure, the co-distribution of

Table 1 Sets of residues unrelated to the crosstalk between phosphorylation and sulfoxidation do not show coevolutionary signal

Group	Set at 218	Set at 222	LRT	p-value
Positive	{S,T}	{M}	15.9	0.003
Control-1	{S,T}	{V}	2.4	0.668
Control-1	{S,T}	{C}	3.1	0.542
Control-1	{S,T}	{N}	4.6	0.336
Control-1	{S,T}	{L}	0.0	1.000
Control-2	{N,Q}	{M}	3.7	0.449
Control-2	{G,A}	{M}	0.0	1.000
Control-2	{D,E}	{M}	0.0	1.000
Control-2	{H,K}	{M}	0.0	1.000
Control-3	{Q,N}	{V}	0.2	0.993
Control-3	{G,A}	{V}	6.2	0.181
Control-3	{D,E}	{V}	0.0	1.000
Control-3	{H,K}	{V}	3.5	0.475
Control-4	{Q,P}	{G}	0.0	1.000
Control-4	{W,V}	{A}	0.0	1.000
Control-4	{L,T}	{I}	0.0	1.000
Control-4	{I,Y}	{H}	0.0	1.000

The existence of correlated evolution between the sites 218 and 222 was examined using different sets of residues to define the model states. Thus, beside the sets formed by phosphorylatable residues at 218 and sulfoxidable methionine at 222, used as positive reference, four other types of set combinations were analyzed. In the first type (Control-1, rows 2–5), the residues providing X = 1 are still serine and threonine, but the residue making Y = 1 is now different to methionine (as indicated in the table). Val, Cys, Asn and Leu have been selected because of their frequencies (from higher to lower) at position 222 in the eukaryotic species examined. The second group of control analyses (Control-2, rows 6–9) always included methionine at position 222 as the residue providing Y = 1, while the set leading to X = 1 was now formed by a pair of non-phosphoacceptor amino acids of similar physicochemical properties. The third group (Control-3, rows 10–13) was formed by sets of residues that were neither phosphorylatable nor sulfoxidable. Finally, the fourth group (Control-4, last four rows), was established by randomly taking the sets, among the twenty proteinogenic amino acids. The LRTs obtained, and their respective p-values, are shown

phosphorylatable and sulfoxidable residues throughout eukaryotes does not seem to fit in Darwin's scenario.

In an attempt to numerically support this graphic result, we resorted to pairwise comparisons of terminal taxa to test for character correlation [25]. Only pairs of taxa that differed in both characters and were phylogenetically separate were considered. A pair is said to be phylogenetically separate if the path between the members of the pair, along the branches of the tree, do not touch the path of any other pair (Fig. 6). Since we are examining pairs contrasting in both characters, there are two types of possible pairs: a pair is considered to be positive when the presence of a phosphorylatable residue in a member of the pair is accompanied by a methionine in the same member of the pair ({(1,1), (0,0)}), otherwise the pair is referred to as negative ({(1,0), (0,1)}. The idea underlying this analysis is simple. When pairs of species contrasting in the state of a particular character are examined, the member of a pair with a particular state might be more likely than the other member to exhibit a particular state in the second character. In other words, the null hypothesis states that the number of positive pairs is equal to the number of negative pairs. When we subjected our data to such an analysis, we found that many different maximal pairings could be defined (a maximal pairing is a set of phylogenetically separate pairs of terminal taxa that contains the most pairs possible for the given tree). In all the cases, these pairings contained five positive pairs and one negative pair (Fig. 6). Pairwise comparison analysis can avoid the pitfalls of being influenced by a single origin of a character state by choosing pairs of taxa that contrast in the states of both variables. However, the method uses only a small subset of taxa, discarding much of the data, and so it has a very low power to detect correlations [33]. That seems to be the case with our data. Despite that the number of positive pairs overtook the number of negative ones (five to one), the difference did not reach statistical significance due to the low number of total phylogenetically separate pairs (only six). Nevertheless, the fact that among contrasting pairs the probability of being positive is five times greater than that of being negative allows, at least, to rule out a single origin of the phosphorylatable-sulfoxidable co-distribution among eukaryotes.

Stochastic character mapping also supports coevolution

To tackle the potential problem of non-independence in phylogenies using a different approach, we took advantage of the idea underlying Ridley's method, which was

Table 2 Influence of the phylogeny on the likelihood ratio test between the models of dependent and independent changes

Tree	Ln(L)	Fitch score	Topology difference	Branch score	LRT	p-Value
eggNOG	−43,009	7499	0	0	15.9	0.0030
NJ	−43,677	7596	198	1.664	20.8	0.0003
MP	−60,127	7472	28	426.745	14.1	0.0070
ML	−37,544	7511	111	1.118	17.1	0.0020

Besides the pre-computed tree from eggNOG, trees reconstructed using the methods of neighbor-joining (NJ), maximum parsimony (MP) and maximum likelihood (ML), were used to test the hypothesis of correlated evolution between Ser-218 and Met-222. In addition to the LRT and its related p-value, the table shows the natural logarithm of the likelihood, Ln(L), and the Fitch score for each tree. The distance between each tree and that from eggNOG was assessed using either the metric proposed by [53] (topology difference) or the branch score proposed by [54]. The former, is defined as twice the number of internal branches that differ in their splits, while the latter is defined as the sum of squares of the differences between each branch's length in both trees

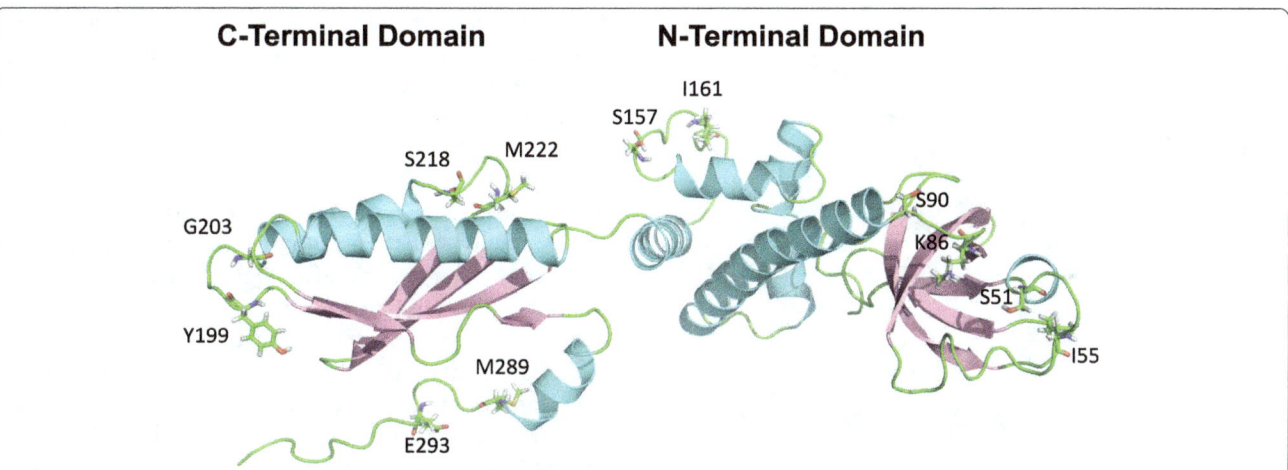

Fig. 3 Spatial distribution of structurally comparable residue pairs from eIF2α. Ribbon cartoon of the three-dimensional structure of the α subunit of human eIF2 (PDB 1Q8K). The modifiable residues (either Ser, Thr, Tyr or Met) located outside helices and strands are displayed using stick representation, as well as those amino acids found four positions away from them

specifically tailored to get around the problem of pseudo-replications [34]. In the original description of the method, the character states of internal nodes are reconstructed using parsimony. Once the internal nodes have been assigned, one works through the phylogeny keeping a tally of the number of transitions in the tree. More concretely, the method scores transitions only in those branches for which the beginning and end-states differ. That is, branches along which no change occurs are not included. By not including such branches, the method avoids counting species or internal nodes that share character states with an immediate

Table 3 Sites from non-regulatory pairs evolve independently despite their spatial proximity

Group	Sites	Regulatory	Loop	Distance	ΔΔG	LTR	p-value
A	S51-I55	Yes	β3/β4	6.2	−0.6 ± 1.2	27.7	1.5 10^{-5}
A	S90-K86	No	β5/α1	9.2	2.1 ± 1.3	5.4	0.242
A	S157-I161	No	α4/α5	9.7	0.0 ± 0.8	3.9	0.420
A	Y199-G203	No	β6/α6	8.8	9.7 ± 4.8	0.0	1.000
A	S218-M222	Yes?	α6/β7	6.3	2.2 ± 2.0	15.9	0.003
A	M289-E293	No	CT	9.5	−0.3 ± 0.8	5.9	0.207
B	C217-M222	No	α6/β7	9.9	1.2 ± 2.1	2.1	0.712
B	S218-M222	Yes?	α6/β7	6.3	2.2 ± 2.0	15.9	0.003
B	T219-M222	No	α6/β7	4.9	0.6 ± 1.6	0.0	1.000
B	E220-M222	No	α6/β7	6.7	1.4 ± 1.6	6.5	0.164
B	N221-M222	No	α6/β7	5.3	1.5 ± 1.9	6.6	0.158
B	P223-M222	No	α6/β7	5.3	2.1 ± 1.5	10.5	0.033
C	S218-C217	No	α6/β7	4.4	1.6 ± 1.9	8.5	0.075
C	S218-T219	?	α6/β7	4.4	0.5 ± 1.4	17.9	0.001
C	S218-E220	No	α6/β7	8.4	1.7 ± 1.3	12.9	0.012
C	S218-N221	No	α6/β7	9.7	2.0 ± 1.8	8.1	0.088
C	S218-M222	Yes?	α6/β7	6.3	2.2 ± 2.0	15.9	0.003
C	S218-P223	No	α6/β7	7.7	2.6 ± 1.4	10.5	0.033

The LRTs between the models of correlated and uncorrelated evolution were computed for the indicated site pair, as well as their associated p-values. The distances, in ångströms, between residues are also given. The pairs shown in the upper part of the table (Group A) are those whose members are four residues away from each other and they are outside helices and strands. Three of these tested site pairs were found within the N-terminal domain (NTD), while the other three were located in the C-terminal domain (CTD) of the eIF2α protein. In the middle (Group B) and lower (Group C) parts of the table, the relationships between M222 and its neighbors and S218 and its neighbors, respectively, are analyzed. For each pair of sites, the thermodynamic stability change (ΔΔG) for the 400 possible double-mutants was computed and the mean ± standard deviation is shown in kcal/mol

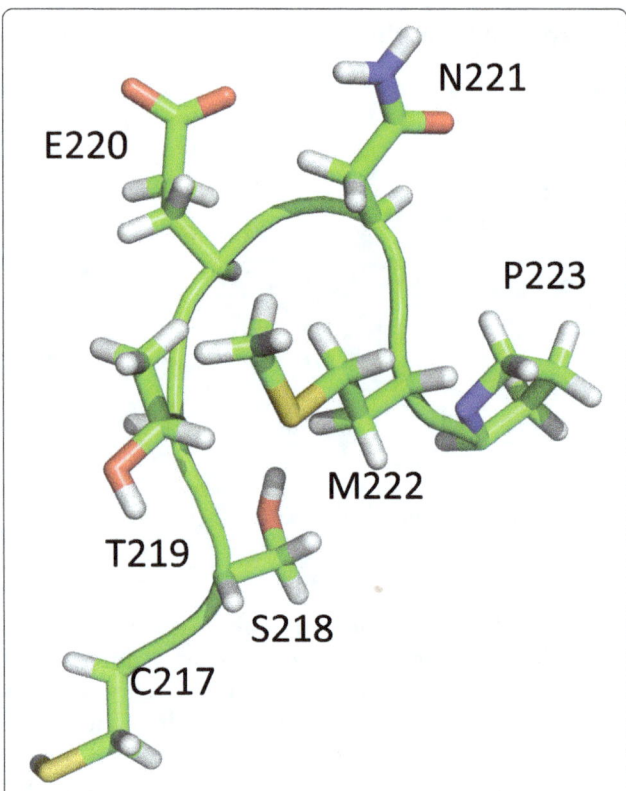

Fig. 4 Structure of the α6/β7 loop from eIF2α. The spatial disposition of the residues forming α6/β7 is shown using stick representation

common ancestor, and which thereby cannot be considered independent data points. Herein, instead of reconstructing the states of both characters at internal nodes using parsimony, which have a number of serious limitations [26], we rather carried out stochastic character mapping using a Markov chain Monte Carlo approach to sample character histories from their posterior probability distribution. In this way, after simulating 10 stochastic character histories, a contingency table showing the number of branches along which transitions occurred that ended in each of the four possible states, {(0,0), (0,1), (1,0), (1,1)}, could be computed (173, 22, 285 and 68, respectively). Making use of such contingency table, the null hypothesis that changes at position 218 (character X) and 222 (character Y) were independent was rejected (p-values = 0.021 and 0.016, for Yates' chi-squared and Fisher's exact tests, respectively).

Phosphorylation-sulfoxidation relationship in other stress-related proteins

For most protein kinases, the selection of target substrates is strongly influenced by the amino acid sequence surrounding the phospho-acceptor site [35]. The amino acids within these environments that either promote or compromise the phosphorylation are referred to as specificity determinants. In a recent work, we investigated those phosphorylation motifs where methionine may play a role as specificity determinant, finding that the reversible oxidation of methionines located at one (P + 1) or four (P + 4) positions carboxyl-terminal to the phosphosite was a process highly selective among stress-related proteins, which may couple oxidative signals with changes in protein phosphorylation [9]. In this way, besides eIF2α other three proteins constituents of SGs, such as ataxin-2, ataxin-2-like and pumilio homolog 1, were identified as potential targets for crosstalk between sulfoxidation and phosphorylation [9]. In humans, each of these proteins contains a serine that has been proved to be phosphorylatable [36] and that is accompanied by a methionine either at P + 1 or P + 4. In addition, these methionine residues are known to be oxidized in vivo after an oxidative stimulus [7]. Herein, the coevolution of these PTM site pairs (phosphorylatable serine and sulfoxidable methionine) was evaluated by the markovian-likelihood method described above for eIF2α. Table 4 summarizes the results of such analyses. As it can be observed, the evolution of the PTM site pairs was significantly better explained, in all the cases, by the dependent evolution model when compared to the model that assumed independent evolution.

Discussion

Many of the cellular responses triggered by oxidative stress are known to be mediated by signaling cascades involving protein phosphorylation [37, 38]. Despite the enormous research effort that has been devoted to the study of protein phosphorylation, the molecular mechanisms coupling oxidative signals to changes in phosphorylation remain poorly understood. A direct way through which oxidants may be sensed and transduced into biological responses involves reversible oxidation of protein-bound methionine to MetO [8]. Like phosphorylation, methionine oxidation is a reversible covalent PTM that can impact protein function in different ways. Thus, it has been shown that sulfoxidation of specific methionine residues can determine the subcellular distribution and activity of the target protein [39–41]. In yet another parallelism with phosphorylation, methionine oxidation can lead to either down-regulation [42, 43] or up-regulation [44, 45] of protein activity. In addition, both methionine oxidation and methionine sulfoxide reduction are reactions that can be enzyme-catalyzed [46, 47]. In this context, it has been proposed that oxidation of methionine, that converts the side chain of this amino acid from hydrophobic to hydrophilic [48], may provide the basis for regulating the specificity of protein kinase-substrate

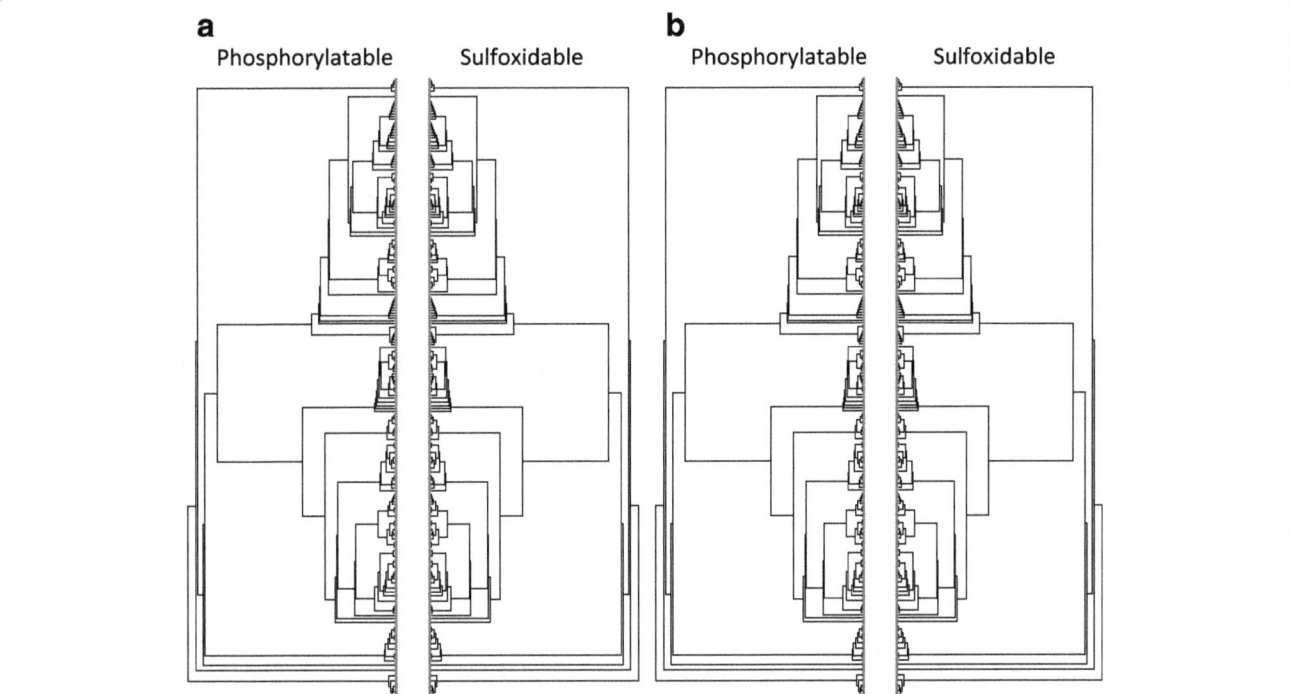

Fig. 5 Co-distribution of phosphorylatable and sulfoxidable residues through eukaryote evolution. **a**. The same phylogeny of eIF2α is mirrored to show the pattern of co-distribution of phosphorylatable and sulfoxidable residues. In the rightwards tree, the presence/absence of a phosphorylatable residue at position 218 of eIF2α in that eukaryotic species is indicated by a *red/blue dot*, respectively. Similarly, the presence/absence of a sulfoxidable residue at position 222 is indicated by a *red/blue dot* in the leftwards tree. **b**. Hypothetical co-distribution of both characters, made up to illustrate Darwin's scenario: when the co-distribution of specific states of two characters is the result of sharing a pair of synapomorphic characters, each emerged independently and then maintained each by its own motives in the descendants

interactions [49, 50]. In a previous study, we addressed the potential for crosstalk between sulfoxidation and phosphorylation at the proteome scale, reaching the conclusion that the interplay between serine phosphorylation and methionine oxidation was most prevalent among proteins involved in the control of translation during stress response. However, no study, hitherto, has developed an evolutionary approach to address the potential crosstalk between these two PTM types. To this respect, the original rationale for the current work was that if two PTM sites are functionally related and they are modified in a coordinated fashion, then they might have been coevolving over time.

Among the wide range of computational methods that have been proposed to detect coevolving residues (reviewed in [11, 51]), several exist that attempt to model coevolution in a phylogenetic context [29, 52]. However, far more popular methods are those that search for covariation between sites in a tree-independent manner. Although the need to account for the phylogenetic relationship is a well-acknowledged fact among evolutionary biologists, it has been poorly addressed, or simply ignored, by those authors more biased toward functional and/or structural biology. However, because molecular sequences share common ancestries and are therefore not independent from each other, the underlying evolutionary history

of sequences should be taken into account if we pretend to properly extract the coevolutionary signal out of the noise. Therefore, even though less popular, conceptually more complex and computationally expensive, we resorted to a phylogenetic method to assess whether the two eIF2α PTM sites of interest have been coevolving along the eukaryotic tree. To this end, the method proposed by Pagel to detect correlated evolution on phylogenies [22] was tailored to meet the requirements imposed by the molecular model. According to this method, a likelihood ratio test was used to discriminate between two models that were fitted to the data. Both models were based on continuous-time Markov processes, one allowing only for independent evolution of the two PTM sites, the other allowing for correlated changes. The results of this analysis convincingly favored the model where both characters evolve influencing each other. To further strengthen the conclusion that these PTM sites are bona fide coevolving sites, we accounted for the uncertainty in the phylogeny by repeating the analysis on different trees obtained using diverse approaches. To this respect, we can be confident that the described coevolution of these two PTM sites is a robust conclusion with respect to slight differences in the phylogeny employed (Table 2).

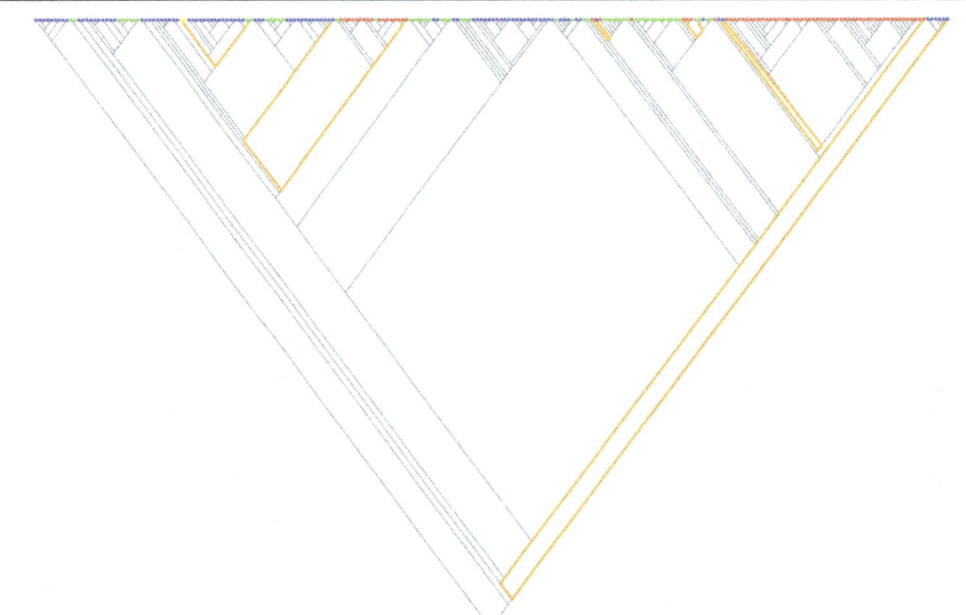

Fig. 6 Separate evolutionary origins of the co-occurrence of modifiable residues at positions 218 and 222 of eIF2α. The combined character states were encoded with colors as follows. Blue (state 1: (0,0)); yellow (state 2: (0,1)); green (state 3: (1,0)); red (state 4: (1,1)). Those pairs of species that differed in both characters and were phylogenetically separate (the path between them along the branches of the tree do not touch the path of any other pair) are shown connected by a thick orange line. Since we are examining pairs contrasting in both characters, there are two types of possible pairs: a pair is considered to be positive when the presence of a phosphorylatable residue in a member of the pair is accompanied by a methionine in the same member of the pair {red, blue}, otherwise the pair is referred to as negative {yellow, green}. In addition, to show that the co-occurrence of both PTMs has multiple and independent evolutionary origins, it can also be noted that the number of positive pairs is greater than the number of negative pairs

The statistical evidence that these two traits (PTM sites) coevolve across a range of species suggests that common selective pressures have been acting on the traits, which may point to a functional or adaptive relationship between them. Such conclusion was further supported by two additional observations. Firstly, the co-localization of a phosphorylatable residue at position 218 and a sulfoxidable methionine at 222 has emerged several times over the evolution of eukaryotes (Figs. 5 and 6), and secondly, something more than a mere structural effect seems to be behind the observed coevolution of these sites (Table 3 and Additional file 1: Figures S2 and S3). Indeed, it is expected that residues

that are close in the spatial structure of the protein will mutually influence their evolution. At such sites, a substitution that partly destabilizes the protein structure could be compensated by a subsequent change at an adjacent site to restore the stability. However, the results summarized in Table 3 suggest a functional, rather than structural, relationship between Ser-218 and Met-222. Hence, we hypothesize that the oxidation status of Met-222 allows protein kinases to monitor oxidative stress and subsequently to code this information in terms of Ser-218 phosphorylation. At this juncture, we wondered whether the presence/absence of a phosphorylatable residue at position 218 would influence the strength of the selective pressure acting on position 222, or, alternatively, whether possessing a sulfoxidable methionine at the position 222 promotes the gain/maintenance of a phospho-acceptor at 218. To examine these potential scenarios, we performed a number of analyses using reduced models. For instance, the hypothesis that the presence of a phosphorylatable residue favors the gain of methionine, was investigated by testing whether the rate of the transition parameter q_{34}, $(1,0) \rightarrow (1,1)$, differed from the rate of the transition parameter q_{12}, $(0,0) \rightarrow (0,1)$. Unfortunately, with the data at hand, there was not sufficient evidence to reject the null hypothesis for any of the examined reduced models (results not

Table 4 Testing for correlated evolution between PTM sites from SG

Protein	Sites	$l(I)$	$l(D)$	LRT	p-Value	N
eIF2α	S218-M222	−108.2	−100.2	15.9	0.003	233
Ataxin-2	S814-M815	−95.3	−89.8	10.9	0.027	215
Ataxin-2-like	S211-M215	−135.0	−120.8	28.3	10^{-5}	215
Pumilio homolog 1	S124-M125	−38.9	−23.3	31.2	2.8×10^{-6}	51

The column 'Sites' gives the residue (S: serine and M: methionine) found at the indicated position in the human ortholog sequence. LRT stands for the likelihood ratio test statistics and N for the number of species included in the analyses. $l(I)$: natural logarithm of the likelihood value for the model of independent evolution. $l(D)$: natural logarithm of the likelihood value for the model of dependent evolution shown). Therefore, although the results from the current work strongly support the conclusion that both PTM sites have been coevolving, we failed to identify the probable temporal ordering of changes in these two traits.

Methionine, a relatively hydrophobic amino acid, can be found as a specificity determinant in a number of protein kinase substrate motifs [35]. Although methionine can occupy any position within these canonical recognition motifs, it is most often found at 1 or 4 positions carboxyl-terminal to the phosphorylatable serine (P + 1 and P + 4, respectively). On the other hand, oxidation of methionine at these positions has been described as a process highly selective, as opposite to random [9]. Since the oxidation of methionine to methionine sulfoxide converts the side chain of this amino acid from hydrophobic to polar and increases the capacity for H-bonding [48], the redox state at positions P + 1 and P + 4 can impact the recognition of these protein substrates by their cognate protein kinase and/or phosphatase, providing, in this way, a mechanistic coupling between oxidative signals and phosphorylation status. Interestingly, when, in a previous study, we carried out GO analysis to gain insight into the processes that may be regulated by crosstalk between Ser/Thr phosphorylation and sulfoxidation of methionine at P + 1 or P + 4, it turned out that the occurrence of MetO near phosphoserine was more prevalent in proteins related to control of translation and stress related proteins. In addition, a small set of proteins related to SGs was identified as potential target for crosstalk between sulfoxidation and phosphorylation. Therefore, in the current study we extended the coevolutionary analysis described for eIF2α to this set of stress-related proteins. For the four analyzed proteins, their respective pairs of PTM sites (phosphorylatable serine and sulfoxidable methionine at either P + 1 or P + 4) were found to be evolving in a correlated fashion (Table 4), which again suggests a relevant role for methionine sulfoxidation and serine phosphorylation crosstalk in response to oxidative stress. Overall, the findings described in this study should encourage further systematic biochemical and genetic studies aimed at understanding the role of methionine sulfoxide in the control of protein translation.

Conclusions

Protein-bond methionine sulfoxidation was initially perceived as an inevitable damage derived from aerobic metabolism. However, this view of methionine as a vulnerable residue representing the Achilles' heel of proteins has been gradually changing since in the 1990s Levine and coworkers proposed a role in the antioxidant defense for methionine residues as ROS scavengers More recently, the sulfoxidation of certain specific methionine residues is emerging as a posttranslational modification capable of regulating protein

activity during stress conditions. In this line, we have shown that the oxidation of methionines housed within phosphorylation motifs is a highly selective process among stress-related proteins. In the current study, using evolutionary models based on continuous-time Markov chains, we have addressed the interrelationship between phosphorylation and sulfoxidation in four proteins related with the SGs. We have found their respective pairs of phosphorylatable-sulfoxidable PTM sites to be evolving in a correlated fashion through the eukaryotic lineage, which suggests a relevant role for serine/threonine phosphorylation and methionine sulfoxidation crosstalk in the control of protein synthesis during stress conditions.

Abbreviations
CTD: C-terminal domain; eIF: Eukaryotic initiation factor; GO: Gene ontology; LRT: Likelihood ratio test; MetO: Methionine sulfoxide; ML: Maximum likelihood; MP: Maximum parsimony; MSA: Multiple sequence alignment; NJ: Neighbor joining; NTD: N-terminal domain; PTM: Post-translational modification; SGs: Stress granules

Acknowledgements
The author thanks Alicia Esteban del Valle for her invaluable help with the manuscript preparation. The author is also grateful to two anonymous referees who have helped to improve the original manuscript.

Funding
Publishing cost were partially covered by the Universidad de Málaga.

Competing interests
The author declares that he has no competing interests.

References
1. Sonenberg N, Hinnebusch AG. Regulation of translation initiation in eukaryotes: mechanisms and biological targets. Cell. 2009;136:731–45.
2. Rowlands AG, Panniers R, Henshaw EC. The catalytic mechanism of guanine nucleotide exchange factor action and competitive inhibition by phosphorylated eukaryotic initiation factor 2. J Biol Chem. 1988;263(12): 5526–33.
3. Knutsen JHJ, Rødland GE, Bøe CA, Håland TW, Sunnerhagen P, Grallert B, et al. Stress-induced inhibition of translation independently of eIF2α phosphorylation. J Cell Sci. 2015;128:4420–7.
4. Shenton D, Smirnova JB, Selley JN, Carroll K, Hubbard SJ, Pavitt GD, et al. Global translational responses to oxidative stress impact upon multiple levels of protein synthesis. J Biol Chem. 2006;281(39):29011–21.
5. Hamanaka RB, Bennett BS, Cullinan SB, Diehl JA. PERK and GCN2 contribute to eIF2 alpha phosphorylation and cell cycle arrest after activation of the unfolded protein response pathway. Mol Biol Cell. 2005;16:5493–501.
6. Kettenbach AN, Schweppe DK, Faherty BK, Pechenick D, Pletnev AA, Gerber SA. Quantitative phosphoproteomics identifies substrates and functional modules of aurora and polo-like kinase activities in mitotic cells. Sci Signal. 2011;4(179):rs5.
7. Ghesquière B, Jonckheere V, Colaert N, Van Durme J, Timmerman E, Goethals M, et al. Redox proteomics of protein-bound methionine oxidation. Mol Cell Proteomics. 2011;10:M110.006866.
8. Kim G, Weiss SJ, Levine RL. Methionine oxidation and reduction in proteins. Biochim Biophys Acta. 1840;2014:901–5.
9. Veredas FJ, Cantón FR, Aledo JC. Methionine residues around phosphorylation sites are preferentially oxidized in vivo under stress conditions. Scientific Rep. 2017;7:40403.
10. Ochoa D, Pazos F. Practical aspects of protein co-evolution. Front Cell Dev Biol. 2014;2:1–9.

11. de Juan D, Pazos F, Valencia A. Emerging methods in protein co-evolution. Nat Rev Genet. 2013;14:249–61.

12. Pazos F, Helmer-Citterich M, Ausiello G, Valencia A. Correlated mutations contain information about protein-protein interaction. J Mol Biol. 1997;271: 511–23.

13. Dunn SD, Wahl LM, Gloor GB. Mutual information without the influence of phylogeny or entropy dramatically improves residue contact prediction. Bioinformatics. 2007;24:333–40.

14. Weigt M, White R, Szurmant H, Hoch J, Hwa T. Identification of direct residue contacts in protein–protein interaction by message passing. Proc Natl Acad Sci U S A. 2009;106:67–72.

15. Morcos F, Pagnani A, Lunt B, Bertolino A, Marks D, Sander C, et al. Direct-coupling analysis of residue coevolution captures native contacts across many protein families. Proc Natl Acad Sci U S A. 2011;108:E1293–301.

16. Jones DT, Buchan DWA, Cozzetto D, Pontil M. PSICOV: precise structural contact prediction using sparse inverse covariance estimation on large multiple sequence alignments. Bioinformatics. 2012;28:184–90.

17. Talavera D, Lovell SC, Whelan S. Covariation is a poor measure of molecular coevolution. Mol Biol Evol. 2015;32:2456–68.

18. Huerta-Cepas J, Szklarczyk D, Forslund K, Cook H, Heller D, Walter MC, et al. eggNOG 4.5: a hierarchical orthology framework with improved functional annotations for eukaryotic, prokaryotic and viral sequences. Nucl Ac Res. 2016;44:D286–93.

19. Saitou N, Nei M. The neighbor-joining method: a new method for reconstructing phylogenetic trees. Mol Biol Evol. 1987;4:406–25.

20. Schliep KP. Phangorn: phylogenetic analysis in R. Bioinformatics. 2011;27:592–3.

21. Paradis E, Claude J, Strimmer K. APE: analyses of phylogenetics and evolution in R language. Bioinformatics. 2004;20:289–90.

22. Pagel M. Detecting correlated evolution on phylogenies: a general method for the comparative analysis of discrete characters. Proc R Soc Lond B. 1994; 255:37–45.

23. Harmon LJ, Weir JT, Brock CD, Glor RE, Challenger W. GEIGER: investigating evolutionary radiations. Bioinformatics. 2007;24:129–31.

24. Maddison WP. Testing character correlation using pairwise comparisons on a phylogeny. J Theor Biol. 2000;202:195–204.

25. Huelsenbeck JP, Nielsen R, Bollback JP. Stochastic mapping of morphological characters. Syst Biol. 2003;52:131–58.

26. Revell LJ. Phytools: an R package for phylogenetic comparative biology (and other things). Methods Ecol Evol. 2012;3:217–23.

27. Schymkowitz J, Borg J, Stricher F, Nys R, Rousseau F, Serrano L. The FoldX web server: an online force field. Nucl Ac Res. 2005;33:W382–8.

28. Guerois R, Nielsen JE, Serrano L. Predicting changes in the stability of proteins and protein complexes: a study of more than 1000 mutations. J Mol Biol. 2002;320:369–87.

29. Pollock DD, Taylor WR, Goldman N. Coevolving protein residues: maximum likelihood identification and relationship to structure. J Mol Biol. 1999;287:187–98.

30. Dar AC, Dever TE, Sicheri F. Higher-order substrate recognition of eIF2α by the RNA-dependent protein kinase PKR. Cell. 2005;122:887–900.

31. Hornbeck PV, Zhang B, Murray B, Kornhauser JM, Latham V, Skrzypek E. PhosphoSitePlus, 2014: mutations. PTMs and Recalibrations Nucl Ac Res. 2015;43:D512–20.

32. Landry CR, Levy ED, Michnick SW. Weak functional constraints on phosphoproteomes. Trend Gen. 2009;25:193–7.

33. Maddison WP, FitzJohn RG. The unsolved challenge to phylogenetic correlation tests for categorical characters. Syst Biol. 2014;64:127–36.

34. Ridley M. The explanation of organic diversity: the comparative method and adaptations for mating. Oxford: Clarendon Press; 1983.

35. Amanchy R, Periaswamy B, Mathivanan S, Reddy R, Tattikota SG, Pandey A. A curated compendium of phosphorylation motifs. Nat Biotechnol. 2007;25:285–6.

36. Hornbeck PV, Kornhauser JM, Tkachev S, Zhang B, Skrzypek E, Murray B, et al. PhosphoSitePlus: a comprehensive resource for investigating the structure and function of experimentally determined post-translational modifications in man and mouse. Nucl Ac Res. 2011;40:D261–70.

37. Burgoyne JR, Oka S-I, Ale-Agha N, Eaton P. Hydrogen peroxide sensing and signaling by protein kinases in the cardiovascular system. Antioxid Red Signal. 2013;18:1042–52.

38. Schieber M, Chandel NS. ROS function in redox signaling and oxidative stress. Curr Biol. 2014;24:R453–62.

39. Gallmetzer A, Silvestrini L, Schinko T, Gesslbauer B, Hortschansky P, Dattenböck C, et al. Reversible oxidation of a conserved methionine in the nuclear export sequence determines subcellular distribution and activity of the fungal nitrate regulator NirA. PLoS Gen. 2015;11:e1005297–27.

40. Allu PK, Marada A, Boggula Y, Karri S, Krishnamoorthy T, Sepuri N. Methionine sulfoxide reductase 2 reversibly regulates Mge1, a cochaperone of mitochondrial Hsp70, during oxidative stress. Mol Biol Cell. 2015;26:406–19.

41. Lee BC, Péterfi Z, Hoffmann FW, Moore RE, Kaya A, Avanesov A, et al. MsrB1 and MICALs regulate actin assembly and macrophage function via reversible stereoselective methionine oxidation. Mol Cell. 2013;51:397–404.

42. Hämdahl U, Kokke BP, Gustavsson N, Berggren K, Tjerneld F, Boelens WC, et al. The chaperone-like activity of a small heat shock protein is lost after sulfoxidation of conserved methionines in a surface amphipathic alpha-helix. Biochim Biophys Acta. 2001;1545(1–2):227–37.

43. Taggart C, Cervantes-Laurean D, Kim G, McElvaney NG, Wehr N, Moss J, et al. Oxidation of either methionine 351 or methionine 358 in alpha 1-antitrypsin causes loss of anti-neutrophil elastase activity. J Biol Chem. 2000; 275:27258–65.

44. Tang XD, Daggett H, Hanner M, Garcia ML, McManus OB, Brot N, et al. Oxidative regulation of large conductance calcium-activated potassium channels. J Gen Physiol. 2001;117:253–73.

45. Drazic A, Miura H, Peschek J, Le Y, Bach N, Kriehuber T, et al. Methionine oxidation activates a transcription factor in response to oxidative stress. Proc Natl Acad Sci U S A. 2013;110:9493–8.

46. Lim JC, You Z, Kim G, Levine RL. Methionine sulfoxide reductase a is a stereospecific methionine oxidase. Proc Natl Acad Sci U S A. 2011;108:10462–77.

47. Manta B, Gladyshev VN. Regulated methionine oxidation by monooxygenases. Free Radic. Biol. Med. Elsevier. 2017;17:30072.

48. Black SD, Mould DR. Development of hydrophobicity parameters to analyze proteins which bear post- or cotranslational modifications. Anal Biochem. 1991;193:72–82.

49. Hardin SC, Larue CT, Oh M-H, Jain V, Huber SC. Coupling oxidative signals to protein phosphorylation via methionine oxidation in Arabidopsis. Biochem J. 2009;422:305–12.

50. Miernyk JA, Johnston ML, Huber SC, Tovar Méndez A, Hoyos E, Randall DD. Oxidation of an adjacent methionine residue inhibits regulatory seryl-phosphorylation of pyruvate dehydrogenase. Proteomics Insights. 2009;2:15–22.

51. Dutheil JY. Detecting coevolving positions in a molecule: why and how to account for phylogeny. Brief Bioinform. 2011;13:228–43.

52. Yeang C-H, Haussler D. Detecting coevolution in and among protein domains. PLoS Comput Biol. 2007;3:e211–3.

53. Penny D, Hendy MD. The use of tree comparison metrics. Syst Biol. 1985;34: 75–82.

54. Kuhner MK, Felsenstein J. A simulation comparison of phylogeny algorithms under equal an unequal evolutionary rates. Mol Biol Evol. 1994;11:459–68.

9

Expression and phylogenetic analyses reveal paralogous lineages of putatively classical and non-classical MHC-I genes in three sparrow species (*Passer*)

Anna Drews[*] ⓘ, Maria Strandh, Lars Råberg and Helena Westerdahl

Abstract

Background: The Major Histocompatibility Complex (MHC) plays a central role in immunity and has been given considerable attention by evolutionary ecologists due to its associations with fitness-related traits. Songbirds have unusually high numbers of MHC class I (MHC-I) genes, but it is not known whether all are expressed and equally important for immune function. Classical MHC-I genes are highly expressed, polymorphic and present peptides to T-cells whereas non-classical MHC-I genes have lower expression, are more monomorphic and do not present peptides to T-cells. To get a better understanding of the highly duplicated MHC genes in songbirds, we studied gene expression in a phylogenetic framework in three species of sparrows (house sparrow, tree sparrow and Spanish sparrow), using high-throughput sequencing. We hypothesize that sparrows could have classical and non-classical genes, as previously indicated though never tested using gene expression.

Results: The phylogenetic analyses reveal two distinct types of MHC-I alleles among the three sparrow species, one with high and one with low level of polymorphism, thus resembling classical and non-classical genes, respectively. All individuals had both types of alleles, but there was copy number variation both within and among the sparrow species. However, the number of highly polymorphic alleles that were expressed did not vary between species, suggesting that the structural genomic variation is counterbalanced by conserved gene expression. Overall, 50% of the MHC-I alleles were expressed in sparrows. Expression of the highly polymorphic alleles was very variable, whereas the alleles with low polymorphism had uniformly low expression. Interestingly, within an individual only one or two alleles from the polymorphic genes were highly expressed, indicating that only a single copy of these is highly expressed.

Conclusions: Taken together, the phylogenetic reconstruction and the analyses of expression suggest that sparrows have both classical and non-classical MHC-I genes, and that the evolutionary origin of these genes predate the split of the three investigated sparrow species 7 million years ago. Because only the classical MHC-I genes are involved in antigen presentation, the function of different MHC-I genes should be considered in future ecological and evolutionary studies of MHC-I in sparrows and other songbirds.

Keywords: MHC class I, Passer, sparrows, Classical genes, Non-classical genes, gene expression

* Correspondence: anna.drews@biol.lu.se
Department of Biology, Lund University, Ecology Building, 223 62 Lund, Sweden

Background

The major histocompatibility complex (MHC) is a key component of adaptive immunity and holds the most polymorphic genes known in the vertebrate genome [1]. MHC class I (MHC-I) proteins are expressed on all nucleated cells whereas MHC-II proteins are expressed only on antigen presenting cells [2]. Typically, animals have a handful of functional MHC-I genes, as exemplified by humans (six genes), swine (*Sus scrofa domesticus*; six genes) and domestic chicken (*Gallus gallus*; four genes) [3–5]. On the contrary, songbirds of the order Passeriformes have a larger number of MHC-I genes than most other species investigated to date [6–8]. O'Connor et al. (2015) reported between four and 20 MHC-I genes per individual across Passerida (i.e. genomic MHC-I exon 3 sequences in open reading frame), and Biedrzycka et al. (2017) found 65 alleles per individual, i.e. at least 33 MHC-I genes, in the sedge warbler *Acrocephalus schoenobaenus* [8, 9]. The functional significance of all these MHC-I gene copies in songbirds is not known.

In most species studied to date — for example humans and other primates, swine, mice and chicken — MHC-I genes are categorized as classical or non-classical MHC-I genes [3, 5, 10, 11]. In mammals, classical MHC-I genes (MHC-Ia) are highly polymorphic and highly expressed, whereas non-classical (MHC-Ib) are less polymorphic and have low expression [12, 13]. The non-classical genes do not appear to have a common origin among distantly related mammals but seem to have arisen independently from recent duplications of classical MHC-I genes within species [14]. MHC-Ia molecules play an important role in adaptive immunity by presenting peptides to T-cells [1], whereas MHC-Ib molecules have other immune functions [12, 13, 15, 16]. Humans have three MHC-Ia genes (HLA-A, -B and -C) and three MHC-Ib genes (HLA-E, -F and -G) [3]. HLA-A, -B and -C have a much larger number of alleles (>9000 world-wide) than HLA-E, -F and -G (<90 world-wide) [17]. HLA-A, -B and -C genes are expressed in most tissues, but there are gene-specific expression differences among the genes; HLA-C is expressed to a lower degree than HLA-A and -B resulting in high variation in expression levels among classical genes [18].

Classical and non-classical class I genes have also been reported in birds of the order Galliformes, e.g. in chicken, turkey (*Meleagris gallopavo*) and golden pheasant (*Chrysolophus pictus*) and here the MHC-Ia and MHC-Ib genes are referred to as MHC-B and MHC-Y, respectively [4, 19–21]. The chicken has two classical MHC-I genes at the MHC-B locus and two non-classical MHC-I genes at the MHC-Y locus [4]. Both genes at the classical B-locus are expressed, but the 'major' gene is highly expressed compared to the 'minor' gene [22–24]. Only one of the non-classical Y-locus genes has been shown to be expressed and then specifically in spleen [25, 26]. The presence of classical and non-classical MHC-I genes is less established in non-galliform birds, but has been suggested in species of the orders Anseriformes, Charadriiformes and Pelecaniformes [27–30]. Moreover, in Anseriformes and Pelecaniformes there seem to be one putatively classical gene that is highly expressed [30, 31].

MHC genes evolve by frequent gene duplications, but there is also gene loss when gene copies become nonfunctional, and the evolution of MHC therefore fit a birth-and-death model of molecular evolution [32–34]. The large number of MHC-I genes in the genomes of songbirds indicates either a higher rate of gene duplications in songbirds compared to other animals, or that several copies of their MHC genes have been duplicated simultaneously [6–8]. Very little is known about neofunctionalization among the multiple gene copies in songbirds, though it seems likely that some gene copies have evolved different functions, like the classical and non-classical genes found in other species [27–30]. It is important to distinguish non-classical and classical MHC genes in evolutionary and ecological studies since only the latter are subject to balancing selection and expected to be associated with disease resistance and fitness.

Karlsson and Westerdahl [35] showed that house sparrows (*Passer domesticus*) have two types of MHC-I alleles that exhibit some of the hallmarks of classical and non-classical genes. The putatively non-classical house sparrow MHC-I alleles have low levels of polymorphism, few positively selected sites and form a distinct phylogenetic cluster, whereas the putatively classical MHC-I alleles have high polymorphism, many positively selected sites and do not form a supported phylogenetic cluster [35]. The putatively non-classical alleles in house sparrows are easily identified by a six base pair deletion in exon 3 [35, 36]. However, the expression pattern of these putatively classical and non-classical genes in house sparrow has not been investigated. Hence, a way to further establish the occurrence of classical and non-classical MHC genes in house sparrows would be to measure their relative expression. Classical genes are often highly expressed compared to non-classical genes, and certain genes within each category might be more highly expressed, e.g. if 'major' and 'minor' loci are present as indicated in species of the bird orders Galliformes, Anseriformes and Pelecaniformes [25–30].

Previous studies on the evolution of MHC genes have shown that orthologous classical MHC genes often survive longer in the genome—over speciation events—than non-classical genes [10]. A next step to continue studying putatively non-classical genes in Passerines is therefore to investigate their occurrence in species that are closely related to house sparrows. Finding putatively classical and non-classical genes in several species would not only make the finding more solid but also give an indication of their evolutionary age.

We set out to investigate the structure, number and expression patterns of MHC-I genes among three sparrow species, the house sparrow, the Spanish sparrow (*Passer hispaniolensis*) and the tree sparrow (*Passer montanus*), in order to identify neo-functionalization of MHC-I genes, in particular whether there are both putatively classical and non-classical genes. We firstly reconstruct allelic phylogenies where we classify the sparrow MHC-I alleles as putatively classical or non-classical genes and then test if these putatively classical and non-classical alleles differ in i) number of genomic alleles between species, ii) number of expressed alleles between species and iii) relative gene expression within species.

Methods
Study species
We study three species of sparrows in the Passer clade; house sparrow (*Passer domesticus*) (*n* = 5), Spanish sparrow (*P. hispaniolensis*) (*n* = 3) and tree sparrow (*P. montanus*) (*n* = 5). The native range of the house sparrow and tree sparrow covers most of Eurasia whereas the Spanish sparrow has a more restricted distribution around the Mediterranean Sea and in south-west Asia [37]. All three species live in both urban and rural environments [37]. In order to determine when house sparrow, Spanish sparrow and tree sparrow separated phylogenetically a maximum clade credibility tree was constructed based on data from 23 Passer species and an outgroup (*Cyanistes caeruleus*) from the Bird Tree website [38, 39], for details see Additional file 1: Method S1. House sparrows and Spanish sparrows split 3 million years ago, while the tree sparrows are more distantly related and split from the other two species 7 million years ago.

Sample collection and extractions
House sparrows and tree sparrows were caught, with mist nets, in Löberöd, Skåne, Sweden. The Spanish sparrows were kept and caught in aviaries at University of Oslo, Norway. All samples were collected during the autumn of 2012. Blood samples (20-40 µl) were taken from the brachial vein and then stored either at −20 °C in SET buffer (150 mM NaCl, 50 mM TRIS, 1 mM EDTA, pH 8.0), for DNA extraction, or at 4 °C in 100 µl K_2EDTA and 500 µl TRIzol LS (Life Technologies, Carlsbad, CA, USA), for RNA extraction. DNA was extracted with ammonium acetate extraction [40]. RNA was extracted with a combination of the TRIzol LS protocol (Life Technologies, Carlsbad, CA, USA) and the RNeasy Mini kit (QIAGEN, Hilden, Germany). Briefly, the homogenization and phase separation was done according

to the TRIzol LS protocol, resulting in an aqueous phase. One volume of 70% EtOH was added to the aqueous phase and from this step the RNeasy protocol was followed, including an on-column DNase treatment [41]. The RNA (mRNA) was reverse transcribed to complementary DNA (cDNA) using the RETROscript kit (Life Technologies, Carlsbad, CA, USA) according to the manufacturer's protocol.

High-throughput amplicon sequencing
We sequenced partial MHC-I exon 3 amplicons (185–226 bp) obtained from genomic DNA (gDNA) and cDNA (to examine gene expression) from house sparrows, tree sparrows and Spanish sparrows using 454 amplicon sequencing. Four different primer combinations were used to amplify MHC-I exon 3 alleles in each species to minimize the effects of amplification bias and allelic dropouts which is often a problem when using only one primer pair (Additional file 1: Table S1, Figure S1). Primer combinations 1 and 2 amplify both putatively classical and non-classical alleles whereas primer combination 3 and 4 exclusively amplifies putatively classical and non-classical alleles, respectively [8, 35, 42]. Each individual was represented by one gDNA and one cDNA sample and samples were technically duplicated (two PCRs from 40% of the samples). We performed PCR with individually tagged 454 fusion primers (6-bp tag on forward and reverse primer) [43]. Each 15 µl PCR reaction contained either 25 ng gDNA or 10 ng cDNA, 0.2 µM of each primer and 1× QIAGEN Multiplex PCR Master Mix (QIAGEN, Hilden, Germany). The cycling conditions for primer combination 1 and 2 were set to 35 cycles at 95 °C (30s), 60 °C (90s), 72 °C (60s) followed by 72 °C for 10 min. For primer combination 3 and 4 the cycling conditions were 30 cycles at 95 °C (30s), 65 °C (60s), 72 °C (60s) followed by 72 °C for 10 min. The PCR products were verified on a 2% agarose gel and products were pooled semi-equimolarly based on the strength of the bands with a maximum of eight products per pool. These pools were purified on MinElute PCR purification columns (QIAGEN, Hilden, Germany) according to the manufacturer's protocol and quantified on a NanoDrop 2000/2000c (Thermo Fisher Scientific, Wilmington, DE, USA). The purified pools were pooled in equimolar DNA amounts in one final pool per primer combination. Amplicons were sequenced in two separate 454 sequencing runs (one for primer combination 1 and 2 amplicons and another for primer combination 3 and 4 amplicons) at the Lund University DNA Sequencing facility, Faculty of Science, Sweden.

Filtering of high-throughput sequencing data
There are errors associated with high-throughput sequencing techniques that will result in artefactual alleles

(AA). The AAs were distinguished from the true alleles (TA) and removed from the dataset by a number of filtering steps, (for details see Additional file 1: Methods S2). Briefly, the first filtering step handled AA originating from homopolymer errors, these were identified by eye and if a possible AA always occurred with its parental sequences the read depth from the AA was added to that of the parental sequence. As a second step all amplicons with insufficient coverage was removed (the threshold was set to 110 for gDNA (and to 140 for cDNA, values in brackets) for primer combination 1, 70 (100) for primer combination 2, 100 (180) for primer combination 3 and 100 (140) for primer combination 4, see Additional file 1: Methods S2 for more details). Next, all sequences that had too few reads within an amplicon were deleted, this read depth was measured as percentage of the total read depth (varying from 1.1% to 3.0% depending on species and primer combination, see Additional file 1: Methods S2 for more details). All sequences that varied by 1–2 bp were identified and the read depth of the possible AA was only added to the parental sequences if the AA occurred only once in the entire data set, together with the parental sequence and when the read depth of the possible AA was less than half of the parental sequence. Chimeras and non-functional sequences were identified by eye and deleted from the data set. Possible chimeras were deleted from the data set only when they occurred with both parental sequences and when the read depth of the putative chimera was less than half of both parental sequences. Sequences that only occurred in one amplicon in the entire data set or in a single amplicon within an individual were deleted. Lastly low frequency sequences that were only amplified in cDNA were deleted since all cDNA sequences should also be found in the corresponding gDNA sample.

All sequences that remained after the strict filtering were considered TA. BLAST was used to determine which alleles had previously been published. When an allele was 100% identical to a previously published sequence and of the same length the allele was named according to the published sequence. If an allele was 100% identical to a previously published sequence but of different length the allele was given the name of the published sequence followed by 'a'. Alleles that had not previously been published were given species specific names according to the recommended guide lines for naming MHC alleles [44]. All new sequences were deposited in GenBank (GenBank Acc nr KY303944-KY304003). TA in the cDNA samples, i.e. transcribed alleles, are hereafter called expressed alleles. Thirty-six samples were run in duplicates and the repeatability between duplicates was calculated as the percentage of total number of alleles amplified (i.e. the concatenated

number of alleles using both duplicates as 'total number').

Number and expression of MHC-I alleles

The use of multiple primer combinations enabled a detailed characterization of the total number of classical and non-classical MHC-I alleles per individual since the possibility of amplifying all alleles in an individual increases when different primer combinations are used [8, 42]. The number of alleles per individual was determined by combining the result from all four primer combinations. The amplification range of the four different primer combinations was calculated as the proportion of the total number of alleles amplified with all primer combinations. This was done separately for each individual and an average was calculated for classical and non-classical alleles in each species (Additional file 1: Table S2). We used primer combination 1 in the expression analysis since this primer combination amplified the majority of all the classical and non-classical alleles simultaneously. With the concatenated result from all four primer combinations we identified which alleles were expressed in each individual and in the expression analysis we only included individuals where primer combination 1 amplified the majority of all expressed alleles. Primer combination 1 fulfilled these strict criteria in six individuals (house sparrow, $n = 3$, tree sparrow, $n = 3$). The relative expression of each allele was estimated as the proportion of the total number of reads per individual. These six individuals expressed up to four classical and up to four non-classical alleles per individual, the maximum total number of expressed alleles was eight. In order to get an estimate of how many alleles that were highly expressed we set a custom threshold to define and separate highly expressed alleles from the remaining alleles based on the following reasoning: Given that there was a maximum of eight alleles per individual, each allele would with an even distribution contribute with 12.5% of the reads. We set the threshold for 'high allelic expression' to twice as high; hence highly expressed alleles should contribute with more than 25% of the reads in an individual.

Statistical analysis and phylogenetic relationship

Statistical analyses were performed in SPSS (IBM SPSS Statistics 22). The differences between species regarding number of alleles and number of expressed alleles were determined with one-way ANOVAs. The differences in variation in expression between classical and non-classical genes were determined in two ways. First, when only expressed alleles were included in the model, Levine's test of equal variance was used. Second, when also non-expressed alleles (i.e. alleles with zero read depth in cDNA sample) were included in the model the data was no longer normally distributed and hence the Brown-Forsythe variance test was used. In order to determine the phylogenetic relationship between MHC-I alleles both a maximum likelihood tree and a neighbor-net network were construed. The maximum likelihood tree was constructed with the RAxML software (version 7.0.4) using the GTRGAMMA model and 1000 bootstraps and illustrated with iTOL (version 3.4.3) [45]. The network was constructed with SplitsTree v. 4.14.4 [46] using a GTR model with the α parameter for gamma distribution set to 0.3450, which was recommended by jModelTest 2.1.10 [47] and 1000 bootstraps. Figures were produced in R, (version 2.15.3), using the built in packages barplot and plot [48].

Results

Phylogenetic relationship of putatively classical and non-classical MHC-I genes

MHC-I alleles were genotyped in 13 individuals, five house sparrows, three Spanish sparrows and five tree sparrows, using high-throughput sequencing of both gDNA (genomic DNA) and cDNA (complementary DNA, i.e. reverse transcribed RNA, as a measure of gene expression). The average read depth of true alleles (alleles that remained after filtering the HTS data) per individual varied between 413 and 990 reads for gDNA and between 528 and 1108 reads for cDNA (combining three or four primer combinations, for further details on read depth see Additional file 1: Table S3). The repeatability in genotyping between duplicates varied between 91% and 100% across primers (Additional file 1: Table S4). Putatively classical and non-classical MHC-I genes were found in all three Passer species, and the putatively non-classical alleles were identified by a 6 bp deletion in exon 3, as described previously for house sparrows. In total, in all three species, 129 alleles were identified (Additional file 1: Figure S2).

The phylogenetic relationship of the putatively classical and non-classical MHC-I genes placed all putatively non-classical alleles in a distinct cluster, in the maximum likelihood tree and the neighbor-net network with high bootstrap support, 92 and 93 respectively (Fig. 1, Additional file 1: Figure S3). This shows that the separation of putatively classical and non-classical MHC-I genes predates the speciation of the investigated sparrow species. The putatively non-classical alleles have short branches and are hence highly similar, whereas the putatively classical alleles are much more variable, and this is further supported by the higher nucleotide diversity and amino acid sequence per nucleotide sequence for classical alleles (Additional file 1: Table S5). Moreover, no clear phylogenetic separation based on expression could be seen since expressed alleles were found across the phylogenetic tree (Fig. 1, Additional file 1: Figure S3).

Number of putatively classical and non-classical MHC-I alleles among sparrow species

The number of putatively classical gDNA alleles per individual varied significantly between species ($F = 8.418$, $p = 0.007$), as did the number of putatively non-classical gDNA alleles ($F = 7.003$, $p = 0.013$, Fig. 2, Additional file 1: Table S6, Table S7). The highest number of gDNA alleles in a house sparrow was seven putatively classical and 13 putatively non-classical alleles, in a Spanish sparrow six and 12 and in a tree sparrow 14 and seven. The number of expressed putatively non-classical alleles varied significantly between the three species ($F = 13.018$, $p = 0.002$, Fig. 2, Additional file 1: Table S6, Table S7), whereas no such difference was seen for the number of expressed putatively classical alleles. The highest number of expressed alleles in house sparrows was four putatively classical and five putatively non-classical, in Spanish sparrows four and six and in tree sparrows four and three.

Variance in expression of classical and non-classical MHC-I genes in house sparrows and tree sparrows

Putatively classical MHC-I genes had a significantly higher variance in expression, measured as relative read depth, than putatively non-classical genes (Levine's test: $F = 5.20$, $p = 0.005$; Fig. 3a). This difference between putatively classical and non-classical genes is still present when non-expressed alleles are included in the model (Brown-Forsythe variance test: $F = 6.62$, $p = 0.012$). The large variance in expression among putatively classical genes suggests that only a subset of these genes is highly expressed. The variance in relative read depth was not significantly different between putatively classical and non-classical genes in gDNA (Levine's test: $F = 0.883$, $p = 0.455$, Fig. 3b), and hence there is no bias in amplification efficiency of the primers between the genes. The difference in variance seen for expressed putatively classical and non-classical genes can therefore not be a result of biased amplification of certain alleles. Two tree sparrow individuals were run in duplicates and the relative read depth is highly similar between duplicates, (Additional file 1: Table S8). Only three house sparrows and three tree sparrows fulfilled the criteria to be included in the expression analyses, i.e. that the majority of all the identified expressed alleles were amplified with primer combination 1 (for further details see Additional file 1: Table S9, Figure S4). In these six individuals at most two putatively classical alleles were highly expressed per individual suggesting that only a single putatively classical gene is highly expressed in sparrows (assuming heterozygosity). None of the putatively non-classical alleles were highly expressed.

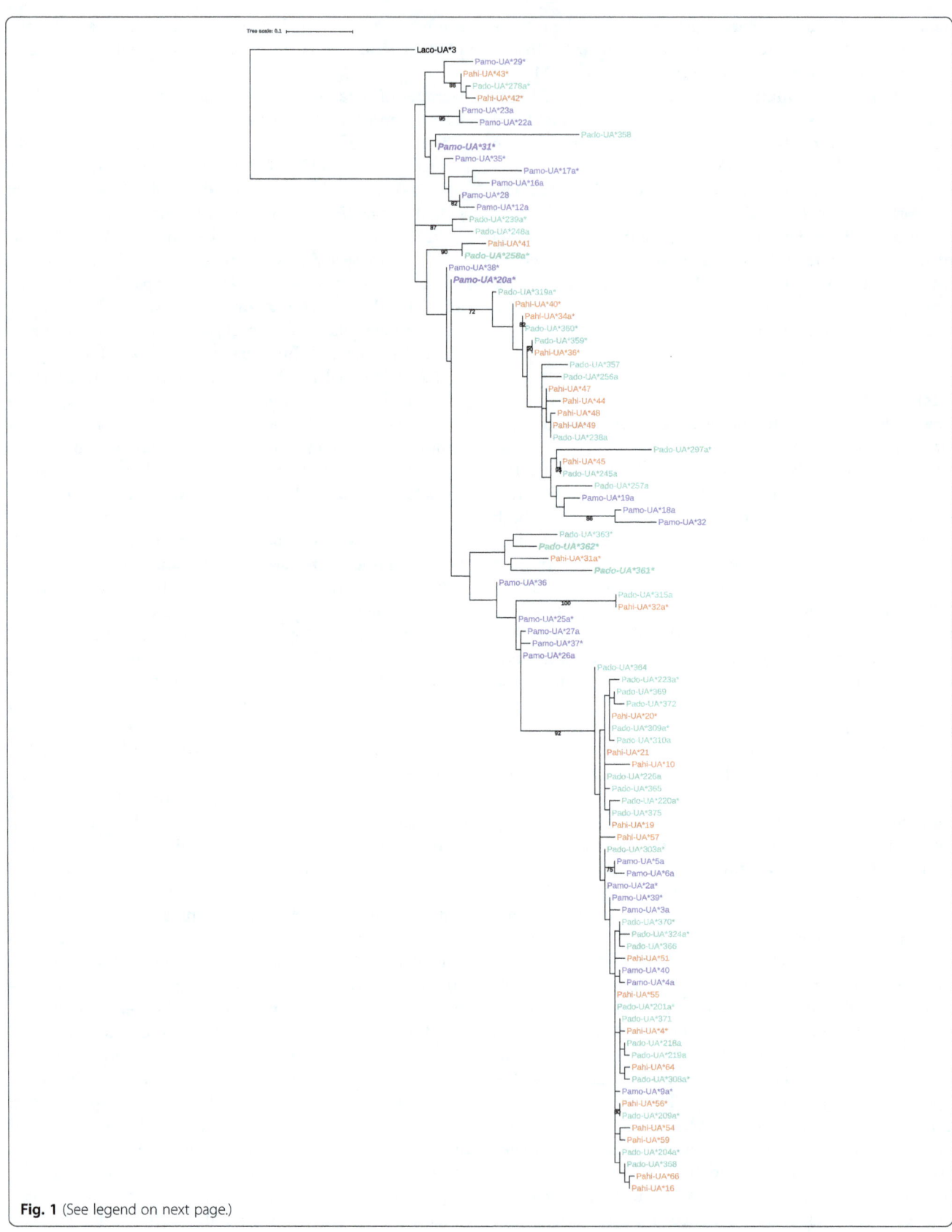

Fig. 1 (See legend on next page.)

(See figure on previous page.)

Fig. 1 Maximum likelihood tree based on 94 MHC class I exon 3 nucleotide sequences from house sparrows (Pado; indicated in green), Spanish sparrows (Pahi; indicated in orange) and tree sparrows (Pamo; indicated in purple) amplified with primer combination 1. One MHC class I sequence (Acc nr KU169762) from *Lanius collaris* was used as outgroup. The tree was constructed with the RAxML software (version 7.0.4) using the GTRGAMMA model and 1000 bootstraps, displaying bootstrap values larger than 70%. *Stars* (*) indicates alleles that were found in both gDNA and cDNA (i.e. expressed alleles). The classical alleles that were identified as highly expressed are marked in bold and italic. All putatively non-classical alleles are found in the lower cluster, with no clustering based on species, whereas the putatively classical alleles do not form a distinct cluster

Discussion

The subdivision of MHC-I genes into classical, highly polymorphic genes that present peptides to T-cells, and non-classical genes, less polymorphic genes that do not present peptides to T-cells, have been reported in many mammal species and also in several bird species [3–5, 10, 11, 19–21, 27–30]. However, classical and non-classical genes have not yet been confirmed in songbirds in the largest bird order Passeriformes, though such subdivision is likely since non-classical genes have been

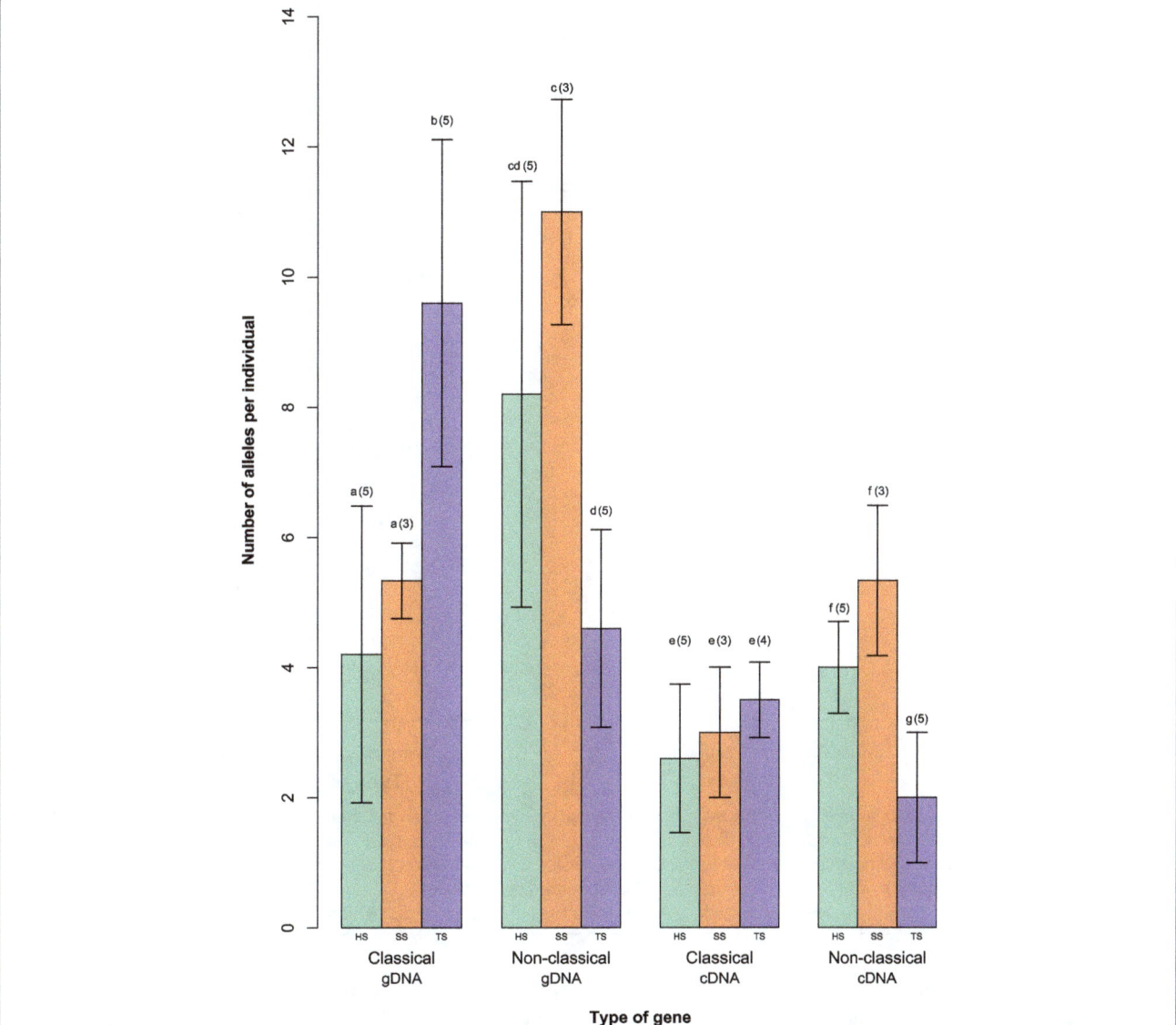

Fig. 2 Average numbers of putatively classical and non-classical MHC-I alleles per individual in gDNA (genomic DNA) and cDNA (i.e. expressed alleles) in house sparrows (HS; indicated in *green*), Spanish sparrows (SS; indicated in orange) and tree sparrows (TS; indicated in *purple*), for further details see Additional file 1: Table S4. Tukey posthoc test was used to determine differences between the groups, significance is indicated by letters (a-g; $p < 0.05$) and number of individuals is reported within brackets

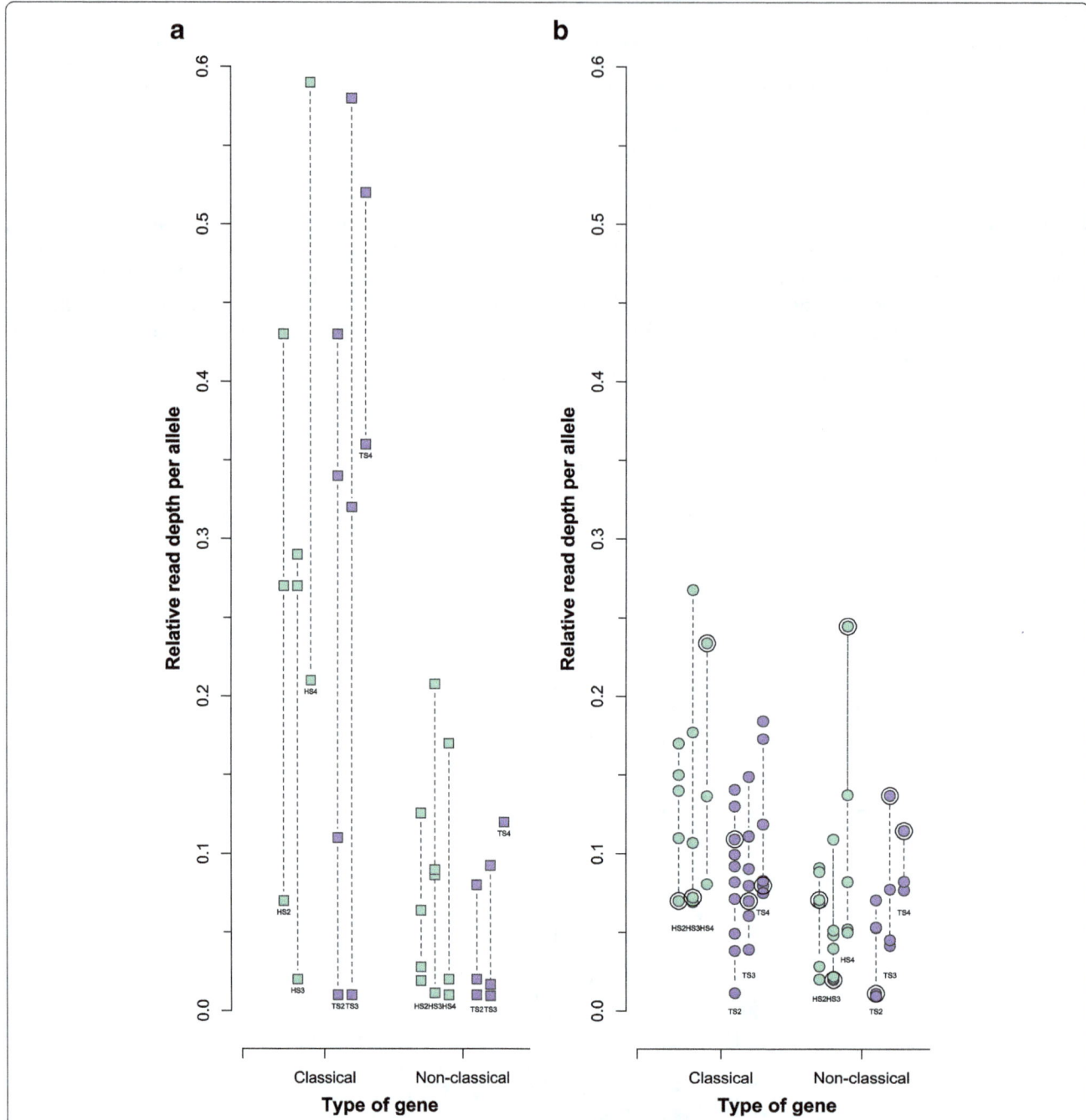

Fig. 3 Variance in relative read depth per allele of putatively classical and non-classical alleles in three house sparrow (HS2, HS3, and HS4; indicated in *green*) and three tree sparrow individuals (TS2, TS3 and TS4; indicated in *purple*). **a** In cDNA (*squares*) there is a significant difference in variance in expression (measured as relative read depth per allele) between putatively classical and non-classical alleles (Levine's test: $F = 5.20$, $p = 0.005$). **b** In gDNA (*circles*) there is no difference in relative read depth per allele. The highest expressed allele in cDNA does not correspond to the allele with highest relative read depth in gDNA. These highly expressed alleles are marked with a *circle* in the gDNA plot

identified in other bird species from several different orders [25–30]. In the present study we found high numbers of MHC-I gene copies in sparrows; the maximum number of MHC-I alleles per individual that we identified was 21 (i.e sparrows have eleven or more MHC-I gene copies), though only about 50% of these alleles were expressed. In the maximum likelihood tree we identified one distinct strongly supported cluster (bootstrap = 92 and for the neighbor-net network the corresponding the bootstrap = 93) containing alleles (from all three species) with low polymorphism and a 6 bp deletion (putatively non-classical genes). The remaining alleles were more polymorphic and found in non-significantly supported groups (putatively classical genes). Several previous studies have reported considerably lower diversity estimates and lower rates of non-synonymous substitutions in the peptide binding region of putatively non-classical alleles compared with putatively classical alleles in house sparrows [35, 36, 49], and we found similar results in our data from three different sparrow species considering nucleotide diversity. Moreover, the analyses of gene expression showed that the polymorphic group with the putatively classical genes had variable expression, that is, some alleles were highly expressed while others had low expression. Strikingly, at most two alleles among these putatively classical MHC-I alleles were ever highly expressed in each individual. In contrast, the group with putatively non-classical genes had more uniformly low expression. Taken together, these results strongly indicate that we have identified classical and non-classical MHC-I genes in sparrows.

Phylogenetic ages of classical and non-classical genes

The phylogenetic reconstruction of sparrow MHC-I alleles places the non-classical genes in a single well-supported cluster, confirming that the subdivision of these putatively classical and non-classical genes predates the separation of the investigated sparrow species. This shows that orthologous gene copies of classical and non-classical genes have persisted in sparrows over several speciation events in a time frame of at least 7 million years. Classical (MHC-B) and non-classical (MHC-Y) genes have also been reported among a wide range of birds in the order Galliformes, including chicken, turkey and golden pheasant, species that split 28–40 million years ago, and non-classical alleles in chicken and turkey form a gene specific cluster [4, 20, 21, 50–52]. It is possible that these different sets of paralogous genes are even older and evolved in an ancient common ancestor of galliforms more than 65 million years ago [53]. However, non-classical genes in sparrows of the order Passeriformes and species within the order Galliformes are not likely to be orthologous, presently available data suggest that the non-classical genes originate from more recent duplications of classical genes, a pattern seen also among distantly related mammals [14, 30]. Though, orthologous clusters of classical and non-classical MHC-I genes among relatively closely related mammals has been seen in hominids and here the classical and non-classical genes even cluster by locus [10]. Classical (HLA-A, -B and -C) and non-classical (HLA-E, -F and -G) MHC-I genes in human and chimpanzee (*Pan troglodytes*), species that diverged 6 to 7 million years ago, form gene specific clusters at each of these six loci (A-G).

Number of MHC alleles in gDNA and cDNA in three sparrow species

We found no difference in the copy number of classical and non-classical alleles in the genome between house sparrows and Spanish sparrows (species that split 3 million years ago) but there was a difference in allele copy number relative to the tree sparrows, which diverged earlier (7 million years ago). This difference in gene copy number could have originated in two different ways, first the divergence time of house sparrows and Spanish sparrows may be too short for gene copy number to evolve while tree sparrows that are more distantly related have evolved further. Tree sparrows have a higher number of classical gDNA alleles and a lower number of non-classical gDNA alleles than house sparrows and Spanish sparrows. Alternatively, house sparrow and Spanish sparrow could have experienced different selection and lost classical gene copies. Without knowing the proportion of classical and non-classical genes in ancestral sparrows it is impossible to determine how this gene copy variation occurred. Moreover, there are certain problems associated with co-amplifying multiple genes at the same time which makes it more difficult to determine the exact number of genes and possible copy number variation [54, 55]. Here we have tried to overcome this problem by using several primer combinations.

The number of expressed alleles (cDNA) varied less between species than the number of alleles in the genome and, interestingly, there was no significant difference in any species comparisons in number of expressed classical alleles. One explanation could be that the number of expressed genes are more conserved than the total number of genes in the genome and there may be selection for expressing a certain number of genes, e.g. expressing an optimal number of classical genes [56]. There was a significant difference in the number of expressed non-classical alleles among species; tree sparrows expressed fewer non-classical alleles than both house sparrows and Spanish sparrows. It is interesting to note that tree sparrows, which express significantly lower numbers of non-classical alleles, also have a lower number of non-classical alleles in the genome.

Variation in expression of classical and non-classical MHC-I alleles

The variance in gene expression was larger in putatively classical than non-classical MHC-I genes in sparrows. This finding is consistent with the existence of 'major' (highly expressed) and 'minor' (low expressed) loci for classical but not for non-classical MHC-I genes in sparrows, as previously reported for species within the order Galliformes [22, 23]. The highest numbers of classical alleles per individual in sparrows were 14 in gDNA but out of these only two alleles were highly expressed. Two highly expressed classical MHC-I alleles have been reported for several bird species in the order Galliformes (e.g. chicken and Japanese quail (*Coturnix japonica*)) and also in mallards belonging to the order Anseriformes. In chicken, Japanese quail and mallard the MHC genomic regions have been characterized and each species has a single major classical MHC-I locus, that is, one classical gene that is highly expressed [27, 57–59]. In all these species, the highly expressed gene is located next to the TAP gene (Transporter associated with antigen processing) and co-evolutionary processes between TAP and MHC is thought to explain why only a single MHC-I gene is highly expressed [4, 31, 57, 60–62]. Our findings on classical MHC-I genes in house sparrows and tree sparrows agree well with these previous findings from other birds, though with our data we cannot determine with certainty if sparrows have one single major classical MHC-I gene. It is possible that the six sparrows are homozygous for two classical MHC-I genes that are highly expressed, even though this is unlikely since heterozygosity is much more common than homozygosity at classical MHC-I loci. Alternatively, different genes could be highly expressed in the six individuals or the two alleles of one gene could be differently expressed, meaning that two genes could be highly expressed. It would be interesting to study this further and to determine if the highly expressed alleles belong to the same gene and if this gene is located next to TAP. Since we set strict criteria for including individuals in the expression analysis we could only investigate six of the individuals. Future analysis of more individuals would help determining how general our results are.

In our study of expression of MHC-I genes we only estimated gene expression in blood. Blood to some extent represent gene expression in different tissues but we do not claim that our results should be extrapolated to be representative for expression in all tissues. Non-expressed genes in our study could for example have specific expression under other conditions, in other tissues or in birds of different age classes. Chen et al. [30] recently characterized the genomic MHC region in the crested ibis (*Nipponia nippon*) and reported considerable differences in gene expression of five different MHC-I genes between tissues. Interestingly one particular MHC-I gene was highly expressed in all tissues in the crested ibis and this gene was the only gene situated in the core MHC genomic region [30]. Expression in blood was however not reported in the crested ibis.

Conclusions

We have studied the highly duplicated MHC-I gene family in three species of sparrows and based on phylogeny and gene expression patterns found strong indications for the existence of classical and non-classical MHC-I genes. This

subdivision of genes has previously been reported in many groups of vertebrates, for example in galliform birds and hominids, but never in songbirds. A majority of the sparrow MHC-I genes are putatively non-classical; hence they are presumably not involved in T-cell mediated immunity. Such a distinctly separated phylogenetic cluster of putatively non-classical genes is rarely found among songbirds, and within songbirds non-classical genes could be a unique feature for sparrows. However, we find it more likely that there are groups of non-classical MHC-I genes in most songbirds but that they often are missed. Therefore, it would be valuable if future studies of MHC-I in songbirds investigated the existence of putatively non-classical MHC-I genes, preferably using gene expression. Future ecological and evolutionary studies of MHC-I in wild birds would gain from considering the existence of classical and non-classical genes, since these two types of MHC-I genes have different functions.

Additional file

Additional file 1: Method S1. Extended methods for creating the Passer maximum clade credibility tree. **Method S2.** Extended filtering protocol for treating the high-throughput amplicon data. **Table S1.** Detailed information regarding the primers used for high-throughput amplicon sequencing. **Table S2.** Comparison of how efficient the four different primer combinations used in this study amplified MHC-I alleles. **Table S3.** Read depth per individual before and after filtering of the high-throughput amplicon data. **Table S4.** Comparison of the reputability between duplicated samples sequenced with 454 amplicon sequencing. **Table S5.** Diversity measurements of putatively classical and non-classical alleles. Calculated for all gDNA alleles, expressed and non-expressed alleles separately. **Table S6.** The number of putatively classical and non-classical MHC-I alleles identified, per individual, in gDNA and cDNA, for the 13 sparrow individuals used in this study. **Table S7.** List of the different alleles, both classical and non-classical, amplified in each individual. **Table S8.** Comparison of the relative read depth per allele between two duplicated tree sparrow individuals that were used for the expression analysis. **Table S9.** Number of expressed MHC-I alleles identified in the three house sparrow and three tree sparrow individuals used for the expression analysis. **Figure S1.** Schematic overview of MHC-I exon 3 displaying the different locations for all primers used in this study. **Figure S2.** Alignmnet of the 129 MHC-I concatenated exon 3 alleles identified. **Figure S3.** Neighbor-net network displaying the 94 MHC-I alleles amplified with primer combination 1. **Figure S4.** Comparison of the proportion of reads per allele between gDNA and cDNA in the three house sparrow and three tree sparrow individuals selected for the expression analysis. (DOCX 5357 kb)

Acknowledgements

We are grateful to Fredrik Haas for assistance during field work and to Emily O'Connor for assistance during lab work and data analysis.

Funding

This work was supported by the Swedish Research Council (grant 621–2011-3674 and 2015–05149) and by the Crafoord Foundation to H.W.

Authors' contributions

HW and AD designed the study and conducted the field work. AD did the lab work. The data was analyzed by AD, HW and LR. All authors discussed the results and contributed to the writing of the manuscript. All authors read and approved the final manuscript.

Competing interests

The authors declare that they have no competing interests.

References

1. Murphy K, Travers P, Walport M. Janeway's immunobiology. 7th ed. New York: Garland Science; 2008.
2. Neefjes J, Jongsma ML, Paul P, Bakke O. Towards a systems understanding of MHC class I and MHC class II antigen presentation. Nat Rev Immunol. 2011;11:823–36.
3. Shiina T, Hosomichi K, Inoko H, Kulski JK. The HLA genomic loci map: expression, interaction, diversity and disease. J Hum Genet. 2009;54:15–39.
4. Kaufman J, Milne S, Göbel TW, Walker BA, Jacob JP, Auffray C, et al. The chicken B locus is a minimal essential major histocompatibility complex. Nature. 1999;401:923–5.
5. Lunney JK, Ho CS, Wysocki M, Smith DM. Molecular genetics of the swine major histocompatibility complex, the SLA complex. Dev Comp Immunol. 2009;33:362–74.
6. Westerdahl H. Passerine MHC: genetic variation and disease resistance in the wild. J Ornithol. 2007;148:469–77.
7. Sepil I, Moghadam HK, Huchard E, Sheldon BC, Kuduk K, Babik W, et al. Characterization and 454 pyrosequencing of major histocompatibility complex class I genes in the great tit reveal complexity in a passerine system. BMC Evol Biol. 2012;12:68.
8. O'Connor EA, Strandh M, Hasselquist D, Nilsson J, Westerdahl H. The evolution of highly variable immunity genes across a passerine bird radiation. Mol Ecol. 2016;25:977–89.
9. Biedrzycka A, O'Connor E, Sebastian A, Migalska M, Radwan J, Zając T, et al. Extreme MHC class I diversity in the sedge warbler (Acrocephalus schoenobaenus); selection patterns and allelic distributions suggest that different genes have different functions. BMC Evol Biol. 2017. In press.
10. Adams E, Parham P. Species-specific evolution of MHC class I genes in the higher primates. Immunol Rev. 2001;183:41–64.
11. Velten F, Rogel-Gaillard C, Renard C, Pontarotti P, Tazi-Ahnini R, Vaiman M, et al. A first map of the porcine major histocompatibility complex class I region. Tissue Antigens. 1998;51:183–94.
12. Shawar S, Vyas J. Antigen presentation by major histocompatibility complex class IB molecules. Annu Rev Immunol. 1994;12:839–80.
13. Rodgers JR, Cook RG. MHC class Ib molecules bridge innate and acquired immunity. Nat Rev Immunol. 2005;5:459–71.
14. Hughes ALL, Nei M. Evolution of the major histocompatibility complex: independent origin of nonclassical class I genes in different groups of mammals. Mol Biol Evol. 1989;6:559–79.
15. Ishitani A, Sageshima N, Lee N, Dorofeeva N, Hatake K, Marquardt H, et al. Protein expression and peptide binding suggest unique and interacting functional roles for HLA-E, F, and G in maternal-placental immune recognition. J Immunol. 2003;171:1376–84.
16. Diefenbach A, Raulet DH. The innate immune response to tumors and its role in the induction of T-cell immunity. Immunol Rev. 2002;188:9–21.
17. Robinson J, Halliwell J, Hayhurst JD, Flicek P, Parham P, Marsh SGE. The IPD and IMGT/HLA database: Allele variant databases. Nucleic Acids Res. 2015;43:D423–31.
18. Apps R, Meng Z, Del Prete GQ, Lifson JD, Zhou M, Carrington M. Relative Expression Levels of the HLA Class-I Proteins in Normal and HIV-Infected Cells. J Immunol. 2015;194:3594–600.
19. Briles WE, Goto RM, Auffray C, Miller MM. A polymorphic system related to but genetically independent of the chicken major histocompatibility complex. Immunogenetics. 1993;37:408–14.
20. Chaves LD, Krueth SB, Reed KM. Characterization of the turkey MHC chromosome through genetic and physical mapping. Cytogenet Genome Res. 2007;117:213–20.
21. Zeng Q, Zhong G, He K, Sun D, Wan Q. Molecular characterization of classical and nonclassical MHC class I genes from the golden pheasant (Chrysolophus pictus). Immunogenetics. 2016;43:8–17.
22. Kaufman J, Jacob J, Shaw I, Walker B, Milne S, Beck S, et al. Gene organisation determines evolution of function in the chicken MHC. Immunol Rev. 1999;167:101–17.
23. Kaufman J. Co-evolving genes in MHC haplotypes: the "rule" for nonmammalian vertebrates? Immunogenetics. 1999;50:228–36.

24. Wallny H-J, Avila D, Hunt LG, Powell TJ, Riegert P, Salomonsen J, et al. Peptide motifs of the single dominantly expressed class I molecule explain the striking MHC-determined response to Rous sarcoma virus in chickens. Proc Natl Acad Sci U S A. 2006;103:1434–9.
25. Afanassieff M, Goto RM, Ha J, Sherman MA, Zhong L, Auffray C, et al. At least one class I gene in restriction fragment pattern-Y (Rfp-Y), the second MHC gene cluster in the chicken, is transcribed, polymorphic, and shows divergent specialization in antigen binding region. J Immunol. 2001;166:3324–33.
26. Hunt HD, Goto RM, Foster DN, Bacon LD, Miller MM. At least one YMHCI molecule in the chicken is alloimmunogenic and dynamically expressed on spleen cells during development. Immunogenetics. 2006;58:297–307.
27. Moon DA, Veniamin SM, Parks-Dely JA, Magor KE. The MHC of the duck (Anas platyrhynchos) contains five differentially expressed class I genes. J Immunol. 2005;175:6702–12.
28. Cloutier A, Mills JA, Baker AJ. Characterization and locus-specific typing of MHC class I genes in the red-billed gull (Larus scopulinus) provides evidence for major, minor, and nonclassical loci. Immunogenetics. 2011;63:377–94.
29. Buehler DMDM, Verkuil YIYI, Tavares ESES, Baker AJAJ. Characterization of MHC class i in a long-distance migrant shorebird suggests multiple transcribed genes and intergenic recombination. Immunogenetics. 2013;65:211–25.
30. Chen L-C, Lan H, Sun L, Deng Y-L, Tang K-Y, Wan Q-H. Genomic organization of the crested ibis MHC provides new insight into ancestral avian MHC structure. Sci Rep. 2015;5:7963.
31. Mesa CM, Thulien KJ, Moon D a, Veniamin SM, Magor KE. The dominant MHC class I gene is adjacent to the polymorphic TAP2 gene in the duck, Anas platyrhynchos. Immunogenetics. 2004;56:192–203.
32. Ohno S. Evolution by Gene Duplication. Berlin, Heidelberg: Springer Berlin Heidelberg; 1970.
33. Eirin-Lopez JM, Rebordinos L, Rooney AP, Rozas J. The Birth-and-Death Evolution of Multigene Families Revisited. Genome Dyn. 2012;7:170–96.
34. Nei M, Gu X, Sitnikova T. Evolution by the birth-and-death process in multigene families of the vertebrate immune system. Proc Natl Acad Sci National Acad Sciences. 1997;94:7799–806.
35. Karlsson M, Westerdahl H. Characteristics of MHC Class I Genes in House Sparrows Passer domesticus as Revealed by Long cDNA Transcripts and Amplicon Sequencing. J Mol Evol. 2013;77:8–21.
36. Bonneaud C, Sorci G, Morin V, Westerdahl H, Zoorob R, Wittzell H. Diversity of Mhc class I and IIB genes in house sparrows (Passer domesticus). Immunogenetics. 2004;55:855–65.
37. Mullarney K, Svensson L, Zetterström D, Grant PJ. Bird guide. The most complete field guide to the birds of britain and europe. London: HarperCollins Publishers Ltd; 2006.
38. Jetz W, Thomas GH, Joy JB, Hartmann K, Mooers AO. The global diversity of birds in space and time. Nature. 2012;491:444–8.
39. A Global Phylogeny of Birds. http://birdtree.org. Accessed 25 Nov 2015.
40. Sambrook J, Fritsch EFMT. Molecular cloning: a laboratory manual. 2nd ed. Cold Spring Harbour: Cold Spring Harbour Laboratory Press; 1989.
41. Chiari Y, Galtier N. RNA extraction from sauropsids blood: evaluation and improvement of methods. Amphibia-Reptilia. 2011;32:136–9.
42. Westerdahl H, Wittzell H, von Schantz T, Bensch S. MHC class I typing in a songbird with numerous loci and high polymorphism using motif-specific PCR and DGGE. Heredity (Edinb). 2004;92:534–42.
43. Kloch A, Babik W, Bajer A, Siński E, Radwan J. Effects of an MHC-DRB genotype and allele number on the load of gut parasites in the bank vole Myodes glareolus. Mol Ecol. 2010;19 Suppl 1:255–65.
44. Klein J, Bontrop RE, Dawkins RL, Erlich HA, Gyllensten UB, Heise ER, et al. Nomenclature for the major histocompatibility complexes of different species: a proposal. Immunogenetics. 1990;31:217–9.
45. Letunic I, Bork P. Interactive tree of life (iTOL) v3: an online tool for the display and annotation of phylogenetic and other trees. Nucleic Acids Res. 2016;44:W242–W245.
46. Huson DH, Bryant D. Application of phylogenetic networks in evolutionary studies. Mol Biol Evol. 2006;23:254–67.
47. Posada D. jModelTest: Phylogenetic model averaging. Mol Biol Evol. 2008;25:1253–6.
48. R Core Team. In: RDC T, editor. R: A Language and Environment for Statistical Computing. Vienna: R Foundation for Statistical Computing; 2014.

49. Borg AA, Pedersen SA, Jensen H, Westerdahl H. Variation in MHC genotypes in two populations of house sparrow (*Passer domesticus*) with different population histories. Ecol Evol. 2011;1:145–59.

50. Reed KM, Bauer MM, Monson MS, Benoit B, Chaves LD, O'Hare TH, et al. Defining the turkey MHC: identification of expressed class I- and class IIB-like genes independent of the MHC-B. Immunogenetics. 2011;63:753–71.

51. Dimcheff DE, Drovetski SV, Mindell DP. Phylogeny of Tetraoninae and other galliform birds using mitochondrial 12S and ND2 genes. Mol Phylogenet Evol. 2002;24:203–15.

52. Van Tuinen M, Dyke GJ. Calibration of galliform molecular clocks using multiple fossils and genetic partitions. Mol Phylogenet Evol. 2004;30:74–86.

53. Jarvis E, Mirarab S, Aberer A, Li B, Houde P, Li C, et al. Whole-genome analyses resolve early branches in the tree of life of modern birds. Science. 2014;346:1320–31.

54. Burri R, Promerová M, Goebel J, Fumagalli L. PCR-based isolation of multigene families: lessons from the avian MHC class IIB. Mol Ecol Resour. 2014;14:778–88.

55. Gaigher A, Burri R. Family-assisted inference of the genetic architecture of major histocompatibility complex variation; 2016. p. 1353–64.

56. Milinski M. The Major Histocompatibility Complex, Sexual Selection, and Mate Choice. Annu Rev Ecol Evol Syst. 2006;37:159–86.

57. Kaufman J, Völk H, Wallny HJ. A "minimal essential Mhc" and an "unrecognized Mhc": two extremes in selection for polymorphism. Immunol Rev. 1995;143:63–88.

58. Shiina T, Hosomichi K, Hanzawa K. Comparative genomics of the poultry major histocompatibility complex. Anim Sci J. 2006;77:151–62.

59. Fleming-canepa X, Jensen SM, Christine M, Diaz-satizabal L, Roth AJ, Parks-dely JA, et al. Extensive Allelic Diversity of MHC Class I in Wild Mallard Ducks. J Immunol. 2016;197:783–94.

60. Shiina T, Oka A, Imanishi T, Hanzawa K, Gojobori T, Watanabe S, et al. Multiple class I loci expressed by the quail Mhc. Immunogenetics. 1999;49:456–60.

61. Shiina T, Shimizu S, Hosomichi K, Kohara S, Watanabe S, Hanzawa K, et al. Comparative genomic analysis of two avian (quail and chicken) MHC regions. J Immunol. 2004;172:6751–63.

62. Walker BA, Hunt LG, Sowa AK, Skjødt K, Göbel TW, Lehner PJ, et al. The dominantly expressed class I molecule of the chicken MHC is explained by coevolution with the polymorphic peptide transporter (TAP) genes. Proc Natl Acad Sci U S A. 2011;108:8396–401.

63. Drews A, Strandh M, Råberg L, Westerdahl H. Data from: Expression and phylogenetic analyses reveal paralogous lineages of putatively classical and non-classical MHC-I genes in three sparrow species (*Passer*). Dryad Digital Repository. http://dx.doi.org/10.5061/dryad.79t4b.

10

Phylogenetic relationships of cone snails endemic to Cabo Verde based on mitochondrial genomes

Samuel Abalde[1], Manuel J. Tenorio[2], Carlos M. L. Afonso[3], Juan E. Uribe[1], Ana M. Echeverry[1] and Rafael Zardoya[1*]

Abstract

Background: Due to their great species and ecological diversity as well as their capacity to produce hundreds of different toxins, cone snails are of interest to evolutionary biologists, pharmacologists and amateur naturalists alike. Taxonomic identification of cone snails still relies mostly on the shape, color, and banding patterns of the shell. However, these phenotypic traits are prone to homoplasy. Therefore, the consistent use of genetic data for species delimitation and phylogenetic inference in this apparently hyperdiverse group is largely wanting. Here, we reconstruct the phylogeny of the cones endemic to Cabo Verde archipelago, a well-known radiation of the group, using mitochondrial (mt) genomes.

Results: The reconstructed phylogeny grouped the analyzed species into two main clades, one including *Kalloconus* from West Africa sister to *Trovaoconus* from Cabo Verde and the other with a paraphyletic *Lautoconus* due to the sister group relationship of *Africonus* from Cabo Verde and *Lautoconus ventricosus* from Mediterranean Sea and neighboring Atlantic Ocean to the exclusion of *Lautoconus* endemic to Senegal (plus *Lautoconus guanche* from Mauritania, Morocco, and Canary Islands). Within *Trovaoconus*, up to three main lineages could be distinguished. The clade of *Africonus* included four main lineages (named I to IV), each further subdivided into two monophyletic groups. The reconstructed phylogeny allowed inferring the evolution of the radula in the studied lineages as well as biogeographic patterns. The number of cone species endemic to Cabo Verde was revised under the light of sequence divergence data and the inferred phylogenetic relationships.

Conclusions: The sequence divergence between continental members of the genus *Kalloconus* and island endemics ascribed to the genus *Trovaoconus* is low, prompting for synonymization of the latter. The genus *Lautoconus* is paraphyletic. *Lautoconus ventricosus* is the closest living sister group of genus *Africonus*. Diversification of *Africonus* was in allopatry due to the direct development nature of their larvae and mainly triggered by eustatic sea level changes during the Miocene-Pliocene. Our study confirms the diversity of cone endemic to Cabo Verde but significantly reduces the number of valid species. Applying a sequence divergence threshold, the number of valid species within the sampled *Africonus* is reduced to half.

Keywords: Mitochondrial genomes, *Africonus*, *Trovaoconus*, *Kalloconus*

* Correspondence: rafaz@mncn.csic.es
[1]Museo Nacional de Ciencias Naturales (MNCN-CSIC), José Gutiérrez Abascal 2, 28006 Madrid, Spain
Full list of author information is available at the end of the article

Background

The cone snails (Conidae, Gastropoda) endemic to the archipelago of Cabo Verde in West Africa represent one of the few textbook examples of a well-documented insular species radiation involving marine organisms [1–3]. Cone snails, which are found in tropical and subtropical marine waters throughout the world, show a hotspot of species diversity in the Cabo Verde archipelago with up to 95 endemic species (roughly 10% of cone species diversity worldwide) narrowly confined to about 4000 km^2 [4]. As in other parts of the world, cone snails endemic to Cabo Verde constitute a key component of the intertidal and subtidal ecosystems associated to rocky shores, coral reefs, and sandy bottoms. All cones endemic to Cabo Verde feed on marine annelid worms [1] and use a sophisticated venom apparatus (including a venom gland that produces conotoxins and a specialized harpoon-like radular tooth) to capture their preys [5]. Another interesting biological feature common to all these endemic species is that they have direct development. Their larvae lack a pelagic stage, and thus show a considerably reduced dispersal capacity [1]. Survival rate is higher for this type of larvae since they are less likely to be eaten by predators and are not dependent on plankton for feeding (i.e, non-planktotrophic).

The origin and evolutionary history of cones endemic to Cabo Verde has been the subject of several recent phylogenetic studies [1, 2, 6, 7]. Molecular phylogenies demonstrated that two different ancestors reached the archipelago independently and subsequently diversified following recurrent biogeographic patterns [1, 2, 7]. The existence of two clades led to the classification of cone species endemic to Cabo Verde into two genera, *Africonus* and *Trovaoconus* [8]. The question of which species are the closest living sister groups to *Africonus* and *Trovaoconus* remains open [1, 2]. According to a previous study, the ancestor of *Africonus* colonized the archipelago in the Miocene, about 16.5 million years ago (mya; [1]), and spread to all islands (except Fogo, the youngest, with steep slopes in the coast and ongoing volcanic activity). Most (95%) of the currently described species endemic to Cabo Verde belong to *Africonus*, and are normally referred to as restricted to a single island and in some cases even to single bays within an island [3]. The ancestor of *Trovaoconus* arrived at Cabo Verde archipelago in the Pliocene, about 4.6 mya, and diversified only in four islands (Sal, Boa Vista, Maio, and possibly Santiago), which are the closest to the continent [1]. These cones are significantly larger in size than those belonging to *Africonus* and show wider distributions extending in some cases to more than one island. It has been hypothesized that diversification within each genus was in allopatry and followed recurrent eustatic sea level changes during the Neogene that intermittently connected and disconnected the islands [1, 7]. However, sea level fluctuations alone do not fully explain the extraordinary diversity of cones in Cape

Verde since nearby archipelagos in the Macaronesia biogeographic region such as the Canary Islands subjected to similar trends since the Miocene do not have endemic cone species [6]. A larger distance to the mainland, which enhances isolation and restricts gene flow combined with a higher mean sea surface temperature and the presence of more suitable habitats may have promoted a significant increase in diversification rates in the Cabo Verde archipelago [6].

The rate of description of new cone species endemic to Cabo Verde has accelerated more than expected during the last years (Fig. 1). After the early descriptions in the eighteenth and nineteenth centuries based on samples brought to Europe by naturalists [9], the main contribution to the cataloguing of cone species endemic to Cabo Verde was due to the work of Emilio Rolán [10], who drew attention to this singular radiation. Hence, around year 2000, there were about 50 species recognized [11] and remarkably this number has almost doubled in the last 2-3 years [12–19]. However, it is important to note that many of the recent species diagnoses in cones are mainly based on the shape, color, and banding patterns of the shell. These phenotypic characters are highly variable at the population level and prone to local adaptation and convergence, making species assignment problematic and sometimes, misleading [7]. In many cases, distinguishing whether different shell morphotypes of cone snails represent valid species or ecotypes of the same species is challenging [20]. Therefore, determination of genetic variation and inference of phylogenetic relationships based on DNA sequence data are timely as part of a multidisciplinary approach [21] to identify and delimit species and to understand evolutionary processes underlying diversification within cones, in general, and within those endemic to Cabo Verde, in particular.

Here, we used nearly complete mitochondrial (mt) genomes, which have proven to successfully reconstruct robust phylogenies of Conidae [22] and of particular groups such as the cones endemic to Senegal [23]. In this study, we sequenced the nearly complete mt genomes of 88 individuals representing different populations and species of *Africonus* and *Trovaoconus* endemic to Cabo Verde. We aimed to: (1) reconstruct a highly resolved phylogeny of cones endemic to Cabo Verde; (2) determine the closest living sister groups of *Africonus* and *Trovaoconus*; (3) date major cladogenetic events and analyze biogeographical patterns; (4) study radular tooth evolution within the two genera; and (5) provide a first genetic hypothesis of species delimitation in the radiation of Cabo Verde endemic cones.

Results

Sequencing, assembly, and genome organization

The nucleotide sequences of the near-complete mt genomes of 75 specimens of *Africonus*, 13 specimens of

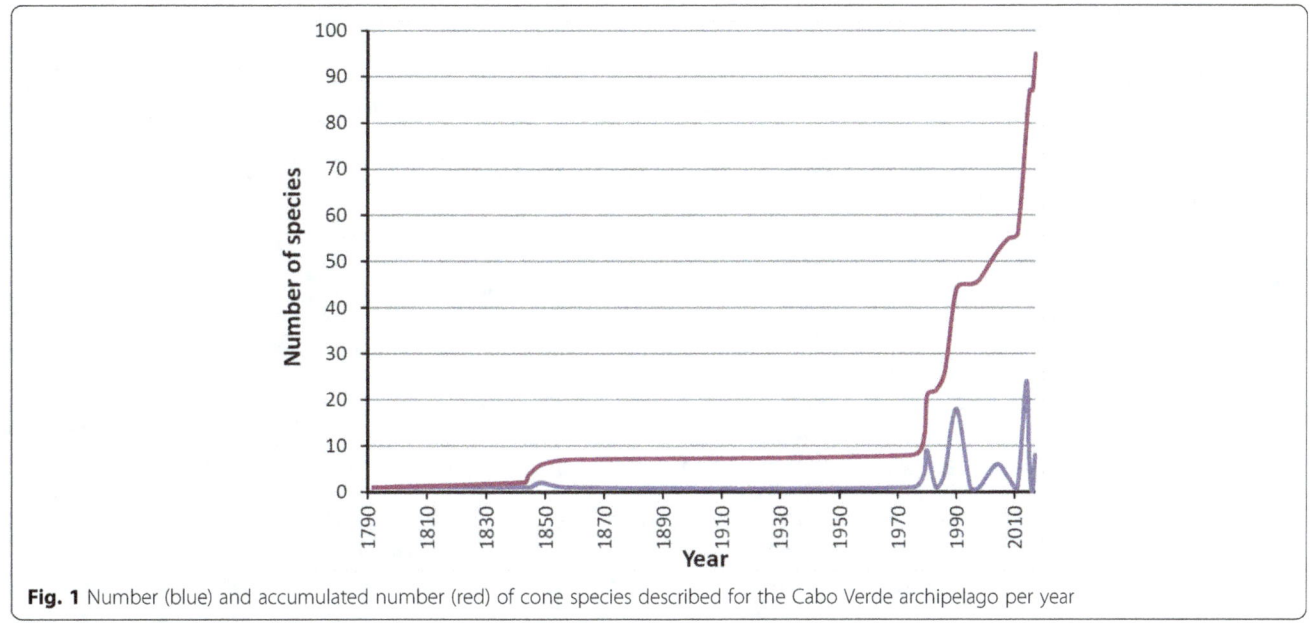

Fig. 1 Number (blue) and accumulated number (red) of cone species described for the Cabo Verde archipelago per year

Trovaoconus, and one specimen of *Lautoconus ventricosus* were determined (Table 1). These mt genomes lacked the *trnF* gene, the control region, and the start of the *cox3* gene because the corresponding fragment was not PCR amplified. The number of reads, mean coverage, and length of each mt genome are provided in Table 1. The mt genomes of *Africonus boavistensis* and *Africonus denizi* received the minimum (42,021) and maximum (906,765) number of reads, respectively. The same samples received the minimum (412×) and maximum (8,885×) mean coverage, respectively (Table 1). All sequenced mt genomes encode for 13 protein-coding, 2 rRNA and 21 tRNA genes (but note that the *trnF* gene could not be determined; see above). They all share the same genome organization: the major strand encodes all genes, except those forming the cluster MYCWQGE (*trnM*, *trnY*, *trnC*, *trnW*, *trnQ*, *trnG*, *trnE*) and the *trnT* gene.

Phylogenetic relationships and sequence divergences between clades

Phylogenetic relationships of cones endemic to Cabo Verde were reconstructed based on the nucleotide sequences of the concatenated 13 mt protein-coding and two rRNA genes using probabilistic methods and *Chelyconus ermineus* as outgroup. The final matrix was 13,572 positions in length. According to the AIC, the best partition scheme for the protein-coding genes was the one combining all these genes but analyzing each codon position separately. The best substitution model for each of the three codon positions was GTR + I + G. For the rRNA genes, the best scheme had both genes combined under the GTR + I + G model. Both, ML (−*lnL* =

75,600.18) and BI (−*lnL* = 76,002.71 for run 1; −*lnL* = 76,288.44 for run 2) arrived at almost identical topology (Figs. 2 and 3). Most nodes received high statistical support and differences in topology were restricted exclusively to three relatively shallow nodes that had low support in ML and were unresolved in BI. Two of these nodes involved almost identical sequences and corresponded to *Africonus bernardinoi*/ *Africonus pseudocuneolus* and *Africonus teodorae*/ *Africonus fiadeiroi*, respectively. The third unresolved node corresponded to a trichotomy involving *Africonus felitae*, *Africonus regonae* and *Africonus longilineus*/ *Africonus cagarralensis*/ *Africonus melissae*.

The reconstructed phylogeny (Fig. 2) grouped the analyzed species into two main clades, one including *Kalloconus* from mainland West Africa sister to *Trovaoconus* from Cabo Verde and the other having paraphyletic *Lautoconus* due to the sister group relationship of *Africonus* from Cabo Verde and *Lautoconus ventricosus* from Mediterranean Sea and neighboring Atlantic Ocean to the exclusion of *Lautoconus* endemic to Senegal (plus *Lautoconus guanche* from Mauritania, Morocco, and Canary Islands). Within *Trovaoconus*, up to three main lineages could be distinguished (Fig. 2). The first one included two specimens from Sal initially identified as *Trovaoconus ateralbus*, which were sister to a clade including one lineage with specimens from Maio and Boa Vista identified as *Trovaoconus venulatus* and another lineage having mostly specimens of *Trovaoconus pseudonivifer* from Maio and Boa Vista but also one specimen of *Trovaoconus trochulus* from Boa Vista and one of *Trovaoconus atlanticoselvagem* from Baixo João Valente (Fig. 2).

The clade of *Africonus* from Cabo Verde included four main lineages (named I to IV), each further subdivided

Table 1 Mitochondrial (mt) genomes analyzed in this study

New mt genomes

ID CV	Initial species identification	Location	Coordinates	Coverage		Length (bp)	GenBank Acc. No	Voucher DNA (MNCN/ADN)	Voucher shell (MNCN 15.05/)	New species proposed[a]
				n° reads	mean depth					
1020	*Africonus antoniaensis*	Água Doce, Boa Vista, Cabo Verde	16°12'29"N, 22°44'7"W	151104	1476.8	15332	MF491587	95072	79889	—
0885	*Africonus antoniomonteiroi*	Pedra Lume, Sal, Cabo Verde	16°45'44"N, 22°53'2"W	232069	2273.4	15328	MF491578	95063	79794	—
0927	*Africonus bernardinoi*	Pedra Lume, Sal, Cabo Verde	16°45'44"N, 22°53'2"W	59799	583.3	15328	MF491582	95067	79835	*Africonus cuneolus*
0520	*Africonus boavistensis*	Baía do Ervatão (North), Boa Vista, Cabo Verde	16°12'3"N, 22°54'43"W	42021	412.8	15217	MF491563	95045	80413	—
1135	*Africonus cabraloi*	Estancinha, Boa Vista, Cabo Verde	16°13'12"N, 22°55'9"W	74446	730.4	15329	MF491598	95083	80004	*Africonus crotchii*
0895	*Africonus cagarralensis*	Pedra Lume, Sal, Cabo Verde	16°45'44"N, 22°53'2"W	161290	1367.2	15320	MF491579	95064	79804	*Africonus longilineus*
0173	*Africonus calhetae*	Praia da Soca, Maio, Cabo Verde	15°15'8"N, 23°13'4"W	55433	544.7	15242	MF491534	95016	78798	—
0920	*Africonus* cf. *anthonyi*	Ilhéus do Chano, Sal, Cabo Verde	16°41'37"N, 22°52'47"W	172336	1678.6	15315	MF491581	95066	79828	*Africonus cuneolus*
0162	*Africonus* cf. *claudiae*	Praia da Soca, Maio, Cabo Verde	15°15'8"N, 23°13'4"W	87407	858.6	15326	MF491533	95015	78787	*Africonus calhetae*
0465	*Africonus* cf. *delanoyae*	Ponta Antónia, Boa Vista, Cabo Verde	16°13'24"N, 22°46'59"W	382817	3736.1	15335	MF491559	95041	80409	*Africonus fuscoflavus*
0207	*Africonus* cf. *galeao*	Ponta Pipa, Maio, Cabo Verde	15°19'30"N, 23° 9'48"W	81447	797.3	15325	MF491536	95018	78832	*Africonus galeao*
0135	*Africonus* cf. *gonsaloi*	Praia Gonçalo, Maio, Cabo Verde	15°16'13"N, 23°6'15"W	148032	1455.5	15250	MF491529	95011	78760	*Africonus gonsaloi*
0380	*Africonus* cf. *miguelfiaderoi*	Jorrita, Baía da Gata, Boa Vista, Cabo Verde	16°12'9"N, 22°42'22"W	358342	3507.9	15328	MF491548	95030	80398	*Africonus vulcanus*
1400	*Africonus* cf. *miruchae*	Calhau, São Vicente, Cabo Verde	16°51'7"N, 24°51'59"W	523002	5104.9	15321	MF491601	95088	78562	*Africonus* sp. nov. 1
0223	*Africonus claudiae*	Ponta Pipa, Maio, Cabo Verde	15°19'30"N, 23° 9'48"W	148508	1434.5	15337	MF491537	95019	78848	*Africonus galeao*
0303	*Africonus condei*	Baía Grande, Derrubado, Boa Vista, Cabo Verde	16°13'31"N, 22°47'17"W	253863	2472.2	15248	MF491542	95024	80392	*Africonus crotchii*
0045	*Africonus crioulus*	Praia Santana, Maio, Cabo Verde	15°18'13"N, 23°11'49"W	255019	2502	15247	MF491521	95003	78670	*Africonus maioensis*
1075	*Africonus crotchii*	Morro de Areia, Boa Vista, Cabo Verde	16°5'24"N, 22°57'7"W	332385	3237.6	15329	MF491591	95076	79944	—
0803	*Africonus cuneolus*	Calheta Funda, Sal, Cabo Verde	16°39'6"N, 22°56'53"W	184181	1791.6	15329	MF491569	95053	79712	—
0936	*Africonus cuneolus*	Santa Maria, Sal, Cabo Verde	16°35'38"N, 22°53'36"W	80472	787.4	15328	MF491583	95068	79844	—
1420	*Africonus curralensis*	Praia de Palmo Tostão, Santa Luzia, Cabo Verde	16°45'19"N, 24°45'24"W	857123	8358.6	15329	MF491602	95089	78581	—
1017	*Africonus damioi*	Água Doce, Boa Vista, Cabo Verde	16°12'29"N, 22°44'7"W	76477	745.5	15326	MF491586	95071	79886	*Africonus roeckeli*
0405	*Africonus damottai*	Baía da Gata (center), Boa Vista, Cabo Verde	16°11'50"N, 22°42'32"W	315488	2914.3	15358	MF491551	95033	80401	—
1428	*Africonus decoratus*	Curral, Santa Luzia, Cabo Verde	16°46'23"N, 24°47'13"W	566822	5540.2	15326	MF491603	95090	78589	—
0370	*Africonus delanoyae*	Jorrita, Baía da Gata, Boa Vista, Cabo Verde	16°12'9"N, 22°42'22"W	158489	1543.7	15323	MF491547	95029	80397	—
1471	*Africonus denizi*	Praia Grande, São Vicente, Cabo Verde	16°51'40"N, 24°52'30"W	906765	8885.2	15326	MF491605	95092	78621	—
0315				214173	2089.2	15243	MF491543	95025	80393	

Table 1 Mitochondrial (mt) genomes analyzed in this study *(Continued)*

	Africonus derrubado	Baía Grande, Derrubado, Boa Vista, Cabo Verde	16°13'31"N, 22°47'17"W							*Africonus damottai*
0565	*Africonus diminutus*	Ilhéu de Sal Rei, Boa Vista, Cabo Verde	16°9'50"N, 22°55'31"W	840424	8204.1	15330	MF491566	95049	80416	—
1025	*Africonus docensis*	Água Doce, Boa Vista, Cabo Verde	16°12'29"N, 22°44'7"W	47313	464.8	15329	MF491588	95073	79894	*Africonus crotchii*
0385	*Africonus evorai*	Zebraca (near Ilhéu do Galeão), Boa Vista, Cabo Verde	16°12'6"N, 22°42'40"W	226416	2218	15243	MF491549	95031	80399	*Africonus crotchii*
0070	*Africonus fantasmalis*	Porto Cais, Maio, Cabo Verde	15°19'15"N, 23°11'10"W	97527	954.7	15330	MF491524	95006	78695	*Africonus fuscoflavus*
0835	*Africonus felitae*	Rabo de Junco, Sal, Cabo Verde	16°41'44"N, 22°58'35"W	344190	3343.9	15404	MF491573	95057	79744	—
1437	*Africonus fernandesi*	Porto Novo, Santo Antão, Cabo Verde	17°1'4"N, 25°3'22"W	742414	7244.2	15324	MF491604	95091	78598	—
0332	*Africonus fiadeiroi*	Derrubado (bay West), Boa Vista, Cabo Verde	16°13'22"N, 22°47'41"W	205910	2016.5	15243	MF491545	95027	80395	*Africonus crotchii*
0855	*Africonus fontonae*	Baía da Fontona, Sal, Cabo Verde	16°44'22"N, 22°58'46"W	156259	1523.9	15328	MF491575	95059	79764	*Africonus cuneolus*
0945	*Africonus fontonae*	Regona, Sal, Cabo Verde	16°48'5"N, 22°59'33"W	56310	549.8	15327	MF491584	95069	79853	*Africonus regonae*
0450	*Africonus fuscoflavus*	Derrubado (bay East), Boa Vista, Cabo Verde	16° 13'331"N, 22°47'3"W	151904	1478.6	15331	MF491557	95039	80407	—
0052	*Africonus galeao*	Navio Quebrado, Terras Salgadas, Maio, Cabo Verde	15°18'54"N, 23°11'2"W	117940	1139.2	15326	MF491522	95004	78677	—
0134	*Africonus gonsaloi*	Praia Gonçalo, Maio, Cabo Verde	15°16'13"N, 23°6'15"W	188174	1835.9	15339	MF491528	95010	78759	—
1390	*Africonus grahami*	Calhau, São Vicente, Cabo Verde	16°51'7"N, 24°51'59"W	464704	4536.1	15325	MF491599	95086	78552	—
0140	*Africonus irregularis*	Porto Cais (North), Maio, Cabo Verde	15°19'45"N, 23°10'57"W	202254	1937.9	15321	MF491530	95012	78765	*Africonus maioensis*
0317	*Africonus irregularis*	Baía Grande, Derrubado, Boa Vista, Cabo Verde	16°13'31"N, 22°47'17"W	170523	1668.2	15331	MF491544	95026	80394	*Africonus maioensis*
0392	*Africonus irregularis*	Baía da Gata, Boa Vista, Cabo Verde	16°11'50"N, 22°42'32"W	252126	2454.3	15324	MF491550	95032	80400	*Africonus crotchii*
1084	*Africonus irregularis*	Morro de Areia, Boa Vista, Cabo Verde	16°5'24"N, 22°57'7"W	125264	1225.1	15330	MF491593	95078	79953	*Africonus crotchii*
1128	*Africonus irregularis*	Estancinha, Ponta do Sol, Boa Vista, Cabo Verde	16°13'12"N, 22°55'9"W	469101	4597.1	15313	MF491597	95082	79997	*Africonus crotchii*
0225	*Africonus isabelarum*	Ponta do Pau Seco, Maio, Cabo Verde	15°15'26"N, 23°13'16"W	247567	2431.7	15244	MF491538	95020	78850	—
0085	*Africonus josephinae*	Lage Branca, Maio, Cabo Verde	15°18'32"N, 23°8'17"W	224495	2204.6	15239	MF491525	95007	78710	*Africonus* sp. nov. 2
0555	*Africonus josephinae*	Ilhéu de Sal Rei, Boa Vista, Cabo Verde	16°9'50"N, 22°55'31"W	169723	1611.8	15330	MF491565	95048	80415	—
0830	*Africonus longilineus*	Serra Negra, Sal, Cabo Verde	16°38'17"N, 22°53'56"W	148726	1453	15316	MF491572	95056	79739	—
0847	*Africonus longilineus*	Rabo de Junco, Sal, Cabo Verde	16°41'44"N, 22°58'35"W	308057	3000.9	15333	MF491574	95058	79756	*Africonus miruchae*
0410	*Africonus luquei*	Praia Canto, Boa Vista, Cabo Verde	16°11'10"N, 22°42'28"W	83198	815.1	15244	MF491552	95034	80402	*Africonus delanoyae*
0064	*Africonus maioensis*	Porto Cais, Maio, Cabo Verde	15°19'15"N, 23°11'10"W	143797	1402.9	15327	MF491523	95005	78689	—
0510	*Africonus marckeppensi*	Ervatao Norte, Boa Vista, Cabo Verde	16°12'3"N, 22°54'43"W	254030	2480.3	15330	MF491562	95044	80412	*Africonus josephinae*
0102				80254	783.1	15326	MF491527	95009	78727	

Table 1 Mitochondrial (mt) genomes analyzed in this study *(Continued)*

	Africonus marcocastellazzii	Lage Branca, Maio, Cabo Verde	15°18'32"N, 23°8'17"W							*Africonus maioensis*
0870	*Africonus melissae*	Baía da Parda, Sal, Cabo Verde	16°45'7"N, 22°53'56"W	195531	1907.9	15328	MF491577	95061	79779	*Africonus longilineus*
0455	*Africonus messiasi*	Derrubado (bay East), Boa Vista, Cabo Verde	16°13'33"N, 22°47'3"W	250171	2447.7	15260	MF491558	95040	80408	*Africonus fuscoflavus*
0426	*Africonus miguelfiaderoi*	Praia Canto, Boa Vista, Cabo Verde	16°11'10"N, 22°42'28"W	130100	1270.5	15328	MF491554	95036	80404	*Africonus vulcanus*
0905	*Africonus mordeirae*	Baía do Roucamento, Sal, Cabo Verde	16°41'20"N, 22°56'24"W	99337	971.4	15241	MF491580	95065	79814	*Africonus cuneolus*
1091	*Africonus morroensis*	Morro de Areia, Boa Vista, Cabo Verde	16°5'24"N, 22°57'7"W	172039	1515.1	15337	MF491594	95079	79960	*Africonus diminutus*
1395	*Africonus navarroi*	Calhau, São Vicente, Cabo Verde	16°51'7"N, 24°51'59"W	665250	6509.2	15331	MF491600	95087	78557	—
0250	*Africonus nelsontiagoi*	Tarrafal, Santiago, Cabo Verde	15°16'50"N, 23°45'15"W	173873	1687.2	15339	MF491541	95023	78875	*Africonus verdensis*
0820	*Africonus pseudocuneolus*	Serra Negra, Sal, Cabo Verde	16°38'17"N, 22°53'56"W	131838	1288.7	15337	MF491571	95055	79729	*Africonus cuneolus*
0036	*Africonus raulsilvai*	Praia da Soca, Maio, Cabo Verde	15°15'8"N, 23°13'4"W	345872	1699.6	15534	MF491520	95002	78661	—
0865	*Africonus regonae*	Baía da Fontona, Sal, Cabo Verde	16°44'22"N, 22°58'46"W	246041	2411	15328	MF491576	95060	79774	—
0950	*Africonus regonae*	Regona, Sal, Cabo Verde	16°48'5"N, 22°59'33"W	88673	864.1	15337	MF491585	95070	79858	—
0586	*Africonus roeckeli*	Praia Canto, Boa Vista, Cabo Verde	16°11'10"N, 22°42'28"W	141600	1385.7	15320	MF491567	95050	80417	—
0549	*Africonus salreiensis*	Ilhéu de Sal Rei, Boa Vista, Cabo Verde	16°9'50"N, 22°55'31"W	349070	3402.9	15331	MF491564	95047	80414	*Africonus crotchii*
0810	*Africonus serranegrae*	Serra Negra, Sal, Cabo Verde	16°38'17"N, 22°53'56"W	182124	1777.6	15335	MF491570	95054	79719	*Africonus cuneolus*
1078	*Africonus silviae*	Morro de Areia, Boa Vista, Cabo Verde	16°5'24"N, 22°57'7"W	293568	2876.1	15336	MF491592	95077	79947	*Africonus fuscoflavus*
0445	*Africonus swinneni*	Porto Ferreira, Boa Vista, Cabo Verde	16°7'45"N, 22°40'17"W	221672	2169.1	15244	MF491556	95038	80406	*Africonus delanoyae*
1125	*Africonus teodorae*	Estancinha, Ponta do Sol, Boa Vista, Cabo Verde	16°13'12"N, 22°55'9"W	326654	3185.5	15334	MF491596	95081	79994	*Africonus crotchii*
1035	*Africonus umbelinae*	Espingueira, Boa Vista, Cabo Verde	16°12'55"N, 22°47'49"W	110526	1085.5	15333	MF491589	95074	79904	*Africonus damottai*
0240	*Africonus verdensis*	Tarrafal, Santiago, Cabo Verde	15°16'50"N, 23°45'15"W	53400	519.8	15339	MF491540	95022	78865	—
0435	*Africonus vulcanus*	Porto Ferreira, Boa Vista, Cabo Verde	16°7'45"N, 22°40'17"W	284766	2787.4	15242	MF491555	95037	80405	—
1110	*Africonus zinhoi*	Curral Velho, Boa Vista, Cabo Verde	15°58'4"N, 22°47'42"W	388828	3785.5	15331	MF491595	95080	79979	*Africonus maioensis*
7036	*Trovaoconus atlanticoselvagem*	Baixo João Valente, Cabo Verde	15°44'27"N, 23°5'26"W	95264	933.8	15352	MF491606	7036	—	*Kalloconus trochulus*
0616	*Trovaoconus* cf. *ateralbus*	Serra Negra, Sal, Cabo Verde	16°38'17"N, 22°53'56"W	87550	853.7	15344	MF491568	95052	79664	*Kalloconus* sp. nov. 1
0010	*Trovaoconus pseudonivifer*	Ponta do Pau Seco, Maio, Cabo Verde	15°15'26"N, 23°13'16"W	56486	555.5	15351	MF491519	95000	78635	*Kalloconus trochulus*
0094	*Trovaoconus pseudonivifer*	Lage Branca, Maio, Cabo Verde	15°18'32"N, 23°8'17"W	454415	4429.8	15351	MF491526	95008	78719	*Kalloconus trochulus*
0154	*Trovaoconus pseudonivifer*	Porto Cais (north), Maio, Cabo Verde	15°19'45"N, 23°10'57"W	199736	1954.2	15352	MF491532	95014	78779	*Kalloconus trochulus*
0420	*Trovaoconus pseudonivifer*	Praia Canto, Boa Vista, Cabo Verde	16°11'10"N, 22°42'28"W	111182	1085	15347	MF491553	95035	80403	*Kalloconus pseudonivifer*
0500				223915	2177.2	15351	MF491561	95043	80411	

Table 1 Mitochondrial (mt) genomes analyzed in this study *(Continued)*

ID	Species	Location	Coordinates			Length (bp)	GenBank Acc No	Voucher (MNCN/ADN)	Voucher shell (MNCN 15.05/)	New species proposed[a]
	Trovaoconus trochulus	Baía do Ervatão (North), Boa Vista, Cabo Verde	16°12'3"N, 22°54'43"W							*Kalloconus trochulus*
0149	*Trovaoconus venulatus*	Lage Branca, Maio, Cabo Verde	15°18'32"N, 23°8'17"W	105732	1014.9	15276	MF491531	95013	78774	*Kalloconus venulatus*
0187	*Trovaoconus venulatus*	Praia Real, Maio, Cabo Verde	15°19'45"N, 23°10'40"W	144651	1415	15330	MF491535	95017	78812	*Kalloconus venulatus*
0234	*Trovaoconus venulatus*	Ponta do Pau Seco, Maio, Cabo Verde	15°15'26"N, 23°13'17"W	67867	661.5	15320	MF491539	95021	78859	*Kalloconus venulatus*
0347	*Trovaoconus venulatus*	Derrubado (bay West), Boa Vista, Cabo Verde	16°13'22"N, 22°47'41"W	48636	475.9	15326	MF491546	95028	80396	*Kalloconus venulatus*
0475	*Trovaoconus venulatus*	Ponta Antónia, Boa Vista, Cabo Verde	16°13'24"N, 22°46'59"W	409041	3966.1	15340	MF491560	95042	80410	*Kalloconus venulatus*
1038	*Trovaoconus venulatus*	Praia Canto, Boa Vista, Cabo Verde	16°11'10"N, 22°42'28"W	143644	2403.8	15336	MF491590	95075	79907	*Kalloconus venulatus*
IB001	*Lautoconus ventricosus*	Estani des Peix, Formentera, Balearic Islands, Spain	38°43'49"N, 1°24'42"E	86290	842.5	15341	MF491607	95094	80426	*Lautoconus* sp. nov 1

GenBank mt genomes

ID	Species	Location	Coordinates	Reference	Length (bp)	GenBank Acc No	Voucher (MNCN/ADN)	Voucher shell (MNCN 15.05/)	New species proposed[a]
6990	*Africonus borgesi*	Porto Ferreira, Boa Vista, Cabo Verde	16°7'45"N, 22°40'170"W	Cunha et al., (2009)	15536	NC_013243	6990	—	—
0025	*Africonus infinitus*	Ponta do Pau Seco, Maio, Cabo Verde	15°15'26"N, 23°13'17"W	Abalde et al., (in prep.)	15522	KY864967	95001	78650	—
0875	*Africonus miruchae*	Terrinha Fina, Palhona, Sal, Cabo Verde	16°49'12"N, 22°59'12"W	Abalde et al., (in prep.)	15336	KY864971	95062	79784	—
0534	*Trovaoconus pseudonivifer*	Estancinha, Ponta do Sol, Boa Vista, Cabo Verde	16°13'12"N, 22°55'9"W	Abalde et al., (in prep.)	15351	KY864969	95046	80418	*Kalloconus trochulus*
0550	*Trovaoconus venulatus*	Ilhéu de Sal Rei, Boa Vista, Cabo Verde	16°9'56"N, 22°55'23"W	Uribe et al., (2017)	15524	KX263250	86741	80419	*Kalloconus venulatus*
0601	*Trovaoconus ateralbus*	Calheta Funda, Sal, Cabo Verde	16°39'6"N, 22°56'53"W	Abalde et al., (in prep.)	15327	KY864970	95051	79649	*Kalloconus ateralbus*
1375	*Kalloconus* cf. *byssinus*	North Senegal	unknown	Abalde et al., (in prep.)	15348	KY864973	95085	78536	*Kalloconus pulcher*
1253	*Kalloconus pulcher*	Les Almadies, Dakar, Senegal	14°44'40"N, 17°31'442"W	Abalde et al., (in prep.)	15332	KY864972	95084	78414	*Kalloconus pulcher*
1343	*Lautoconus belairensis*	Terrou-Bi. Dakar, Senegal	14°40'29"N, 17°28'12"W	Abalde et al., (2017)	15321	KY801849	91293	78504	*Gen. nov. belairensis*
1338	*Lautoconus bruguieresi*	Île de Gorée, Dakar, Senegal	14°40'16"N, 17°23'58"W	Abalde et al., (2017)	15340	KY801851	91291	78499	*Gen. nov. bruguieresi*
1296	*Lautoconus cloveri*	Ndayane, Senegal	14°33'45"N, 17°7'34"W	Abalde et al., (2017)	15323	KY801859	91283	78457	*Gen. nov. cloveri*
CG13	*Lautoconus guanche*	Lanzarote, Canary Islands, Spain	28°57'16"N, 13°34'22"W	Abalde et al., (2017)	15506	KY801847	91295	—	*Gen. nov. guanche*
1266	*Lautoconus hybridus*	NGor, Dakar, Senegal	14°45'67"N, 17°30'36.33"W	Abalde et al., (2017)	15507	KY801863	91279	78427	*Gen. nov. hybridus*
1278	*Lautoconus mercator*	NGor, Dakar, Senegal	14°45'6"N, 17°30'36"W	Abalde et al., (2017)	15329	KY801862	91280	78439	*Gen. nov. mercator*
CV13	*Lautoconus ventricosus*	Ria Formosa, Faro, Portugal	36°58'0"N, 7°53'2"W	Uribe et al., (2017)	15534	KX263251	86742	—	—
CVERM1	*Chelyconus ermineus*	Praia Gonçalo, Maio, Cabo Verde	15°16'13"N, 23°6'15"W	Abalde et al., (in prep.)	15365	KY864977	95095	78876	—

[a]hyphen indicates that original species name is maintained and considered valid

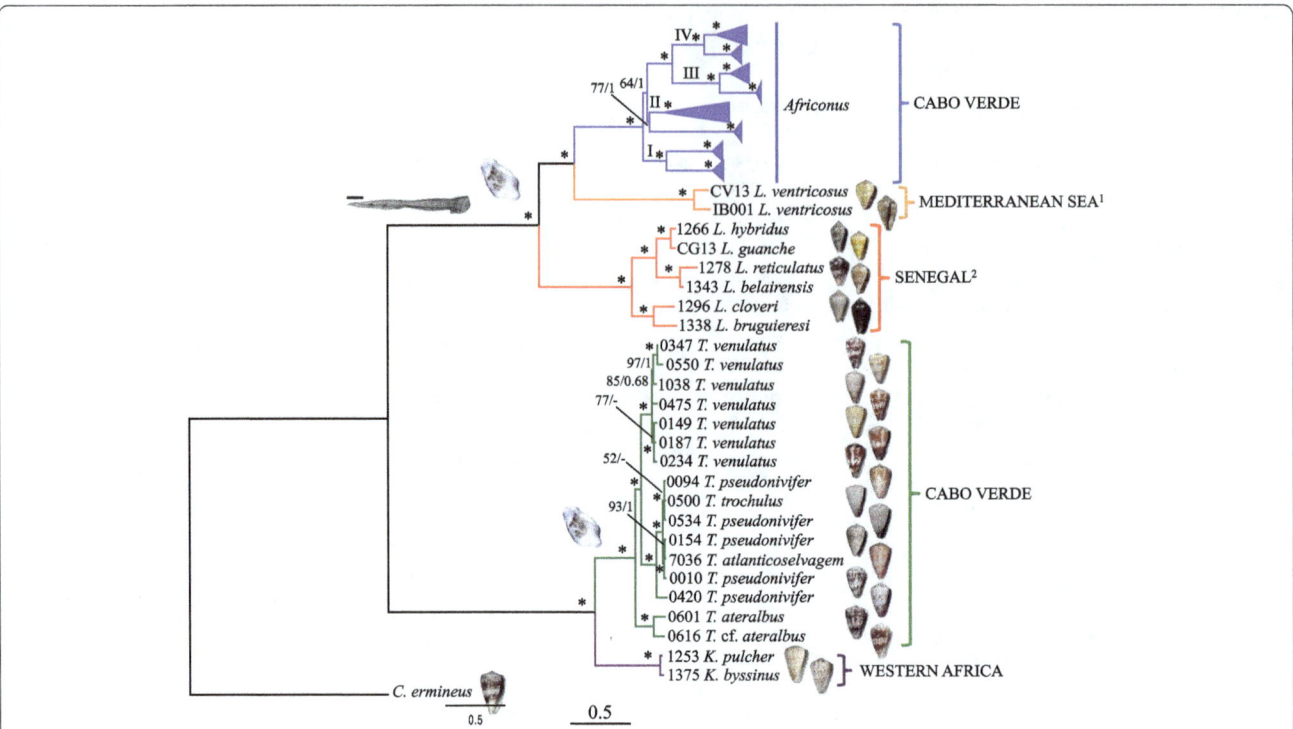

Fig. 2 Closest sister groups of cone snails endemic to Cabo Verde based on mt genomes (concatenated protein coding plus rRNA genes analyzed at the nucleotide level). The reconstructed ML tree using *C. ermineus* as outgroup is shown. Number of specimen, initial species assignment, and a ventral picture of the shell are provided. Numbers at nodes are statistical support values for ML (bootstrap proportions)/BI (posterior probabilities). Scale bar indicates substitutions/site. Two major clades are recovered: the first one includes *Kalloconus* from Western Africa (purple) and *Trovaoconus* from Cabo Verde (green) whereas the second one includes *Lautoconus* from the Mediterranean Sea (and neighboring Atlantic Ocean[1]; orange), *Lautoconus* from Senegal (and *L. guanche* from Canary Islands[2]; red), and *Africonus* (blue) from Cabo Verde. Lineages within *Africonus* are expanded in Fig. 3. The "robust" type of radular tooth is shown as the ancestral character state for the second clade (scale bar equals 0.1 mm; see diversity of radular teeth of the first clade in Additional file 1). The two transitions from planktonic to non-planktonic larvae are indicated by a sac of eggs

into two monophyletic groups (Figs. 2 and 3). Lineage I was the sister group of the remaining *Africonus* and its two lineages had each species from Maio sister to species from Boa Vista (Fig. 3a). Lineage II included species from Santiago and Maio sister to species endemic to the westernmost islands (Santo Antão, São Vicente and Santa Luzia). These latter species could be grouped into three main lineages, one containing species endemic to São Vicente, another containing species distributed both in Santa Luzia and São Vicente, and the third one including species from the three islands (Fig. 3a). Lineage III included species from Maio sister to species from Boa Vista (Fig. 3a). Lineage IV contained specimens representing most of the described species of *Africonus*. One monophyletic group included species endemic to Sal whereas the other clade included *Africonus isabelarum* from Maio as sister to four lineages, two containing exclusively species from Boa Vista, one having species from Maio sister to *Africonus irregularis* from Boa Vista, and one having species from Boa Vista and *Africonus fantasmalis* from Maio (Fig. 3b).

Pairwise uncorrected sequence divergences were estimated based on the alignment including the nucleotide sequences of the 13 mt protein-coding and two rRNA genes. Pairwise uncorrected sequence divergences between *C. ermineus* and ingroup taxa averaged 18%. The average pairwise uncorrected sequence divergence between the two main ingroup clades (genera *Kalloconus* + *Trovaoconus* versus genera *Lautoconus* + *Africonus*) was 16%. Pairwise uncorrected sequence divergences between *Lautoconus* endemic to Senegal (plus *L. guanche*) and *L. ventricosus* plus *Africonus* averaged 11%. The average pairwise uncorrected sequence divergence between the sister groups *L. ventricosus* and *Africonus* was 10% whereas between *Kalloconus* and *Trovaoconus*, it was 5%. The pairwise uncorrected sequence divergences between the four main lineages within *Africonus* averaged 6%. The corresponding values for the pairwise divergences between the two major clades defined within each of the lineages I–IV were 4%, 6%, 3%, and 3%, respectively. Pairwise uncorrected sequence divergence comparisons between sister species level were distributed into two different ranges, one closer to 1% (0.5–

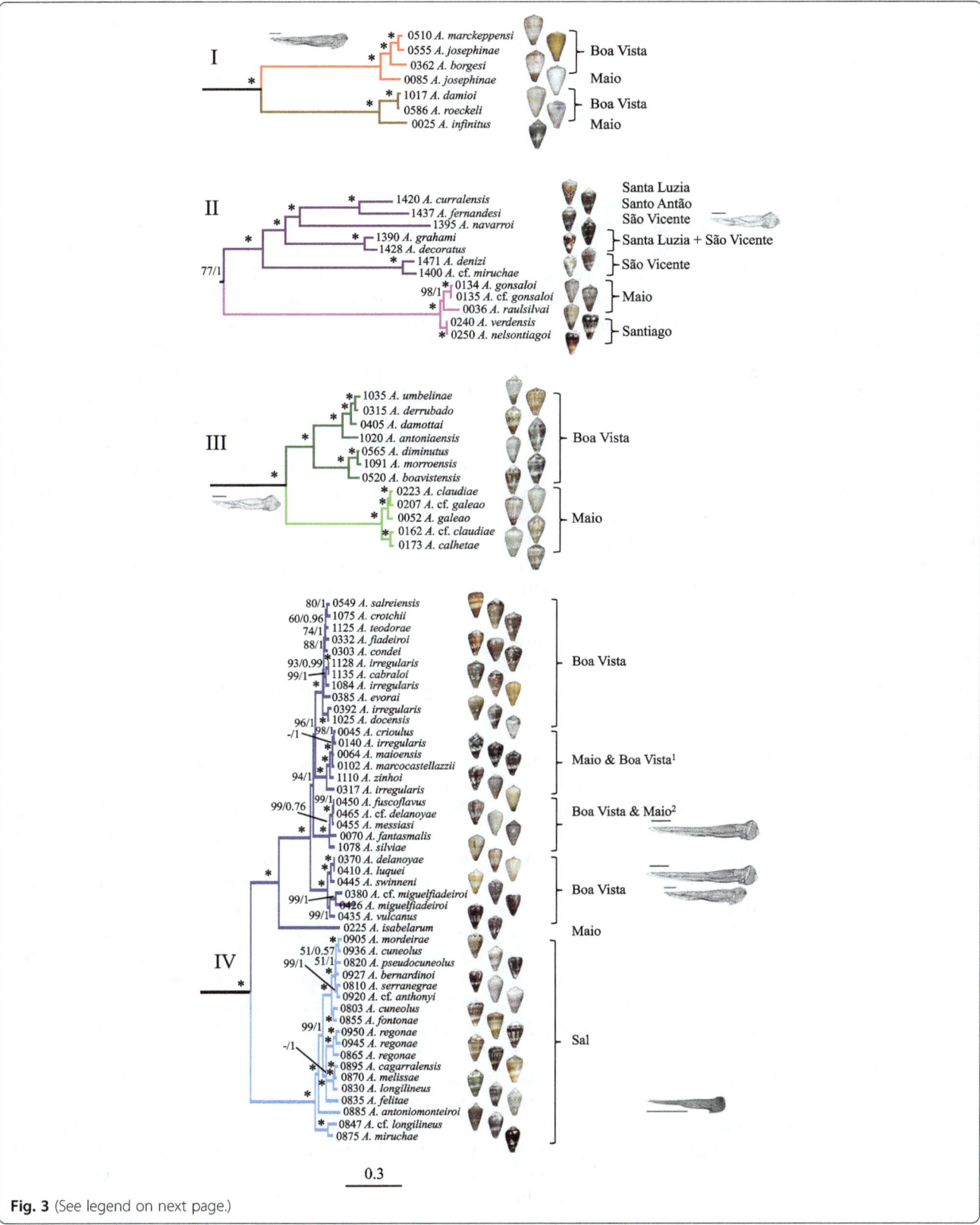

Fig. 3 (See legend on next page.)

(See figure on previous page.)

Fig. 3 Phylogeny of *Africonus* based on mt genomes (concatenated protein coding plus rRNA genes analyzed at the nucleotide level). Number of specimen, initial species assignment, island, and a ventral picture of the shell are provided. Numbers at nodes are statistical support values for ML (bootstrap proportions)/BI (posterior probabilities). Hyphen indicates a bootstrap value below 50%. Scale bar indicates substitutions/site. Four major lineages (I-IV) are recovered and indicated with different colors. All *Africonus* have the "robust" type of radular tooth except when indicated (scale bar equals 0.1 mm). [1]All taxa endemic to Maio except *A. irregularis* endemic to Boa Vista. [2]All taxa endemic to Boa Vista except *A. fantamalis* endemic to Maio

1.5%) and the other closer to 0% (0-0.5%). The latter divergences were particularly abundant among sister species comparisons within Maio, Boa Vista and Sal. Several mt genomes of different species were almost identical (<0.05%) in sequence including (1) *Africonus delanoyae* and *Africonus luquei*, (2) *Africonus fuscoflavus*, *Africonus* cf. *delanoyae*, and *Africonus messiasi*, (3) *Africonus irregularis* (#1128) and *Africonus cabraloi*, (4) *Africonus verdensis* and *Africonus nelsontiagoi*, and (5) *Africonus gonsaloi* and *Africonus* cf. *gonsaloi*.

Evolution of radular types

The different lineages within *Africonus* exhibit distinct radular types (Fig. 3). Most lineages and species showed the "robust" type, which is of medium relative size, with a short, pointed barb and a basal spur (see Additional file 1). The anterior section of the tooth is equal or slightly shorter than the posterior section, and the blade covers most of the anterior section (80% – 85%). There are usually 19 to 30 denticles in the serration, arranged in one row (occasionally two). Several species within lineage IV (*Africonus delanoyae*, *Africonus luquei*, *Africonus swinneni*, *Africonus fuscoflavus*, *Africonus messiasi*, *Africonus silviae* and *Africonus* cf. *delanoyae* from Boa Vista island, and *Africonus fantasmalis* from Maio island) exhibited radular teeth of the "elongated" type, similar to the "robust" type but characterized by an anterior section which is longer than the posterior section, a blade covering 40 to 50% of the anterior section, and more numerous denticles in the serration (usually more than 30) often arranged in two rows. Several species (*Africonus borgesi*, *Africonus josephinae* and *Africonus marckeppensi* in lineage I, *Africonus navarroi* in lineage II), all species in lineage III, plus *Africonus vulcanus*, *Africonus miguelfiadeiroi* and *Africonus* cf. *miguelfiadeiroi* in lineage IV) displayed radular tooth of the "broad" type, which is characterized by a medium-sized (Shell Length/Tooth Length = 32-45) and very broad radular tooth (Shell Length/Anterior section Width = 7-12), with an anterior section which is shorter than the posterior section (Tooth Length/Anterior section Length = 2.1-2.9), a blade covering most of the anterior section, and with a variable number of denticles (8 to 30) in the serration arranged in two or more rows. The radular morphology of *Africonus felitae* may represent a special case with a small relative size (Shell Length/Tooth Length = 63-67),

narrow (Tooth Length/Anterior section Width = 20-23), the anterior section shorter than the posterior section (Tooth Length/Anterior section Length = 2.2-2.4), and characterized by the total absence of denticles in the serration. The base of this tooth is relatively large and broad.

The species of *Kalloconus* and *Trovaoconus* exhibit essentially two kinds of radular morphologies (Additional file 1). The teeth in *K. pulcher*, and also in *Trovaoconus trochulus*, *T. pseudonivifer* and *Trovaoconus atlanticoselvagem* are narrow and elongated; the blade is moderately short being about one third to almost one-half the length of the anterior section of the tooth, which is distinctly longer than the posterior section of the tooth. There are many denticles (25 to 45 or more) in the long serration, arranged usually in multiple rows with a major row flanked by numerous smaller serrations. In the case of *T. venulatus*, *Trovaoconus ateralbus*, and *Trovaoconus* cf. *ateralbus* the teeth are broader, and the anterior and posterior sections are almost equal in length. There are 16 to 33 denticles in the serration, often coarse and hook-shaped in the middle portion, arranged initially in one row becoming two rows below.

Dating of major cladogenetic events

Major cladogenetic events within the reconstructed phylogeny were dated using an uncorrelated relaxed molecular clock model, which was calibrated using the age of Sal (28 mya; the oldest island of the archipelago) for the node separating *Africonus* from its sister group, *L. ventricosus*, and the age of the origins of São Vicente, Santo Antão, and Santa Luzia (7.5 mya) for the node splitting the lineage including the endemics to these islands from its sister group lineage including endemics to Maio and Santiago islands [24]. The first divergence event involving *Kalloconus* + *Trovaoconus* versus (paraphyletic) *Lautoconus* + *Africonus* was dated at 34 mya (Fig. 4; note that genera and species labels in the chronogram take into account proposed synonymizations, see Discussion). The divergence between the clade containing cones endemic to Senegal (+ *L. guanche*) and the clade including *L. ventricosus* plus *Africonus* was dated at 26 mya. The split between the latter two lineages was dated at 23 mya. The diversification of the crown group of *Africonus* into its four main lineages (I-IV) was estimated to have occurred between 9.4 - 6.9 mya (Fig. 4). The separation of *Kalloconus* and *Trovaoconus* was dated

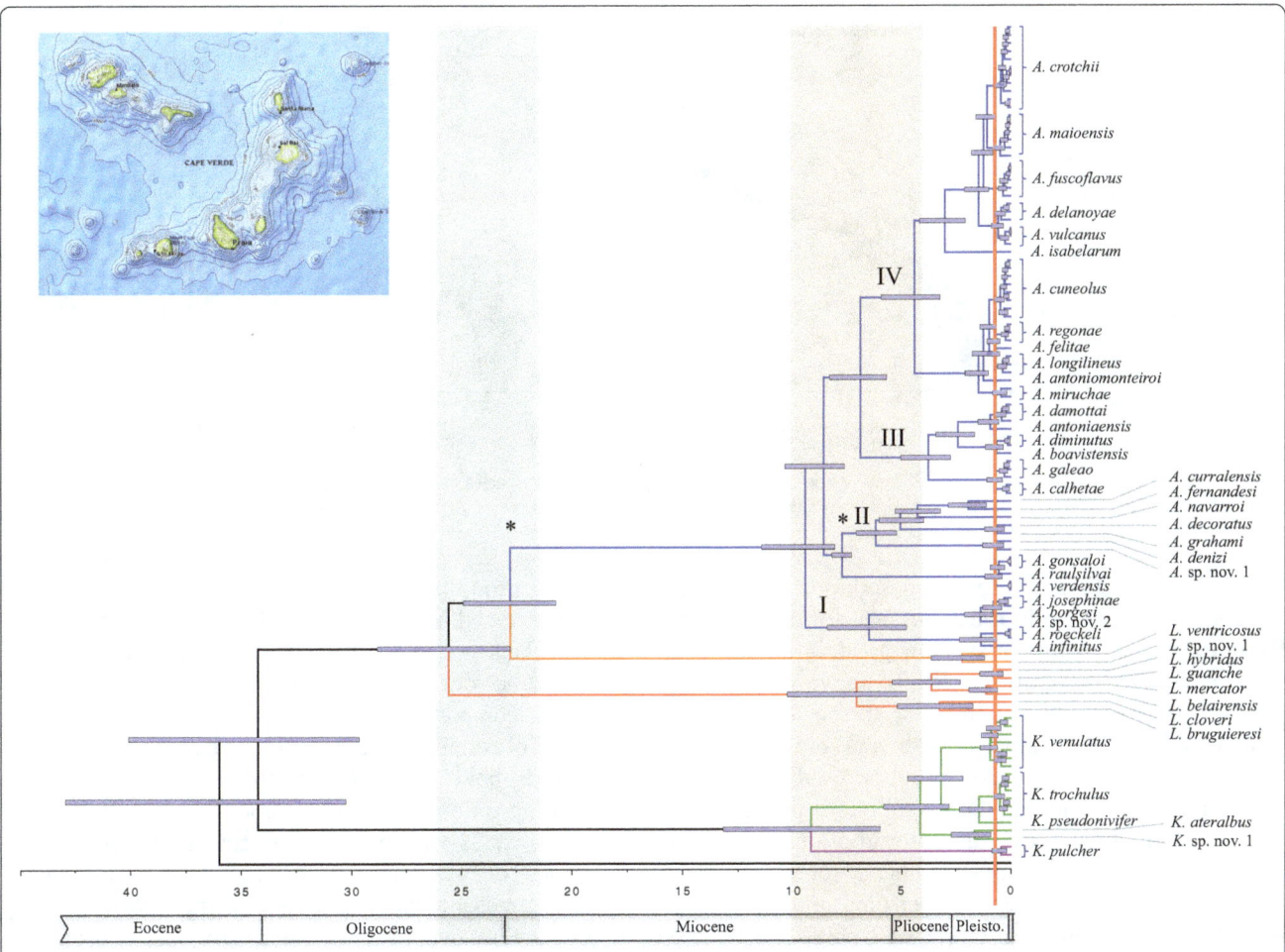

Fig. 4 Chronogram based on mt genomes (concatenated protein coding plus rRNA genes analyzed at the nucleotide level) and using the fixed topology of the ML tree shown in Figs. 2 and 3. Lineages are colored as in Fig. 2. Synonymizations proposed (see text) are taken into account. A Bayesian uncorrelated relaxed lognormal clock with geographic-based calibration priors (denoted by asterisks) was used in BEAST. Horizontal bars represent 95% credible intervals for time estimates; dates are in millions of years. Geological periods are indicated. Brown and orange bands indicate main divergence periods around the Oligocene-Miocene and Miocene-Pliocene transitions, respectively. A red line indicates the threshold for species delimitation. The bathymetry of Cabo Verde archipelago is shown in an inset

9 mya and the diversification of the crown group of *Trovaoconus* was established at 4 mya (Fig. 4).

Diversification rates through time

Variations in the diversification rates through time were estimated for *Africonus* and the hypothesis of a radiation during the evolutionary history of the clade was tested. The gamma-statistic, which measures departures from a constant rate of diversification, had values of 7.19 ($p <$ 0.05) and 3.12 (p < 0.05) when considering the currently named (based on phenotypic traits) or only the here-proposed (considering genetic evidence) species for the genus, respectively. In both cases, the hypothesis of a radiation is accepted. According to the lineage through time plots (Fig. 5), the initial rate of increase in the number of species slowed down between six and one and a half million years ago regardless of the species delimitation hypothesis

tested. Afterwards, the diversification rate accelerated considerably, and the increase in number of species either continued or abandoned a normal Yule process of speciation when considering the species delimitation hypothesis here proposed or the currently number of named species, respectively.

Discussion

Cone snails are marine gastropods well known to evolutionary biologists due to their extraordinary species and ecological diversity [25], but also to molecular biologists and pharmacologists due to their sophisticated venom cocktails [26], as well as to amateur naturalists due to their brightly colored and highly appreciated shells [27]. Therefore, they are the subject of intensive research across disciplines and additionally have received wide attention from the general public. There are more than 800 described species and this number increases steadily

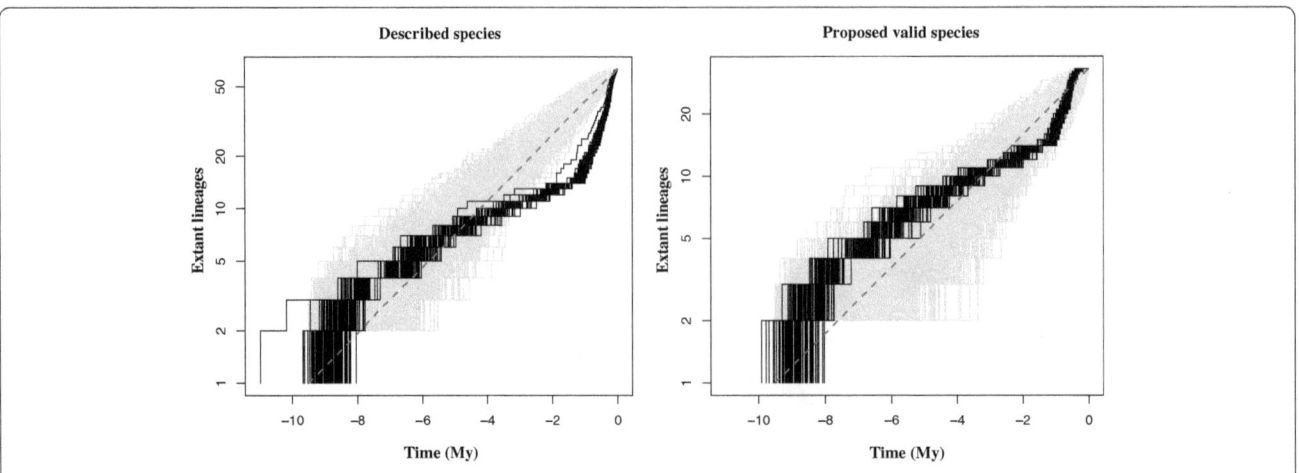

Fig. 5 Logarithmic lineage through time (LTT) plots of described (**a**) and proposed valid (**b**) species. The red bar indicates the pure Yule process of speciation, the grey shadow shows 1000 simulated trees and the black lines represent 100 trees randomly chosen among the trees generated by BEAST

every year [28]. Thus far, species description and identification of cones heavily relies on shell form, color and banding patterns, which may show great variety at local scale leading to important levels of synonymy within the family Conidae [7, 20]. In this regard, species delimitation could greatly improve with the aid of robust molecular phylogenies, which could be used in addition as framework to uncover the evolutionary patterns and processes underlying the diversification of the group. While reconstructing a robust phylogeny for all described cone species worldwide is cumbersome and at present unrealistic within the framework of a single study, it is possible, however, to accomplish a proof-of-concept study in a particular region [23].

We have here reconstructed a molecular phylogeny of cones endemic to Cabo Verde and allied species in the Macaronesian region, continental West Africa, and the Mediterranean region. These cones are particularly interesting from an evolutionary perspective as they have radiated in an oceanic archipelago and constitute a natural experiment to gain insights onto the processes governing diversification and adaptation [29]. Phylogenetic analyses were based on nearly complete mt genomes (only missing the control region and neighboring sequences) and included 105 specimens comprising most of the cone species diversity of the analyzed regions. Probabilistic methods of phylogenetic inference arrived at a robust and highly resolved phylogeny (virtually all nodes received high statistical support, which in most cases was maximal). To our knowledge, this is the first wide application of mt genomes to the resolution of a phylogeny within mollusks (but see [30, 31] for comparable examples in fish and insects, respectively). Previous studies in gastropods were restricted in the number of taxa analyzed (e.g., [22] for the family Conidae) but here we were able to achieve a lineage representation of the

reference group (in this case, Cabo Verde cones) only previously attained by studies using few concatenated partial gene sequences (see e.g., [25] for the family Conidae or [1] for the cones of Cabo Verde). Previous phylogenetic studies using complete mt genomes have demonstrated that the level of resolution of these molecular markers is compromised above the superfamily level due to saturation, base composition biases, and among-lineage rate heterogeneity [32, 33]. Here, we show that phylogenetic performance of mt genomes achieves best results when analyzing closely related genera (and their corresponding species). Moreover, results were particularly promising taking into account that *Africonus* diversity in Cabo Verde was originated through radiation processes, which normally lead to relatively short tree nodes (often difficult to disentangle).

A thorough sampling of closely-related outgroup taxa allowed us to tackle key questions on the origin of the cones endemic to Cabo Verde and on their closest living sister groups. As previously reported, there are two independent origins of Cabo Verde cones, leading to the genera *Africonus* and *Trovaoconus*, respectively [1]. The closest living sister group of *Africonus* is *L. ventricosus* from the Mediterranean Sea and neighboring Atlantic Ocean. Therefore, the origin of this clade is clearly Macaronesian/Mediterranean and these cones are only distantly related to the geographically closer cones endemic to Senegal. These latter cones were ascribed to the genus *Lautoconus* (as was the case of *L. guanche* from Canary islands, deeply nested within the clade of Senegal cones; [23]). However, the closest sister group relationship of *Africonus* and *L. ventricosus* requires formal description of a different genus for Senegal cones (plus *L. guanche*), which will be done elsewhere. We could not include any representative of cones endemic to Angola (genus *Varioconus*) but a recent phylogeny based on partial *cox1* gene sequences recovered all these cones (including *Varioconus jourdani* from Saint Helena Island) as a monophyletic group sister to Senegal cones [34]. Alternatively, all previously mentioned genera could be merged into genus *Lautoconus* [35]. However, the relatively high levels of sequence divergence (using the sequence divergence of genus *Chelyconus* as reference) and the restricted (endemic) distribution of the clades fit better with the former taxonomic proposal. The closest living sister group of *Trovaoconus* is genus *Kalloconus* from West Africa. Therefore, the origin of this clade is clearly related to neighboring regions of the continent. Actually, the sequence divergence between *Trovaoconus* and *Kalloconus* is much lower than that estimated between *Lautoconus* and *Africonus*. This observation argues against maintaining the generic status of *Trovaoconus*, and supports the inclusion of their species within genus *Kalloconus*, as some authors have proposed [35]. Hence, *Kalloconus* would be a genus that is present throughout the coast of West Africa from Morocco to Angola as well as in Canary Islands and Cabo Verde.

Altogether, the two main clades in the reconstructed phylogeny show very distinct patterns of distribution. One clade includes a single genus with widespread distribution in Macaronesia and West Africa whereas the other clade, which occupies the same geographical regions, is divided into several valid genera (*Africonus*, *Lautoconus*, *Varioconus*, Gen. nov. for Senegal endemics). These distinct patterns could be explained partly taking into account differences in larval dispersal capabilities between the two clades [1]. According to the phylogeny, the ancestor of the *Kalloconus* clade was inferred to have planktotrophic larvae, capable of long dispersals whereas the ancestor of the other clade would have non-planktotrophic larvae, and thus a limited dispersal capability leading to restricted gene flow and higher rates of diversification [36, 37]. Interestingly, the ancestor of *Kalloconus* species endemic to Cabo Verde (former *Trovaoconus* species) lost planktotrophy, which is a common evolutionary pattern in insular species [38].

According to the reconstructed phylogeny, cones belonging to genus *Africonus* are divided into four main lineages (I-IV; with each further subdivided into two distinct clades). Species endemic to Maio and Boa Vista are found in all four lineages whereas species endemic to Sal form a clade within lineage IV, species from the westernmost islands (Santo Antão, São Vicente and Santa Luzia) form a clade within lineage II, and the single species from Santiago is recovered within lineage II. Unfortunately, we could not sample specimens of *Africonus furnae* from Brava and *Africonus kersteni* from São Nicolau, and cannot determine whether they could be ascribed to any of the above-mentioned four lineages or form their own independent lineages. The single origin of cones endemic to Sal, Santiago, and westernmost islands could be explained by the deep slopes separating these islands

whereas the multiple origins of the cones found in Maio and Boa Vista could be associated to the relatively shallow seamount (Baixo João Valente) connecting both islands [1]. These differences in bathymetry in connection with past eustatic sea level changes could be determinant in preventing or promoting dispersal in *Africonus* species, whose larvae are all non-planktotrophic.

Diversification events among main lineages were concentrated in three major periods. The first one, around the Oligocene-Miocene boundary (23 mya), includes the divergence of cones endemic to Senegal (and Angola) from their sister clade, and the posterior separation within this sister clade of cones endemic to Cabo Verde and those endemic to the Mediterranean Sea and neighboring Atlantic Ocean. During Oligocene-Miocene transition, there was a global cooling event [39, 40], the ice sheet of Antarctica greatly expanded, and a sea level drop of ~50 m occurred [41]. The second period corresponds to a sustained global cooling in the Late Miocene starting 12 mya [42] that produced an eustatic sea level drop between −10 and −30 m from 6.26 to 5.50 Mya [43] and culminated with the Messinian Salinity crisis and the desiccation of the Mediterranean Sea at the end of the Miocene from 5.96 to 5.33 Mya [44]. During this period, the divergence of the main lineages within *Africonus* (I-IV), the cones endemic to Senegal, and *Kalloconus* occurred. Finally, a burst of speciation events is inferred during the Pleistocene when another cooling period characterized by extreme climate oscillations and drastic eustatic sea level changes concurring with glacial-interglacial periods [45]. Global cooling has been recently proposed to be a driver of diversification of marine species [46] in agreement with our results. The reconstructed phylogeny, the chronogram, and the current geographical distribution of the species altogether support that allopatry is the main mode of speciation for cone snails with non-planktotrophic larvae, as previously suggested [1]. The complex geology of the island of Boa Vista with several eruptions at >16, 15-12.5, and 9.5-4.5 mya [47], involving different parts of the island may have also contributed to creating additional niches along the coast and could explain that this island harbors the highest number of endemic cones.

The reconstructed phylogeny also allows inferring the evolution of the radula in the studied lineages [8, 48]. All analyzed ingroup taxa are vermivorous [8]. Studies documenting potential specialization of the vermivore radular type to prey on specific worm species are scarce and restricted thus far to cone species preying on amphinomids [49]. Here, we show that most *Africonus* species show a "robust" radular type, which is shared also with *L. ventricosus* and a lineage of Senegal cones represented by *Lautoconus cloveri* and *Lautoconus bruguieresi* in the phylogeny [23]. Therefore, the common ancestor of cones endemic to Senegal (plus *L. guanche*), *Africonus*, and *L.*

ventricosus was inferred to have a "robust" type radula. The "elongated" type of radular tooth, which was found in several species within lineage IV of *Africonus*, also appears in a lineage of Senegal cones that is represented by *Lautoconus hybridus* and *L. guanche* in the phylogeny [23]. The radular tooth of *A. felitae* resembles the "small" type observed in a lineage of Senegal cones represented by *Lautoconus reticulatus* and *Lautoconus belairensis* in the phylogeny [23]. The "broad" type of radular tooth that appeared independently in several lineages of *Africonus* has not been observed in any cone from Senegal. While shifts in radular type could be correlated with early clado-genesis in cones endemic to Senegal [23], the evolution of different types of radular tooth within *Africonus* was restricted to few specific cases. Thus, future studies are needed to determine whether in such cases there has been a dietary shift to prey on specific worms. The radula teeth identified in *Kalloconus* resemble the types "elongated" and "robust" observed within *Lautoconus* and *Africonus*, although are clearly distinct. This might indicate instances of convergence, and that only a discrete number of different main types of radula could be found in a given clade.

During the last few years, the number of new cone species described from Cabo Verde has increased at an astonishing rate (e.g., [12]). These new species are identified based on differences (often subtle) in shell shape and color, and their status needs to be contrasted with genetic data to uncover cases of local phenotypic variation within species due to either genetic polymorphism or phenotypic plasticity that may be producing overestimations of the number of species in the group [21]. In addition, genetic data could help identify cases of phenotypic convergence due to adaptation of genetically distinct populations (ecotypes) or species (sibling or cryptic) to similar environments [50–52], also affecting the total number of valid species. Comparative analyses of pairwise uncorrected sequence divergences taking into account the reconstructed phylogeny showed that some described species shared almost identical mt genomes with levels of sequence divergence normally considered to be associated to genetic variation at the population level. Clades comprising these sets of closely related sequences indicate that an uncorrected sequence divergence threshold around 1% could be associated to the species status. This threshold lies well within the so-called grey zone of speciation between 0.5-2% [53]. Of course, these results need to be further confirmed with genomic nuclear data that discard potential events of incomplete lineage sorting and hybridization [54]. In addition, the present study could be further improved in the future by increasing the number of individuals analyzed per original species. Importantly, the comparative analyses on variation of diversification rates through time support the here proposed hypothesis of species delimitation as it concurs with a Yule process of speciation whereas the number of currently named species clearly exceed expectations and would imply an extraordinary recent acceleration of speciation rates.

Our study confirms the diversity of cone endemic to Cabo Verde but significantly reduces the number of valid species. Applying the threshold in a conservative manner (i.e., maintaining described species as valid in case of doubt due to closeness to the threshold) to cones endemic to Cabo Verde would reduce the number of valid species within the sampled *Africonus* from 65 to 32 (see Table 1 and Fig. 4). The proposed nomenclatural changes follow standard ICZN recommendations maintaining the most senior (oldest) name. Among the species not sampled, two correspond to São Nicolau and Brava islands, four of them are from the islands of São Vicente and Santa Luzía, and most likely represent valid species (*Africonus bellulus*, *Africonus lugubris*, *Africonus saragasae*, *Africonus santaluziensis*, *A. kersteni* and *A. furnae*) given the relative high sequence divergences found among species endemic to these islands. The 19 remaining ones were recently described, mostly from Boa Vista, and are expected to fall in most cases into some of the clades already discussed in the present work, and therefore may correspond to morphs of other described species. A direct consequence of synonymization is that some previously described species of rather restricted distribution are merged as populations into the new species, which considerably increase their range of distribution (Additional file 1). For instance, *A. crotchii*, which was reported as endemic from Southwest Boa Vista, would be now distributed also in the whole north half of the island. This increase in range of distribution of several species has important effects on their IUCN conservation status [3]. In the case of *Kalloconus*, some morphotypes attributed to *Kalloconus pseudonivifer* are now assigned to *Kalloconus trochulus*, and *Kalloconus atlanticoselvagem* is synonymized with *K. trochulus*. Our specimen of *Kalloconus* cf. *byssinus* is from North Senegal and has little sequence divergence compared to *Kalloconus pulcher*. In this case, it would be important to study *K. byssinus* from Mauritania or Morocco before considering synonymization. In the opposite direction, there are three clear instances of morphological convergence and thus, of the existence of cryptic species. Those are the cases of *Africonus josephinae* from Maio, *Africonus* cf. *miruchae* from São Vicente, and *Kalloconus* cf. *ateralbus* from Sal, which will be described as new species in due course.

Conclusions

We reconstructed a robust phylogeny based on mitochondrial genomes of cone snails endemic to Cabo Verde, which provides the necessary framework for future evolutionary studies focused on this radiation. The double origin of Cabo Verde endemic cones was supported. The ancestor of *Africonus* separated from L. ventricosus during the Oligocene-Miocene boundary (about 23 mya) and diversified into four main lineages (I to IV) in the Late Miocene (about 9.4-6.9 mya). The divergence of the ancestor of *Kalloconus* endemic to Cabo Verde from those inhabiting mainland occurred also in the Late Miocene whereas its diversification into three main lineages was dated in the Pliocene (4 mya). Main cladogenetic events within cones endemic to Cabo Verde coincide with global cooling periods, which were characterized by radical climate oscillations and eustatic sea level changes. Recurrent cycles of island connection/ disconnection likely favored speciation in allopatry in these cones, which lack a pelagic larval stage, and thus have limited dispersal capacity. Direct development evolved in the ancestor of *Kalloconus* endemic to Cabo Verde, likely associated to the colonization of the archipelago by a cone with a planktotrophic larval stage. However, in the case of *Africonus*, the ancestor that arrived to Cabo Verde was already non-planktotrophic as the corresponding independent evolutionary shift to direct development predated the separation of cones endemic to Senegal (and Canary Islands) from *L. ventricosus* plus *Africonus*. Radular types were modified during the diversification of *Africonus* from an ancestral "robust" type, although correlation with diet specializations await better knowledge of the specific worm species preyed by the different species of cones. Sequence divergence comparisons and reconstructed phylogenies supported the diversity of cone species endemic to Cabo Verde but significantly reduced its number, which was likely overestimated in the past due to important homoplasy in shell morphology, the, thus far, main discriminant character used for species description and identification.

Methods
Samples and DNA extraction
The complete list of specimens analyzed in this study corresponding to different populations and species of *Africonus* and *Trovaoconus* from Cabo Verde is shown in Table 1, as well as details on the respective sampling localities and museum vouchers. As outgroup taxa, we also sampled and analyzed one specimen of *L. ventricosus* from Formentera Island (Spain). Specimens were collected by snorkel at 1-3 m depth, or picked by hand at low tide. All samples were stored in 100% ethanol. The initial species identification (see corresponding column in Table 1) was based on comparison with type material (mostly deposited in the MNCN) or consulting the original publications. Total DNA was isolated from 5 to 10 mg of foot tissue following a standard phenol-chloroform extraction [55].

Radular tooth preparation
The radular sac was dissected from the main body and soft parts were digested in concentrated aqueous potassium hydroxide for 24 h. The resulting mixture was then placed in a petri dish and examined with a binocular microscope. The entire radula was removed with fine tweezers and rinsed with distilled water, then mounted on a slide using Aquatex (Merck, Germany) mounting medium, and observed under a compound microscope. Photographs were taken with a charge-coupled device (CCD) camera attached to the microscope. Terminology for radular morphology follows [8], with abbreviations following [48]. Names of radular types follow [23].

PCR amplification and sequencing
Near-complete (without the control region) mt genomes were amplified through a combination of standard and long PCRs using the primers and following the protocols of [22]. Standard-PCR products were sequenced using Sanger technology. Long-PCR products were subjected to next-generation sequencing. Briefly, PCR amplified fragments from the same mt genome were pooled together in equimolar concentrations. For each cone mt genome a separate indexed library was constructed using the NEXTERA XT DNA library prep kit (Illumina, San Diego, CA, USA). The average size of the Nextera libraries varied between 307 and 345 bp. Libraries were pooled and run in an Illumina MiSeq platform (v.2 chemistry; 2 × 150 paired-end) at Sistemas Genómicos (Valencia, Spain).

Genome assembly and annotation
The reads corresponding to each mt genome were sorted using the corresponding library indices, and read assembly was performed in the TRUFA webserver [56]. Briefly, adapters were removed using SeqPrep [57], quality of the reads was checked using FastQC v.0.10.1 [58], and raw sequences were trimmed and filtered out according to their quality scores using PRINSEQ v.0.20.3 [59]. Filtered reads were used for de novo assembly of each mt genome using default settings (minimum contig length: 200; sequence identity threshold: 0.95) of Trinity r2012-06-08 [60] in TRUFA, and only retaining contigs with a minimum length of 3 kb. These contigs were used as starting point to assemble the mt genomes using Geneious® 8.0.3. First, the (raw) reads with a minimum identity of 99% were mapped against the contigs to correct possible sequence errors. Then, successive mapping iterations using a 100% identity as threshold were performed to elongate the contigs.

The mt genomes were annotated with the option "Annotate from Database" in Geneious® 8.0.3, using published mt genomes of Conidae as references. Annotations of the 13mt protein-coding genes were refined manually identifying the corresponding open reading frames using the invertebrate mitochondrial code. The transfer RNA (tRNA) genes were further identified with tRNAscan-SE 1.21 [61], which infer cloverleaf secondary structures (with a few exceptions that were determined manually). The ribosomal RNA (rRNA) genes were identified by sequence comparison with other Conidae mt genomes [22], and assumed to extend to the boundaries of adjacent genes [62]. GenBank accession numbers of each mt genome are provided in Table 1.

Sequence alignment and phylogenetic analyses

The newly sequenced mt genomes were aligned with the mt genomes of *A. borgesi, Africonus infinitus, Africonus miruchae, T. ateralbus, T. pseudonivifer, T. venulatus,* and *C. ermineus* from Cabo Verde, *L. hybridus, L. mercator, L. belairensis, L. cloveri, L. bruguieresi, K. pulcher,* and *K.* cf. *byssinus* from Senegal, *L. guanche* from Canary Islands, and *L. ventricosus* from Portugal, which were downloaded from GenBank (Table 1). A sequence data set was constructed concatenating the nucleotide sequences of the 13 mt protein-coding and two rRNA genes. The deduced amino acid sequences of the 13 mt protein-coding genes were aligned separately and used to guide the alignment of the corresponding nucleotide sequences with Translator X [63]. Nucleotide sequences of the mt rRNA genes were aligned separately using MAFFT v7 [64] with default parameters. Ambiguously aligned positions were removed using Gblocks, v.0.91b [65] with the following settings: minimum sequence for flanking positions: 85%; maximum contiguous non-conserved positions: 8; minimum block length: 10; gaps in final blocks: no. Finally, the different single alignments were concatenated using Geneious® 8.0.3. Sequences where format converted for further analyses using the ALTER webserver [66]. The concatenated alignment is available at http://purl.org/phylo/treebase/phylows/study/TB2:S21557.

Phylogenetic relationships were inferred using maximum likelihood (ML, [67]) and Bayesian inference (BI, [68]). For ML, we used RAxML v8.1.16 [69] with the rapid hill-climbing algorithm and 10,000 bootstrap pseudoreplicates (BP). BI analyses were conducted with MrBayes v3.1.2 [70], running four simultaneous Markov chains for 10 million generation, sampling every 1000 generations, and discarding the first 25% generations as burn-in (as judged by plots of ML scores and low SD of split frequencies) to prevent sampling before reaching stationarity. Two independent Bayesian inference runs were performed to increase the chance of adequate mixing of the Markov chains and to increase the chance of detecting failure to converge, as determined using Tracer v1.6 [71]. The effective sample size (ESS) of all parameters was checked to be above 200. Node support

was assessed based on Bayesian Posterior Probabilities (BPP). A node was considered highly supported with BP and BPP values above 70% and 0.95, respectively. The ML and BI phylogenetic trees are available at http://purl.org/phylo/treebase/phylows/study/TB2:S21557.

The best partition schemes and best-fit models of substitution for the data set were identified using PartitionFinder2 [72] with the Akaike information criterion [73]. For the protein-coding genes, the partitions tested were: all genes grouped; all genes separated (except *atp6-atp8* and *nad4-nad4L*); and genes grouped by subunits (*atp, cob, cox,* and *nad*). In addition, these three partitions schemes were tested taking into account separately the three codon positions. The rRNA genes were tested with two different schemes, genes separated or combined.

Estimation of divergence times

The program BEAST v.1.8.0 [74] was used to perform a Bayesian estimation of divergence times. An uncorrelated relaxed molecular clock was used to infer branch lengths and nodal ages. The tree topology was fixed using the one recovered by the ML analysis. For the clock model, the lognormal relaxed-clock model was selected, which allows rates to vary among branches without any a priori assumption of autocorrelation between adjacent branches. For the tree prior, a Yule process of speciation was employed. Concatenated protein coding plus rRNA genes were analyzed at the nucleotide level. The partitions and models selected by PartitionFinder2 were applied (see results). The final Markov chain was run twice for 100 million generations, sampling every 10,000 generations, and the first 1000 trees were discarded as part of the burn-in process, according to the convergence of chains checked with Tracer v.1.5. [71]. The ESS of all parameters was above 200.

Despite the fact that there are many fossils of Conidae, it is difficult in many instances to be certain about species identifications given the important levels of homoplasy in shell shape [75]. Hence, although there are fossils attributed to *L. ventricosus* [76] and *L. mercator* [77], which could be applied to the reconstructed phylogeny, we opted to calibrate the clock using biogeographical events (i.e., the age of the islands of Cabo Verde). We run a preliminary analysis in which the posterior distribution of the estimated divergence times was obtained by specifying one calibration point as prior for the divergence time of the split between *L. ventricosus* and the genus *Africonus*. This genus is endemic to Cabo Verde, and we used the age of formation of the oldest island, Sal (28 Mya; [24]), as biogeographical calibration point. We applied a log-normal distribution as the prior model for the calibration and enforced the median divergence time to equal 25 (s.d. = 0.05, offset = 0.7). According to the results of the preliminary analysis, we found that only in the case of São Vicente, Santo Antão, and Santa Luzia, the early divergence of living cone endemic lineages followed the origin of the corresponding island, and therefore, we used a second calibration point corresponding to the origin of these islands about 7.5 mya [24]. We applied a log-normal distribution as the prior model for the calibration and enforced the median divergence time to equal 7.5 (s.d. = 0.03, offset = 0). The BEAST tree is available at http://purl.org/phylo/treebase/phylows/study/TB2:S21557.

Diversification rate through time

The chronogram was used to determine diversification rate through time of genus *Africonus* under alternative (phenotypic versus genetic) species delimitation hypotheses. A lineage through time (LTT) plot analysis was conducted using the APE 4.1 R package [78]. A random sample of 100 trees was selected and mapped over a simulation of 1000 trees following a Yule process of speciation (net diversification rate = 0.4). The phytools R package [79] was used to calculate the Gamma-Statistic [80].

Acknowledgements
We are indebted to Cabo Verde biology students Paulo Vasconcelos and Stiven Pires, as well as to our colleagues Sara Rocha, David Posada and Julio Rozas for their valuable help during sampling in Cabo Verde. We thank Dr. Rui Freitas from the Universidade de Cabo Verde (UniCV) for his continuous support and for granting us access to the facilities of the UniCV in Mindelo, São Vicente. We also thank Dr. Iderlindo Silva dos Santos and Dra. Sonia Monteiro de Pina Araujo from the Direcçao Nacional do Medio Ambiente of the Ministério do Ambiente, Habitação e Ordenamento do Território (MAHOT) of the Republic of Cabo Verde for their help with collecting permits (Autorizações 07/2013, 26/2013, 01/2104, 04/2015, and 03/2016). The sample of *L. ventricosus* from Formentera Island was obtained by Paula C. Rodríguez Flores. We are grateful to Jesús Marco and Aida Palacios, who provided access to the supercomputer Altamira at the Institute of Physics of Cantabria (IFCA-CSIC), member of the Spanish Supercomputing Network, for performing assembling and phylogenetic analyses.

Funding
This work was supported by the Spanish Ministry of Science and Innovation (CGL2013-45211-C2-2-P and CGL2016-75255-C2-1-P (AEI/FEDER, UE)) to RZ; BES-2011-051469 to JEU; BES-2014-069575 to SA; Doctorado Nacional-567, Colciencias-Universidad Nacional de Colombia to AME.

Authors' contributions
MJT, CMLA, and RZ collected the material. MJT prepared the radula. SA, JEU, and AME generated the molecular data. SA analyzed the data. RZ wrote the first draft of the manuscript and all authors contributed to writing the final version. All authors read and approved the final manuscript.

Competing interests
The authors declare that they have no competing interests.

Author details
[1]Museo Nacional de Ciencias Naturales (MNCN-CSIC), José Gutiérrez Abascal 2, 28006 Madrid, Spain. [2]Departamento CMIM y Q. Inorgánica-INBIO, Facultad de Ciencias, Universidad de Cádiz, 11510 Puerto Real, Cádiz, Spain. [3]Centre of Marine Sciences (CCMAR), Universidade do Algarve, Campus de Gambelas, 8005 - 139 Faro, Portugal.

References
1. Cunha RL, Castilho R, Rüber L, Zardoya R. Patterns of cladogenesis in the venomous marine gastropod genus *Conus* from the Cape Verde Islands. Syst Biol. 2005;54(4):634–50.
2. Duda TF, Rolán E. Explosive radiation of Cape Verde *Conus*, a marine species flock. Mol Ecol. 2005;14(1):267–72.
3. Peters H, O'Leary BC, Hawkins JP, Roberts CM. The cone snails of Cape Verde: marine endemism at a terrestrial scale. Global Ecology and Conservation. 2016;7:201–13.
4. Tucker JK, Tenorio MJ. Illustrated catalog of the living cone shells. Wellington: MDM Publishing; 2013.
5. Olivera BM, Watkins M, Bandyopadhyay P, Imperial JS, de la Cotera EPH, Aguilar MB, Vera EL, Concepcion GP, Lluisma A. Adaptive radiation of venomous marine snail lineages and the accelerated evolution of venom peptide genes. Ann N Y Acad Sci. 2012;1267(1):61–70.
6. Cunha RL, Lima FP, Tenorio MJ, Ramos AA, Castilho R, Williams ST. Evolution at a different pace: distinctive phylogenetic patterns of cone snails from two ancient oceanic archipelagos. Syst Biol. 2014;63(6):971–87.
7. Cunha RL, Tenorio MJ, Afonso C, Castilho R, Zardoya R. Replaying the tape: recurring biogeographical patterns in Cape Verde *Conus* after 12 million years. Mol Ecol. 2008;17(3):885–901.
8. Tucker JK, Tenorio MJ. Systematic classification of recent and fossil conoidean gastropods: with keys to the genera of cone shells. Hackenheim: Conchbooks; 2009.
9. Röckel D, Rolán E, Monteiro M. Cone shells from Cape Verde Islands - a difficult puzzle. I. Feito: Vigo, Spain; 1980.
10. Rolán E: Descripción de nuevas especies y subespecies del género *Conus* (Mollusca, Neogastropoda) para el archipiélago de Cabo Verde. Iberus 1990, Suppl. 2:5 - 70.
11. Monteiro A, Tenorio MJ, Poppe GT. The family Conidae. The west African and Mediterranean species of *Conus*. A Conchological Iconography. Hackenheim: ConchBooks; 2004.
12. Cossignani T. Dieci nuovi coni da Capo Verde. Malacol Mostra Mond. 2014;82:18–29.
13. Cossignani T, Fiadeiro R. Cinque nuovi coni da Capo Verde. Malacol Mostra Mond. 2014;84:21–7.
14. Tenorio MJ, Afonso CML, Cunha RL, Rolán E. New species of *Africonus* (Gastropoda, Conidae) from boa vista in the Cape Verde archipelago: molecular and morphological characterization. Xenophora Taxonomy. 2014;2:5–21.
15. Afonso CML, Tenorio MJ. Recent findings from the islands of Maio and boa vista in the Cape Verde archipelago, West Africa: description of three new *Africonus* species (Gastropoda: Conidae). Xenophora Taxonomy. 2014;3:47–60.
16. Cossignani T, Fiadeiro R. Quattro nuovi coni da Capo Verde. Malacol Mostra Mond. 2014;83:14–9.
17. Cossignani T, Fiadeiro R. Otto nuovi coni da Capo Verde. Malacol Mostra Mond. 2017;94:26–36.
18. Cossignani T, Fiadeiro R. Tre nuovi coni da Capo Verde. Malacol Mostra Mond. 2015;86:17–21.
19. Cossignani T, Fiadeiro R. Due nuovi coni da Capo Verde. Malacol Mostra Mond. 2015;87:3–5.
20. Duda TF, Palumbi SR. Developmental shifts and species selection in gastropods. Proc Natl Acad Sci U S A. 1999;96(18):10272–7.
21. Dayrat B. Towards integrative taxonomy. Biol J Linn Soc. 2005;85(3):407–15.
22. Uribe JE, Puillandre N, Zardoya R. Beyond *Conus*: phylogenetic relationships of Conidae based on complete mitochondrial genomes. Mol Phylogenet Evol. 2017;107:142–51.
23. Abalde S, Tenorio MJ, CML A, Zardoya R. Mitogenomic phylogeny of cone snails endemic to Senegal. Mol Phylogenet Evol. 2017;112:79–87.
24. Holm PM, Grandvuinet T, Friis J, Wilson JR, Barker AK, Plesner S. An 40Ar-39Ar study of the Cape Verde hot spot: temporal evolution in a semistationary plate environment. J. Geophys. Res. Solid Earth. 2008;113(B8):n/a-n/a.

25. Puillandre N, Bouchet P, Duda TF Jr, Kauferstein S, Kohn AJ, Olivera BM, Watkins M, Meyer C. Molecular phylogeny and evolution of the cone snails (Gastropoda, Conoidea). Mol Phylogenet Evol. 2014;78:290–303.

26. Olivera BM. Conus peptides: biodiversity-based discovery and exogenomics. J Biol Chem. 2006;281(42):31173–7.

27. Filmer RM. A catalogue of nomenclature and taxonomy in the living Conidae 1758 – 1998. Leiden: Backhuys Publishers; 2001.

28. Bouchet P, Gofas S. Conidae Fleming, 1822. In: MolluscaBase (2017). 2010. Accessed through: World Register of Marine Species at https://marinespeciesorg/aphiaphp?p=taxdetails&id=14107. Accessed 11 Nov 2017.

29. Warren BH, Simberloff D, Ricklefs RE, Aguilée R, Condamine FL, Gravel D, Morlon H, Mouquet N, Rosindell J, Casquet J, et al. Islands as model systems in ecology and evolution: prospects fifty years after MacArthur-Wilson. Ecol Lett. 2015;18(2):200–17.

30. Crampton-Platt A, Timmermans MJTN, Gimmel ML, Kutty SN, Cockerill TD, Vun Khen C, Vogler AP. Soup to tree: the phylogeny of beetles inferred by mitochondrial metagenomics of a Bornean rainforest sample. Mol Biol Evol. 2015;32(9):2302–16.

31. Saitoh K, Sado T, Mayden RL, Hanzawa N, Nakamura K, Nishida M, Miya M. Mitogenomic evolution and interrelationships of the Cypriniformes (Actinopterygii: Ostariophysi): the first evidence toward resolution of higher-level relationships of the world's largest freshwater fish clade based on 59 whole mitogenome sequences. J Mol Evol. 2006;63(6):826–41.

32. Stöger I, Schrödl M. Mitogenomics does not resolve deep molluscan relationships (yet?). Mol Phylogenet Evol. 2013;69(2):376–92.

33. Osca D, Irisarri I, Todt C, Grande C, Zardoya R. The complete mitochondrial genome of Scutopus ventrolineatus (Mollusca: Chaetodermomorpha) supports the Aculifera hypothesis. BMC Evol Biol. 2014;14(1):1–10.

34. Tenorio MJ, Lorenz F, Dominguez M. New insights into Conus jourdani da Motta, 1984 (Gastropoda, Conidae), an endemic species from Saint Helena Island. Xenophora Taxonomy. 2016;11:32–42.

35. Puillandre N, Duda TF, Meyer C, Olivera BM, Bouchet P. One, four or 100 genera? A new classification of the cone snails. J Molluscan Stud. 2015;81:1–23.

36. Jablonski D. Larval ecology and macroevolution in marine invertebrates. Bull Mar Sci. 1986;39(2):565–87.

37. Scheltema RS. Planktonic and non-planktonic development among prosobranch gastropods and its relationship to the geographic range of species. In: Ryland JS, Tyler RA, editors. Reproduction, genetics and distribution of marine organisms. Fredensborg: Olsen & Olsen; 1989. p. 183–8.

38. Swearer SE, Shima JS, Hellberg ME, Thorrold SR, Jones GP, Robertson DR, Morgan SG, Selkoe KA, Ruiz GM, Warner RR. Evidence of self-recruitment in demersal marine populations. Bull Mar Sci. 2002;70(1):251–71.

39. Miller KG, Kominz MA, Browning JV, Wright JD, Mountain GS, Katz ME, Sugarman PJ, Cramer BS, Christie-Blick N, Pekar SF. The Phanerozoic record of global sea-level change. Science. 2005;310(5752):1293.

40. Zachos JC, Flower BP, Paul H. Orbitally paced climate oscillations across the Oligocene/Miocene boundary. Nature. 1997;388(6642):567–70.

41. Beddow HM, Liebrand D, Sluijs A, Wade BS, Lourens LJ. Global change across the Oligocene-Miocene transition: high-resolution stable isotope records from IODP site U1334 (equatorial Pacific Ocean). Paleoceanography. 2016;31(1):81–97.

42. Herbert TD, Lawrence KT, Tzanova A, Peterson LC, Caballero-Gill R, Kelly CS. Late Miocene global cooling and the rise of modern ecosystems. Nat Geosci. 2016;9(11):843–7.

43. Hodell DA, Curtis JH, Sierro FJ, Raymo ME. Correlation of late Miocene to early Pliocene sequences between the Mediterranean and North Atlantic. Paleoceanography. 2001;16(2):164–78.

44. Krijgsman W, Hilgen FJ, Raffi I, Sierro FJ, Wilson DS. Chronology, causes and progression of the Messinian salinity crisis. Nature. 1999;400(6745):652–5.

45. Lisiecki LE, Raymo ME. A Pliocene-Pleistocene stack of 57 globally distributed benthic δ18O records. Paleoceanography. 2005;20:PA1003.

46. Davis KE, Hill J, Astrop TI, Wills MA. Global cooling as a driver of diversification in a major marine clade. Nat Commun. 2016;7:13003.

47. Dyhr CT, Holm PM. A volcanological and geochemical investigation of boa vista, Cape Verde Islands; 40Ar/39Ar geochronology and field constraints. J Volcanol Geotherm Res. 2010;189(1–2):19–32.

48. Kohn AJ, Nishi M, Pernet B. Snail spears and scimitars: a character analysis of Conus radular teeth. J Molluscan Stud. 1999;65:461–81.

49. Duda TF, Kohn AJ, Palumbi SR. Origins of diverse feeding ecologies within Conus, a genus of venomous marine gastropods. Biol J Linn Soc. 2001;73(4):391–409.

50. Dowle EJ, Morgan-Richards M, Brescia F, Trewick SA. Correlation between shell phenotype and local environment suggests a role for natural selection in the evolution of Placostylus snails. Mol Ecol. 2015;24(16):4205–21.

51. Hollander J, Butlin RK. The adaptive value of phenotypic plasticity in two ecotypes of a marine gastropod. BMC Evol Biol. 2010;10(1):333.

52. Knowlton N. Sibling species in the sea. Annu Rev Ecol Syst. 1993;24(1):189–216.

53. Roux C, Fraïsse C, Romiguier J, Anciaux Y, Galtier N, Bierne N. Shedding light on the Grey zone of speciation along a continuum of genomic divergence. PLoS Biol. 2016;14(12):e2000234.

54. Gerard D, Gibbs HL, Kubatko L. Estimating hybridization in the presence of coalescence using phylogenetic intraspecific sampling. BMC Evol Biol. 2011;11(1):291.

55. Sambrook J, Fritsch EF, Maniatis T. Molecular cloning: a laboratory manual. 2nd ed. New York: Cold Spring Harbor Laboratory Press; 1989.

56. Kornobis E, Cabellos L, Aguilar F, Frías-López C, Rozas J, Marco J, Zardoya R. TRUFA: a user-friendly web server for de novo RNA-seq analysis using cluster computing. Evol Bioinforma. 2015; 11(Supplementary Material 23873):97–104.

57. StJohn J. SeqPrep. 2011. https://githubcom/jstjohn/SeqPrep.

58. Andrews S. FastQC. 2010. http://wwwbioinformaticsbabrahamacuk/projects/fastqc/.

59. Schmieder R, Edwards R. Quality control and preprocessing of metagenomic datasets. Bioinformatics. 2011;27(6):863–4.

60. Grabherr MG, Haas BJ, Yassour M, Levin JZ, Thompson DA, Amit I, Adiconis X, Fan L, Raychowdhury R, Zeng Q, et al. Full-length transcriptome assembly from RNA-Seq data without a reference genome. Nat Biotech. 2011;29(7):644–52.

61. Schattner P, Brooks AN, Lowe TM. The tRNAscan-SE, snoscan and snoGPS web servers for the detection of tRNAs and snoRNAs. Nucleic Acids Res. 2005;33(suppl 2):W686–9.

62. Boore JL, Macey JR, Medina M. Sequencing and comparing whole mitochondrial genomes of animals. Methods Enzymol. 2005;395:311–48.

63. Abascal F, Zardoya R, Telford MJ. TranslatorX: multiple alignment of nucleotide sequences guided by amino acid translations. Nucleic Acids Res. 2010;38(Web Server issue):W7–13.

64. Katoh K, Standley DM. MAFFT multiple sequence alignment software version 7: improvements in performance and usability. Mol Biol Evol. 2013; 30(4):772–80.

65. Castresana J. Selection of conserved blocks from multiple alignments for their use in phylogenetic analysis. Mol Biol Evol. 2000;17(4):540–52.

66. Glez-Peña D, Gómez-Blanco D, Reboiro-Jato M, Fdez-Riverola F, Posada D. ALTER: program-oriented conversion of DNA and protein alignments. Nucleic Acids Res. 2010;38(suppl 2):W14–8.

67. Felsenstein J. Evolutionary trees from DNA sequences: a maximum likelihood approach. J Mol Evol. 1981;17(6):368–76.

68. Huelsenbeck J, Ronquist F. MrBayes: Bayesian inference of phylogenetic trees. Bioinformatics. 2001;17:754–5.

69. Stamatakis A. RAxML-VI-HPC: maximum likelihood-based phylogenetic analyses with thousands of taxa and mixed models. Bioinformatics. 2006; 22(21):2688–90.

70. Ronquist F, Huelsenbeck JP. MrBayes 3: Bayesian phylogenetic inference under mixed models. Bioinformatics. 2003;19(12):1572–4.

71. Rambaut A, Drummond AJ. Tracer v1.4. 2007. Available from http://beast.bio.ed.ac.uk/Tracer.

72. Lanfear R, Frandsen PB, Wright AM, Senfeld T, Calcott B. PartitionFinder 2: new methods for selecting partitioned models of evolution for molecular and morphological phylogenetic analyses. Mol Biol Evol. 2017;34(3):772–3.

73. Akaike H. Information theory and an extension of the maximum likelihood principle. In: Petrov BN, Csaki F, editors. 2nd international symposium on information theory. Budapest: Akademiai Kiado; 1973. p. 267–81.

74. Drummond A, Rambaut A. BEAST: Bayesian evolutionary analysis by sampling trees. BMC Evol Biol. 2007;7:214.

75. Duda TF, Bolin MB, Meyer CP, Kohn AJ. Hidden diversity in a hyperdiverse gastropod genus: discovery of previously unidentified members of a Conus species complex. Mol Phylogenet Evol. 2008;49(3):867–76.

76. Sacco F: I molluschi dei terreni terziarii del Piemonte e della Liguria. Conidae e Conorbidae, vol. XIII. Torino. Italy: Stamperia Reale; 1893.

77. Glibert M. Les Conacea fossiles du Cénozoïque étranger des collections de l'Institut Royal des Sciences Naturelles de Belgique. Institut Royal des Sciences Naturelles de Belgique Mémoire, série 2. 1960;64:1–132.

11

Phylogenetic analysis of the SINA/SIAH ubiquitin E3 ligase family in *Metazoa*

Ian J. Pepper, Robert E. Van Sciver and Amy H. Tang[*]

Abstract

Background: The RAS signaling pathway is a pivotal developmental pathway that controls many fundamental biological processes including cell proliferation, differentiation, movement and apoptosis. *Drosophila* Seven-IN-Absentia (SINA) is a ubiquitin E3 ligase that is the most downstream signaling "gatekeeper" whose biological activity is essential for proper RAS signal transduction. Vertebrate SINA homologs (SIAHs) share a high degree of amino acid identity with that of *Drosophila* SINA. SINA/SIAH is the most conserved signaling component in the canonical EGFR/RAS/RAF/MAPK signal transduction pathway.

Results: Vertebrate SIAH1, 2, and 3 are the three orthologs to invertebrate SINA protein. SINA and SIAH1 orthologs are found in all major taxa of metazoans. These proteins have four conserved functional domains, known as RING (Really Interesting New Gene), SZF (SIAH-type zinc finger), SBS (substrate binding site) and DIMER (Dimerization). In addition to the *siah1* gene, most vertebrates encode two additional *siah* genes (*siah2* and *siah3*) in their genomes. Vertebrate SIAH2 has a highly divergent and extended N-terminal sequence, while its RING, SZF, SBS and DIMER domains maintain high amino acid identity/similarity to that of SIAH1. But unlike vertebrate SIAH1 and SIAH2, SIAH3 lacks a functional RING domain, suggesting that SIAH3 may be an inactive E3 ligase. The SIAH3 subtree exhibits a high degree of amino acid divergence when compared to the SIAH1 and SIAH2 subtrees. We find that SIAH1 and SIAH2 are expressed in all human epithelial cell lines examined thus far, while SIAH3 is only expressed in a limited subset of cancer cell lines.

Conclusion: Through phylogenetic analyses of metazoan SINA and SIAH E3 ligases, we identified many invariant and divergent amino acid residues, as well as the evolutionarily conserved functional motifs in this medically relevant gene family. Our phylomedicinal study of this unique metazoan SINA/SIAH protein family has provided invaluable evolution-based support towards future effort to design logical, potent, and durable anti-SIAH-based anticancer strategies against oncogenic K-RAS-driven metastatic human cancers. Thus, this method of evolutionary study should be of interest in cancer biology.

Keywords: RAS signal transduction, SINA/SIAH E3 ligases, Ubiquitin-mediated proteolysis, Phylogenetic analysis, Invariant and divergent amino acid residues, And conserved functional domains in SINA, SIAH1, SIAH2 and SIAH3

Background

The RAS signaling pathway in metazoans is one of the most fundamental and evolutionarily conserved signaling pathways controlling cell proliferation, motility, and stem cell renewal during normal development, tissue regeneration, and pathogenesis [1–5]. Aberrant EGFR/HER2/K-RAS pathway hyperactivation is known to promote hyperplastic growth, tumorigenesis, and metastasis

[6, 7]. Therefore, inhibiting oncogenic K-RAS pathway activation is a logical strategy to stop tumor progression and prevent metastasis in human cancer [8–10]. However, the design of long-lasting anti-K-RAS therapies has remained elusive for past 40 years due to unique RAS kinetics, extensive pathway cross-talk, signal bifurcation, compensatory activation, feedback control, context-dependent adaptation and dynamic rewiring of the ERBB/K-RAS signaling network [11, 12]. Hence, alternative anti-K-RAS strategies are urgently needed to shutdown "undruggable" oncogenic K-RAS activation and eradicate oncogenic K-RAS-driven metastatic cancer in the clinic [10].

* Correspondence: TangAH@evms.edu
Department of Microbiology and Molecular Cell Biology, Eastern Virginia Medical School, Leroy T. Canoles Jr. Cancer Research Center, Harry T. Lester Hall, Room 454-457, 651 Colley Avenue, Norfolk, VA 23501, USA

Many key RAS signaling components were identified successfully in *Drosophila* via genetic modifier screens, including Sevenless (SEV, the *Drosophila* homolog of mammalian EGFR membrane receptor), rat sarcoma viral oncogene (RAS), RAF serine/threonine kinase, Mitogen-activated Protein Kinase (MAPK), Son of Sevenless (SOS), and Seven-In-Absentia (SINA) [2, 13–19]. *Drosophila* SINA was identified as the most downstream signaling component in the *Drosophila* RAS signaling pathway, playing a critical "gatekeeper" role in R7 photoreceptor cell fate determination [2, 10, 14, 20]. Interestingly, the loss of the R7 photoreceptor mutant phenotypes observed in *sina* [loss-of-function] mutant flies is identical to that observed in *sev* [loss-of-function] mutant flies [13, 14]. SINA-mediated degradation of a neuronal repressor, Tramtrack (TTK[88]), is required to unleash active RAS signaling to initiate the neuronal cell differentiation program in Drosophila R7 precursor cells [20]. Among all the signaling components identified thus far in the RAS pathway, Drosophila SINA and human SIAH1/2 share the highest level of evolutionary conservation and amino acid identities [14, 21]. Extensive genetic epistasis analyses have demonstrated that proper SINA function is critical for RAS signal transduction, and that active EGFR, RAS, RAF, and MAPK signals cannot be transmitted properly without functional SINA. This suggests that SINA is the most downstream signaling "gatekeeper" identified thus far in the RAS signaling pathway and it is a key signaling hub critical for transmitting EGFR/RAS/RAF/MAPK activation signals in vivo [2, 17, 22].

Drosophila SINA and two of its human homologs, SIAH1 and SIAH2, belong to a highly evolutionarily conserved family of RING domain E3 ligases [20, 21, 23]. SINA and SIAH function as homo- and hetero-dimers [24–27]. The family of the SINA/SIAH1/SIAH2 E3 ligases has four highly conserved and distinct functional domains: (1) the Really Interesting New Gene (RING) domain is the catalytic active site for the E3 ligase activity, (2) the SIAH-type zinc finger (SZF) domain contains a dual zinc-finger motif, (3) the substrate-binding site (SBS) recognizes substrates, and (4) the dimerization (DIMER) domain allows for homo- and heterodimer formation between SINA/SIAH proteins [23–31]. The substrate-binding domain (SBD), composed of SZF, SBS and DIMER domains, is responsible for substrate recognition, targeting, interaction, and degradation [23, 25–31]. As an E3 ligase, SINA/SIAH is known to interact, modify, and target a multitude of substrates/partners/regulators to orchestrate ubiquitin-mediated proteolysis and regulate protein stability, protein complex assembly, protein subcellular localization, and other cellular functions in normal development and human diseases [10, 32, 33]. SIAH3 is a newly identified member of the vertebrate SIAH family [34, 35]. Although SIAH3 has been reported to function as a negative

regulator of Parkin protein translocation within the mitochondria, SIAH3 biochemical function remains largely understudied in mammalian systems [23, 34].

Controlling oncogenic K-RAS-driven metastatic cancer remains an unmet need in medicine [8, 12]. Based on the highly conserved molecular principles and regulatory mechanisms learned from the *Drosophila* RAS signaling pathway, we have proposed and demonstrated the efficacy of a novel antitumor strategy to inhibit the "undruggable" oncogenic K-RAS signal at its most downstream signaling hub. This was achieved by inhibiting SIAH1/2 E3 ligases, using both in vitro and in vivo tumor models of human pancreatic and lung cancer [36, 37]. We and others have shown that blocking SIAH1/2 activity is a promising and logical strategy to inhibit oncogenic K-RAS/B-RAF activation and to impede oncogenic K-RAS/B-RAF-driven tumorigenesis in preclinical xenograft models [32, 36–38]. In the current study, we conducted a phylogenetic analysis of the SINA/SIAH family of E3 ligases across the entire animal kingdom to further delineate SINA/SIAH biological function by focusing on the evolutionary conservation and mutational constraints observed in this family of SINA/SIAH E3 ligases. We identified invariant and divergent amino acid residues, as well as several highly conserved functional motifs in the SINA/SIAH family. This provides an evolutionary, structural, functional, and translational basis with which to design more potent and long-lasting anti-SIAH-based anti-K-RAS strategies against oncogenic K-RAS-driven metastatic human cancers in the future.

Results
Phylogenetic analysis of the SINA/SIAH family of E3 ligases
We retrieved all currently existing SINA/SIAH amino acid sequences from the NCBI Refseq protein database by conducting separate BLAST searches using *Drosophila* SINA along with human SIAH1, SIAH2, and SIAH3 full-length proteins as the query sequences. The E-value cutoff was 10^{-60} for each of the four separate searches. A subset of 70 unique SINA/SIAH sequences, spanning the entire known taxonomy of the animal kingdom, was used for the phylogenetic analyses as shown in this study (Table 1 and Fig. 1). This collection contained SINA/SIAH sequences from 19 vertebrate species and 20 invertebrate species. Among the 19 vertebrate species, all four major classes of tetrapod vertebrates were represented, i.e. mammals (1–6), birds (7–8), reptiles (9–12), and amphibians (13) (these numeric numbers represent their positions as shown in Table 1). The three major groups of jawed fish were represented, including coelacanth (14), teleost fish (15–17), and cartilaginous fish (18). Additionally, lamprey (19) was utilized as a jawless vertebrate representative. Among the 20 invertebrate species, all major divisions of invertebrates were represented, including invertebrate deuterostomes

Table 1 List of sequences (*n* = 70) that were utilized in all evolutionary analyses as conducted in this study

	Taxonomy	Species	SIAH1	SIAH2	SIAH3
		Vertebrates			
1	Mammalia	*Homo sapiens*	NP_003022.3	NP_005058.3	NP_942146.2
2	Mammalia	*Pan troglodytes*	NP_001233288.1	XP_516819.2	XP_522672.3
3	Mammalia	*Mus musculus*	NP_033198.1	NP_033200.2	NP_001121565.1
4	Mammalia	*Bos taurus*	XP_005218736.1	NP_001193983.1	NP_001192350.1
5	Mammalia	*Tursiops truncatus*	XP_019788031.1	XP_019789908.1	XP_019792000.1
6	Mammalia	*Monodelphis domestica*	XP_007474893.1	XP_001363407.1	XP_007501643.1
7	Aves	*Gallus gallus*	XP_015147897.1	XP_426719.2	XP_417044.1
8	Aves	*Sturnus vulgaris*	XP_014738058.1	XP_014741604.1	XP_014750862.1
9	Reptilia	*Pogona vitticeps*	XP_020643987.1	XP_020637543.1	
10	Reptilia	*Thamnophis sirtalis*	XP_013913965.1	XP_013927960.1	
11	Reptilia	*Alligator mississippiensis*	XP_019352994.1	XP_006259223.1	XP_019356083.1
12	Reptilia	*Chrysemys picta*	XP_008165194.1	XP_005286979.1	XP_005287220.1
13	Amphibia	*Xenopus tropicalis*	NP_001015836.1	NP_001095281.1	XP_002941011.1
14	Sacropterygii	*Latimeria chalumnae*	XP_005998413.1	XP_006008925.1	XP_014352065.1
15	Neopterygii	*Danio rerio*	NP_955815.1	NP_956721.2	
16	Neopterygii	*Salmo salar*	XP_013979721.1	XP_014052924.1	
17	Neopterygii	*Oreochromis niloticus*	XP_019217515.1	XP_003459581.3	
18	Chondrichthyes	*Callorhinchus milii*	XP_007887616.1	XP_007889716.1	XP_007889930.1
19	Cyclostomata	*Petromyzon marinus*	S4R9G1_PETMA		
		Invertebrates	SINA		
20	Cephalochordata	*Branchiostoma floridae*	XP_002609562.1		
21	Echinodermata	*Strongylocentrotus purpuratus*	XP_797311.2		
22	Arthropoda	*Drosophila melanogaster*	NP_476725.1		
23	Arthropoda	*Anopheles gambiae*	XP_001688791.1		
24	Arthropoda	*Apis mellifera*	XP_394284.2		
25	Arthropoda	*Acyrthosiphon pisum*	XP_008186354.1		
26	Nematoda	*Caenorhabditis elegans*	NP_500409.1		
27	Nematoda	*Necator americanus*	XP_013306181.1		
28	Nematoda	*Brugia malayi*	XP_001898781.1		
29	Nematoda	*Trichinella spiralis*	XP_003379392.1		
30	Spiralia	*Octopus bimaculoides*	XP_014779248.1		
31	Spiralia	*Crassostrea gigas*	XP_011434753.1		
32	Spiralia	*Helobdella robusta*	XP_009028819.1		
33	Spiralia	*Schistosoma mansoni*	XP_018646300.1		
34	Cnidaria	*Nematostella vectensis*	XP_001637064.1		
35	Cnidaria	*Orbicella faveolata*	XP_020623250.1		
36	Cnidaria	*Acropora digitifera*	XP_015766642.1		
37	Cnidaria	*Hydra vulgaris*	XP_002162099.1		
38	Porifera	*Amphimedon queenslandica*	XP_019850617.1		
39	Placozoa	*Trichoplax adhaerens*	XP_002108034.1		

Taxonomic designation for representative species (*n* = 39) is listed to the left of its name. With the exception of the sequence from *Petromyzon marinus*, which was acquired from UniProtKB, all amino acid sequence identifiers presented in this table refer to their NCBI Genbank accession numbers

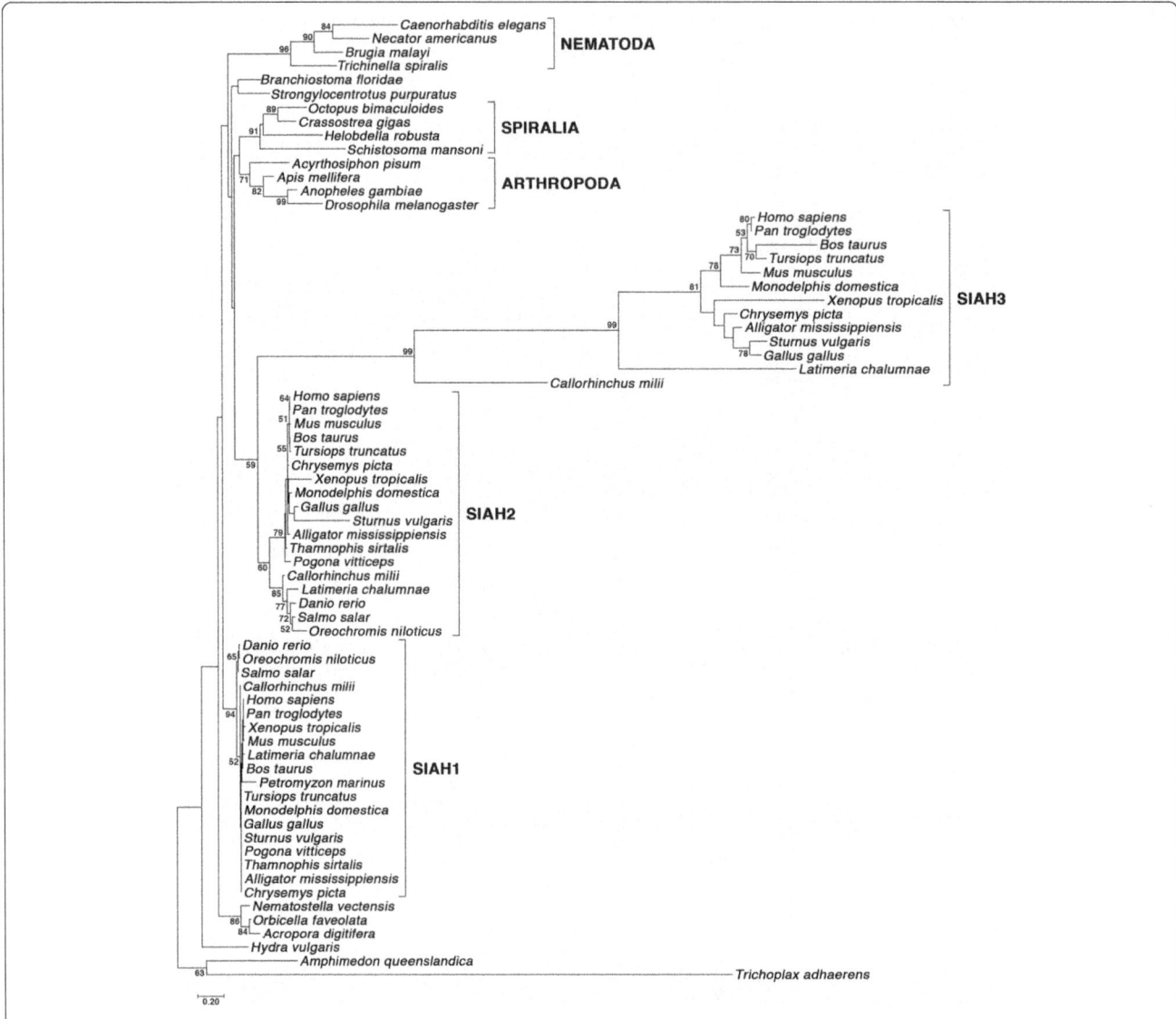

Fig. 1 Phylogenetic tree of the evolutionarily conserved SINA/SIAH family across metazoan species. The phylogenetic tree was constructed to illustrate the evolutionary relationships of SINA/SIAH family using the representative species from all the major taxa across the entire metazoan kingdom. The LG + G4 + F model was utilized for construction of the tree. The numbers listed on each node represent the bootstrap support value associated with that node after running 100 replicates. All bootstrap values < 50 were eliminated from the tree display. The tree was manually rooted at the node containing the outgroup sequences *A. queenslandica* and *T. adhaerens*. Major clades that were recovered by the analysis are indicated by the brackets on the right side

(20–21), arthropods (22–25), nematodes (26–29), mollusks (30–31), annelids (32), platyhelminths (33), cnidarians (34–37), sponges (38), and placozoans (39). Using this manual selection of 70 sequences from 39 unique metazoan species, we constructed a master phylogenetic tree of the entire SINA/SIAH protein family (Fig. 1). Invertebrates and jawless vertebrates possess a single seven-in-absentia (*sina*) gene. Jawed vertebrates (gnathostomes) possess at least two SINA orthologs/homologs: *siah1* and *siah2* genes. The majority of gnathostomes possess three SINA orthologs/homologs: *siah1*, *siah2* and *siah3* genes within their genomes.

Jawed vertebrate taxa for which BLAST searches did not retrieve a SIAH3 sequence include teleost fish and the squamate division of reptiles (includes lizards and snakes, excludes crocodiles and turtles). Thus, the loss of the *siah3* gene from these specific vertebrate lineages in gnathostomes appears to be a recent evolutionary event.

We utilized the maximum-likelihood method (ML) to construct the phylogenetic tree. The chosen model of sequence evolution was LG + G4 + F, and 100 bootstrap replicates were performed to assess the validity of the groupings within the generated ML tree (log likelihood = –9417.10). The sequences for *Amphimedon queenslandica* and *Trichoplax adhaerens* were utilized as the phylogenetic outgroup. This phylogenetic analysis does not resolve any ancestral evolutionary relationships among the three vertebrate SIAH subfamilies, nor does it recover them as a monophyletic grouping. There was confident support for the individual subtrees of vertebrate SIAH1 (94) and SIAH3 (99). By contrast, a considerably weaker bootstrap value (60) was obtained for monophyly of the SIAH2 subtree, and a value of similar magnitude (59) was obtained for a sister-group relationship between the SIAH2 and SIAH3 sequences (Fig. 1). Among the three vertebrate SIAH proteins, the SIAH3 subfamily exhibited the highest rate of amino acid substitution (measured in number of substitutions per site). These values were determined by summing branch lengths along the path starting from the node adjacent to the subtree's root and ending at the tip of the *H. sapiens* branch within each subtree (designated "root-to-tip distance"). The root-to-tip distance for SIAH3 subtree is the greatest by a large margin (2.58 substitutions/site). The SIAH2 subfamily, with a root-to-tip distance of 0.15 substitutions/site, exhibits a mutation rate that is over 2-fold higher than that of the SIAH1 subfamily (0.07 substitutions/site). Moreover, SIAH3's mutational rate is 17-fold higher across the vertebrate lineage when compared with SIAH2's mutational rate, and 34-fold higher when compared with SIAH1's mutational rate.

As the first identified member of this evolutionarily conserved family of RING E3 ligases, *Drosophila* SINA shares an extensive degree of amino acid sequence identity/similarity to vertebrate SIAH1, SIAH2 and SIAH3 proteins [14, 21]. The phylogenetic analyses presented here demonstrate that vertebrate SIAH1, SIAH2, and SIAH3 are all equally orthologous to invertebrate SINA (Fig. 1). There are two possible scenarios for the emergence of the three SIAH paralogs from an ancestral SINA/SIAH gene that existed in the vertebrate last common ancestor (LCA): (1) tandem gene duplications or (2) successive whole genome duplications. The details of these *siah* gene duplication events are not yet fully understood, as the phylogenetic analysis presented here does not yield enough resolution (low node supports) to reveal their lineage relationships. Thus, SIAH1, SIAH2 and SIAH3 are three paralogous lineages in vertebrates, but their exact evolutionary history remains unclear.

Invertebrate SINA sequences have evidently undergone significant divergent evolution within their various lineages since the bifurcation of bilaterians into protostomes and deuterostomes. This is demonstrated by the relatively strong bootstrap values obtained for the Arthropoda (71), Nematoda (96), and Spiralia (91) subtrees (Fig. 1). The analysis also contained SINA sequences from Cnidaria, a clade that is a sister-group to the entire bilaterian lineage. Their

overall placement within the tree topology was correct; however, the exclusion of *Hydra vulgaris* from the well-supported Cnidaria clade suggests a significant sequence dissimilarity among the hydrozoan and anthozoan branches of this lineage. The phylogenetic analysis presented in Fig. 1 also helped to resolve ambiguity regarding the classification of the sequence retrieved from *Petromyzon marinus*. It was previously suspected that this was a SINA protein since the *P. marinus* genome only possesses one sequence belonging to the SINA/SIAH combined family, much like the other invertebrates with a single SINA protein. However, the inclusion of *P. marinus* within the well-supported SIAH1 clade provides support for an alternative view. *Petromyzon marinus* contains a SIAH1 sequence with high similarity to SIAH1 of gnathostomes. There may have once been SIAH2 and SIAH3 genes in the genomes of ancestral jawless vertebrates that were subsequently lost over evolutionary time and are now absent from extant species.

The internal node supports within the vertebrate SIAH paralog subtrees also show great discrepancy. For the SIAH1 subtree, the only clade with some degree of bootstrap support is teleost fish (65), with all other SIAH1 sequences lumped into a sister-group to teleosts. SIAH2 sequences produce a bifurcating subtree with a correct tetrapod clade (79) and an incorrect grouping containing all fish SIAH2 sequences (84). The SIAH3 subtree exhibits the most robust recovery of an accurate vertebrate phylogeny. The *C. milii* and *L. chalumnae* branches are correctly placed in their basal position to tetrapods with maximal bootstrap support. The monophyletic tetrapod clade was recovered with a similar support value (81) as the SIAH2 tree. In contrast to the SIAH1 and SIAH2 trees, moderately strong support was obtained for a monophyletic mammalian clade (78) as well as a placental mammal group within this clade (73). Only the SIAH3 subfamily is congruent with the established phylogeny of vertebrate species as reported in the literature. The extraordinary amino acid sequence conservation in the vertebrate SIAH1 and SIAH2 subfamilies likely leads to a lack of demarcating phylogenetic signals, hampering the proper reconstruction of the expected species tree within the SIAH family.

Structural motifs and functional topology of the invertebrate SINA subfamily

To identify the conserved amino acid residues and structural motifs of the invertebrate SINA family, we conducted a functional domain analysis from 20 invertebrate SINA sequences as selected in this study (Fig. 2). A very noticeable observation upon obtaining the raw alignment is the high degree of the length variance in the highly diverse and variable N-terminal sequences of invertebrate SINAs. For invertebrate SINA proteins, the general length of the evolutionarily conserved C-terminal SINA sequence

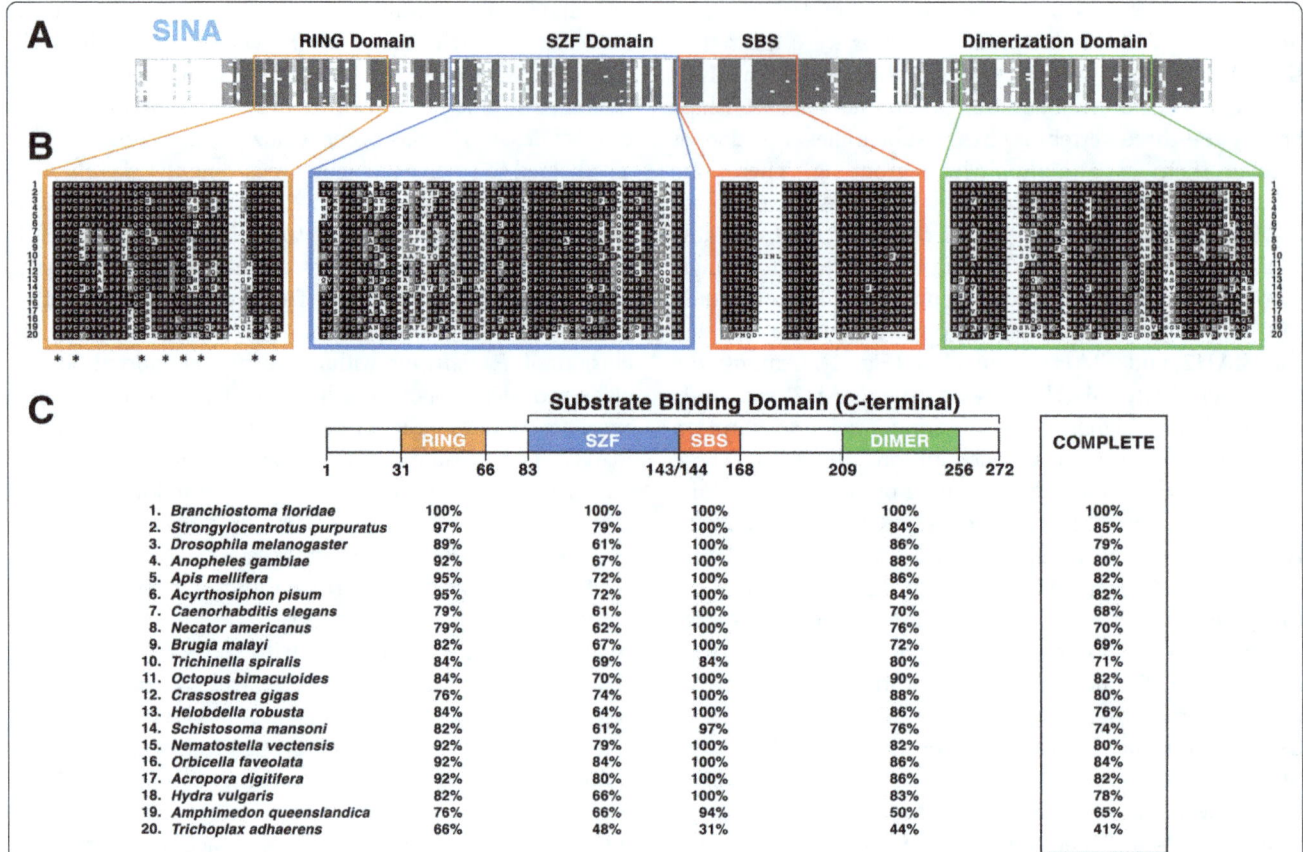

Fig. 2 Sequence alignment of the invertebrate SINA subfamily reveals its invariant amino acid residues, and the four conserved structural motifs. Sequence comparison of SINA proteins from 20 representative invertebrate species (#1-#20) is shown. **a** Overview of the entire alignment produced by the 20 invertebrate SINA sequences. Four key functional domains are marked in four distinct colors: RING domain (*orange*), SZF domain (*blue*), SBS (*red*), and DIMER domain (*green*). **b** Schematic illustration of amino acid conservation within the 4 domains of the SINA sequences is shown. Amino acid identity is shown as white letters in a *black box*, amino acid similarity is shown as white letters in a *grey box*, and amino acid divergence is shown as black letters in a *white box*. The asterisks located below the RING domain alignment indicate unanimous conservation of the cysteine/histidine zinc-binding residues. **c** The percentages of amino acid conservation in each distinct domain and the entire SINA sequence between *Branchiostoma floridae* and each of the representative invertebrate species are shown. The diagram of the domain architecture was based on *B. floridae* SINA

(comprised of the 4 known functional domains, RING, SZF, SBS and DIMER) was fairly consistent among the 20 diverse invertebrate species in our analysis (Fig. 2a and b). By contrast, the length of the N-terminal sequence preceding the conserved RING domain in invertebrate SINA sequences was quite variable and diverse, ranging anywhere between 26 and 155 amino acids in length (see Additional file 1).

For the purposes of standardizing and refining the invertebrate SINA sequencing alignment, *Branchiostoma floridae* (a basal chordate with ≥ 90% identity to human SIAH1 in all functional domains except SZF) was chosen as a reference sequence, and all positions that resulted in gaps within this sequence exclusive of the four functional domains were eliminated from the refined alignment (Fig. 2). Sequence alignments indicated a considerable degree of amino acid diversity in the RING, SZF, and DIMER domains of invertebrate SINA

sequences (Fig. 2). The SBS exhibits a higher level of conservation, but also possesses noticeable species-specific insertions that produce gaps in the alignment (Fig. 2a and b). Quantification of the percent similarity in invertebrate SINA amino acid sequences as compared to the *B. floridae* reference SINA sequence support this observation (Fig. 2c). Several invertebrates share 100% amino acid identity with the SBS domain of *B. floridae*; however, there are notably reduced amino acid identities observed within the RING, SZF, and DIMER domains of this SINA family (Fig. 2c). Quantification of the two outgroup sequences indicates that *Trichoplax adhaerens* SINA completely lacks the evolutionary conservation observed in all other metazoan SINA/SIAH sequences. It is essentially a unique "outlier" sequence within the family, as evidenced by the fact that its SBS shares just 31% identity with the highly conserved SBS motif found in SINA, SIAH1, and SIAH2 sequences (Fig. 2c).

Structural motifs and functional topology of the vertebrate SIAH1 subfamily

Here, we focused on identifying the invariant amino acid residues and highly conserved structural motifs among the SIAH1 subfamily in all vertebrate species (Fig. 3). The alignment of the vertebrate SIAH1 sequences demonstrates an extraordinary degree of amino acid sequence identity among SIAH1 orthologs (Fig. 3a), even within an N-terminal sequence that was quite divergent in the invertebrate SINA family (Fig. 2). The RING and SZF domains in the SIAH1 subfamily possess 8 immutable zinc-coordinating histidine (His) and cysteine (Cys) amino acid residues [39–41]. In fact, all aligned gnathostome SIAH1 sequences possess 100% sequence identity to each other within their RING domains (Fig. 3b and c). The SIAH1 sequence for *Petromyzon marinus* only possesses two amino acid differences from this conserved RING domain sequence. Similarly, the SIAH1 sequences from jawed vertebrates demonstrated 100% interspecies conservation within the SBS and DIMER domains (Fig. 3b and c). *Petromyzon marinus* SIAH1 is also identical to the vertebrate SIAH1 SBS domain, and only bears a single divergent amino acid compared to the DIMER domain sequence. The SZF contains the lowest degree of amino acid conservation out of the four domains, with *P. marinus* having 92% identity with human SIAH1 (Fig. 3b and c). Additionally, it is the only functional domain to exhibit any degree of amino acid divergence amongst all the gnathostome SIAH1 sequences (Fig. 3c). Together, this data shows that SIAH1 sequences have maintained an extraordinarily high degree of amino acid conservation

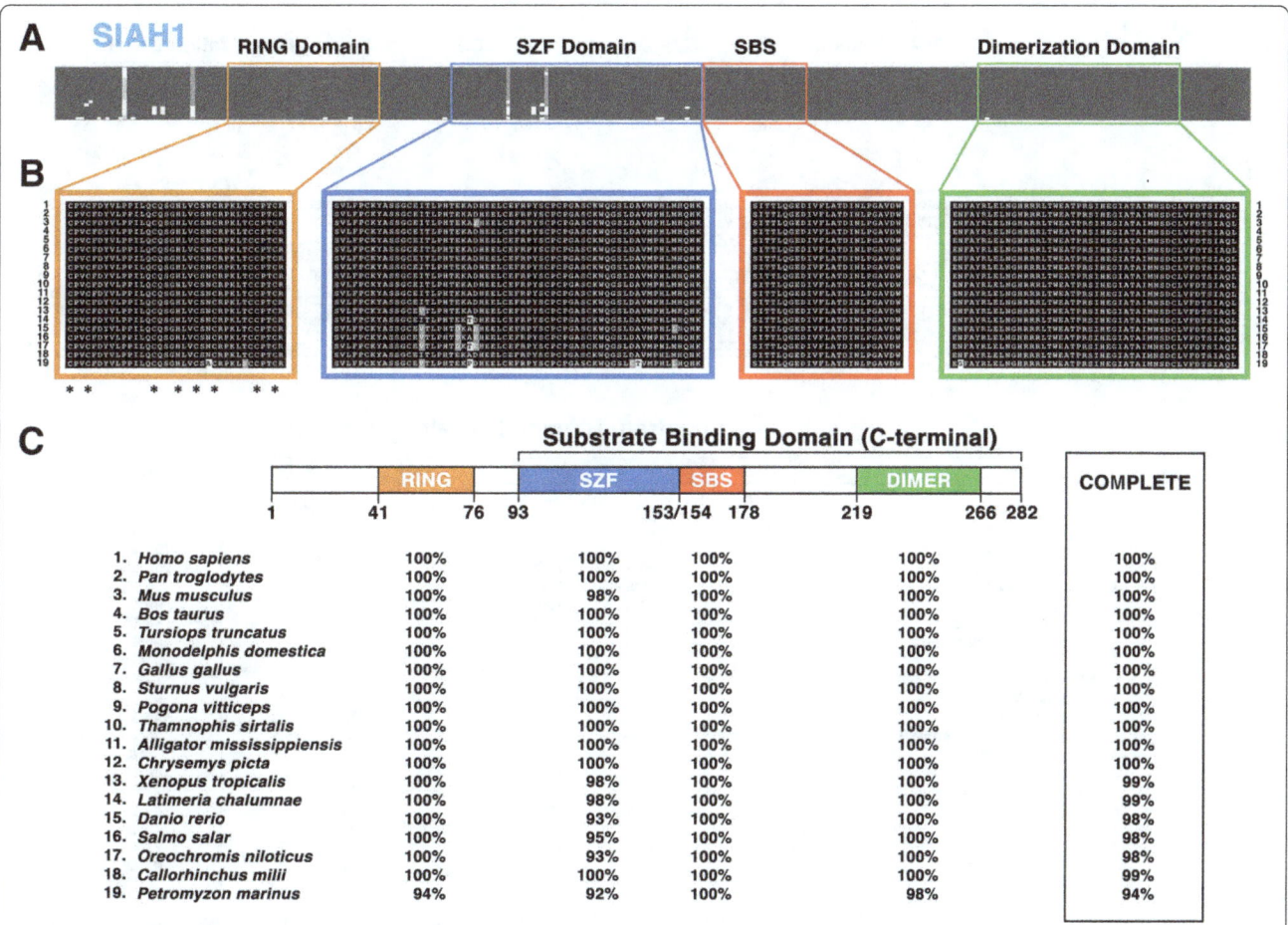

Fig. 3 Sequence alignment of the vertebrate SIAH1 subfamily reveals its invariant amino acid residues, and the four conserved structural motifs. Sequence comparison of SIAH1 proteins from 19 representative vertebrate species (#1-#19) is shown. **a** The level of amino acid conservation in the N-terminal portion (#1 to #40) is high among SIAH1 sequences. Four key functional domains are marked in four distinct colors: RING domain (*orange*), SZF domain (*blue*), SBS (*red*), and DIMER domain (*green*). **b** Schematic illustration of amino acid conservation within the 4 domains of the SIAH1 sequences is shown. Amino acid identity is shown as white letters in a *black box*, amino acid similarity is shown as white letters in a *grey box*, and amino acid divergence is shown as black letters in a *white box*. The asterisks located below the RING domain alignment indicate unanimous conservation of the cysteine (Cys)/histidine (His) zinc-binding residues. **c** The percentages of amino acid conservation in each distinct domain and the entire SIAH1 sequence between human and each of the representative vertebrate species are shown. The diagram of the domain architecture was based on *Homo sapiens* SIAH1

ever since the vertebrate SIAH1 paralog first originated from the vertebrate LCA's SIAH protein (Fig. 3).

Structural motifs and functional topology of the vertebrate SIAH2 subfamily

To identify the invariant amino acid residues and highly conserved structural motifs in the SIAH2 subfamily, we aligned vertebrate SIAH2 sequences from 18 diverse species of jawed vertebrates. The sampling of these species used in this analysis of the SIAH2 subtree includes representatives from all the major taxa within the gnathostome clade, with an emphasis on mammals (Fig. 4). Vertebrate SIAH2 has an extended N-terminal fragment that is 40 amino acids longer than that of SIAH1 (Fig. 4). The unique 80 amino acid N-terminal

fragments in the SIAH2 subfamily are quite diverse, while the SIAH2 core sequences (#80-#324) share a high level of amino acid identity with the SIAH1 core sequence (#41-#282) (Fig. 4). Like SIAH1 orthologs, SIAH2 orthologs have 4 essential functional domains: the RING, SZF, SBD, and DIMER domains (Fig. 4a and b). The evolutionary conservation in the SIAH2 subfamily is illustrated by the extraordinarily high level of amino acid identities observed in all four functional domains (Fig. 4).

Sequence homology analysis of the SIAH2 subfamily shows that the four distinct SIAH2 functional domains exhibit a high degree of evolutionary conservation (Fig. 4b and c). Like SIAH1 orthologs, the SBS motif exhibits the highest degree of conservation in these SIAH2 orthologs (Fig. 4b). Sixteen of the eighteen metazoan

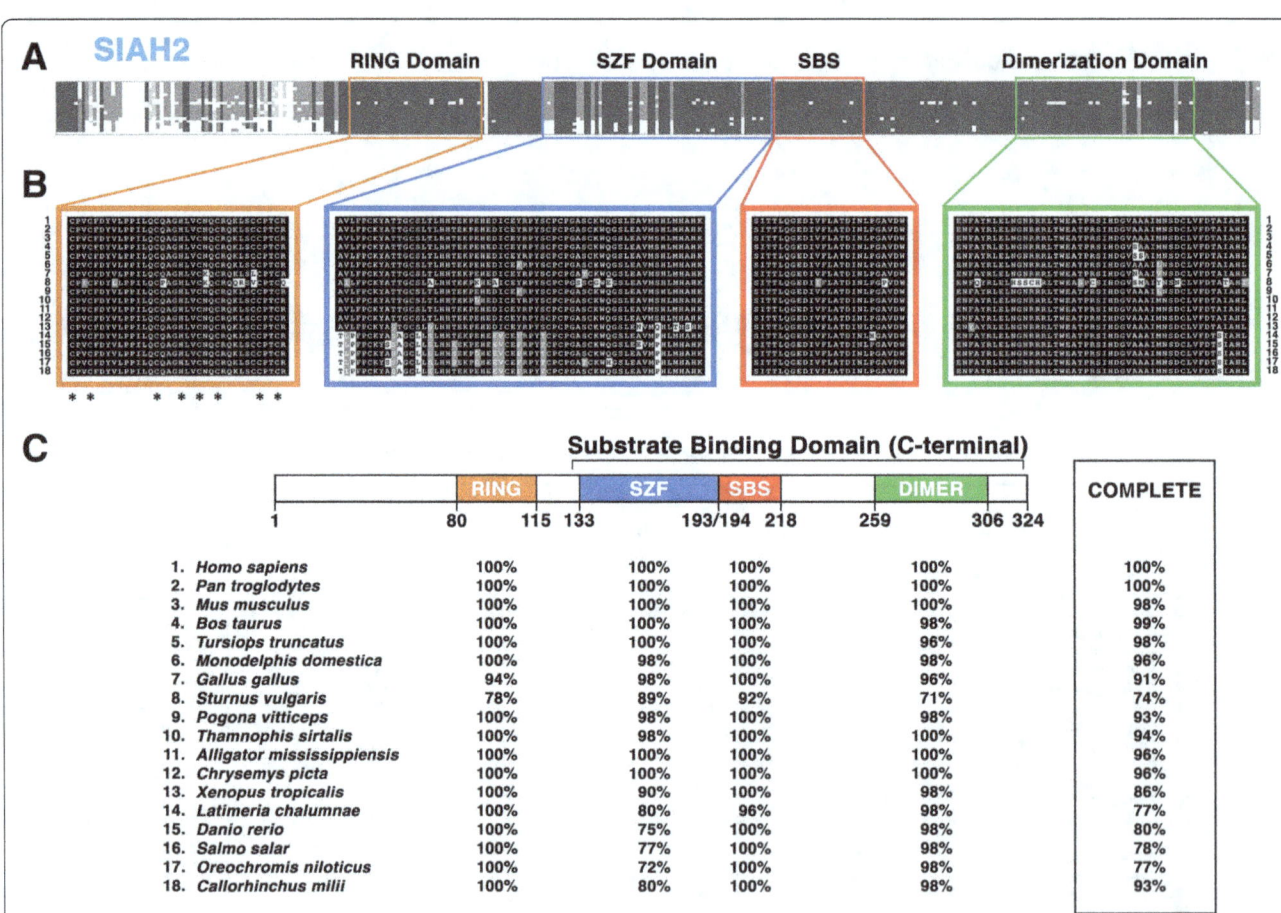

Fig. 4 Sequence alignment of the vertebrate SIAH2 subfamily reveals its invariant amino acid residues, and the four conserved structural motifs. Sequence comparison of SIAH2 proteins from 18 representative vertebrate species (#1–#18) is shown. **a** The level of amino acid divergence in the N-terminal fragments (#1 to #80) is high. Four key functional domains are marked in four distinct colors: RING domain (*orange*), SZF domain (*blue*), SBS (*red*), and DIMER domain (*green*). **b** Schematic illustration of amino acid conservation within the 4 domains of the SIAH2 sequences is shown. Amino acid identity is shown as white letters in a *black box*, amino acid similarity is shown as white letters in a *grey box*, and amino acid divergence is shown as black letters in a *white box*. The asterisks located below the RING domain alignment indicate unanimous conservation of the cysteine (Cys)/histidine (His) zinc-binding residues. **c** The percentages of amino acid conservation in each distinct domain and the entire SIAH2 sequence between human and each of the representative vertebrate species are shown. The diagram of the domain architecture was based on *Homo sapiens* SIAH2. The SIAH2 sequence for *C. milii* was incomplete, and these gaps induced by the incompleteness of the sequence were disregarded when calculating conservation across the whole protein

SIAH2 SBS sequences are 100% identical to the human SIAH2 SBS, with two species (*S. vulgaris*, and *L. chalumnae*) evolving only two and one amino acid substitutions, respectively. The extraordinarily high level of amino acid conservation was also observed throughout the RING (100%), SZF (98%) and DIMER (94%) domains of vertebrate SIAH2 orthologs (Fig. 4b and c, and Table 2). When comparing SIAH2 amino acid sequences from all gnathostomes, the SIAH2 sequence conservation observed in the SBS domain is significantly higher compared to that of the other three domains (Fig. 4). For example, 22/25 (88%) of amino acid residues in the SBS are identical among all 18 vertebrate SIAH2 sequences as aligned (Fig. 4). In contrast, just 32/48 (67%) of amino acid residues in the DIMER domain are unanimously conserved (Fig. 4c). The RING domains share 67% identity and the SZF domains share 64% identity (Fig. 4b and c, and Table 2). Similar to vertebrate SIAH1, all eight metal-coordinating cysteine (Cys) and histidine (His) residues are immutable amino acids in both the RING and SZF domains of the SIAH2 subfamily (Fig. 4b). Interestingly, the majority of the divergent amino acids in the domains belong to *Sturnus vulgaris*, suggesting that there has been accelerated SIAH2 amino acid sequence evolution in birds of the Passeriformes family in comparison to other tetrapod species.

Structural motifs and functional topology of the vertebrate SIAH3 subfamily

Vertebrate SIAH3 is the most divergent member of the vertebrate SIAH family E3 ligases. Unlike its vertebrate SIAH1 and SIAH2 counterparts, SIAH3 orthologs do not possess a functional RING domain, suggesting that SIAH3 is an enzymatically-inactive E3 ligase (Fig. 5a and b). In addition to the loss of the RING domain, SIAH3 contains only a single Zinc finger in the SZF domain compared to the double Zinc finger motif found in vertebrate SIAH1 and SIAH2 [23]. SIAH3 orthologs, apart from *Latimeria chalumnae*, possess SIAH 3-unique N-terminal sequences (S3UNS) that are highly conserved and evolutionarily unique to the SIAH3 subfamily, but are completely absent from the SIAH1 and SIAH2 subfamilies (Fig. 5). SIAH3 has 4 distinct functional motifs, including S3UNS, SZF, SBS and DIMER domains (Fig. 5a–c). Among the evolutionarily

conserved SZF, SBS and DIMER domain in vertebrates, the SIAH3 subfamily exhibits a higher mutation rate and additional amino acid divergence when compared to SIAH1 and SIAH2 subfamilies (Fig. 5c).

Comparison of SINA, SIAH1, SIAH2 and SIAH3 consensus amino acid sequences in metazoans

To identify the invariant amino acid residues and conserved functional domain sequences in the SINA/SIAH family, we compared the core consensus sequences of invertebrate SINA with its three vertebrate SIAH orthologs: SIAH1, SIAH2 and SIAH3. We aligned the 4 core consensus sequences of full-length SINA, SIAH1, SIAH2 and SIAH3 extracted from Figs. 2, 3, 4 and 5 together (Fig. 6 and Table 2). SIAH1 and SIAH2 are two functional E3 ligases with high amino acid identity/similarity observed in their core consensus sequences, while SIAH3 is a nonfunctional and inactive E3 ligase that is missing a catalytically active RING domain (Fig. 5). The invertebrate SINA consensus sequence bears the highest degree of similarity to vertebrate SIAH1, as the majority of amino acids within the functional domains of these two consensus sequences are conserved (Table 2). Despite the conservation of these core consensus sequences, invertebrate SINA is still considered equally orthologous to each of the three vertebrate SIAH proteins. These orthologous relationships exist because all extant SIAH paralogs originated from duplication events involving the ancestral vertebrate SIAH protein. This ancestral vertebrate SIAH protein diverged from invertebrate SINA via a speciation event (the emergence of the vertebrate lineage). The SBS and DIMER domains are fully intact in all three SIAH paralogs, and exhibit the highest degree of amino acid conservation among the entire vertebrate SIAH family (Fig. 6). By contrast, SIAH3 lacks the high amino acid identity in the RING and SZF domains as reported in SIAH1 and SIAH2 proteins (Figs. 5 and 6). Additionally, the SIAH3 consensus sequence contains a histidine-rich region within the first half of the SZF observed in SIAH1 and SIAH2 consensus sequences (Figs. 5 and 6). Our phylogenetic analysis did not support the previous conclusion that the *siah3* gene is derived from a duplication of the *siah2* gene [23].

SIAH1, SIAH2 and SIAH3 mRNA expression in human cancer cells

RT-PCR was performed to examine the expression levels of *siah1*, *siah2* and *siah3* mRNA transcripts in 15 human epithelial cell lines with a range of tumorigenicities (Fig. 7). These cell lines include human pancreatic, prostate, breast, lung and cervical cancer cell lines. Specific and unique PCR primers were used to synthesize mRNA products from each distinct SIAH subfamily. Our findings show that *siah1* and *siah2* mRNA transcripts are

Table 2 Comparison of amino acid identities observed in the three SIAH paralogs

Consensus sequence	RING	SZF	SBS	DIMER	WHOLE
SINA	100%	100%	100%	100%	100%
SIAH1	100%	89%	100%	90%	88%
SIAH2	89%	83%	100%	85%	72%
SIAH3	28%	25%	60%	63%	45%

Percentage identity values for each paralog's domains were calculated relative to the consensus sequence of invertebrate SINA, the mutual ortholog of all three SIAH paralogs

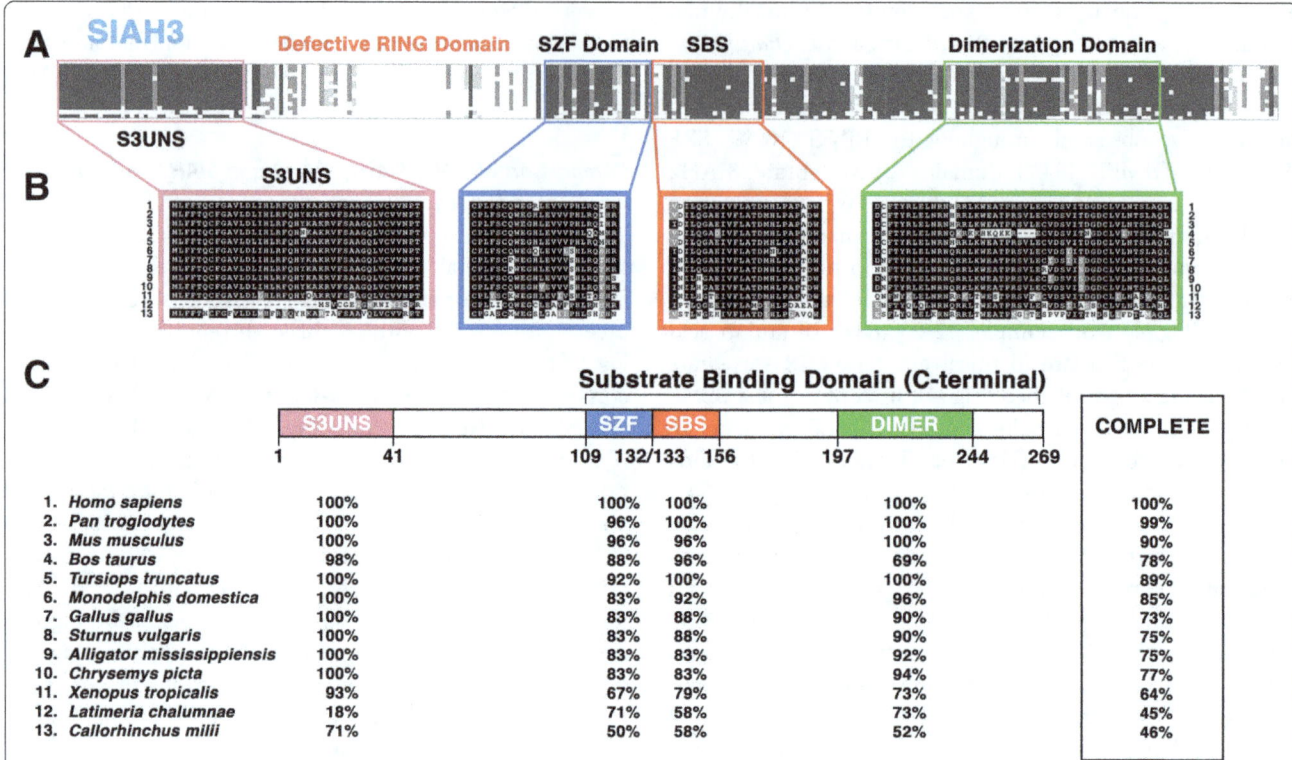

Fig. 5 Sequence alignment of the vertebrate SIAH3 subfamily reveals its invariant amino acid residues, absence of RING domain, and conserved structural motifs. Sequence comparison of SIAH3 proteins from 13 representative vertebrate species (#1-#13) is shown. **a** The level of amino acid conservation in the N-terminal fragments is high, while the portion of the protein sequences that would be expected to contain the RING domain is highly divergent compared to SIAH1 and SIAH2. Four key functional domains are marked in four distinct colors: SIAH3 unique N-terminal sequence (S3UNS, *pink*), SZF domain (*blue*), SBS (*red*) and DIMER domain (*green*). **b** Schematic illustration of amino acid conservation within the 4 distinct domains of the SIAH3 sequences is shown. Amino acid identity is shown as white letters in a *black box*, amino acid similarity is shown as white letters in a *grey box*, and amino acid divergence is shown as black letters in a *white box*. **c** The percentages of amino acid conservation in each distinct domain and the entire SIAH3 sequence between human and each of the representative vertebrate species are shown. The diagram of the domain architecture was based on *Homo sapiens* SIAH3

universally expressed in all human epithelial cell lines examined so far [36, 37], whereas *siah3* mRNA transcript is expressed in a small subset of human tumor cell lines. This result suggests a biological function of SIAH3 that is distinct from those of SIAH1 and SIAH2 in human tumor biology (Fig. 7).

Discussion

The field of cancer biology desperately needs a more effective method for controlling and conquering oncogenic K-RAS hyperactivation in metastatic human cancer. With our new strategy of targeting SINA/SIAH1/SIAH2, the most conserved and the most downstream "signaling gatekeeper" identified thus far in the RAS signaling pathway, we aim to bypass obstacles such as the extensive bifurcation, cross-talk and dynamic feedback controls downstream of several major compensatory K-RAS effector pathways to impede and block oncogenic K-RAS-driven malignant tumors [10]. To determine the biological function and evolutionary constraints of the SINA/SIAH family

of E3 ligases, we conducted a phylogenetic analysis to identify the invariant amino acids and conserved functional motifs in the SINA/SIAH family among all major taxa of metazoans. The co-existence of the three unique SINA orthologs, SIAH1, SIAH2, and SIAH3, may represent an interesting event of genome evolution, gene duplication, and gene divergence in vertebrates. The three vertebrate SIAH paralogs were not recovered as a monophyletic grouping, which could be a result of long-branch attraction between the Cnidaria SINA sequences and vertebrate SIAH1 sequences. Given the high amino acid identity observed in one-to-one alignments of human SIAH1 with cnidarian SINA sequences, this possibility may be a logical deduction. Future phylogenetic analyses with greater taxon sampling and density, combined with dN/dS calculations on the SINA/SIAH protein-coding nucleotide sequences to detect residues under positive selection, will be required. They will help to further elucidate the overall evolutionary relationship between the three vertebrate SIAH paralogs and where they fit within

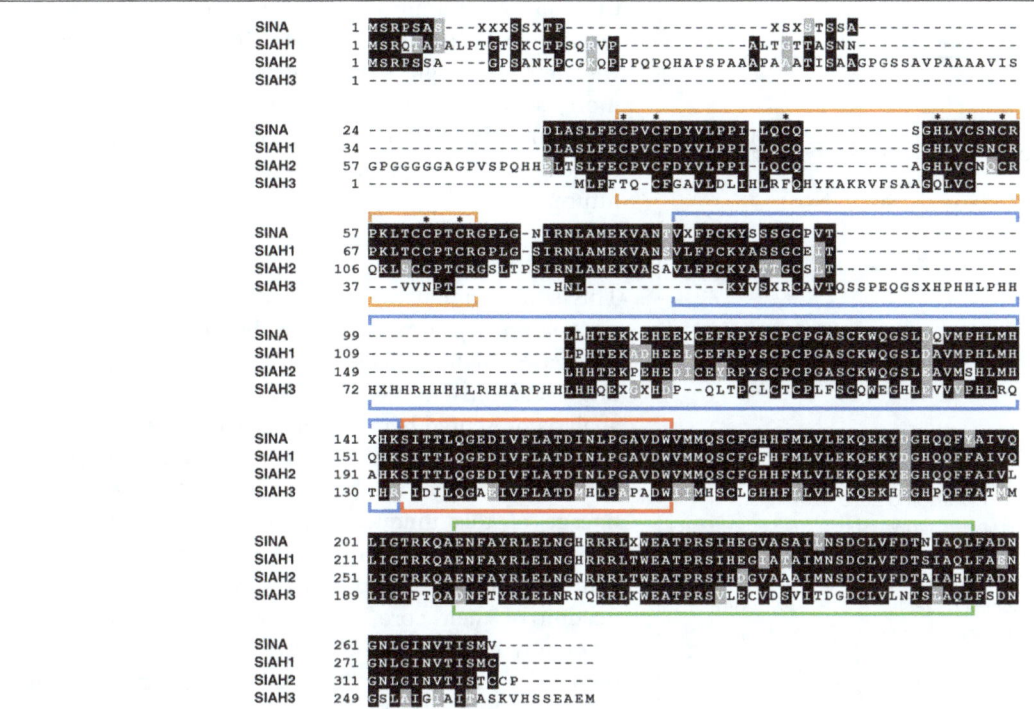

Fig. 6 The consensus sequences of SINA, SIAH1, SIAH2, and SIAH3 were aligned to identify the invariant and divergent amino acid residues in this evolutionarily highly conserved SINA/SIAH E3 ligase family. There is a high level of amino acid conservation in the SBD domain in the SINA, SIAH1, SIAH2, and SIAH3 core consensus sequences in the SIAH family. The RING domain is marked by an *orange bracket*, the SZF domain by a *blue bracket*, the SBS by a *red bracket*, and the DIMER domain by a *green bracket*. Asterisks within the RING domain indicate the position of the invariant cysteine (Cys)/histidine (His) residues in SINA, SIAH1 and SIAH2. Amino acid positions marked with an "X" (instead of a valid one-letter amino acid abbreviation) indicate that the consensus at this site could not be resolved unambiguously. SINA, SIAH1, and SIAH2 share extensive sequence homology between each other in their core consensus sequences, whereas SIAH3 shows dramatic sequence divergence in the corresponding RING and SZF domains to those of SINA, SIAH1, and SIAH2 proteins. SINA, SIAH1, SIAH2, and SIAH3 share high levels of amino acid conservation in the SBS domains

the context of the metazoan evolution of this medically important family of RING-domain E3 ligases in RAS signal transduction in all metazoan cells.

The vertebrate SIAH1 and SIAH2 core consensus sequences exhibit the highest degree of amino acid identities within this RING domain E3 ligase family (Fig. 6). The invertebrate SINA underwent significant divergent evolution in the protostome lineage after the divergence of bilaterians (Fig. 1). Unlike vertebrate SIAH1 and SIAH2 orthologs, we report that SIAH3 orthologs exhibit the highest rates of amino acid substitution and sequence divergence within the gnathostome lineage. Based on phylogenetic analyses and amino acid substitution patterns, we speculate that the emergence of this SIAH3 subfamily, unique to jawed vertebrates, may have suppressor effects on its sibling SIAH1 and SIAH2 subfamilies. This finding may have possible medical relevance in cancer biology. *Siah3* mRNA transcript expression in several human cancer cell lines as shown in Fig. 7, raised the interesting possibility that nonfunctional SIAH3 may function as an endogenous SIAH inhibitor that regulates and antagonizes SIAH1

and SIAH2 biological activity through its DIMER domain to suppress and inhibit SIAH1 and SIAH2 biological activities to control cell proliferation, tissue growth, pattern formation, and homeostasis. This idea is supported by several previous studies that demonstrated that the RING domain-deleted SIAH1 or SIAH2 mutants (termed SIAH1/2 dominant-negative) functionally ablated endogenous activity of SIAH1 and SIAH2 in cancer biology [23, 27, 32, 38, 42, 43].

The molecular phylogenetic analysis of the SINA/SIAH family provides valuable insights into immutable amino acid residues and conserved functional motifs across the entire metazoan kingdom. Previous biomedical studies have narrowly focused on SINA, SIAH1, and SIAH2 in *Drosophila*, mice, and humans [14, 21, 32, 44–47]. Based on the phylogenetic analyses in metazoans, we suggest that SIAH3 is a new member of the SINA/SIAH E3 ligase family that lacks functional E3 ligase activity. The phylogenetic analysis conducted in this study provides a new framework and a novel evolutionary perspective with which we can identify and dissect the invariant and divergent amino acid residues and the conserved functional

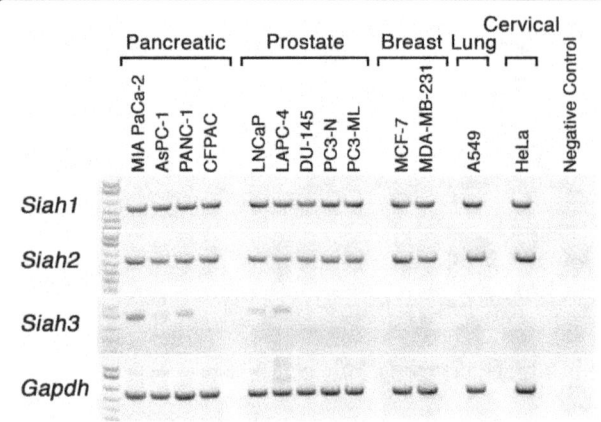

Fig. 7 SIAH1, SIAH2 and SIAH3 mRNA expression in human cancer cell lines. Semi-quantitative reverse transcription – polymerase chain reaction (RT-PCR) analysis of siah1, siah2 and siah3 mRNA transcript expression in human cancer cell lines is shown. The relative expression levels of siah1, siah2 and siah2 mRNA transcripts in 13 epithelial cancer cells including pancreatic cancer cells (MiaPaCa, AsPC-1, PANC-1 and CFPAC), prostate tumor cell lines (LNCaP, LAPC-4. PC-3-N (normal), PC-3-ML (metastatic)), breast tumor cell lines (MCF-7 and MDA-MB-231), non-small cell lung cancer cell line (A549) and cervical cancer cell line (HeLa) were estimated semi-quantitatively for serial dilutions of the complementary DNA templates. Glyceraldehyde-3-phosphate dehydrogenase (GAPDH) mRNA transcript was used as an internal control. The RT-PCR mixture without cDNA template was used as a negative control

domains of SINA/SIAH proteins at the molecular level for phylomedicine. This is especially useful in the context of the SINA/SIAH1/2 gatekeeper function required for proper K-RAS activation and context-dependent RAS signal transmission [10, 36, 37]. This phylogenetic analysis of SINA/SIAH evolution in the animal kingdom is likely to provide valuable insights into the logical design of effective anti-K-RAS drugs that selectively target and specifically inhibit human SIAH1/2 proteins to rapidly shut down oncogenic K-RAS-driven malignant tumor growth and block metastatic cancer cell dissemination.

The discovery of SIAH3's increased expression in a subset of human cancer cells presents an interesting opportunity for novel drug discovery. By taking advantage of nonfunctional SIAH3 as a putative endogenous SIAH inhibitor, we may be able to develop a novel anti-SIAH1 and anti-SIAH2 strategy by utilizing SIAH3 expression to antagonize oncogenic K-RAS-driven metastatic human cancer cells. SIAH3 shares a common ancestry with SIAH1 and SIAH2, as well as three conserved structural motifs (i.e., SBS, SZF and DIMER domains). It is conceivable that SIAH3 may function as a highly specific inhibitor of endogenous SIAH1 and SIAH2 activity by binding to SIAH1/2 via its DIMER domain. Additional work will be conducted to examine the interplay between SIAH3 and SIAH1/SIAH2 in oncogenic K-RAS-driven human cancers in the future. By focusing on

these invariant amino acid residues and conserved functional motifs identified in the SINA/SIAH superfamily, we aim to design a phylogenetic-based, targeted and more specific anti-SIAH-based anticancer strategy to both impede and eradicate oncogenic K-RAS-driven metastatic human cancers for clinical translation in the future.

Conclusions

This study demonstrates the extraordinarily high degree of evolutionary conservation in the SINA/SIAH family of E3 ligases in metazoans. SINA/SIAH proteins evidently originated early in metazoan evolution. The phylogenetic analysis presented here indicates that invertebrate SINA is a mutual ortholog of the three vertebrate SIAH paralogs, and future analyses will help resolve the exact evolutionary lineage of this unique RING-domain E3 ligase family. These ancestral SINA/SIAH E3 ligases occur under stringent evolutionary selection pressure that prevents diversification of their core sequences in all major metazoan taxonomy groups, as shown in the highly conserved SBS domain, as well as all immutable Cys/His zinc-binding residues within the RING and SZF domains of SIAH1, SIAH2, and SINA proteins. SIAH3 orthologs lack the conserved Zinc-binding Cys/His residues, suggesting a loss of the functional E3 ligase activity. Together, the phylogenetic analysis of the SINA/SIAH family can be utilized to pinpoint the invariable amino acid residues and conserved structural domains that are absolutely critical for their enzymatic functions and biological activities in transmitting active RAS signal in metazoa. By analyzing the evolutionary relationship between invertebrate SINA and its vertebrate SIAH paralogs (SIAH1, SIAH2, and SIAH3), we have gained an in-depth understanding of the extraordinarily high degree of amino acid conservation in this medically significant gene family. This knowledge will promote a phylogenetic-based SIAH-centered drug design toward generating useful SIAH-specific peptides and SIAH small molecule inhibitors as new and more efficacious therapeutics to eradicate oncogenic K-RAS-driven metastatic cancer in the future.

Methods
Sequence database search and data partitioning
The putative vertebrate paralogs of human SIAH1 were determined using the Ensembl gene tree associated with the protein (GeneTree ENSGT00390000005434). PSI-BLAST searches were performed on the NCBI protein database using the three SIAH paralogs (SIAH1, SIAH2 and SIAH3) from *Homo sapiens* and SINA from *Drosophila melanogaster* as the query sequences. The default algorithm parameters were utilized for each query, except for Max target sequences (1000) and the E-value threshold (10^{-60}). BLAST hits returned from the

SIAH1 and SIAH2 searches were included in the master sequence collection. Table 1 is a subset of the master sequence collection if (a) they were >240 amino acids in length, and (b) produced an alignment with the functional domain-containing region of their human orthologs. The same length criteria were applied for hits from the SINA search, and the *Drosophila* query sequence was used as the reference for evaluating whether each hit produced a functional domain region alignment. In the case of the SIAH3 search, the length criteria and alignment-based selection filter were relaxed, and BLAST hits were included for species which returned results >200 amino acids in length.

A total of 70 sequences (20 SINA sequences from invertebrates; and 50 vertebrate SIAH sequences (19 SIAH1, 18 SIAH2, and 13 SIAH3 sequences respectively) were manually selected from the master collection of BLAST hits across all four searches for utilization in the phylogenetic analysis ("Main dataset") (Table 1 and Additional file 2: Table S1). The sequence selection was targeted so that a balanced sampling of vertebrate and invertebrate species, in addition to adequate representation of the major metazoan taxonomy groupings, was achieved. The *Petromyzon marinus* SIAH1 sequence was manually downloaded from UniProtKB and added to the main dataset to obtain a total of 70 SINA/SIAH family sequences. For each vertebrate species within the Main dataset, all SIAH paralogs encoded by their genomes were included. The only vertebrate species within the Main dataset without three SIAH paralogs detected in their genomes by BLAST searches are the teleosts (*Danio rerio, Salmo salar, Oreochromis niloticus*), squamates (*Pogona vitticeps, Thamnophis sirtalis*), and the jawless vertebrate representative *Petromyzon marinus* (Table 1).

To conduct subsequent functional domain analyses, the Main dataset was subdivided into four smaller datasets. A SINA dataset consisting of all 20 invertebrate SINA sequences was created, as well as three distinct datasets for each vertebrate SIAH paralog. The SIAH1, SIAH2, and SIAH3 datasets consisted of 19, 18, and 13 sequences, respectively (Table 1 and Additional file 2: Table S1).

Construction of multiple sequence alignments

Protein sequences were aligned using the MAFFT algorithm in all instances. The SINA/SIAH family alignment utilized the Main dataset, resulting in an alignment of 70 sequences that was 534 positions in length (Additional file 3). To eliminate positions with less than 30% sequence coverage (i.e. gaps present in >70% of sequences), the alignment was trimmed to 305 positions for usage in phylogenetic analysis (Additional file 4).

Three individual alignments were also constructed for each vertebrate SIAH paralog using their respective datasets. Following completion of MAFFT alignment on each of these smaller datasets, each alignment was manually refined to eliminate positions that resulted in gaps within a designated "reference sequence". In the case of each SIAH paralog alignment, the respective *Homo sapiens* amino acid sequence was chosen as the reference. The gap-trimming procedure caused reductions in alignment length as follows: 287 to 282 positions for SIAH1, 391 to 324 positions for SIAH2, 283 to 269 positions for SIAH3. It should be noted that the large majority of the trimmed positions across all three paralog alignments were within the N-terminal portion, and not the conserved functional domains identified in each distinct SIAH subtree.

Additionally, an alignment for the mutual SINA ortholog of the three vertebrate SIAH paralogs was built using the "SINA dataset". For the SINA alignment, *Branchiostoma floridae* was chosen because it is considered a basal chordate. This taxonomic status puts it closer phylogenetically to vertebrates than any other species within the SINA dataset. The gap-trimming procedure was executed differently for the SINA alignment due to the presence of more significant gaps within the functional domain region. Instead of trimming gaps present within the entire reference sequence, the manual refinement procedure was restricted to the alignment's N-terminal portion (i.e. all positions upstream of the *B. floridae* RING domain start that resulted in gaps were cut). Additionally, an overhang at the C-terminal that was only present in *C. elegans* was eliminated from the alignment. These procedures cut down the alignment length from 495 (original) to 287 (refined). The unprocessed, original alignments for all four proteins are available in Additional files 1, 5, 6 and 7, while the refined alignments used in the figures are contained in Additional files 8, 9, 10 and 11.

Phylogenetic analysis of SINA/SIAH protein family

To select an optimal amino acid substitution model within the Maximum-likelihood (ML) framework, the model selection tool within MEGA7 was utilized. The refined SINA/SIAH family alignment was used as the input data. The output returned results from 56 models, and we narrowed our selection to 16 results from the general amino acid replacement matrices (Dayhoff, JTT, WAG, LG) which included a gamma parameter (+G) (see Additional file 12 for output spreadsheet). The LG + G + F model was selected for ML-based phylogenetic analysis.

All gathered SINA/SIAH sequences were aligned and subjected to phylogenetic reconstruction using MEGA7 software [48]. The MEGA7 analysis involved the refined family alignment (70 amino acid sequences with 305 positions) as the input data. The evolutionary history was inferred by using the Maximum Likelihood method based on the LG + G + F model, and a tree with log likelihood = −9417.10) was obtained. Bootstrapping with 100 replicates was applied as a test of phylogeny. Subtree-Pruning-Regrafting (SPR) level 5 was chosen as the ML heuristic method. Initial tree(s) for the heuristic search were obtained automatically by applying Neighbor-Join and BioNJ algorithms to a matrix of pairwise distances estimated using a JTT model, and then selecting the topology with superior log likelihood value. A discrete gamma distribution was used to model evolutionary rate differences among sites [4 categories (+G, parameter = 0.5852)]. The tree was drawn to scale, with branch lengths measured in the number of substitutions per site.

Functional domain analysis

The locations and length of functional domains for all three human SIAH paralogs were derived from their respective entries in NCBI protein database. Using these domain sequences as the reference in each of the three individual paralog alignments, sub-alignments for each of the four domains were extracted. Sequence identity calculations for each domain were performed using the "Calculate distance matrix" function within UniPro UGENE. These values were reported as percentage identity (not excluding gaps), and were calculated using human SIAH1/SIAH2/SIAH3 sequence as the reference. For the SINA alignment, the *Branchiostoma floridae* sequence was utilized as the reference for calculating identity values.

Generation and comparison of SINA/SIAH core consensus sequences

Consensus sequences were extracted and downloaded from the four individual sub-alignments using the default settings within UGENE. Any amino acid position that was ambiguous (i.e. two or more amino acids are prevalent with equal frequency at a given position) was replaced with an "X" character. The four resulting consensus sequences were aligned using MAFFT and the results of this alignment are presented in Additional file 13.

RNA isolation, cDNA synthesis, and RT-PCR amplification of siah1, siah2, and siah3 mRNA transcripts in human cancer cell lines

Total RNA was isolated from cancer cell lines using RNeasy Mini Kit per the manufacturer's instructions and extraction protocol (Qiagen. Germantown, MD). cDNA synthesis was carried out using AMV First Strand cDNA synthesis kit following the manufacturer's protocol (New England BioLabs. Ipswich, MA). PCR amplification was performed using Expand High Fidelity PCR System (Roche. Indianapolis, IN). All primers were purchased from Integrated DNA Technologies (Coralville, IA).

The forward and reverse primers for PCR amplification of the *siah1* cDNA transcript were 5′-ATGAGCCGTCA GACTGCTACAG-3′ and 5′-CAGGACTGCATCATCAC

CCAGT-3′, respectively. The forward and reverse primers for PCR amplification of the *siah2* cDNA transcript were 5′-GCCATCGTCCTGCTCATTGGCA-3′ and 5′-ACCAA TATGGGAAGGCAGGCAGGAAGGGGC-3′, respectively. The forward and reverse primers for PCR amplification of the *siah3* cDNA transcript were 5′-ATGCTTT TCTTTACCCAGTGCT-3′ and 5′-TCACATTTCAGCTT CTGAGGGGA-3′, respectively. The forward and reverse primers for PCR amplification of the *gapdh* cDNA transcript were 5′-AAAGGGTCATCATCTCTGCC-3′ and 5′-TGACAAAGTGGTCGTTGAGG-3′, respectively.

Additional files

> **Additional file 1:** Original alignment for SINA subtree. 20 invertebrate SINA sequences, no trimmed positions. (FASTA 11 kb)
>
> **Additional file 2: Table S1.** Each SINA/SIAH sequence is linked with online accession by one click. List of SINA/SIAH protein sequences (n = 70) that were utilized in the phylogenetic analyses as presented in this study. Taxonomic designation for representative species (n = 39) is listed to the left of its name. With the exception of the sequence from *Petromyzon marinus*, which was acquired from UniProtKB, all amino acid sequence identifiers presented in this table refer to their NCBI Genbank accession numbers. Additional file 2: Table S1 is identical to Table 1, except that the online accession version of these SINA/SIAH sequences was included in the Additional file 2: Table S1. (DOCX 24 kb)
>
> **Additional file 3:** Original alignment of entire SINA/SIAH database. 70 SINA/SIAH protein sequences, no trimmed positions. (FASTA 37 kb)
>
> **Additional file 4:** Refined alignment of entire SINA/SIAH database. 70 SINA/SIAH sequences, trimmed according to Methods. (FASTA 21 kb)
>
> **Additional file 5:** Original alignment for SIAH1 subtree. 19 vertebrate SIAH1 sequences, no trimmed positions. (FASTA 6 kb)
>
> **Additional file 6:** Original alignment for SIAH2 subfamily. 18 vertebrate SIAH2 sequences, no trimmed positions. (FASTA 8 kb)
>
> **Additional file 7:** Original alignment for SIAH3 subtree. 13 vertebrate SIAH3 sequences, no trimmed positions. (FASTA 4 kb)
>
> **Additional file 8:** Refined alignment for SIAH1 subtree. 19 vertebrate SIAH1 sequences, trimmed according to Methods. (FASTA 6 kb)
>
> **Additional file 9:** Refined alignment for SIAH2 subtree. 18 vertebrate SIAH2 sequences, trimmed according to Methods. (FASTA 7 kb)
>
> **Additional file 10:** Refined alignment for SIAH3 subtree. 13 vertebrate SIAH3 sequences, trimmed according to Methods. (FASTA 4 kb)
>
> **Additional file 11:** Refined alignment for SINA subtree. 20 invertebrate SINA sequences, trimmed according to Methods. (FASTA 7 kb)
>
> **Additional file 12:** Output from MEGA7 Model Selection to generate Fig. 1. Excel spreadsheet related to selection of phylogenetic models for Fig. 1 tree, sorted by AICc values. (XLS 525 kb)
>
> **Additional file 13:** Sequence alignment of the four consensus sequences (SINA, SIAH1, SIAH2, SIAH3). Consists of 4 sequences generated from the refined subalignments (see Methods). (FASTA 1 kb)

Abbreviations

DIMER: Dimerization; EGFR: Epidermal growth factor receptor; HER2: Human epidermal growth factor receptor 2; K-RAS: Kirsten rat sarcoma viral oncogene; LCA: Last common ancestor; MAPK: Mitogen-activated protein kinase; MEK: Mitogen-activated protein kinase kinase/extracellular signal-regulated kinase kinase (MAPK/ERK kinase).; RAF: Rapidly Accelerated Fibrosarcoma; RAS: Rat sarcoma viral oncogene; RING: Really Interesting New Gene; S3UNS: SIAH3-unique N-terminal sequences; SBD: Substrate binding domain; SBS: Substrate binding site; SEV: Sevenless; SIAH: SINA homolog; SINA: Seven-In-Absentia; SOS: Son of Sevenless; SZF: SIAH-type zinc finger

Acknowledgments

We would like to thank the two anonymous reviewers for their invaluable critiques and high-quality comments that have helped to strengthen, streamline and improve this manuscript significantly. We are indebted to Dr. Andy G. Clark at Cornell University, and Dr. Kent E. Carpenter at Old Dominion University for guiding us with conducting phylogenetic analyses, and providing expert advice and suggestions in data analyses. We thank Drs. Stephen I. Deutsch, Edward M. Johnson, Jeffrey L. Platt, Julie A. Kerry, Ms. Lauren L. Siewertsz van Reesema and Ms. Angela M. Tang-Tan for critical proofreading of this manuscript, their valuable critiques and positive comments. Correspondence should be addressed to AHT.

Funding

AHT and this work are supported by National Institutes of Health (R01-CA140550), and National Institutes of Health Grant (GM069922-06S1).

Authors' contributions

AHT conceived the idea, designed the Figures, interpreted the results, and wrote this manuscript. IP conducted phylogenetic analyses, performed all the sequence alignments of SINA/SIAH family of E3 Ligases (Figs. 1, 2, 3, 4 and 5), generated Tables (Tables 1 and 2), and was a major contributor in writing the manuscript. REVS conducted the RT-PCR analyses of *siah1*, *siah2* and *siah3* mRNA transcript expression in a panel of human cancer cells (Fig. 6). REVS generated and refined all the finalized and revised Figures and Tables. All authors read and approved this final manuscript for publication.

Competing interests

The authors declare that they have no competing interests.

References

1. Greenwald I, Rubin GM. Making a difference: the role of cell-cell interactions in establishing separate identities for equivalent cells. Cell. 1992;68(2):271–81.
2. Zipursky SL, Rubin GM. Determination of neuronal cell fate: lessons from the R7 neuron of drosophila. Annu Rev Neurosci. 1994;17:373–97.
3. Downward J. Targeting RAS signalling pathways in cancer therapy. Nat Rev Cancer. 2003;3(1):11–22.
4. Schubbert S, Shannon K, Bollag G. Hyperactive Ras in developmental disorders and cancer. Nat Rev Cancer. 2007;7:295–308.
5. Xu C, Liu R, Zhang Q, Chen X, Qian Y, Fang W. The diversification of evolutionarily conserved MAPK cascades correlates with the evolution of fungal species and development of lifestyles. Genome Biol Evol. 2016;
6. Barbacid M. ras genes. Annu Rev Biochem. 1987;56:779–827.
7. Bos JL. Ras oncogenes in human cancer: a review. Cancer Res. 1989;49(17): 4682–9.
8. Malumbres M, Barbacid M. RAS oncogenes: the first 30 years. Nat Rev Cancer. 2003;3(6):459–65.
9. Pylayeva-Gupta Y, Grabocka E, Bar-Sagi D. RAS oncogenes: weaving a tumorigenic web. Nat Rev Cancer. 2011;11(11):761–74.
10. Van Sciver RE, Njogu MM, Isbell AJ, Odanga JJ, Bian M, Svyatova E, van Reesema LLS, Zheleva V, Eisner JL, Bruflat JK et al.: Blocking SIAH proteolysis, an important K-RAS vulnerability, to control and eradicate K-RAS-driven metastatic cancer. In: Conquering RAS. Edited by Azmi A: Elsevier, Inc.; 2016.
11. Cox AD, Fesik SW, Kimmelman AC, Luo J, Der CJ. Drugging the undruggable RAS: mission possible? Nat Rev Drug Discov. 2014;13(11):828–51.
12. McCormick F. KRAS as a therapeutic target. Clin Cancer Res. 2015;21(8): 1797–801.
13. Simon MA, Bowtell DD, Rubin GM. Structure and activity of the sevenless protein: a protein tyrosine kinase receptor required for photoreceptor development in drosophila. Proc Natl Acad Sci U S A. 1989;86(21):8333–7.
14. Carthew RW, Rubin GM. Seven in absentia, a gene required for specification of R7 cell fate in the drosophila eye. Cell. 1990;63(3):561–77.
15. Simon MA, Bowtell DD, Dodson GS, Laverty TR, Rubin GM. Ras1 and a putative guanine nucleotide exchange factor perform crucial steps in signaling by the sevenless protein tyrosine kinase. Cell. 1991;67(4):701–16.
16. Dickson B, Sprenger F, Morrison D, Hafen E. Raf functions downstream of Ras1 in the Sevenless signal transduction pathway. Nature. 1992; 360(6404):600–3.
17. Fortini ME, Simon MA, Rubin GM. Signaling by the sevenless protein tyrosine kinase is mimicked by Ras1 activation. Nature. 1992;355(6360): 559–61.
18. Brunner D, Oellers N, Szabad J, Biggs WH 3rd, Zipursky SL, Hafen E. A gain-of-function mutation in drosophila MAP kinase activates multiple receptor tyrosine kinase signaling pathways. Cell. 1994;76(5):875–88.
19. Biggs WH 3rd, Zavitz KH, Dickson B, van der Straten A, Brunner D, Hafen E, Zipursky SL. The drosophila rolled locus encodes a MAP kinase required in the sevenless signal transduction pathway. EMBO J. 1994;13(7):1628–35.
20. Tang AH, Neufeld TP, Kwan E, Rubin GM. PHYL acts to down-regulate TTK88, a transcriptional repressor of neuronal cell fates, by a SINA-dependent mechanism. Cell. 1997;90(3):459–67.
21. Hu G, Chung YL, Glover T, Valentine V, Look AT, Fearon ER. Characterization of human homologs of the drosophila seven in absentia (sina) gene. Genomics. 1997;46(1):103–11.
22. Wassarman DA, Therrien M, Rubin GM. The Ras signaling pathway in drosophila. Curr Opin Genet Dev. 1995;5(1):44–50.
23. Zhang Q, Wang Z, Hou F, Harding R, Huang X, Dong A, Walker JR, Tong Y. The substrate binding domains of human SIAH E3 ubiquitin ligases are now crystal clear. Biochim Biophys Acta. 2017;1861(1 Pt A):3095–105.
24. Mei Y, Xie C, Xie W, Wu Z, Wu M. Siah-1S, a novel splice variant of Siah-1 (seven in absentia homolog), counteracts Siah-1-mediated downregulation of beta-catenin. Oncogene. 2007;26(43):6319–31.
25. Polekhina G, House CM, Traficante N, Mackay JP, Relaix F, Sassoon DA, Parker MW, Bowtell DD. Siah ubiquitin ligase is structurally related to TRAF and modulates TNF-alpha signaling. Nat Struct Biol. 2002;9(1):68–75.
26. Matsuzawa S, Li C, Ni CZ, Takayama S, Reed JC, Ely KR. Structural analysis of Siah1 and its interactions with Siah-interacting protein (SIP). J Biol Chem. 2003;278(3):1837–40.
27. House CM, Hancock NC, Moller A, Cromer BA, Fedorov V, Bowtell DD, Parker MW, Polekhina G. Elucidation of the substrate binding site of Siah ubiquitin ligase. Structure. 2006;14(4):695–701.
28. Hu G, Fearon ER. Siah-1 N-terminal RING domain is required for proteolysis function, and C-terminal sequences regulate oligomerization and binding to target proteins. Mol Cell Biol. 1999;19(1):724–32.
29. Reed JC, Ely KR. Degrading liaisons: Siah structure revealed. Nat Struct Biol. 2002;9(1):8–10.
30. Santelli E, Leone M, Li C, Fukushima T, Preece NE, Olson AJ, Ely KR, Reed JC, Pellecchia M, Liddington RC, et al. Structural analysis of Siah1-Siah-interacting protein interactions and insights into the assembly of an E3 ligase multiprotein complex. J Biol Chem. 2005;280(40):34278–87.
31. Depaux A, Regnier-Ricard F, Germani A, Varin-Blank N. Dimerization of hSiah proteins regulates their stability. Biochem Biophys Res Commun. 2006;348(3):857–63.
32. Moller A, House CM, Wong CS, Scanlon DB, Liu MC, Ronai Z, Bowtell DD. Inhibition of Siah ubiquitin ligase function. Oncogene. 2009;28(2):289–96.
33. Qi J, Kim H, Scortegagna M, Ronai ZA. Regulators and effectors of Siah ubiquitin ligases. Cell Biochem Biophys. 2013;67(1):15–24.
34. Hasson SA, Kane LA, Yamano K, Huang CH, Sliter DA, Buehler E, Wang C, Heman-Ackah SM, Hessa T, Guha R, et al. High-content genome-wide RNAi screens identify regulators of parkin upstream of mitophagy. Nature. 2013;504(7479):291–5.
35. Robbins CM, Tembe WA, Baker A, Sinari S, Moses TY, Beckstrom-Sternberg S, Beckstrom-Sternberg J, Barrett M, Long J, Chinnaiyan A, et al. Copy number and targeted mutational analysis reveals novel somatic events in metastatic prostate tumors. Genome Res. 2011;21(1):47–55.
36. Schmidt RL, Park CH, Ahmed AU, Gundelach JH, Reed NR, Cheng S, Knudsen BE, Tang AH. Inhibition of RAS-mediated transformation and tumorigenesis by targeting the downstream E3 ubiquitin ligase seven in absentia homologue. Cancer Res. 2007;67(24):11798–810.
37. Ahmed AU, Schmidt RL, Park CH, Reed NR, Hesse SE, Thomas CF, Molina JR, Deschamps C, Yang P, Aubry MC, et al. Effect of disrupting seven-in-absentia homolog 2 function on lung cancer cell growth. J Natl Cancer Inst. 2008;100(22):1606–29.
38. Qi J, Nakayama K, Gaitonde S, Goydos JS, Krajewski S, Eroshkin A, Bar-Sagi D, Bowtell D, Ronai Z. The ubiquitin ligase Siah2 regulates tumorigenesis and metastasis by HIF-dependent and -independent pathways. Proc Natl Acad Sci U S A. 2008;105(43):16713–8.
39. Joazeiro CA, Weissman AM. RING finger proteins: mediators of ubiquitin ligase activity. Cell. 2000;102(5):549–52.

40. Budhidarmo R, Nakatani Y, Day CL. RINGs hold the key to ubiquitin transfer. Trends Biochem Sci. 2012;37(2):58–65.

41. Chasapis CT, Spyroulias GA. RING finger E(3) ubiquitin ligases: structure and drug discovery. Curr Pharm Des. 2009;15(31):3716–31.

42. House CM, Moller A, Bowtell DD. Siah proteins: novel drug targets in the Ras and hypoxia pathways. Cancer Res. 2009;69(23):8835–8.

43. Stebbins JL, Santelli E, Feng Y, De SK, Purves A, Motamedchaboki K, Wu B, Ronai ZA, Liddington RC, Pellecchia M. Structure-based design of covalent Siah inhibitors. Chem Biol. 2013;20(8):973–82.

44. Carthew RW, Neufeld TP, Rubin GM. Identification of genes that interact with the sina gene in drosophila eye development. Proc Natl Acad Sci U S A. 1994;91(24):11689–93.

45. Tang AH, Neufeld TP, Rubin GM, Muller HA. Transcriptional regulation of cytoskeletal functions and segmentation by a novel maternal pair-rule gene, lilliputian. Development. 2001;128(5):801–13.

46. Della NG, Senior PV, Bowtell DD. Isolation and characterisation of murine homologues of the drosophila seven in absentia gene (sina). Development. 1993;117(4):1333–43.

47. Neufeld TP, Tang AH, Rubin GM. A genetic screen to identify components of the sina signaling pathway in drosophila eye development. Genetics. 1998;148(1):277–86.

48. Kumar S, Stecher G, Tamura K. MEGA7: molecular evolutionary genetics analysis version 7.0 for bigger datasets. Mol Biol Evol. 2016;33(7):1870–4.

12

Phylogenomic analysis of Copepoda (Arthropoda, Crustacea) reveals unexpected similarities with earlier proposed morphological phylogenies

Seong-il Eyun [ORCID]

Abstract

Background: Copepods play a critical role in marine ecosystems but have been poorly investigated in phylogenetic studies. Morphological evidence supports the monophyly of copepods, whereas interordinal relationships continue to be debated. In particular, the phylogenetic position of the order Harpacticoida is still ambiguous and inconsistent among studies. Until now, a small number of molecular studies have been done using only a limited number or even partial genes and thus there is so far no consensus at the order-level.

Results: This study attempted to resolve phylogenetic relationships among and within four major copepod orders including Harpacticoida and the phylogenetic position of Copepoda among five other crustacean groups (Anostraca, Cladocera, Sessilia, Amphipoda, and Decapoda) using 24 nuclear protein-coding genes. Phylogenomics has confirmed the monophyly of Copepoda and Podoplea. However, this study reveals surprising differences with the majority of the copepod phylogenies and unexpected similarities with postembryonic characters and earlier proposed morphological phylogenies; More precisely, Cyclopoida is more closely related to Siphonostomatoida than to Harpacticoida which is likely the most basally-branching group of Podoplea. Divergence time estimation suggests that the origin of Harpacticoida can be traced back to the Devonian, corresponding well with recently discovered fossil evidence. Copepoda has a close affinity to the clade of Malacostraca and Thecostraca but not to Branchiopoda. This result supports the hypothesis of the newly proposed clades, Communostraca, Multicrustacea, and Allotriocarida but further challenges the validity of Hexanauplia and Vericrustacea.

Conclusions: The first phylogenomic study of Copepoda provides new insights into taxonomic relationships and represents a valuable resource that improves our understanding of copepod evolution and their wide range of ecological adaptations.

Keywords: Copepoda, Crustacea, Arthropoda, Phylogeny, Phylogenomics, Divergence time

Background

Copepods represent the largest biomass of all animals on earth [1–3]. They are aquatic animals, primarily marine, and make up the dominant zooplankton assemblages in nearshore environments [2, 3]. In spite of their critical ecological roles, the taxonomic classification has received poor attention. Copepods exhibit extreme morphological diversity and occupy an enormous range of habitats in the aquatic realm, from freshwater to hypersaline, shallow pool, and cave to deep sea environments [4–6]. Humes [1] described that there are 11,302 species (198 families, 1633 genera; as of the end of 1993) and estimated that a hypothetical total of 75,347 species may exist on the planet [1]. Copepods are also particularly notorious for cryptic speciation [7–10].

Traditionally, there are ten orders of the subclass Copepoda Milne-Edwards, 1840 containing a large different number of families, genera, and species [5]. The morphological phylogenetic analyses of Copepoda have

Correspondence: seyun2@unl.edu
Center for Biotechnology, University of Nebraska-Lincoln, Lincoln, NE 68588, USA

been extensively investigated and there are general agreements such as the monophyletic status of Copepoda [5, 11–14]. Furthermore, copepods can be divided into two infraclasses, Progymnoplea and Neocopepoda [5]. Progymnoplea contains only one order (Platycopioida) and Neocopepoda can be further classified into two superorders, Gymnoplea and Podoplea [5, 12]. For several decades, however, the phylogenetic relationships among the copepod orders have been a matter of controversy [5, 11–17]. Due to an extreme diversity of body forms, the phylogenetic relationships based on traditional morphological data have led to much controversy (see Fig. 1). For example, Ho [11] and Huys and Boxshall [5] analyzed 21 and 54 morphological characters across ten copepod orders [5, 11]. They agreed that Platycopioida and Calanoida were the most basal groups (Fig. 1ab). However, the cladogram from Ho [11] depicted Harpacticoida and Gelyelloida were closely related, but this group was a distinct cluster to the group of Siphonostomatoida, while that of Huys and Boxshall [5] appeared that Harpacticoida had a close affinity to a sister-group of Siphonostomatoida but a discrete to Gelyelloida. Later, some modifications for the morphological phylogenetic models have been proposed [12, 13]. However, as Ho et al. [13] pointed out, the inconsistent position of Harpacticoida that represents an important ecological group in aquatic environments has been still problematical [13].

Furthermore, some molecular-based studies were not congruent with morphological evidence (Fig. 2). Braga et al. [18] focused on the phylogenetic relationships within the copepod family Euchaetidae and also showed the three copepod orders (Harpacticoida, Calanoida, and Poecilostomatoida with a barnacle, *Semibalanus balanoides* as an outgroup) using the large subunit ribosomal RNA (28S rRNA) gene (a total aligned sequence length of 484 bp) [18]. The tree appeared to be markedly inconsistent with

morphological phylogenies; Harpacticoida was closer to Calanoida than to Poecilostomatoida, which was in conflict to the superorder Podoplea (Fig. 2a). Later, other molecular studies recovered and supported the monophyletic podoplean group using the 18S small subunit ribosomal RNA gene (18S rRNA), but still unresolved the phylogenetic position of Harpacticoida (Fig. 2) [19–22]. Recent study using concatenated twelve mitochondrial genes showed that Harpacticoida (*Tigriopus californicus*) was more closely related to Siphonostomatoida (*Lepeophtheirus salmonis* and *Caligus rogercresseyi*) than Calanoida (*Calanus sinicus*) (Fig. 2d) [23]. This mitochondrial phylogenetic hypothesis was generally congruent with the majority of the morphological phylogenies [5, 12, 13] except for the phylogenetic position of Poecilostomatoida (Fig. 2d). Moreover, in the 18S rRNA gene trees of Poecilostomatoida, the Clausidiiform complex and the remaining poecilostomatoid taxa appeared to be paraphyletic (Fig. 2e) [21, 24]. Harpacticoida also may be a paraphyletic taxon with Polyarthra (consisting of the families Canuellidae and Longipediidae) and Oligoarthra (all remaining harpacticoid families) [17, 22, 25]. From the 28S rRNA gene tree (505 bp from the v-x region), two Polyarthra taxa (*Canuella perplexa* and *Longipedia gonzalezi*) were more closely related to other copepods than to Oligoarthra (Fig. 2F) [22]. All these molecular phylogenetic studies used a relatively short length of the sequences (<2,000 bp) or fast evolving genes that were not acceptable for interordinal relationships (Fig. 2; see details in Discussion).

The purpose of the present study was therefore to clarify the phylogenetic relationships among four major orders of copepods using phylogenomics, the inference of phylogenetic relationships using genome-scale data which has increasingly become a powerful tool to resolve difficult phylogenetic questions [26–30]. In particular, the aim was to include the following: 1) an extensive analysis of the

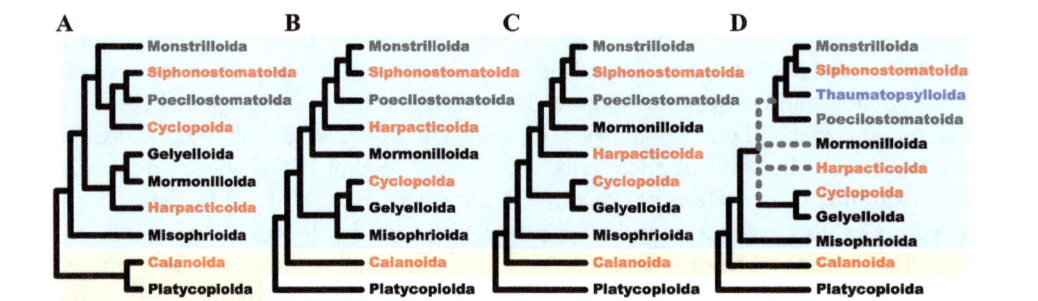

Fig. 1 Major phylogenetic hypotheses based on morphological characters of copepod orders, redrawn from A) Ho [11], B) Huys and Boxshall [5], C) Ho [12], and D) Ho et al. [13]. Cyan and yellow boxes indicate the superorders, Podoplea and Gymnoplea [67]. After Huys and Boxshall [5], Platycopioida is classified as a newly proposed Infraclass, Progymnoplea. A new order, Thaumatopsylloida (indicated by blue) is proposed by Ho et al. [13]. Poecilostomatoida and Monstrilloida (indicated by grey) are considered as the subgroup of Cyclopoida and Siphonostomatoida, respectively [16, 19, 20]. Grey dotted lines depict the ambiguous phylogenetic relationships from Ho et al. [13]. Four copepod orders (indicated by red) are examined in this study

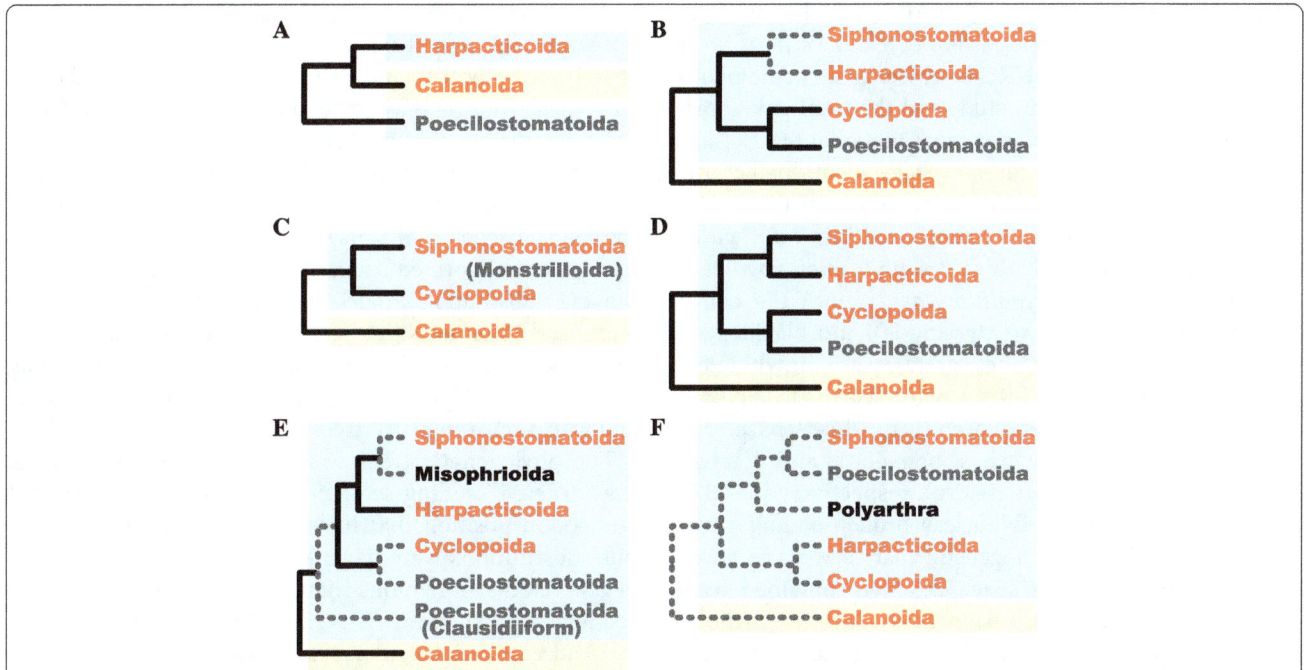

Fig. 2 Phylogenetic hypotheses based on molecular sequence data of copepod orders, redrawn from **a**) Braga et al. [18], **b**) Huys et al. [19], **c**) Huys et al. [20], **d**) Minxiao et al. [23], E) Tung et al. [21], and F) Schizas et al. [22]. Phylogenetic trees using **a**) the large subunit ribosomal RNA (28S rRNA) gene (a total aligned sequence length of 484 bps from the D9/D10 region) [18], **b**) the small subunit ribosomal RNA (18S rRNA) gene (a total aligned sequence length of about 1,882 bp) [19], **c**) 18S rRNA (a total aligned sequence length of about 1,941 bp) [20], **d**) the concatenated twelve mitochondrial genes [23], E) 18S rRNA [21], and F) 28S rRNA (505 bp from the v-x region) [22]. Poecilostomatoida and Monstrilloida (indicated by grey) are considered as the subgroup of Cyclopoida and Siphonostomatoida, respectively [16, 19, 20]. In this study, four copepod orders (indicated by red) are examined. Cyan and yellow boxes indicate Podoplea and Gymnoplea, respectively. Grey dotted lines indicate the low bootstrap values (<60%)

phylogenetic position of Harpacticoida and to evaluate all possible phylogenetic hypotheses; 2) the phylogenetic relationships of copepods among other crustacean groups; and 3) the divergence times of the major copepod orders. Accordingly, the orthologous sequences of 24 nuclear protein-coding genes were retrieved from 18 arthropod species representing four copepod orders (nine species), five other crustaceans (Anostraca, Cladocera, Thecostraca, Amphipoda, and Decapoda), two insects, and two closely related outgroups (Myriapoda and Chelicerata). This study was the first report that provides a rich taxon sampling with genomics-based evidence focusing on the evolution of copepods and their divergence time. Thus, for an ecological perspective, understanding the phylogenetic relationships of copepods would have provided a first step toward elucidating an ecological interaction, habitat colonization, and speciation in Copepoda.

Methods

Taxonomic sampling and identification of orthologous genes

The genome and transcriptome assemblies for 18 arthropod species were obtained from multiple sources (see below). For eight copepod species, five transcriptome

(*Caligus rogercresseyi, Lernaea cyprinacea, Tigriopus californicus, Calanus sinicus*, and *Acartia fossae*) and three genome sequences (*Lepeophtheirus salmonis, Mesocyclops edax*, and *Calanus finmarchicus*) were downloaded from the National Center for Biotechnology Information (NCBI) Sequence Read Archive (SRA) database (http://www.ncbi.nlm.nih.gov/sra) [31, 32]. Three additional crustacean species were also included: the giant tiger prawn *Penaeus monodon* (Malacostraca: Penaeidae), the purple barnacle *Amphibalanus amphitrite* (Thecostraca: Balanidae), and the brine shrimp *Artemia franciscana* (Branchiopoda: Artemiidae) from NCBI SRA. Three other crustacean genome sequences (two copepods and an amphipod; *Eurytemora affinis, Tigriopus californicus*, and *Hyalella azteca*) were downloaded from Baylor College of Medicine Human Genome Sequencing Center (BCM-HGSC), as a part of the pilot project for the i5K arthropod genomes project [33]. Among these crustacean species examined, none of the orthologous sequences for the 24 nuclear protein-coding sequences (see below) was identified in *Calanus finmarchicus* which was excluded from further analysis. Also, the orthologous sequences were further retrieved from the non-redundant (NR) protein database at NCBI (http://www.ncbi.nlm.nih.gov). All orthologous sequences from the copepod *Acanthocyclops vernalis* were obtained from NCBI NR database. All orthologous sequences identified in this study and the GenBank accession numbers were summarized in Additional file 1: Table S1. In addition to the crustacean species mentioned above, five publicly released genomes were added in this study. These sequences of the water flea *Daphnia pulex* (Branchiopoda), the fruit fly (*Drosophila melanogaster*), the red flour beetle (*Tribolium castaneum*), the centipede *Strigamia maritima* (Myriapoda), and blacklegged tick *Ixodes scapularis* (Chelicerata) were downloaded from the wFleaBase (http://wfleabase.org), FlyBase (http://flybase.org), BeetleBase (http://beetlebase.org), BCM-HGSC (https://www.hgsc.bcm.edu), and VectorBase (https://www.vectorbase.org), respectively [34–37].

The previously reported nuclear protein-coding genes that were used for the phylogenetic analysis were retrieved as search queries. These sequences were obtained from Regier et al. [28] and Wiegmann et al. [38]. The orthologous genes were defined by the Basic Local Alignment Search Tool (BLAST, ver. 2.2.30+) programs [39, 40]. The E-value threshold of 1×10^{-30} with the database size 1.4×10^{10} was used to identify orthologous candidates against the genome and transcriptome assemblies. The putative orthologous genes were verified by searches using tblastn against NCBI NR database. After partial sequences or no apparent orthologs were excluded from the analysis, 24 nuclear protein-coding genes were then determined in more than half of the copepod species (Additional file 1: Table S1). All identified copepod protein and nucleotide sequences are provided in Additional files 2 and 3.

Multiple sequence alignments

Multiple alignments of each of the protein gene families were generated using mafft (ver. 7.245) [41] with the L-INS-i algorithm (1,000 maxiterate and 100 retree) which uses a consistency-based objective function and local pairwise alignment with affine gap costs. Alignments were adjusted manually when necessary. Poorly aligned regions with more than 70% of gaps were removed using trimAl (ver. 1.2) [42]. The corresponding coding nucleotide alignments were generated using PAL2NAL [43]. The single gene sequence alignments are available in: http://bioinformatics.unl.edu/eyun/Copepoda_Phylogenomics. All sequences were concatenated using a custom Perl script (ConCat_seq.pl). This Perl script is available upon request from the author. The concatenated dataset used in this study is available in Additional file 4.

Phylogenetic analysis and alternative topology tests

Phylogenetic relationship using the concatenated sequences was reconstructed by the maximum-likelihood (ML) method with the Le and Gascuel (LG) matrix, gamma distributed rates, invariant sites, and the observed amino acid frequencies using PhyML (ver. 3.1) [44–46]. The best-fit model for the concatenated dataset was selected using the Akaike Information Criterion (AIC) as a statistical tool in ProtTest (ver. 3.2) [47]. Nonparametric bootstrapping with 1000 pseudo-replicates was used to estimate the confidence of branching patterns for the ML phylogeny [48]. Bayesian inferences (BI) of phylogeny were performed using MrBayes (ver. 3.2.6) [49] with the LG substitution model, gamma-distributed rate variation, and invariant sites. The Markov chain Monte Carlo search was run for 5×10^6 generations, with a sampling frequency of 10^3, using three heated and one cold chain and with a burn-in of 10^3 trees.

The phylogenetic trees were also reconstructed using the "degen-1" coding sequences, in which nucleotides at any codon position that have the potential of synonymous substitutions were degenerated [28]. To produce the degenerated synonymous matrices (the "degen-1" coding sequences) [28], the Perl script (Degen_v1_4.pl) written by Andreas Zwick and April Hussey was used (http://www.phylotools.com). For the morphological reanalysis, the data matrix of 54 morphological characters was obtained from Ho et al. [13]. This morphological data matrix in Nexus format is available in Additional file 5. Phylogenetic inference of the morphological data was conducted with MrBayes (ver. 3.2.6) [49], using the Mk (Markov K) model [50], a variable rate among characters ("rates = gamma"), and 5×10^8 generations. The Mk model assumes equal state frequencies. In this analysis, trees were sampled every 10^3 generations with the first 25% discarded as burn-in and summarized using a 50% majority rule consensus tree. Presentation of the phylogenies was done with FigTree (ver. 1.4.2) (http://tree.bio.ed.ac.uk/software/figtree).

The Kishino-Hasegawa (KH) [51], the Shimodaira-Hasegawa (SH) [52], and the Approximately Unbiased (AU) [53] tests were used to statistically assess the phylogenetic hypotheses. The site log-likelihood of each tree was calculated in TREE-PUZZLE (ver. 5.3.rc16) [54], and KH, SH, and AU tests were performed in CONSEL (ver. 0.20) with default options [55].

Divergence time estimation

The divergence times of lineages were estimated using BEAST2 (ver. 2.4.3) [56] with Bayesian inference using the calibrated Yule model for the tree prior and the uncorrelated relaxed clock model proposed by Drummond et al. [57]. BEAST2 was using a random tree with 5×10^7 generations and a sample frequency of 5×10^3 generations. Four fossil-based minimum ages were applied for the major splits; 497 MYA for the Diptera-Cladocera divergence, 405 MYA for the Cladocera-Anostraca divergence, 313.7 MYA for the Diptera-Coleoptera divergence, and

358.5 MYA for the Amphipoda-Decapoda divergence [58]. The fossil record of *Wujicaris muelleri* Zhang et al. [59] was also used as the minimum constraint on the crown group of Pancrustacea [58–60].

Results

Monophyly of copepods and their interordinal relationships

24 nuclear protein-coding genes were obtained from 18 arthropod species including nine copepod species (four major orders of copepods) (Additional file 1: Table S1). The common names of species examined with the current taxonomic classification were listed in Table 1. Among 18 arthropods, two non-pancrustacean taxa, the centipede *Strigamia maritima* and blacklegged tick *Ixodes scapularis*, were used as the outgroups [28, 61]. All 24 nuclear protein-coding sequences were concatenated for further phylogenetic analyses (see details in Methods). The data set of the concatenated sequences consisted of 16,710 amino acid sequences (50,106 bp). The phylogenetic relationships obtained from the concatenated sequences were reconstructed by the maximum-likelihood (ML) and Bayesian inferences (BI). The two algorithms confirmed the monophyly of copepods with 100% bootstrap values (Fig. 3). The nine copepod species examined can be classified into two superorders, Gymnoplea (Calanoida) and Podoplea (Siphonostomatoida, Cyclopoida, and Harpacticoida) (Fig. 1; see Discussion). The phylogenomic analyses with ML and BI generated the same topologies supporting the mono-phyly of the podoplean group. The superorder Podoplea was strongly supported with the high maximum likelihood bootstrap value (MLB = 100%) and Bayesian posterior probability (BPP = 1.00) (Fig. 3).

Most notably, within Podoplea, the interordinal relation-ships inferred from the phylogenomic analysis differed from that of the widely accepted hypothesis presented in the majority of the morphological and molecular phylo-genetic studies that Harpacticoida was generally affiliated with Siphonostomatoida rather than with Cyclopoida [5, 12, 19, 21, 23] (Figs. 1 and 2; see Discussion). In addition to order level relationships in this study, all family and genus level relationships were also clearly resolved by high bootstrap values (MLB = 100% and BPP = 1.00) (Fig. 3). This study included three families for Calanoida: ((Acartiidae, Temoridae), Calanidae) and three genera in Cyclopoida: ((*Acanthocyclops*, *Mesocyclops*), *Lernaea*).

Phylogenetic position of Harpacticoida

To confirm the phylogenetic position of Harpacticoida (*Tigriopus californicus*, Oligoarthra), four different phylo-genetic analyses were attempted; 1) Bayesian reestimation of morphological characters, 2) the "degen-1" coding sequences of Regier et al. [28], 3) small-scale phylogeny

Table 1 Taxonomic classification used in this study

[Class] / Species	Order and Family	Common Names
[Copepoda]		
Lepeophtheirus salmonis	Siphonostomatoida, Caligidae	salmon louse
Caligus rogercresseyi	Siphonostomatoida, Caligidae	sea louse
Acanthocyclops vernalis	Cyclopoida, Cyclopidae	
Mesocyclops edax	Cyclopoida, Cyclopidae	freshwater cyclopoid
Lernaea cyprinacea	Cyclopoida, Lernaeidae	anchor worm
Tigriopus californicus	Harpacticoida, Harpacticidae	tide pool copepod
Acartia fossae	Calanoida, Acartiidae	Oceanic shelf copepod
Eurytemora affinis	Calanoida, Temoridae	common estuarine copepod
Calanus sinicus	Calanoida, Calanidae	Asian Pacific copepod
[Thecostraca]		
Amphibalanus amphitrite	Sessilia, Balanidae	purple acorn barnacle
[Malacostraca]		
Hyalella azteca	Amphipoda, Dogielinotidae	
Penaeus monodon	Decapoda, Penaeidae	giant tiger prawn
[Branchiopoda]		
Daphnia pulex	Cladocera, Daphniidae,	water flea
Artemia franciscana	Anostraca, Artemiidae	brine shrimp
[Insecta]		
Drosophila melanogaster	Diptera, Drosophilidae	fruit fly
Tribolium castaneum	Coleoptera, Tenebrionidae	red flour beetle
[Chilopoda]		
Strigamia maritima	Geophilomorpha, Geophilidae	
[Arachnida]		
Ixodes scapularis	Ixodida, Ixodidae	blacklegged tick

dealing with only nine copepod species with two closely related outgroups, and 4) statistical analyses were performed for all possible trees (three topologies here). First, the morphological phylogeny was reconstructed by Bayesian inferences (Additional file 6: Figure S1). Bayesian analysis using 54 morphological characters showed the same topology with that of Ho et al. [13] which was recon-structed by maximum parsimony. This tree yielded mostly congruent results with the majority of the copepod phyl-ogeny above. However, all posterior probabilities with the

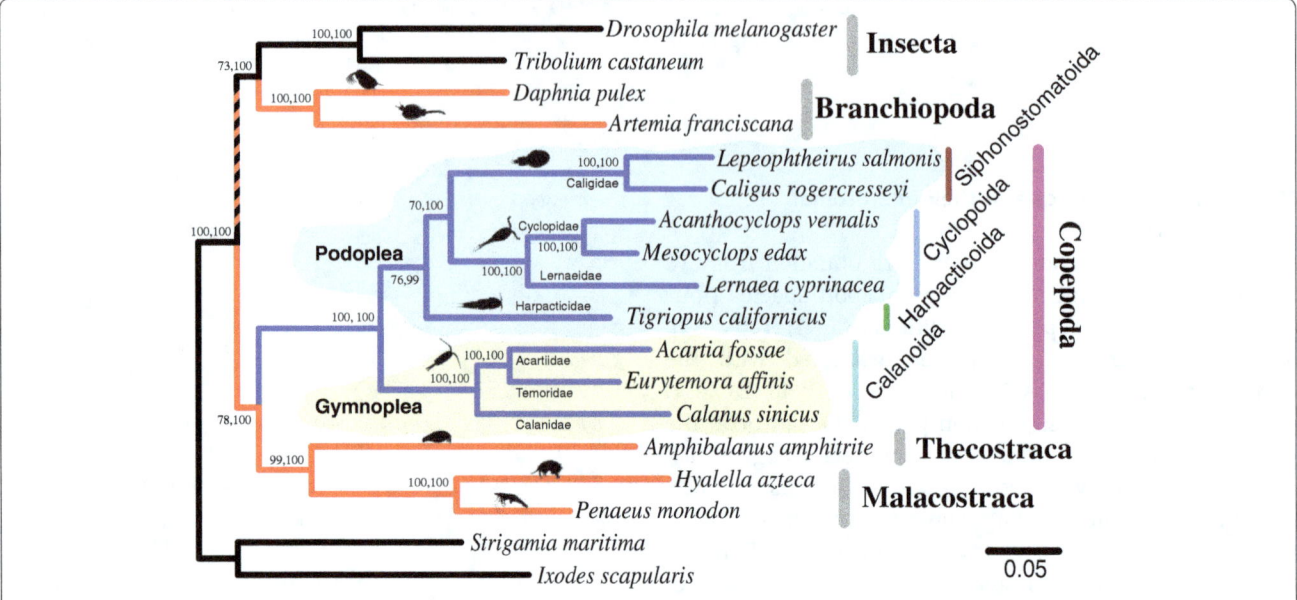

Fig. 3 The maximum-likelihood phylogeny of nine copepod species and nine other arthropod species based on the 24 nuclear protein-coding genes. *Strigamia maritima* (Myriapoda) and *Ixodes scapularis* (Chelicerata) are used as the outgroups. Blue-colored and red-colored branches indicate the copepod groups and all other crustaceans. The numbers at internal branches show the bootstrap support values (%) for the maximum-likelihood phylogeny and the posterior probability (%) for the Bayesian phylogeny in this order. The scale bar represents the number of amino acid substitutions per site

morphology-only data set showed a general lack (BPP < 0.80) of support except for three nodes (indicated by bold-faces in Additional file 6: Figure S1). This suggested that these morphological data might not have sufficient phylogenetic signal. For the second and third attempts, the phylogenetic trees were reconstructed using the "degen-1" coding sequences and from only nine copepod species with *Amphibalanus amphitrite* (Sessilia) and *Penaeus monodon* (Decapoda) as the outgroups. Both the phylogenetic approaches returned the same topology as that obtained from Fig. 3 (Additional files 7 and 8: Figures S2 and S3). The phylogenetic relationships were less resolved (>64% MLB for the clade of Siphonostomatoida and Cyclopoida) using the "degen-1" nucleotide dataset than that of Fig. 3, but better resolved in the small-scale phylogeny by high bootstrap values (>76% MLB for that clade) (Additional files 7 and 8: Figures S2 and S3). Lastly, to evaluate those previously proposed hypotheses shown in Figs. 1 and 2, statistical analyses were performed using TREE-PUZZLE (ver. 5.3.rc16) and CONSEL (ver. 0.20) [54, 55] (see in Methods). In Table 2, the first hypothesis, as mentioned earlier, was obtained from the majority of the copepod phylogeny. The second hypothesis was the best maximum likelihood tree obtained from the concatenated PhyML tree in this study. The third hypothesis was a theoretical tree, in which Harpacticoida was closely related to Cyclopoida. All statistical tests rejected the third hypothesis. Although the KH test ($P = 0.109$) and the SH test ($P = 0.479$) were unable to reject the first hypothesis, the AU topology test was

marginally rejected ($P = 0.085$) at the 0.10 level of significance. This was most likely due to the conservative nature of the KH test and the SH test. The KH test was invalid in this case because the second hypothetical tree was the best ML tree [52]. The SH test is the most conservative estimate and is sensitive to the unlikely tree (i.e., the third hypothesis in Table 2) [62]. Among the three tests, the AU test is known as the best approach to overcome these problems [53]. Thus, the results of the statistical test supported that the most likely phylogenetic scenario is the second hypothesis. Taken together, these results strongly suggested that Siphonostomatoida was closer to Cyclopoida than Harpacticoida.

Copepoda is a sister group to Communostraca

According to the present phylogenomic analysis, the resulting trees revealed that Copepoda was a sister lineage to a group of Thecostraca and Malacostraca but distinct to Branchiopoda (Fig. 3 and Additional file 7: Figure S2), consistent with results from Regier et al. [28] and Oakley et al. [30]. Both ML and BI inferred the following interclass relationships: ((Insecta, Branchiopoda), (Copepoda, (Thecostraca, Malacostraca))). The purple barnacle *A. amphitrite* (Sessilia: Balanidae) was considered to be a sister group to copepods, namely Maxillopoda. In this study, however, this species appeared to be a sister group to the group (Malacostraca) of *H. azteca* (Amphipoda) and *P. monodon* (Decapoda), but distinctly related to copepods (Fig. 3 and Additional file 7: Figure S2). The

Table 2 Statistical comparisons between the best ML tree and alternative phylogenetic hypotheses within podopean copepods

Hypothetical Affinities[a]	References claiming the hypothesis	-lnL[b]	P-values		
			KH[c]	SH[d]	AU[e]
((SI, HA), CY)	Huys and Boxshall (1991) [5], Ho (1994) [12], Huys et al. (2006) [19], Minxiao et al. (2011) [23], and Tung et al. (2014) [21]	177,619.7	0.109	0.479	0.085*
((SI, CY), HA)	Kabata (1979) [68], Ho (1990) [11], and Dahms (2004) [14]	177,399.7	0.377	0.750	0.445
((HA, CY), SI)	none	177,631	<0.001***	<0.001***	<0.001***

[a]SI = Siphonostomatoida, CY = Cyclopoida, and HA = Harpacticoida
[b]lnL = Log-likelihood scores
[c]P-value of the Kishino-Hasegawa (KH) test [51]
[d]P-value of the Shimodaira-Hasegawa (SH) test [52]
[e]P-value of the Approximately Unbiased (AU) test [53]
One (*) and triple (***) asterisks denoted statistical significance at the 0.10 and 0.001 level, respectively

MLB and BPP values for the clade of Sessilia, Amphipoda, and Decapoda were highly supported (MLB > 87% and BPP = 1.00) (Fig. 3 and Additional file 7: Figure S2). Therefore, this study supported the newly proposed clade, Communostraca (common shelled ones) that includes Malacostraca (e.g., crabs or shrimp) and Thecostraca (e.g., barnacles) and the newly proposed clade, Multicrustacea (Copepoda, Malacostraca, and Thecostraca) with high support values (MLB > 71% and BPP = 1.00) [28] (See Discussion).

This study also supported a proposed clade of Insecta and Branchiopoda, consistent with results from Oakley et al. [30] representing the Allotriocarida (Hexapoda/Branchiopoda/Remipedia) clade [30]. A very recent study also supported the monophyly of Allotriocarida [63]. Branchiopoda (a group of Cladocera and Anostraca) was considered to belong to the subphylum Crustacea. However, this group was more closely related to Insecta but distinct to all other crustaceans examined in this study. Although the MLB value for the clade of Allotriocarida (Insecta and Branchiopoda) was not very strong (>73% in Fig. 3 and > 62% in Additional file 7: Figure S2), this hypothesis is often congruent with those obtained from recent studies [27, 30, 64, 65]. Therefore, the phylogenetic trees in this study supported the hypotheses of the three newly proposed clades, Communostraca, Multicrustacea, and Allotriocarida, but challenged the validity of Hexanauplia and Vericrustacea (See Discussion).

Estimation of divergence time in Copepoda
Divergence times were estimated using BEAST2 (ver. 2.4.3) [56] with Bayesian inference. The tree topology was the same as the PhyML tree shown in Fig. 3. Divergence between the groups of podopleans and gymnopleans was estimated to have occurred during the period from the late Cambrian to the Devonian (446.2 ± 47.3 MYA). The origin of *T. californicus* appeared to have occurred in the Devonian (between the late Silurian and the early Carboniferous, 381.4 ± 51.1 MYA) (Fig. 4). The divergence time between the two orders Siphonostomatoida and Cyclopoida occurred in the Carboniferous (351.8 ± 58.1 MYA)

which predated approximately the origin of Harpacticoida (Fig. 4). Seven extant families in this analysis arose before the Cenozoic era and possibly prior to the early stage of breakup of Gondwana [66].

Discussion
The present study provides the first phylogenomic evidence to support the monophyletic origin of four major orders of copepods and the group of podopleans. The monophyletic status of Copepoda has been broadly accepted by both morphological [5, 14] and large-scale phylogenomic analyses [28–30]. Although this study does not include all copepod orders, there can be no doubt of the monophyly of copepods. The subclass Copepoda consists of two infraclasses, Progymnoplea and Neocopepoda, suggested by Huys and Boxshall [5]. The infraclass Neocopepoda can be further divided into two superorder groups, Gymnoplea and Podoplea (Fig. 1). The concept of this classification was proposed by Giesbrecht [67] and became generally accepted [5, 12, 68]. However, the naupliar musculature and the molecular phylogeny using partial nuclear 28S rRNA gene (a total aligned sequence length of 484 bp from the D9/D10 region) (Fig. 2A) showed conflicting results and suggested a possible paraphyletic origin of podopleans [15, 18]. Later, morphological [13, 14] and molecular [19, 20, 23] phylogenetic analyses recovered the monophyly of podopleans. In this study, the phylogenomic analysis shows that three podoplean copepod orders are clearly clustered as a monophyletic clade (supported by high bootstrap values, MLB > 99% and BPP = 1.00) (Fig. 3 and Additional files 7 and 8: Figures S2 and S3).

Unexpectedly, the current phylogenomic evidence is in conflict to the majority of the copepod phylogenies (Figs. 1 and 2; see Results). The present schematic phylogeny resemble those found in the earlier phylogenies and post-embryonic data [11, 14, 68] which show that Calanoida represents the most basal split among the four copepod orders and that Harpacticoida is the basally-branching group of Podoplea. On the basis on postembryonic apomorphies, naupliar characters can be represented by plesiomorphic states because postembryonic stages (both

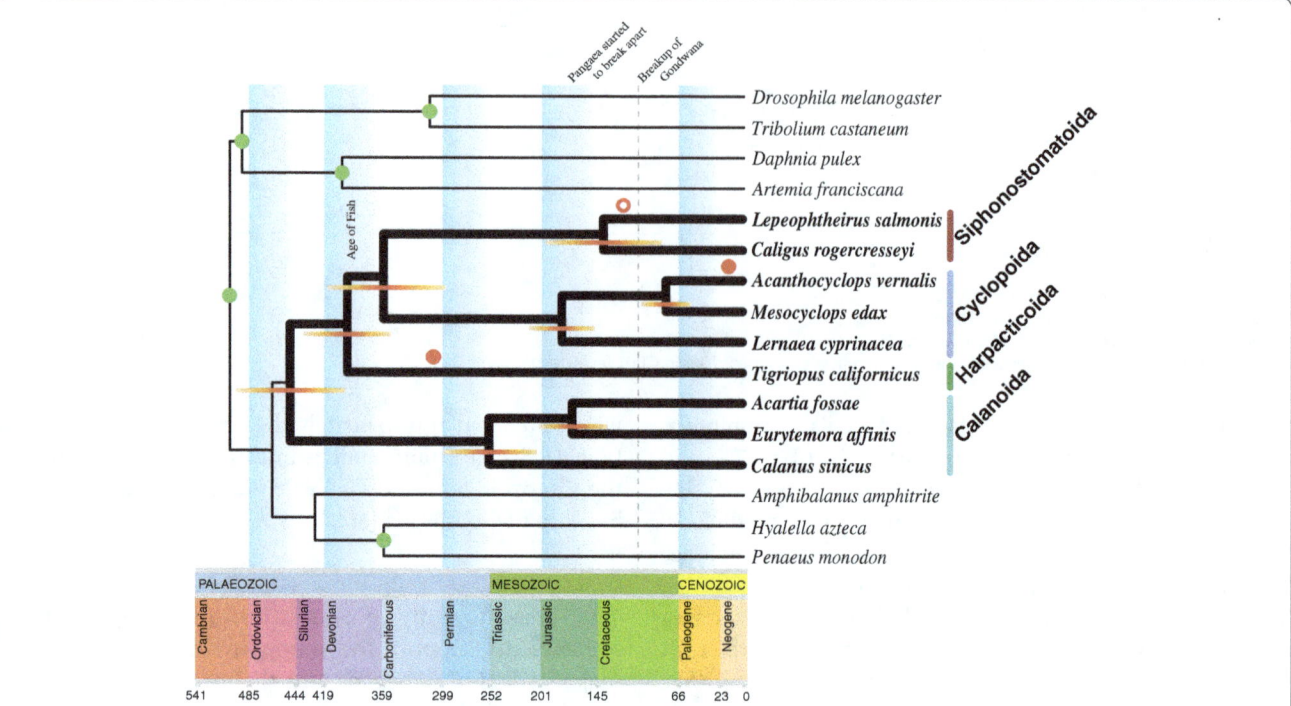

Fig. 4 Estimated divergence times among copepods. BEAST2 (ver. 2.4.3) [56] is used with five calibration points (indicated by green circles, see details in Methods). Orange bars across nodes indicate 95% highest posterior density (HPD) of the Bayesian posterior distribution of molecular time estimates. The open and closed red circles (the parasitic and free-living forms of copepod fossils) above the branches show the oldest fossil records, most likely corresponding to the copepod orders [84–87]. The geologic time scale is according to the International Chronostratigraphic Chart (http://www.stratigraphy.org, v2015/01)

early and later) provide a valuable resource for evolutionary history [14]. His study implied that Harpacticoida is the more basally-branching group than Misophrioida within podopleans, which is hardly reported in previous studies [5, 11–13, 15, 17]. Interestingly, our preliminary survey based on weighted morphological characters after removing the convergent characters appears that Harpacticoida is the most basally-branching podoplean group (Eyun et al., unpublished data). For example, some morphological characters support the current phylogenomic phylogeny; following the characters from Huys and Boxshall [5], character 11 (male antennulary segment XXIII), character 21 (outer seta on basis of maxillule), and character 54 (seta *b* on exopod of male fifth leg). These morphological characters can be the candidates to investigate the order-level relationships of copepods and morphological transitions (e.g., character 21). Based on character 54 which is absent of in Harpacticoida but is present in Misophrioida and many other podopleans, Harpacticoida seems to be the most basally-branching group within Podoplea. Furthermore, as keenly pointed out by Ho [12], some characters such as character 13 (male antennulary segments XXIV and XXV), character 29 (praecoxal seta on maxilliped), and character 39 (number of setae on inner margin of second endopodal segment of first swimming leg) are confirmed as convergent characters in this

study. These implies that differential weighting criteria for the morphological phylogeny [69] and the removal of convergent characters can reduce the phylogenetic noise. In fact, from the preliminary survey removing the convergent characters, the posterior probabilities in Bayesian phylogenetic inference are increased (Eyun et al., unpublished data).

Recent studies have given rise to a new taxonomic classification of Copepoda. Although many progresses have been made toward unraveling the phylogeny and taxonomy of Copepoda, there is so far no consensus of their order-level classification. This should be due to their extreme morphological diversity and a lack of genetic information. Huys and Boxshall [5] summarized ten copepod orders [5]. Ho et al. [13] proposed a new order, Thaumatopsylloida because the family Thaumatopsyllidae was a distinct group from the order Cyclopoida and differed from Monstrilloida and Siphonostomatoida [13]. Boxshall and Halsey [16] suggested that Poecilostomatoida was merged into Cyclopoida [16]. Huys et al. [19], Minxiao et al. [23], and Huys et al. [24] supported this view (but as a sister group) using 18S rRNA and the concatenated twelve mitochondrial genes [19, 23, 24]. Another molecular sequence study using 18S rRNA (a total aligned sequence length of about 1,941 bp) suggested that the order Monstrilloida (indicated by grey in Figs. 1 and 2) was nested within a fish-parasitic clade of the order Siphonostomatoida and thus was considered as the subgroup of Siphonostomatoida [20]. The 18S rRNA gene and 28S rRNA gene trees showed that Poecilostomatoida and Harpacticoida were paraphyletic, respectively (Fig. 2EF) [21, 22, 24].

Some studies have argued that adding more sequences is more important than adding taxa for improved phylogenetic accuracy [70, 71] (but see [72] for the benefits of adding taxa). Indeed, in copepods, insufficient and only partial sequences have been used and showed a limitation for certain order-level [21, 73, 74]. Blanco-Bercial et al. [75] discussed that the use of a single gene at the family or superfamily level of copepods contributed to the disparate results, and the relationships in the superfamily Centropagoidea (Order Calanoida) were still unresolved using the four concatenated genes (18S rRNA, 28S rRNA, cytochrome c oxidase subunit I, and cytochrome b) [75]. Therefore, the phylogenomic approach will make notable contributions to a better resolution of copepod evolution and then can be anchored to certain taxonomic clades. Furthermore, the resulting phylogenomic tree can provide an independent test of morphological character homology and can help to determine the assumptions of plesiomorphic or apomorphic characters and the convergent or homoplastic characters, which are considered as the most difficult issue for copepod taxonomy [5, 11, 12].

The class Maxillopoda (Phylum Arthropoda) is one of the most diverse groups of crustaceans including copepods, barnacles, and a number of related animals (such as a branchiuran fish louse and tongue worms) [6]. However, the monophyly of Maxillopoda seemed increasingly doubtful and the maxillopodan concept became obsolete due to the phylogenetic studies of the Arthropoda [27–30, 61, 76, 77]. These studies appear in the polyphyly of Maxillopoda. In addition, the phylogenetic position of copepods in relationship to other crustacean groups has been controversial, resulting a particularly ambiguous resolution of Copepoda, Thecostraca, Malacostraca, and Branchiopoda. Therefore, the phylogenetic relationships among crustaceans are still far from being resolved [78, 79]. Recent phylogenomic studies advocate a new taxonomic nomenclature for the crustacean groups. Regier et al. [28] and Oakley et al. [30] proposed several crustacean classifications; Communostraca (Malacostraca, Thecostraca), Multicrustacea (Copepoda, Malacostraca, and Thecostraca), and Vericrustacea (Copepoda, Malacostraca, Thecostraca, and Branchiopoda) [28] and Allotriocarida (Hexapoda, Remipedia, Cephalocarida, and Branchiopoda) and Hexanauplia (Copepoda and Thecostraca) [30]. From the currently inferred phylogenies including six crustacean groups (Cladocera, Anostraca, Copepoda, Sessilia, Amphipoda, and Decapoda), the tree supports well the hypothesis of the three newly proposed clades, Communostraca, Multicrustacea, and

Allotriocarida. However, this study challenges the validity of Hexanauplia and Vericrustacea, corroborating those obtained from other phylogenomic analyses (Fig. 3 and Additional file 7: Figure S2) [27, 29, 30].

This study confirms that the rapidly evolving genes tend to generate the phylogenetic noise [30] and that the slower evolving genes contain more informative positions [80]. For instance, the phylogenies using a single gene tree from 6-phosphogluconate dehydrogenase, carbamoylphosphate synthetase, and alanyl-tRNA synthetase show the non-monophyly of Copepoda (Additional file 9: Figure S4). This may be due to incomplete sequences of genes which are not identified to cover the intact region in this study but also to a relatively high level of sequence variation. Regier et al. [28] also categorizes these genes as the fast evolving genes (the gene numbers: 11, 19, and 23) [26]. Therefore, the phylogenetic signals from the fast evolving genes could generate misleading effects in evolutionary studies [81]. Note that, however, the copepod topology after excluding these genes is same as the one shown above (data not shown).

Divergence between the groups of podopleans and gymnopleans is estimated to have occurred in the very late Ordovician. This implies that the origin of copepods may be earlier (probably Cambrian age) than this period [82, 83]. It is because all copepod taxa in this study belong to the Infraclass Neocopepoda, and Platycopioida (the other infraclass Progymnoplea) is known to be the most primitive group of copepods and possibly closer to the ancestral form [5, 12]. Only few fossil records of copepods are available because of their fragile nature and thus having a very low level of potential fossilization. Divergence time estimations in this study are in good agreement with these known fossil records [84–87]. Recently, a new fossil of freshwater harpacticoids (most likely Canthocamptidae) has been found in carboniferous bitumen, dating back to at least 303 MYA [86]. Interestingly, the origin of T. californicus assumed in this study is almost congruent with this fossil record (Fig. 4). The family Canthocamptidae is the largest group (>600 species) of harpacticoids and predominately inhabit fresh water [88]. Boxshall and Jaume [88] speculated that harpacticoids invaded fresh waters on Pangaea based on the pattern of colonization of continental waters. This study supports this hypothesis by molecular sequence analysis. To study the adaptation on the different types of environments (e.g., cave or groundwater) and the timing of colonization events, a strong phylogenetic hypothesis must be established. For the future, comparative genomics of copepod species will help us understanding their evolutionary history and shed light on a wide range of ecological adaptations.

Conclusion

A series of molecular phylogenetic analyses of nine copepod species with five other crustacean groups, two hexapods, and two outgroups (myriapod and spider) is presented using the 24 orthologous nuclear protein-coding genes. Given the phylogeny, this hypothesis provides an overview of the useful directions for future studies and thus will shed a light into new taxonomic investigations. As more sequences become available in the near future, further studies with more comprehensive taxa are essential to evaluate the various hypotheses as well as fully resolve the evolutionary history and taxonomy of Copepoda. Also, some copepod orders (e.g., Thaumatopsylloida, Monstrilloida, and some groups of Poecilostomatoida and Harpacticoida) need to be refined by further phylogenomic studies. The large scale of molecular data such as genomes and transcriptomes of copepods provides us a valuable resource for understanding copepod evolution and a wide range of ecological adaptations.

Additional files

Additional file 1: Table S1. Orthologous sequences used and identified in this study. (XLSX 18 kb)

Additional file 2: This file contains the copepod amino acid sequences in FASTA format. (TXT 71 kb)

Additional file 3: This file contains the copepod nucleotide sequences in FASTA format. (TXT 207 kb)

Additional file 4: This file contains the aligned and concatenated dataset used in this study. (DOCX 102 kb)

Additional file 5: This file contains the data matrix of 54 morphological characters from Ho et al. [13] in Nexus format. (DOCX 17 kb)

Additional file 6: Figure S1. Bayesian phylogenetic analysis of copepod orders with morphological characters taken from Ho et al. [13]. (DOCX 63 kb)

Additional file 7: Figure S2. Bayesian phylogeny using the "degen-1" nucleotide coding sequences. (DOCX 172 kb)

Additional file 8: Figure S3. Bayesian phylogeny of nine copepod species with two outgroups. (DOCX 107 kb)

Additional file 9: Figure S4. Maximum-likelihood phylogenies of arthropods focused on copepod species, based on a single gene region from A) 6-phosphogluconate dehydrogenase, B) carbamoylphosphate synthetase, and C) alanyl-tRNA synthetase. (DOCX 185 kb)

Abbreviations

AU: Approximately unbiased; BI: Bayesian inferences; BPP: Bayesian posterior probability; KH: Kishino-Hasegawa; LG: Le and Gascuel; ML: Maximum-likelihood; MLB: Maximum likelihood bootstrap value; NJ: Neighbor-joining; rRNA: Ribosomal RNA; SH: Shimodaira-Hasegawa

Acknowledgements

The author sincerely thanks to Drs. Hae-Lip Suh and Ho Young Soh (Chonnam National University, Korea) for providing the initial inspiration. Dr. Ju-shey Ho (California State University at Long Beach, USA) provided helpful comments and suggestions on an earlier draft of this manuscript. The author also thanks Susumu Ohtsuka (Hiroshima University, Japan) for critical reading of the manuscript.

Funding

This work was supported by the Nebraska Research Initiative (to SE).

Author's Contributions

SE carried out the data analysis and wrote the manuscript.

Competing interests

The author declares that he has no competing interests.

Data deposition

All identified copepod protein and nucleotide sequences can be found in Additional files 2 and 3 respectively. These sequences are also available from the local server: http://bioinformatics.unl.edu/eyun/ Copepoda_Phylogenomics.
A custom Perl script, ConCat_seq.pl, is available upon request from the author.

References

1. Humes AG. How many copepods? Hydrobiologia. 1994;292/293:1–7.
2. Mauchline J. The Biology of Calanoid Copepods. Adv Mar Biol. 1998;33:1–710.
3. Verity P, Smetacek V. Organism life cycles, predation, and the structure of marine pelagic ecosystems. Mar Ecol Prog Ser. 1996;130:277–93.
4. Hardy A: The Open Sea. It's Natural History: The World of Plankton: Collins. London: Houghton Mifflin Company; 1956.
5. Huys R, Boxshall GA. Copepod Evolution. London: The Ray Society; 1991.
6. Martin JW, Davis GE. An updated classification of the recent Crustacea. Nat Hist Mus Los Angel Cty Sci Ser. 2001;39:1–124.
7. Lee CE. Global phylogeography of a cryptic copepod species complex and reproductive isolation between genetically proximate "populations". Evolution. 2000;54:2014–27.
8. Goetze E. Cryptic speciation on the high seas; global phylogenetics of the copepod family Eucalanidae. Proc R Soc Lond B Biol Sci. 2003;270:2321–31.
9. Eyun S, Lee Y-H, Suh H-L, Kim S, Soh HY. Genetic Identification and Molecular Phylogeny of *Pseudodiaptomus* Species (Calanoida, Pseudodiaptomidae) in Korean Waters. Zoolog Sci. 2007;24:265–71.
10. Chen G, Hare MP. Cryptic diversity and comparative phylogeography of the estuarine copepod *Acartia tonsa* on the US Atlantic coast. Mol Ecol. 2011;20:2425–41.
11. Ho J-S. Phylogenetic Analysis of Copepod Orders. J Crustac Biol. 1990;10:528–36.
12. Ho J-S. Copepod phylogeny: a reconsideration of Huys & Boxshall's 'parsimony versus homology'. Hydrobiologia. 1994;292/293:31–9.
13. Ho J-S, Dojiri M, Gordon H, Deets GB. A New Species of Copepoda (Thaumatopsyllidae) Symbiotic with a Brittle star from California, U.S.A., and Designation of a New Order Thaumatopsylloida. J Crustac Biol. 2003;23:582–94.
14. Dahms H-U. Postembryonic Apomorphies Proving the Monophyletic Status of the Copepoda. Zool Stud. 2004;43:446–53.
15. Dussart BH: A propos du répertoire mondial des Calanoïdes des eaux continentales. *Crustaceana* 1984;(Suppl 7):25–31.
16. Boxshall GA, Halsey SH. An Introduction to Copepod Diversity. London: The Ray Society; 2004.
17. Por FD: Canuellidae Lang (Harpacticoida, Polyarthra) and the Ancestry of the Copepoda. *Crustaceana* 1984;(Suppl 7):1–24.
18. Braga E, Zardoya R, Meyer A, Yen J. Mitochondrial and nuclear rRNA based copepod phylogeny with emphasis on the Euchaetidae (Calanoida). Mar Biol. 1999;133:79–90.
19. Huys R, Llewellyn-Hughes J, Olson PD, Nagasawa K. Small subunit rDNA and Bayesian inference reveal *Pectenophilus ornatus* (Copepoda incertae sedis) as highly transformed Mytilicolidae, and support assignment of Chondracanthidae and Xarifiidae to Lichomolgoidea (Cyclopoida). Biol J Linn Soc. 2006;87:403–25.
20. Huys R, Llewellyn-Hughes J, Conroy-Dalton S, Olson PD, Spinks JN, Johnston DA. Extraordinary host switching in siphonostomatoid copepods and the demise of the Monstrilloida: integrating molecular data, ontogeny and antennulary morphology. Mol Phylogenet Evol. 2007;43:368–78.
21. Tung C-H, Cheng Y-R, Lin C-Y, Ho J-S, Kuo C-H, Yu J-K, Su Y-H. A New Copepod With Transformed Body Plan and Unique Phylogenetic Position Parasitic in the Acorn Worm *Ptychodera flava*. Biol Bull. 2014;226:69–80.

22. Schizas NV, Dahms H-U, Kangtia P, Corgosinho PHC, Galindo Estronza AM. A new species of Longipedia Claus, (Copepoda: Harpacticoida: Longipediidae) from Caribbean mesophotic reefs with remarks on the phylogenetic affinities of Polyarthra. Mar Biol Res. 1863;2015(11):789–803.

23. Minxiao W, Song S, Chaolun L, Xin S. Distinctive mitochondrial genome of Calanoid copepod *Calanus sinicus* with multiple large non-coding regions and reshuffled gene order: Useful molecular markers for phylogenetic and population studies. BMC Genomics. 2011;12:73.

24. Huys R, Fatih F, Ohtsuka S, Llewellyn-Hughes J. Evolution of the bomolochiform superfamily complex (Copepoda: Cyclopoida): New insights from ssrDNA and morphology, and origin of umazuracolids from polychaete-infesting ancestors rejected. Int J Parasitol. 2012;42:71–92.

25. Dahms H-U. Exclusion of the Polyarthra from Harpacticoida and its reallocation as an underived branch of the Copepoda (Arthropoda, Crustacea). Invertebr Zool. 2004;1:29–51.

26. Regier JC, Shultz JW, Ganley ARD, Hussey A, Shi D, Ball B, Zwick A, Stajich JE, Cummings MP, Martin JW, et al. Resolving Arthropod Phylogeny: Exploring Phylogenetic Signal within 41 kb of Protein-Coding Nuclear Gene Sequence. Syst Biol. 2008;57:920–38.

27. Meusemann K, von Reumont BM, Simon S, Roeding F, Strauss S, Kück P, Ebersberger I, Walzl M, Pass G, Breuers S, et al. A Phylogenomic Approach to Resolve the Arthropod Tree of Life. Mol Biol Evol. 2010;27:2451–64.

28. Regier JC, Shultz JW, Zwick A, Hussey A, Ball B, Wetzer R, Martin JW, Cunningham CW. Arthropod relationships revealed by phylogenomic analysis of nuclear protein-coding sequences. Nature. 2010;463:1079–83.

29. von Reumont BM, Jenner RA, Wills MA, Dell'Ampio E, Pass G, Ebersberger I, Meyer B, Koenemann S, Iliffe TM, Stamatakis A, et al. Pancrustacean Phylogeny in the Light of New Phylogenomic Data: Support for Remipedia as the Possible Sister Group of Hexapoda. Mol Biol Evol. 2012;29:1031–45.

30. Oakley TH, Wolfe JM, Lindgren AR, Zaharoff AK. Phylotranscriptomics to Bring the Understudied into the Fold: Monophyletic Ostracoda, Fossil Placement, and Pancrustacean Phylogeny. Mol Biol Evol. 2013;30:215–33.

31. Mojib N, Amad M, Thimma M, Aldanondo N, Kumaran M, Irigoien X. Carotenoid metabolic profiling and transcriptome-genome mining reveal functional equivalence among blue-pigmented copepods and appendicularia. Mol Ecol. 2014;23:2740–56.

32. Eyun S, Soh HY, Posavi M, Munro J, Hughes DST, Murali SC, Qu J, Dugan S, Lee SL, Chao H, et al. Evolutionary history of chemosensory-related gene families across the Arthropoda. Mol Biol Evol. Accepted pending major revision.

33. i5K Consortium. The i5K Initiative: Advancing Arthropod Genomics for Knowledge, Human Health, Agriculture, and the Environment. J Hered. 2013;104:595–600.

34. Chipman AD, Ferrier DEK, Brena C, Qu J, Hughes DST, Schröder R, Torres-Oliva M, Znassi N, Jiang H, Almeida FC, et al. The First Myriapod Genome Sequence Reveals Conservative Arthropod Gene Content and Genome Organisation in the Centipede *Strigamia maritima*. PLoS Biol. 2014;12:e1002005.

35. Colbourne JK, Pfrender ME, Gilbert D, Thomas WK, Tucker A, Oakley TH, Tokishita S, Aerts A, Arnold GJ, Basu MK, et al. The Ecoresponsive Genome of *Daphnia pulex*. Science. 2011;331:555–61.

36. Tribolium Genome Sequencing Consortium. The genome of the model beetle and pest *Tribolium castaneum*. Nature. 2008;452:949–55.

37. Adams MD, Celniker SE, Holt RA, Evans CA, Gocayne JD, Amanatides PG, Scherer SE, Li PW, Hoskins RA, Galle RF, et al. The Genome Sequence of *Drosophila melanogaster*. Science. 2000;287:2185–95.

38. Wiegmann B, Trautwein M, Kim J-W, Cassel B, Bertone M, Winterton S, Yeates D. Single-copy nuclear genes resolve the phylogeny of the holometabolous insects. BMC Biol. 2009;7:34.

39. Altschul SF. Gapped BLAST and PSI-BLAST: a new generation of protein database search programs. Nucleic Acids Res. 1997;25:3389–402.

40. Camacho C, Coulouris G, Avagyan V, Ma N, Papadopoulos J, Bealer K, Madden TL. BLAST+: architecture and applications. BMC Bioinf. 2009;10:1–9.

41. Katoh K, Standley DM. MAFFT Multiple Sequence Alignment Software Version 7: Improvements in Performance and Usability. Mol Biol Evol. 2013;30:772–80.

42. Capella-Gutiérrez S, Silla-Martínez JM, Gabaldón T. trimAl: a tool for automated alignment trimming in large-scale phylogenetic analyses. Bioinformatics. 2009;25:1972–3.

43. Suyama M, Torrents D, Bork P. PAL2NAL: robust conversion of protein sequence alignments into the corresponding codon alignments. Nucleic Acids Res. 2006;34:W609–12.

44. Guindon S, Dufayard J-F, Lefort V, Anisimova M, Hordijk W, Gascuel O. New Algorithms and Methods to Estimate Maximum-Likelihood Phylogenies: Assessing the Performance of PhyML 3.0. Syst Biol. 2010;59:307–21.

45. Yang Z. Maximum likelihood phylogenetic estimation from DNA sequences with variable rates over sites: approximate methods. J Mol Evol. 1994;39:306–14.

46. Le SQ, Gascuel O. An Improved General Amino Acid Replacement Matrix. Mol Biol Evol. 2008;25:1307–20.

47. Abascal F, Zardoya R, Posada D. ProtTest: selection of best-fit models of protein evolution. Bioinformatics. 2005;21:2104–5.

48. Felsenstein J. Confidence limits on phylogenies: an approach using the bootstrap. Evolution. 1985;39:783–91.

49. Ronquist F, Teslenko M, van der Mark P, Ayres DL, Darling A, Höhna S, Larget B, Liu L, Suchard MA, Huelsenbeck JP. MrBayes 3.2: Efficient Bayesian Phylogenetic Inference and Model Choice across a Large Model Space. Syst Biol. 2012;61:539–42.

50. Lewis PO. A likelihood approach to estimating phylogeny from discrete morphological character data. Syst Biol. 2001;50:913–25.

51. Kishino H, Hasegawa M. Evaluation of the maximum likelihood estimate of the evolutionary tree topologies from DNA sequence data, and the branching order in hominoidea. J Mol Evol. 1989;29:170–9.

52. Shimodaira H, Hasegawa M. Multiple Comparisons of Log-Likelihoods with Applications to Phylogenetic Inference. Mol Biol Evol. 1999;16:1114–6.

53. Shimodaira H. An Approximately Unbiased Test of Phylogenetic Tree Selection. Syst Biol. 2002;51:492–508.

54. Schmidt HA, Strimmer K, Vingron M, von Haeseler A. TREE-PUZZLE: maximum likelihood phylogenetic analysis using quartets and parallel computing. Bioinformatics. 2002;18:502–4.

55. Shimodaira H, Hasegawa M. CONSEL: for assessing the confidence of phylogenetic tree selection. Bioinformatics. 2001;17:1246–7.

56. Bouckaert R, Heled J, Kühnert D, Vaughan T, Wu C-H, Xie D, Suchard MA, Rambaut A, Drummond AJ. BEAST 2: A Software Platform for Bayesian Evolutionary Analysis. PLoS Comput Biol. 2014;10:e1003537.

57. Drummond AJ, Ho SYW, Phillips MJ, Rambaut A. Relaxed Phylogenetics and Dating with Confidence. PLoS Biol. 2006;4:e88.

58. Wolfe JM, Daley AC, Legg DA, Edgecombe GD. Fossil calibrations for the arthropod Tree of Life. Earth-Sci Rev. 2016;160:43–110.

59. Zhang X-g, Maas A, Haug JT, Siveter DJ, Waloszek D. A Eucrustacean Metanauplius from the Lower Cambrian. Curr Biol. 2010;20:1075–9.

60. Lee Michael SY, Soubrier J, Edgecombe Gregory D. Rates of Phenotypic and Genomic Evolution during the Cambrian Explosion. Curr Biol. 2013;23:1889–95.

61. Giribet G, Edgecombe GD, Wheeler WC. Arthropod phylogeny based on eight molecular loci and morphology. Nature. 2001;413:157–61.

62. Strimmer K, Rambaut A. Inferring confidence sets of possibly misspecified gene trees. Proc R Soc Lond B Biol Sci. 2002;269:137–42.

63. Lozano-Fernandez J, Carton R, Tanner AR, Puttick MN, Blaxter M, Vinther J, Olesen J, Giribet G, Edgecombe GD, Pisani D: A molecular palaeobiological exploration of arthropod terrestrialization. Philos Trans R Soc Lond B Biol Sci 2016;371(1699). doi:10.1098/rstb.2015.0133.

64. Kashiyama K, Seki T, Numata H, Goto SG. Molecular Characterization of Visual Pigments in Branchiopoda and the Evolution of Opsins in Arthropoda. Mol Biol Evol. 2009;26:299–311.

65. Andrew DR, Brown SM, Strausfeld NJ. The minute brain of the copepod *Tigriopus californicus* supports a complex ancestral ground pattern of the tetraconate cerebral nervous systems. J Comp Neurol. 2012;520:3446–70.

66. Krause DW, O'Connor PM, Rogers KC, Sampson SD, Buckley GA, Rogers RR. Late Cretaceous terrestrial vertebrates from Madagascar: implications for Latin American biogeography. Ann Mo Bot Gard. 2006;93:178–208.

67. Giesbrecht W. Systematik und Faunistik der pelagischen Copepoden des Golfes von Neapel und der angrenzenden Meeres-abschnitte. Fauna Flora Golfes Neapel. 1892;19:1–831.

68. Kabata Z. Parasitic Copepoda of British Fishes. Ray Society: London, England; 1979.

69. Farris JS. The retention index and rescaled consistency index. Cladistics. 1989;5:417–9.

70. Rosenberg MS, Kumar S. Incomplete taxon sampling is not a problem for phylogenetic inference. Proc Natl Acad Sci U S A. 2001;98:10751–6.

71. Rosenberg MS, Kumar S. Taxon sampling, bioinformatics, and phylogenomics. Syst Biol. 2003;52:119–24.

72. Wiens JJ, Tiu J. Highly Incomplete Taxa Can Rescue Phylogenetic Analyses from the Negative Impacts of Limited Taxon Sampling. PLoS ONE. 2012;7:e42925.

73. Wu S, Xiong J, Yu Y. Taxonomic Resolutions Based on 18S rRNA Genes: A Case Study of Subclass Copepoda. PLoS ONE. 2015;10:e0131498.

74. Baek SY, Jang KH, Choi EH, Ryu SH, Kim SK, Lee JH, Lim YJ, Lee J, Jun J, Kwak M, et al. DNA Barcoding of Metazoan Zooplankton Copepods from South Korea. PLoS ONE. 2016;11:e0157307.

75. Blanco-Bercial L, Bradford-Grieve J, Bucklin A. Molecular phylogeny of the Calanoida (Crustacea: Copepoda). Mol Phylogenet Evol. 2011;59:103–13.

76. Rota-Stabelli O, Campbell L, Brinkmann H, Edgecombe GD, Longhorn SJ, Peterson KJ, Pisani D, Philippe H, Telford MJ. A congruent solution to arthropod phylogeny: phylogenomics, microRNAs and morphology support monophyletic Mandibulata. Proc R Soc Lond B Biol Sci. 2011;278:298–306.

77. Jenner RA. Higher-level crustacean phylogeny: Consensus and conflicting hypotheses. Arthropod Struct Dev. 2010;39:143–53.

78. Koenemann S, Jenner RA, Hoenemann M, Stemme T, von Reumont BM. Arthropod phylogeny revisited, with a focus on crustacean relationships. Arthropod Struct Dev. 2010;39:88–110.

79. Stollewerk A. The water flea *Daphnia* - a 'new' model system for ecology and evolution? J Biol. 2010;9:21.

80. Regier JC, Zwick A. Sources of Signal in 62 Protein-Coding Nuclear Genes for Higher-Level Phylogenetics of Arthropods. PLoS ONE. 2011;6:e23408.

81. Philippe H, Brinkmann H, Lavrov DV, Littlewood DTJ, Manuel M, Wörheide G, Baurain D. Resolving Difficult Phylogenetic Questions: Why More Sequences Are Not Enough. PLoS Biol. 2011;9:e1000602.

82. Harvey THP, Vélez MI, Butterfield NJ. Exceptionally preserved crustaceans from western Canada reveal a cryptic Cambrian radiation. Proc Natl Acad Sci U S A. 2012;109:1589–94.

83. Harvey THP, Pedder BE. Copepod mandible palynomorphs from the Nolichucky Shale (Cambrian, Tennessee): Implications for the taphonomy and recovery of small carbonaceous fossils. Palaios. 2013;28:278–84.

84. Cressey R, Boxshall G. *Kabatarina pattersoni*, a Fossil Parasitic Copepod (Dichelesthiidae) from a Lower Cretaceous Fish. Micropaleontol. 1989;35:150–67.

85. Cressey R, Patterson C. Fossil Parasitic Copepods from a Lower Cretaceous Fish. Science. 1973;180:1283–5.

86. Selden PA, Huys R, Stephenson MH, Heward AP, Taylor PN. Crustaceans from bitumen clast in Carboniferous glacial diamictite extend fossil record of copepods. Nat Commun. 2010;1:50.

87. Palmer AR. Miocene Copepods from the Mojave Desert, California. J Paleo. 1960;34:447–52.

88. Boxshall GA, Jaume D. Making waves: The repeated colonization of fresh water by copepod crustaceans. Adv Ecol Res. 2000;31:61–79.

13

Phylogeny, divergence time and historical biogeography of *Laetiporus* (Basidiomycota, Polyporales)

Jie Song and Bao-Kai Cui[*]

Abstract

Background: The aim of this study was to characterize the molecular relationship, origin and historical biogeography of the species in important brown rot fungal genus *Laetiporus* from East Asia, Europe, Pan-America, Hawaii and South Africa. We used six genetic markers to estimate a genus-level phylogeny including (1) the internal transcribed spacer (ITS), (2) nuclear large subunit rDNA (nrLSU), (3) nuclear small subunit rDNA (nrSSU), (4) translation elongation factor 1-α (EF-1α), (5) DNA-directed RNA polymerase II subunit 2 (RPB2), and (6) mitochondrial small subunit rDNA (mtSSU).

Results: Results of multi-locus phylogenetic analyses show clade support for at least seventeen species-level lineages including two new *Laetiporus* in China. Molecular dating using BEAST estimated the present crown group diverged approximately 20.16 million years ago (Mya) in the early Miocene. Biogeographic analyses using RASP indicated that *Laetiporus* most likely originated in temperate zones with East Asia and North America having the highest probability (48%) of being the ancestral area.

Conclusions: Four intercontinental dispersal routes and a possible concealed dispersal route were established for the first time.

Keywords: *Laetiporus*, Wood rot fungi, Phylogeny, Biogeography, Molecular clock

Background

Since the late Tertiary period, severe climatic change and major geological events have played important roles in driving species diversity and in shaping the biogeographic distribution of extant organisms. Benefiting from the development of DNA technology and molecular analysis methods, studies of fungal molecular phylogeny and biogeography have been conducted in recent decades [1–3]. Based on molecular dating, many phylogenetic studies have revealed striking chronological and geographical correlations between evolutionary divergence and geological events [3–8].

Laetiporus Murrill (Fomitopsidaceae, Polyporales) is a cosmopolitan genus, typified by *L. sulphureus* (Bull.) Murrill [9]. Species in this genus grow from cold temperate to tropical zones and are associated with Betulaceae, Burseraceae, Elaeocarpaceae, Fabaceae, Fagaceae, Meliaceae,

Myrtaceae, Oleaceae, Pinaceae, Salicaceae, Sapindaceae and Taxaceae [10–15]. *Laetiporus* spp. have been considered to be forest pathogens and to cause brown cubical heart rot [16, 17], which is implicated in the cycle of the forest ecosystem [13, 15]. Chemical composition research determined that this cultivable mushroom is a potential food due to its rich digestible bioactive substances and lack of detectable levels of poisonous microelements [18]. Some taxa of *Laetiporus* are also valuable sources of medicine, such as ergosterol and acetyl eburicoic acid [19, 20].

Recently, several studies were carried out to clarify the species diversity and phylogeny of *Laetiporus* [11–15]. In these studies, six new species were described, and four new lineages were identified: Clade I, Clade H, Clade L and Clade M. In addition, *L. sulphureus* and *L. versisporus* (Lloyd) Imazeki were shown to each be divisible into three different lineages [11–15], which are represented here as Clade C, Clade E1/E2 and Clade G1/G2/G3, respectively.

* Correspondence: baokaicui2013@gmail.com
Institute of Microbiology, Beijing Forestry University, P.O. Box 6135#, Qinghua East Road, Haidian District, Beijing 100083, People's Republic of China

To date, eleven species and four undescribed taxa of *Laetiporus* have been accepted as belonging to this genus [15]: *L. ailaoshanensis* B.K. Cui & J. Song, *L. cremeiporus* Y. Ota & T. Hatt., *L. versisporus* and *L. zonatus* B.K. Cui & J. Song from East Asia; *L. cincinnatus* (Morgan) Burds., Banik & T.J. Volk, *L. conifericola* Burds. & Banik and *L. huroniensis* Burds. & Banik from North America; *L. caribensis* Banik & D.L. Lindner from Central America; *L. montanus* Černý ex Tomšovský & Jankovský from East Asia and Europe; *L. sulphureus* from North America, South America and Europe; *L. gilbertsonii* Burds. from Pan-America; *L.* sp. 1 from Hawaii; *L.* sp. 2 from South America; *L.* sp. 3 and *L.* sp. 4 from Central America [11–15]. However, the interspecies relationships within *Laetiporus*, as well as the origin and biogeography of the genus, remain unclear.

Here, we present multi-locus phylogenetic analyses using sequences from the internal transcribed spacer (ITS), nuclear large subunit rDNA (nrLSU), nuclear small subunit rDNA (nrSSU), translation elongation factor 1-α (EF-1α), DNA-directed RNA polymerase II subunit 2 (RPB2), and mitochondrial small subunit rDNA (mtSSU) to gain insight into the evolution of species in *Laetiporus*. Our study sought to (1) explore the evolutionary relationships between *Laetiporus* species, and (2) estimate the divergence time and examine hypotheses about the origin and biogeography of *Laetiporus* species.

Results
Phylogenetic analyses
The combined dataset (ITS + nrLSU + nrSSU + mtSSU + EF-1α + RPB2) has an aligned length of 3850 characters, of which 3086 are constant, 247 are variable and parsimony uninformative, and 517 are parsimony informative. The tree obtained from the Maximum likelihood (ML) analysis and the maximum parsimony (MP), maximum likelihood (ML) and Bayesian posterior probability (BPP) values based on the dataset are shown in Fig. 1. The aligned ITS matrix comprises 514 positions, of which 378 are constant, 11 are variable and parsimony uninformative, and 125 are parsimony informative. The tree inferred from the ML analysis and the MP, ML and BPP values are shown in Fig. 2.

The combined dataset and ITS dataset inferred similar topologies (Figs. 1 and 2). The genus *Laetiporus* was supported with low levels of support on the stem branches. Moreover, 21 different phylogenetic lineages were inferred and significantly supported by both datasets.

In the combined dataset topology (Fig. 1), *L. sulphureus* was divided into three different well-supported clades: Clade C (86% MP, 87% ML, 0.99 BPP) and Clade E1 (97% MP, 58% ML, 1.00 BPP) with yellow pore surfaces and Clade E2 (98% MP, 99% ML, 1.00 BPP) with a white pore surface. *L. versisporus* was also divided into three different

clades: Clade G1 (100% MP, 87% ML, 1.00 BPP), Clade G2 and Clade G3 (100% MP, 100% ML, 1.00 BPP). Three sister lineages (*L. montanus*, *L. huroniensis* and *L. conifericola*) that grow on coniferous trees were well supported (Fig. 1). Two novel phylogenetic species from western China formed two significantly supported terminal lineages and were named Clade P (96% MP, 95% ML, 0.99 BPP) and Clade Q (100% MP, 100% ML, 1.00 BPP). Moreover, four groups were recognized (Fig. 1). Group I is well supported by Bayesian inference (BI) (0.99 BPP) and moderately supported by MP and ML analyses (73% MP, 50% ML), it is composed of two cold-temperate to subtropical *Laetiporus* species with white pore surfaces. Group II is well supported by BI (0.99 BPP) but weakly supported by MP and ML analyses, and it contains four North American, Central American and South American *Laetiporus* species. Group III was well supported by BI (0.98 BPP), moderately supported by ML analysis (50% ML) and includes four *Laetiporus* species with a disjunct distribution from Hengduan-Himalayan zones to South Africa. Group IV was supported by MP and ML analyses (77% MP, 75% ML) and only includes the East Asian species *L. versisporus* with a yellow pore surface.

The ITS dataset (Fig. 2) inferred a similar topology despite some existing differences. Clade E1 and Clade E2 clustered together and formed a novel group (Group V) with moderate support from MP and ML analyses (54% MP; 50% ML) and weak support from BI. Notably, this group was weakly supported by BI, MP and ML in the analyses using the combined dataset. The novel phylogenetic species Clade Q clustered together with Clade C and formed a novel group (Group VI) supported by MP and ML analyses (86% MP; 64% ML) but only weakly supported by BI. Moreover, of the 21 lineages identified in the phylogeny, 14 lineages (67%) have temperate distribution, 9 lineages (43%) have subtropical distribution and 9 lineages (43%) have tropical distribution (Fig. 2).

Bayesian estimation of divergence time and the historical biogeography of *Laetiporus*
The alignment of the two concatenated datasets (ITS + nrLSU + nrSSU and EF-1α + RPB2), which were 2172 and 1137 bp in length, respectively, consisted of 44 taxa. The aligned ITS dataset was 514 bp in length and was established to estimate the divergence time and biogeographical history of *Laetiporus*.

Analyses were calibrated using two methods. First, based on the divergence between Ascomycota and Basidiomycota, at 582 million years ago (Mya), *Paleopyrenomycites devonicus* Taylor, Hass, Kerp, M. Krings & Hanlin (Fig. 3) was used to estimate the divergence time of Polyporales at 194.56 ± 0.89 Mya (141.93–247.52 Mya, 95% higher posterior density (HPD)), which is consistent with a previous inference [21]. The initial diversification of

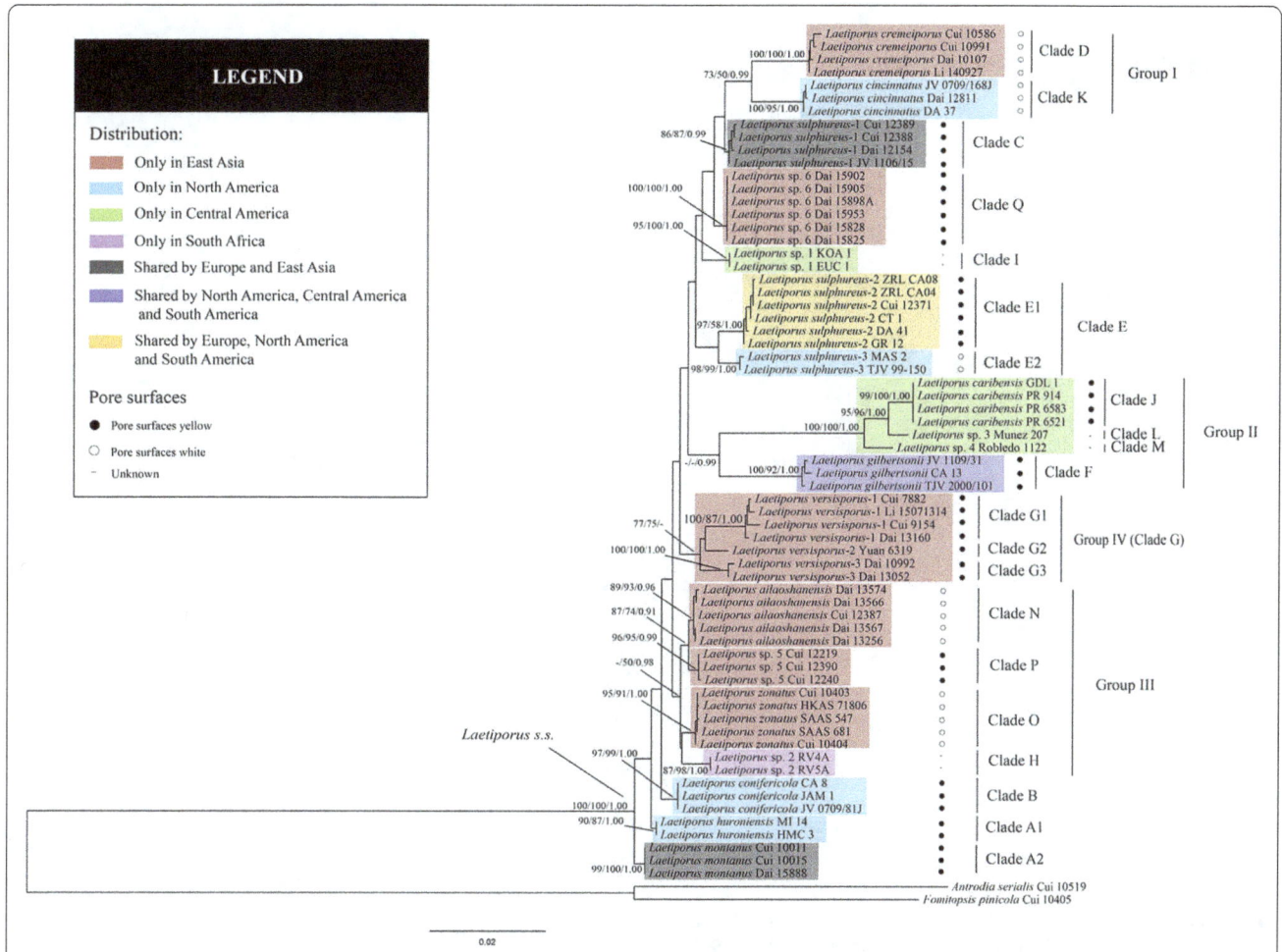

Fig. 1 Phylogenetic consensus tree inferred from the maximum likelihood (ML) analysis based on a concatenated, multi-locus dataset (ITS + nrLSU + nrSSU + mtSSU + EF-1α + RPB2). Branches are labeled where MP/BS support is greater than 75% and collapsed below that support threshold. BPP is labeled where greater than 0.90

Laetiporus occurred during the early Miocene, 20.17 ± 0.12 Mya (12.66–29.09 Mya, 95% HPD), similar to the date of the diversification of the main *Laetiporus* host plants, such as *Quercus, Salix, Populus, Abies, Picea* and *Pinus* [1, 4, 22–25]. Based on the second calibration point, *Quatsinoporites cranhamii* S.Y. Sm., Currah & Stockey, the divergence between Ascomycota and Basidiomycota was estimated to have occurred at 332.93 ± 3.03 Mya (232.23–447.89 Mya, 95% HPD), which was much more recent than the minimal divergence age of the Ascomycota/Basidiomycota (400 Mya). Meanwhile, the crown age of *Laetiporus* estimated based on the calibration point was approximately 12.26 ± 0.13 Mya (7.04–18.48 Mya, 95% HPD), which was also significantly more recent than is required for the estimated divergence time of the main host plants. Thus, the second calibration point seemed to vastly underestimate the divergence time of *Laetiporus*. Therefore, the first calibration

point was used for subsequent analyses, and the divergence times of the main nodes are showed in Fig. 3 and summarized in Additional file 1: Table S1.

The inferred historical biogeographic scenarios from analyses using RASP are shown in Fig. 4. The divergence times of the main groups based on the ITS dating analysis are also showed in Fig. 4 and summarized in Additional file 1: Table S3. The results of the Dispersal-Extinction-Cladogenesis (DEC) analysis suggest a complex biogeographic history for *Laetiporus*. Fifteen dispersal events and six vicariance events were necessary to explain the current distribution of the genus. The ancestral area of *Laetiporus* was ambiguous. In the reconstruction of their ancestral geographic range, several areas contribute to the geography in different proportions: the probability for East Asia and North America was 48%, that for Europe and North America was 42%, and that for North America was 10%. Thus, the geographic range of East Asia and North

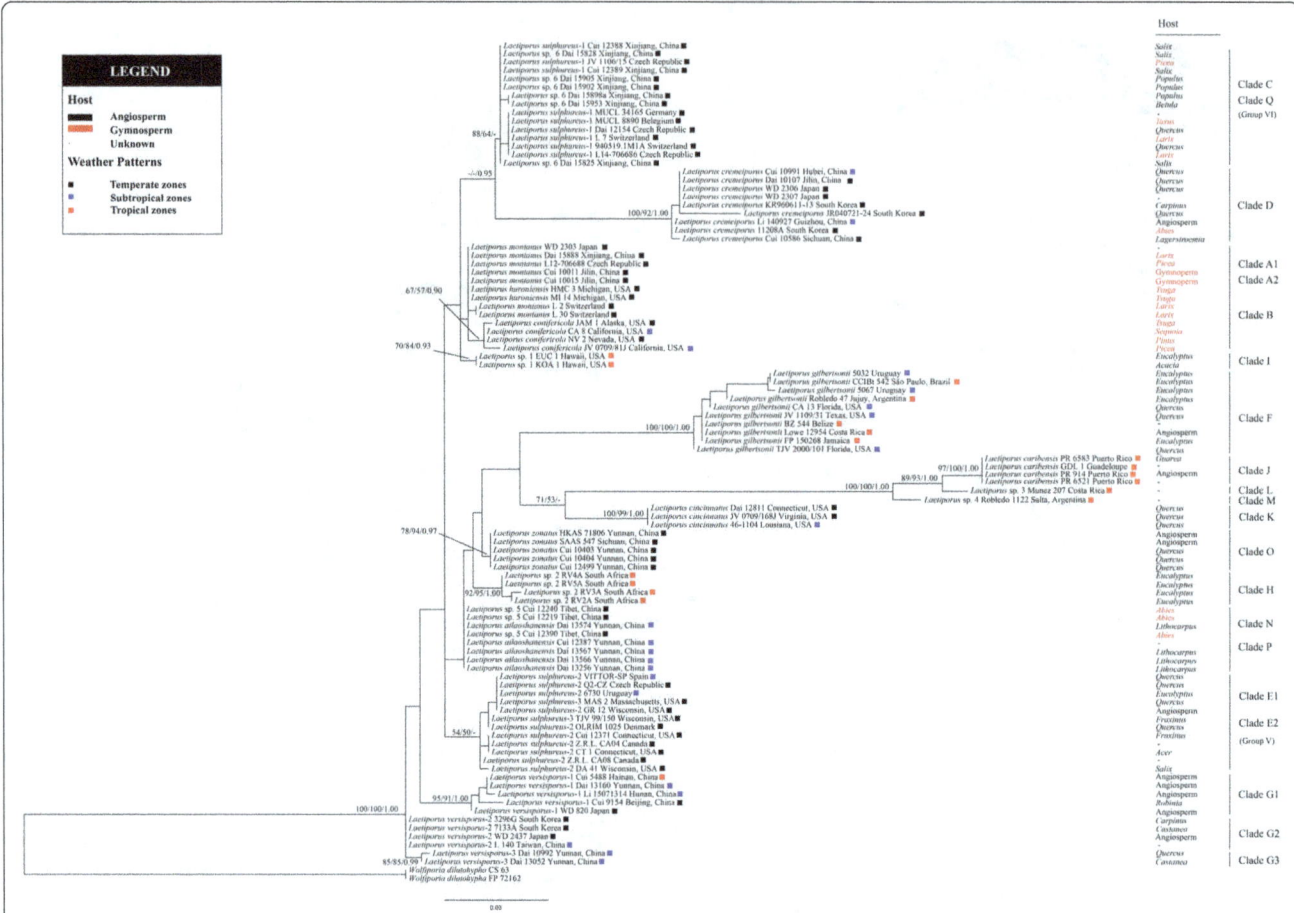

Fig. 2 Phylogenetic tree inferred from the maximum likelihood (ML) analysis based on the ITS sequences. Branches are labeled with MP/BS values if greater than 50% and with BPP values if greater than 0.90

America had the highest probability (48%) of being the ancestral area. The most probable (100%) ancestral area for Group I was East Asia and North America. The most probable (43%) ancestral area for Group II was North America and Central America. East Asia was the most probable ancestral area for Group III and Group IV, at 72% and 85%, respectively. The most probable (100%) ancestral area for Group V was North America. The most probable (74%) ancestral area for Group VI was East Asia. Furthermore, four dispersal routes and a possible concealed dispersal route were inferred: East Asia–eastern North America, North America–Central America–South America, East Asia–South Africa, East Asia–Europe and East Asia–Malay Archipelago–Australia–Hawaii (Fig. 5).

Discussion

Laetiporus has been shown to be a monophyletic group [11–15, 26]. Unexpectedly, despite conducting multi-locus phylogenetic analyses, our study is still unable to entirely resolve the stem relationships within *Laetiporus*. Nevertheless, novel phylogenetic species and certain

clustering tendencies are described. Findings regarding the origin, ancestral area and diversification are also inferred.

Group I contains two sister clades, Clade D and Clade K, with disjunct distribution (Fig. 1). Phylogenetically, this group is supported by the combined dataset analyses (73% MP, 50% ML, 0.99 BPP). However, Clade D and Clade K are distant in the ITS topology. Previous studies showed that both species grow on hardwood with the common cool temperate to subtropical habitat, producing an orange pileal surface and cream pore surface (11, 13, 15). The complete gene information, as well as a similar growth habit and morphology between Clade D and Clade K, suggests that the phylogeny inferred from the analyses of the combined dataset is more reliable.

Group II consists of four North/Central/South American *Laetiporus* clades (Fig. 1). Within this group, Clade F is known to reside in temperate to tropical areas with a Pan-American distribution [10, 12, 27]. This distribution indicates a strong adaptive ability. The other three members of *Laetiporus* that behave as sister species are

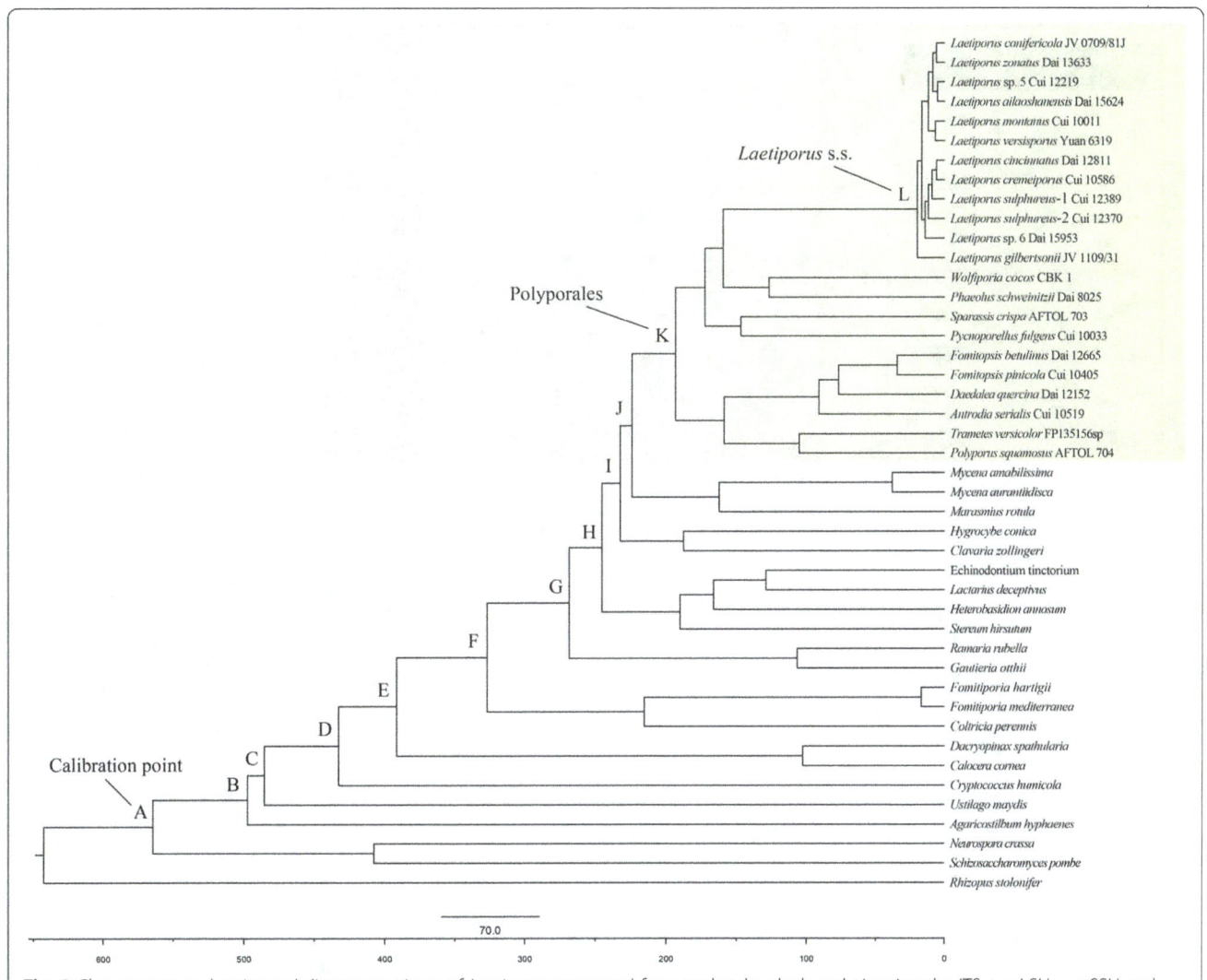

Fig. 3 Chronogram and estimated divergence times of *Laetiporus* generated from molecular clock analysis using the ITS + nrLSU + nrSSU and EF-1α + RPB2 datasets. A chronogram obtained using the Ascomycota–Basidiomycota divergence time of 582 Mya as the calibration point is shown. The calibration point and objects of this study are marked in the chronogram. The lineages in the Polyporales are highlighted in green. The geological time scale is in millions of years ago (Mya)

known to reside only in Central America [14], which is part of the Mesoamerican biodiversity hotspot [28]. Species in this group are found on hardwood and share an orange pileal surface and yellow pore surface, although the characters of Clade L and Clade M are uncertain [15]. Notably, their host plants are usually Fagaceae in North America, tropical plants such as *Guarea* and *Dacryodes* in Central America and mainly *Eucalyptus* in South America [11, 14, 27].

Group III contains four *Laetiporus* clades from East Asia and South Africa, including the novel phylogenetic species Clade P (Fig. 1). Clade P is found on *Abies* in cool temperate areas in the Himalayan region. It acts as a sister species with *L. ailaoshanensis* (Fig. 1), which has been found on *Lithocarpus* and *Castanopsis* in subtropical areas in the Hengduan Mountains [15]. Clade O is the other species

collected from the Hengduan Mountains, and it grows on *Quercus* in temperate areas [15]. Clade H is found on *Eucalyptus* from South Africa, but its characters remain unclear [12]. The relationships between Clade H and the other three species are uncertain due to the low support in the topology of the combined dataset (Fig. 1). Further studies using samples from South Africa are necessary.

Group IV consists of only *L. versisporus* (Clade G), which has a yellow pore surface (Fig. 1). Previous studies have shown that this species is usually divided into two or three clades [13, 15]. In the current study, *L. versisporus* specimens grouped together with significant support from MP and ML analyses. *L. versisporus* covers most parts of East Asia from the Yunnan-Guizhou Plateau, Hainan to Japan and South Korea, and associate with *Robinia*, *Castanea*, *Quercus*, *Elaeocarpus* and

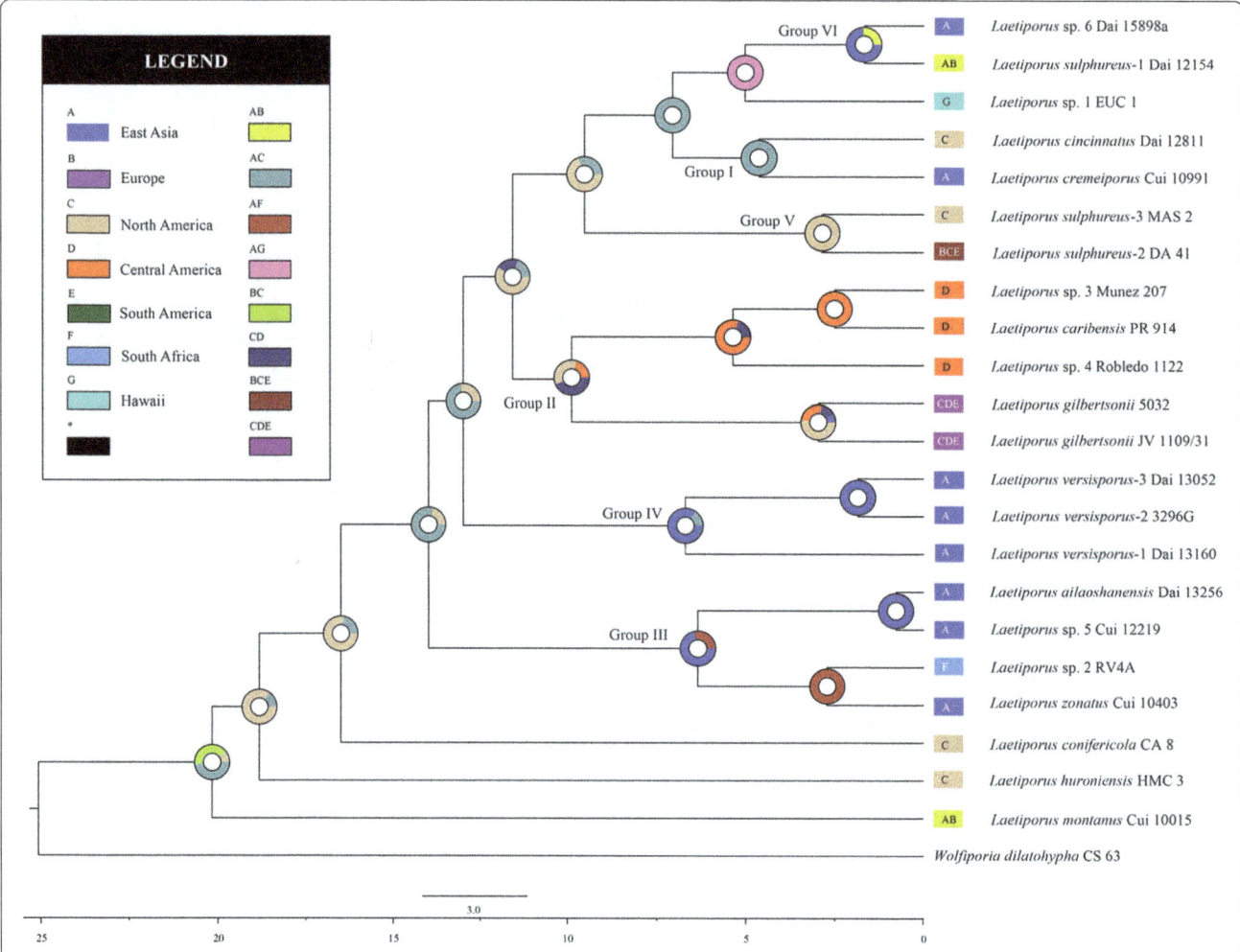

Fig. 4 Divergence time estimation and ancestral area reconstruction of *Laetiporus* using the ITS dataset. The chronogram was obtained via molecular clock analysis using BEAST. A pie chart at each node indicates the possible ancestral distributions inferred from dispersal-extinction-cladogenesis (DEC) analysis implemented in RASP. A black asterisk represents other ancestral ranges

Castanopsis [13, 15]. Infraspecific variation and infra-specific hybridization are considered to occur simultaneously [15].

Group V consists of Clade E1 and Clade E2 (Fig. 1). It is obvious that they are closely related and share similar morphology except for the pore surface [11]. Clade E1 is associated with *Quercus, Eucalyptus, Salix, Acer* and *Fraxinus* and has a disjunct temperate to subtropical areas distribution in North America, South America and Europe. Besides, it produces a yellow pore surface [10, 11]. Clade E2 is distributed in temperate areas of North America, is associated with *Quercus* and *Fraxinus*, and produces a white pore surface [10, 11].

Group VI consists of Clade C and the novel phylogenetic species Clade Q (Fig. 2). This group is only supported by the ITS phylogeny, and the phylogeny analyses do not indicate an obvious species boundary. This suggests a close relationship between Clade C and Clade Q. *Laetiporus* Clade

C has previously been reported only from Europe [11, 13]. Our study presents the first report of Clade C in Xinjiang, China. This species usually grows on hardwoods and conifers such as *Quercus, Sorbus, Populus, Castanea, Prunus, Taxus, Larix* and *Picea* in temperate areas, producing a yellow pore surface. Clade Q is also found in temperate areas in Xinjiang, China, where it is associated with hardwoods such as *Salix, Betula* and *Populus* and produces a yellow pore surface.

The maximum crown age of *Laetiporus* is estimated at the early Miocene (20.17 ± 0.12 Mya) and East Asia and North America are inferred to be the most probable ancestral areas (Figs. 3 and 4). The notable finding is that three coniferous species (*L. montanus, L. huroniensis* and *L. conifericola*) in temperate areas behave as sister species in the analyses of the combined dataset (Fig. 1). Moreover, the temperate host plants are diverse, including *Quercus, Salix, Populus, Picea, Larix, Abies, Tsuga,*

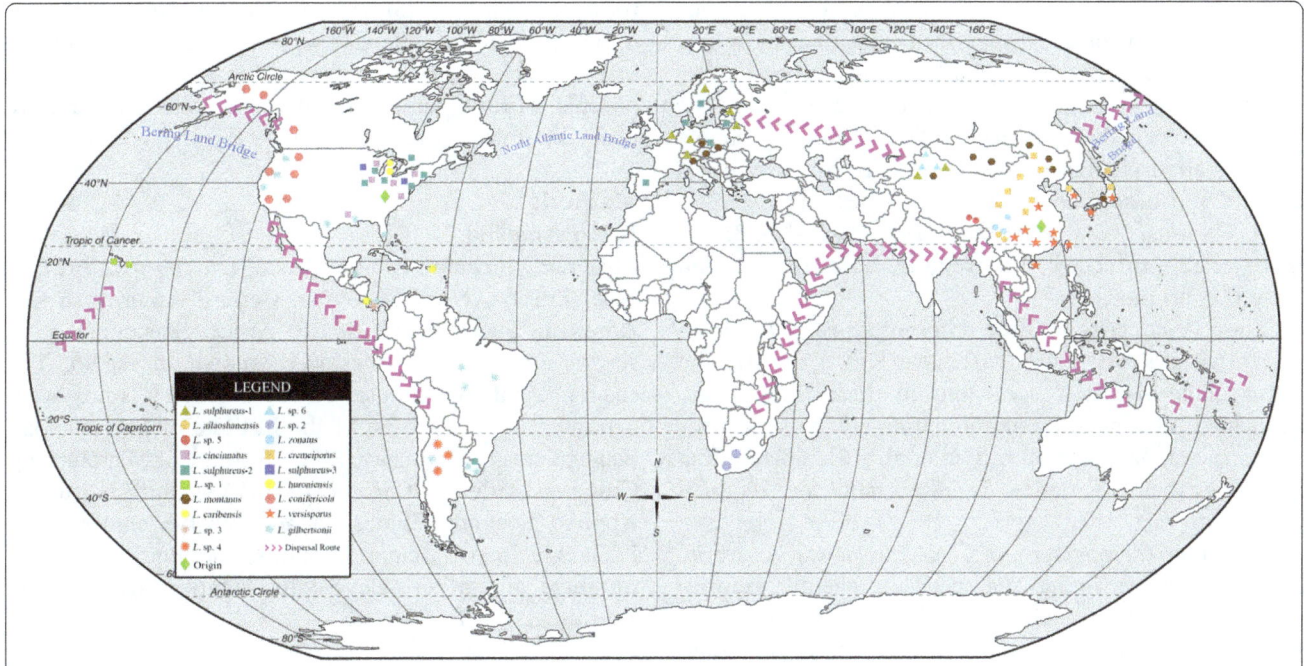

Fig. 5 Map of the geographic distribution of *Laetiporus* and possible dispersal routes generated by ArcGIS v10.1. A hypothetical schematic depiction of the original locations, the migration routes, the rapid radiation and the speciation of *Laetiporus*

Lithocarpus, *Fraxinus* and *Acer*; in contrast, the tropical host plants are limited in variety, including *Eucalyptus* and *Guarea* [10–15]. Based on these findings, an origin in temperate East Asia and North America is proposed.

The independent sister species in Group I indicate an East Asian–eastern North American dispersal route before the estimated divergence time (4.64 Mya) in the early Pliocene (Fig. 4). This divergence time is close to the break time of the Bering Land Bridge (BLB) at approximately 5.4–5.5 Mya [5]. We speculate that their ancestor covered East Asia and North America via the BLB route and that regional speciation after the vicariance emerged due to the disconnection of the BLB and the severe climate change at that time [29–31]. This route is also present in the dispersal of other organisms, especially the common host plant *Quercus* [1]. There may be a strong dispersal and vicariance correlation between *Laetiporus* spp. and their host plants.

Four *Laetiporus* species in Group II with Pan-American distribution exhibit a North American–Central American–South American dispersal route. This group first diverged at approximately 9.88 Mya. North and Central America are inferred to be the most probable ancestral areas. Clade J, Clade L and Clade M are from Central America and the estimated crown age is approximately 5.38 Mya, which coincides with the paleo-elevations that occurred during the late Miocene and early Pliocene [32]. The second intercontinental distribution between North America and South America is exhibited in Group V (Fig. 4). This route has been confirmed by biogeographical research on plants and

animals [1, 33–36]. We speculate that the severe climate change that has occurred since 15 Mya [29] drove the migration from North and Central America to South America and the adaptation to tropical host plants such as *Eucalyptus*, *Guarea* and *Dacryodes*. The vicariance due to tectonic activity is thought to be responsible for the endemism of *Laetiporus* in Central America.

In Group III, four *Laetiporus* species from East Asia and South Africa are closely related (Fig. 1). The estimated divergence time of this group is 6.35 Mya. The DEC analysis inferred East Asia as the most probable ancestral area. However, it is notable that Clade H does not form a robust sister relationship with Clade O (Fig. 1). We speculate that there is incomplete sampling from the Indian Subcontinent to Africa because suitable host plants, such as *Eucalyptus*, are abundant in these areas [12, 37]. Although the estimated divergence time is potentially inaccurate, the dispersal route between East Asia and South Africa is proposed.

The species in Group V also exhibit a continuous distribution in Europe and eastern North America (Fig. 4). The DEC analysis inferred a North American origin for this group, with an estimated divergence time of 2.89 Mya. Clade E1 is found in the eastern North America and Europe with low host-plant specificity. The short-lived North Atlantic Land Bridge acted as a dispersal route until the low Oligocene [6, 36]. Migration to Europe seems unlikely, so the reasonable interpretation is that the human activity introduced this species into new habitats as proposed by Feng et al. [3]. The wind and ocean current could be another driving force and reasonable explanation for the dispersal of fungal basidiospores between Europe and eastern North America [8].

The species in Group VI and Clade A2 have an East Asian-European dispersal route. This route is probable because an exchange of species occurs for *Laetiporus* and its most common host plants such as *Quercus*, *Salix*, *Populus*, *Picea*, *Abies* and *Larix* [1, 22–25, 38]. It is reasonable to accept this route because the Eurasian Plate is continuous.

Group IV consists of three different types of *L. versisporus* that are endemic in East Asia (Figs. 1 and 4). The infraspecific variation is obvious in these three types, but gene exchange and recombination still exist according to the clonal research of Ota et al. [13]. This finding indicates that vicariance is important for regional speciation.

In addition, the migration of Clade I to Hawaii is surprising and worth exploring. We speculate that this example results from an incomplete sampling of molecular data. However, there are many standalone islands in the South Pacific indirectly connecting Hawaii, Australia and Malay Archipelago. The frequent strong winds and continuous ocean currents are potentially responsible for the dispersal of basidiospores between islands. The humid climate and abundant host plants such as *Quercus*, *Castanea* and *Eucalyptus* from the Malay Archipelago to Australia [39, 40] are suitable for *Laetiporus*. A dispersal route of East Asia–Malay Archipelago–Australia–Hawaii seems unlikely. Interestingly, *Eucalyptus*, the host plant of Clade I has been proven to colonize Hawaii via this route [2].

In our study, the samples of *Laetiporus* are scanty in some areas around the world, such as South America, Indian Subcontinent, South Africa and Australia. The taxonomic situation is still unclear, and the evolutionary history of *Laetiporus* remains incompletely understood. A wider range of sampling and further morphological studies, incompatibility tests, and more information of host range and distribution are needed.

Conclusion

The evolutionary history of *Laetiporus* remains incompletely understood. However, this study presents some progress on this topic. (1) Two novel phylogenetic species in East Asia were identified. (2) Our reconstruction and analysis of ancestral areas suggest that *Laetiporus* originated during the early Miocene (20.16 ± 0.13 Mya) in temperate zones and that the combination of East Asia and North America has the highest probability (48%) of being the ancestral area. (3) We also predict that *Laetiporus* may be present in the Indian Subcontinent, in Australia and in the Malay Archipelago. (4) Four intercontinental dispersal routes and a possible concealed dispersal

route are proposed. (5) Vicariance is suggested to play an important role in regional speciation, and recent human activity may render some geographical distribution inexplicable. Further sampling and more molecular data are needed to further clarify the species affinity.

Methods
Taxon sampling

This study included 105 samples of *Laetiporus* from East Asia, Europe, North America, Central America, South America, Hawaii and South Africa. Basidiomata of several *Laetiporus* species were shown in Fig. 6. The sequences of the samples obtained for this study were deposited in the herbaria of the Institute of Microbiology, Beijing Forestry University (BJFC), Institute of Microbiology, Chinese Academy of Sciences (HMAS), and Institute of Applied Ecology, Chinese Academy of Sciences (IFP). Each specimen's scientific name, GenBank accession numbers and other relevant information are listed in Additional file 1: Table S2.

DNA extraction, PCR, and DNA sequencing

Genomic DNA was extracted from dried fruiting bodies using a cetyltrimethylammonium bromide rapid plant genome extraction kit (Aidlab Biotechnologies Co., Ltd., Beijing) according to the manufacturer's instructions with some modifications [26]. Six DNA gene fragments were analyzed, including those coding for RPB2 and EF-1α, along with four non-protein coding regions: ITS, nrLSU, nrSSU and mtSSU. These fragments were actually appropriate in determining the taxonomic status of *Laetiporus*. The primer pairs ITS5/4 [29], LR0R/LR7 [41], PNS1/NS41 [42], MS1/MS2 [29], and 983F/1567R [43] were used to amplify ITS, nrLSU, nrSSU, mtSSU and EF-1α, respectively. Initial attempts to amplify RPB2 using previously published primers that were designed for fungi [44] resulted in weak or non-specific amplification. To improve the success rate of RPB2 amplification, a new primer pair, 6F-1 (CCTCGTCAACTGCACAACA) and 7R-1 (TCTTCCTCGGCATCCAA), was designed based on eleven obtained sequences using Primer-Premier 5 (Premier Biosoft International, Palo Alto, CA, USA).

PCR was performed in a reaction mixture containing 25 μl of 2 × EasyTaq® PCR SuperMix, 2 μl of Forward Primer (10 μM), 2 μl of Reverse Primer (10 μM), and 2 μl of Template DNA. The total volume was adjusted to 50 μl with sterile deionized H_2O. The PCR amplifications were conducted using an Eppendorf Master Cycler (Eppendorf, Netheler-Hinz, Hamburg, Germany), and the cycling conditions were follows: pre-denaturation at 95 °C for 4 min; 35 cycles of denaturation at 94 °C for 40 s, annealing at 50 °C–54 °C for 45 s (ITS, mtSSU, EF-1α and RPB2) or for 60 s (nrLSU and nrSSU), and

Fig. 6 Basidiomata of *Laetiporus* species. **a**–*L. ailaoshanensis*. **b**–*L. cremeiporus*. **c**–*L. montanus*. **d**–*L. sulphurous*. **e**–*L. zonatus*. **f**–*L. versisporus*

elongation at 72 °C for 60 s (ITS, mtSSU, EF-1α and RPB2) or for 90 s (nrLSU and nrSSU); and a final elongation at 72 °C for 10 min. The PCR products were visualized by agarose gel electrophoresis and stored at −20 °C after visualization. The PCR products were purified and sequenced at the Beijing Genomics Institute (China) using the same primers as those used for amplification. Of the 370 sequences of *Laetiporus* used in this paper, 226 sequences of *Laetiporus* were newly generated, including 28 ITS (27% new), 27 nrLSU (40% new), 41 nrSSU (100% new), 47 mtSSU (78% new), 46 EF-1α (85% new), and 37 RPB2 (88% new). All newly generated sequences were deposited in the GenBank database.

Sequence alignments and phylogenetic analyses

To determine the phylogeny of *Laetiporus*, we compiled two datasets: the ITS sequences matrix and a concatenated dataset (ITS + nrLSU + nrSSU + mtSSU + EF-1α + RPB2). In the combined dataset, *Antrodia serialis* (Fr.) Donk and *Fomitopsis pinicola* (Sw.) P. Karst. were used as outgroups; the sequences of ITS, nrLSU, nrSSU, mtSSU, EF-1α and RPB2 were aligned initially by using MAFFT 6 [45] using the "G-INS-I" strategy and then manually optimized in BioEdit [46]. Ambiguously aligned regions were excluded from subsequent analyses. Finally, the six gene fragments were concatenated with SEAVIEW 4 [47] for further phylogenetic analyses. One thousand partition homogeneity test

(PHT) replicates of the ITS, nrLSU, nrSSU, mtSSU, EF-1α and RPB2 sequences were tested using PAUP* version 4.0b10 [48] to determine whether the partitions were homogeneous. The PHT results indicated that all the DNA sequences had a congruent phylogenetic signal (*P* value =0.19). The ITS dataset included more samples compared to the combined dataset. It contained 100 sequences, of which 98 were *Laetiporus* sequences; *Wolfiporia dilatohypha* Ryvarden & Gilb. was used as an outgroup. The sequences were aligned using the same method as that used for the combined dataset. Sequence alignments were deposited at TreeBase (submission ID 20418, 20,419; www.treebase.org).

ML analysis was conducted using RAxML-HPC2 [49] on Abe through the Cipres Science Gateway [50]. To estimate the branch support with an alternative method, we performed BI and MP analyses. For the ML and BI analyses, the optimal substitution models for ITS and the combined dataset were determined using the Akaike information criterion (AIC) as implemented in MrModeltest v2.3 [51, 52]. The selected substitution models for both the combined dataset and ITS dataset were general time reversible + proportion invariant + gamma (GTR + I + G).

In the ML analysis, the concatenated dataset was partitioned into six parts by sequence region, and 1000 ML searches were run under the GTR + GAMMA model with all model parameters estimated using the RAxML-HPC2 program. The best fit maximum likelihood tree from all searches was kept. In addition, 1000 rapid bootstrap replicates were run with the GTR + CAT model to assess the reliability of the nodes.

BI was performed using MrBayes 3.1.2 [53] with 2 independent runs, each beginning from random trees with 4 simultaneous independent chains, performing 5,000,000 replicates each for the concatenated dataset and the ITS dataset, sampling one tree every 1000 generations. Chain convergence was determined using Tracer v1.5 (http://tree.bio.ed.ac.uk/software/tracer/) to confirm sufficiently large ESS values (>200). The first 25% of the sampled trees were discarded as burn-in, and the remaining trees were used to reconstruct a majority rule consensus and calculate BPP of the clades.

MP analysis was performed in PAUP* version 4.0b10 [48]. All characters were equally weighted, and gaps were treated as missing data. Trees were inferred using the heuristic search option with TBR branch swapping and 1000 random sequence additions. Max-trees were set to 5000, branches of zero length were collapsed, and all parsimonious trees were saved. Clade robustness was assessed using a bootstrap (BT) analysis with 1000 replicates [54].

Branches of the consensus tree that received bootstrap support for MP, ML and BPP greater than or equal to 75% (MP/ML) and 0.95 (BPP) were considered to be significantly supported.

Molecular dating analysis

Given that fossil records of fungi are limited, it is difficult to choose a reliable calibration point to estimate the divergence time for any fungal groups. Therefore, extensive sampling of outgroup species for which fossils were available was performed in order to estimate the divergence time of *Laetiporus*. Two primary calibration points were included in our analyses: (1) the divergence between Ascomycota and Basidiomycota, 582 Mya, by placing *P. devonicus* in the subphylum Pezizomycotina [55]; and (2) the divergence between Hymenochaetaceae and Fomitopsidaceae based on a 125 million-year-old fossil of *Q. cranhamii* [56]. The parameter settings for the two calibrations were the same as those used in Feng et al. [3]. As the identifications of the two fossils were fairly ambiguous, the estimated divergence time was constrained by the following two values: the estimated divergence time between Ascomycota and Basidiomycota is at least 400 Mya (the divergence time of *P. devonicus*), and the initial diversification of *Laetiporus* should be close to the divergence times of their host plants as suggested by the co-evolution of the fungi and the plants [57]. The calibration point for which the estimated

results met these two criteria was eventually chosen for our subsequent analyses.

Three nuclear ribosomal RNA genes (ITS, nrLSU and nrSSU) and two protein coding genes (EF-1α and RPB2) were concatenated for molecular dating using the phylogenetic framework described in James et al. [58]. ITS1, ITS2, and the introns in EF-1α and RPB2 were excluded for a conservation analysis. All of the outgroup sequences were retrieved from GenBank and are listed in Additional file 1: Table S2. MrModeltest v2.3 [51, 52] was used to select the best models of evolution using the hierarchical likelihood ratio test. The selected evolutionary models for the two combined datasets were GTR + I + G. The origin time of *Laetiporus* was estimated in BEAST v1.8.0 [59] with the molecular clock and substitution models unlinked but with the trees linked for each gene partition. The BEAST input files were constructed using BEAUti (within BEAST). The lognormal relaxed molecular clock and the Yule speciation prior set were used to estimate the divergence time and the corresponding credibility intervals. The posterior distributions of parameters were obtained using MCMC analysis for 50 million generations with a burn-in percentage of 10%. The convergence of the chains was confirmed using Tracer v1.6. Samples from the posterior distributions were summarized on a maximum clade credibility tree with the maximum sum of posterior probabilities listed on its internal nodes using TreeAnnotator v1.8.0 [59] with the posterior probability limits set to 0.5 to summarize the mean node heights. FigTree v1.4.2 [60] was used to visualize the resulting tree and to obtain the means and 95% HPD [59]. A 95% HPD marks the shortest interval that contains 95% of the values sampled.

We also estimated the divergence time of the main nodes in *Laetiporus* using the ITS dataset, which contained one or two representatives of all of the *Laetiporus* species that were included in our analyses. The estimated crown age of *Laetiporus* based on the combined ITS + nrLSU + nrSSU and EF-1α + RPB2 datasets was used as the calibration point to date the ITS phylogeny by setting the prior to a normal distribution. The other procedures were the same as those applied in the estimation using the combined dataset.

Biogeographic analysis

The reconstruction of ancestral areas in a phylogeny is important for understanding the biogeographic diversification history of a lineage, as this reconstruction makes it possible to infer the original location and dispersal routes of the organisms. To infer ancestral areas, the DEC [61] model was used in RASP 3.2 [62], allowing a maximum of two areas per node. The ancestral area analyses were conducted using the posterior distributions of the dated ITS phylogeny that were estimated from BEAST. The geographic distributions for the *Laetiporus* were delimited into seven areas: (A) East Asia, (B) Europe, (C) North America, (D) Central America, (E) South America, (F) South Africa, and (G) Hawaii. ArcGIS v10.1 [63] was used to visualize the geographic distribution and possible dispersal routes of *Laetiporus*.

Abbreviations

BI: Bayesian inference; BJFC: Herbaria of the Institute of Microbiology, Beijing Forestry University; BPP: Bayesian posterior probabilities; DEC: Dispersal-Extinction-Cladogenesis; EF-1α: Translation elongation factor 1-α; GTR + I + G: General time reversible + proportion invariant + gamma; HMAS: Institute of Microbiology, Chinese Academy of Sciences; HPD: Higher posterior densities; IFP: Institute of Applied Ecology, Chinese Academy of Sciences; ITS: Internal transcribed spacer; ML: Maximum likelihood; MP: Maximum parsimony; mtSSU: Mitochondrial small subunit rDNA; Mya: Million years ago; nrLSU: Nuclear large subunit rDNA; nrSSU: Nuclear small subunit rDNA; RPB2: DNA-directed RNA polymerase II subunit 2

Acknowledgments

The authors extend special thanks to Drs. Shuang-Hui He (BJFC, China) and Xiao-Lan He (SAAS, China) for collecting specimens, as well as to Drs. Mario Rajchenberg (CIEFAP, Argentina) and Michal Tomšovský (BRNU, Czech Republic) for forwarding specimens for our study. We thank Drs. Beatriz Ortiz-Santana (CFMR, USA) and Wanda Daley (PDD, New Zealand), Mr. Pertii Salo (H, Finland) and the curator of the Queensland Herbarium (BRI, Australia) for the loaning of specimens.

Funding

This research was financed by the National Natural Science Foundation of China (Project Nos. 31422001, 31170018) and the Fundamental Research Funds for the Central Universities (No. 2016ZCQ04).

Authors' contributions

BK C and J S designed the experiment; JS conducted the molecular experiments; and J S and BK C analyzed the data and drafted the manuscript. All of the authors approved the manuscript.

Competing interests

The authors declare that they have no competing interests.

References

1. Manos PS, Stanford AM. The historical biogeography of Fagaceae: tracking the tertiary history of temperate and subtropical forests of the northern hemisphere. Int J Plant Sci. 2001;162:S77–93.
2. Percy DM, Garver AM, Wagner WL, James HF. Progressive island colonization and ancient origin of Hawaiian Metrosideros (Myrtaceae). Proc R Soc B. 2008;275:1479–90.
3. Feng B, Xu JP, Wu G, Zeng NK, Li YC, Tolgor B, et al. DNA sequences analyses reveal abundant diversity, endemism and evidence for Asian origin of the porcini. PLoS One. 2012;7:e37567.
4. Eckert AJ, Hall BD. Phylogeny, historical biogeography, and patterns of diversification for *Pinus* (Pinaceae): phylogenetic tests of fossil-based hypotheses. Mol Phylogenet Evol. 2006;40:166–82.
5. Milne RI. Northern hemisphere plant disjunctions: a window on tertiary land bridges and climate change? Ann Bot. 2006;98:465–72.
6. Cai Q, Tulloss RE, Tang LP, Tolgor B, Zhang P, Chen ZH, et al. 2014. Multilocus phylogeny of lethal amanitas: implications for species diversity and historical biogeography. BMC Evol Biol. 2014;14:143.
7. Chen JJ, Cui BK, Zhou LW, Korhonen K, Dai YC. Phylogeny, divergence time estimation, and biogeography of the genus *Heterobasidion* (Basidiomycota, Russulales). Fungal Divers. 2015;71:185–200.
8. Song J, Chen JJ, Wang M, Chen YY, Cui BK. Phylogeny and biogeography of the remarkable genus *Bondarzewia* (Basidiomycota, Russulales). Sci Rep. 2016;6:34568.
9. Murrill WA. The Polyporaceae of North America 9 *Inonotus*, *Sesia* and monotypic genera. Bull Bot Club. 1904;31:593–610.
10. Burdsall HH, Banik MT. The genus *Laetiporus* in North America. Harv Pap Bot. 2001;6:43–55.
11. Lindner DL, Banik MT. Molecular phylogeny of *Laetiporus* and other brown-rot polypore genera in North America. Mycologia. 2008;100:417–30.
12. Vasaitis R, Menkis A, Lim YW, Seok S, Tomšovský M, Jankovský L, et al. Genetic variation and relationships in *Laetiporus sulphureus* s. Lat., as determined by ITS rDNA sequences and in vitro growth rate. Mycol Res. 2009;113:326–36.
13. Ota Y, Hattori T, Banik MT, Hagedorn G, Sotome K, Tokuda S, et al. The genus *Laetiporus* (Basidiomycota, Polyporales) in East Asia. Mycol Res. 2009; 113:1283–300.
14. Banik MT, Lindner DL, Ortiz-Santana B, Lodge DJ. A new species of *Laetiporus* (Basidiomycota, Polyporales) from the Caribbean basin. Kurtziana. 2012;37:15–21.
15. Song J, Chen YY, Cui BK, Liu HG, Wang YZ. Morphological and molecular evidence for two new species of *Laetiporus* (Basidiomycota, Polyporales) from southwestern China. Mycologia. 2014;106:1039–50.
16. Dai YC, Cui BK, Yuan HS, Li BD. Pathogenic wood-decaying fungi in China. Forest Pathol 2007;37: 105–120.
17. Sinclair W A, Lyon HH, Johnson WT. Diseases of trees and shrubs. Cornell University Press; 1987.p. 251–306.
18. Kovács D, Vetter J. Chemical composition of the mushroom *Laetiporus sulphureus* (Bull.) Murill. Acta Aliment Hung. 2015;44:104–10.
19. Saba E, Son Y, Jeon BR, Kim S, Lee I, Yun B, et al. Acetyl eburicoic acid from *Laetiporus sulphureus* var. *miniatus* suppresses inflammation in murine macrophage RAW 264.7 cells. Mycobiology. 2016;43:131–6.
20. Martinez M, Torrez AS, Campi MG, Bravo JA, Vila JL. Ergosterol from the mushroom *Laetiporus* sp.: isolation and structural characterization. Revista Boliviana de Química. 2015;32:90–4.
21. Floudas D, Binder M, Riley R, Barry K, Blanchette RA, Henrissat B, et al. The Paleozonic origin of enzymatic lignin decomposition reconstructed from 31 fungal genomes. Science. 2012;336:1715–9.
22. Ran JH, Wei XX, Wang XQ. Molecular phylogeny and biogeography of *Picea* (Pinaceae): implications for phylogeographical studies using cytoplasmic haplotypes. Mol Phylogenet Evol. 2006;41:405–19.
23. Aguirre-Planter É, Jaramillo-Correa JP, Gómez-Acevedo S, Khasa DP, Bousquet J, Eguiarte LE. Phylogeny, diversification rates and species boundaries of Mesoamerican firs (Abies, Pinaceae) in a genus-wide context. Mol Phylogenet Evol. 2012;62:263–74.
24. Cronk QCB, Needham I, Rudall PJ. Evolution of catkins: inflorescence morphology of selected Salicaceae in an evolutionary and developmental context. Front Plant Sci. 2015;6:1030.
25. Du SH, Wang ZS, Ingvarsson P, Wang DS, Wang JH, Wu ZQ, et al. Multilocus analysis of nucleotide variation and speciation in three closely related *Populus* (Salicaceae) species. Mol Ecol. 2015;24:4994–5005.
26. Han ML, Chen YY, Shen LL, Song J, Vlasák J, Dai YC, et al. Taxonomy and phylogeny of the brown-rot fungi: *Fomitopsis* and its related genera. Fungal Divers. 2016;80:343–73.
27. Pires RM, Motato-Vásquez V, Gugliotta ADM. A new species of *Laetiporus* (Basidiomycota) and occurrence of *L. gilbertsonii* Burds. In Brazil. Nova Hedwigia. 2016;102:477–90.
28. Harvey CA, Komar O, Chazdon RL, Ferguson BG, Finegan B, Griffith DM, et al. Integrating agricultural landscapes with biodiversity conservation in the Mesoamerican hotspot. Conserv Biol. 2008;22:8–15.
29. White TJ, Bruns TD, Lee S, Taylor J. Amplification and direct sequencing of fungal ribosomal RNA genes for phylogenetics. In: Innis MA, Gelfand DH, Sninsky JJ, White TJ, editors. PCR protocols, a guide to methods and applications. California Academic Press: San Diego; 1990. p. 315–22.
30. Tiffney BH, Manchester SR. The use of geological and paleontological evidence in evaluating plant phylogeographic hypotheses in the northern hemisphere tertiary. Int J Plant Sci. 2001;162:S3–S17.

31. Zachos J, Pagani M, Sloan L, Thomas E, Billups K. Trends, rhythms, and aberrations in global climate 65 ma to present. Science. 2001;292:686–93.

32. Burnham RJ, Graham A. The history of neotropical vegetation: new developments and status. Ann Mo Bot Gard. 1999;86:546–89.

33. Cortés-Ortiz L, Bermingham E, Rico C, Rodriguez-Luna E, Sampaio I, Ruiz-Garcia M. Molecular systematics and biogeography of the neotropical monkey genus, Alouatta. Mol Phylogenet Evol. 2003;26:64–81.

34. Dick CW, Abdul-Salim K, Bermingham E. Molecular systematic analysis reveals cyptic tertiary diversification of a widespread tropical rain forest tree. Am Nat. 2003;162:691–703.

35. Morley RJ. Interplate dispersal paths for megathermal angiosperms. Perspect Plant Ecol. 2003;6:5–20.

36. Zamora-Tavares MDP, Martínez M, Magallón S, Guzmán-Dávalos L, Vargas-Ponce O. Physalis and physaloids: a recent and complex evolutionary history. Mol Phylogenet Eovl. 2016;100:41–50.

37. Crame JA, Owen AW. Palaeobiogeography and biodiversity change: the Ordovician and Mesozoic-Cenozoic radiations. Geological society special publication; 2002

38. Cerling TE, Harris JM, MacFadden BJ, Leakey MG, Quade J, Eisenmann V, et al. Global vegetation change through the Miocene/Pliocene boundary. Nature. 1997;389:153–8.

39. Hopper SD, Gioia P. The southwest Australian floristic region: evolution and conservation of a global hot spot of biodiversity. Annu Rev Ecol Evol S. 2004;35:623–50.

40. Swee-Hock S. The population of Malaysia. Institute of Southeast Asian Studies; 2007.

41. Vilgalys R, Hester M. Rapid genetic identification and mapping of enzymatically amplified ribosomal DNA from several Cryptococcus species. J Bacteriol. 1990;172:4238–46.

42. Hibbett DS. Phylogenetic evidence for horizontal transmission of group I introns in the nuclear ribosomal DNA of mushroom-forming fungi. Mol Biol Evol. 1996;13:903–17.

43. Rehner SA, Buckley E. A Beauveria phylogeny inferred from nuclear ITS and EF-1α sequences: evidence for cryptic diversification and links to Cordyceps teleomorphs. Mycologia. 2005;97:84–98.

44. Liu YJ, Wheelen S, Hall BD. Phylogenetic relationships among ascomycetes: evidence from an RNA polymerase II subunit. Mol Biol Evol. 1999;16:1799–808.

45. Katoh K, Toh H. Recent developments in the MAFFT multiple sequence alignment program. Brief Bioinform. 2008;9:286–98.

46. Hall TA. Bioedit: a user-friendly biological sequence alignment editor and analysis program for windows 95/98/NT. Nucleic Acids Symp Ser. 1999;41:95–8.

47. Gouy M, Guidon S, Gascuel O. SeaView version 4: a multiplatform graphical user interface for sequence alignment and phylogenetic tree building. Mol Biol Evol. 2010;27:221–4.

48. Swofford DL. PAUP*: phylogenetic analysis using parsimony (*and other methods), Version 4.0b10. Sunderland: Sinauer Associates; 2002.

49. Stamatakis A. RAxML-VI-HPC: maximum likelihood-based phylogenetic analyses with thousands of taxa and mixed models. Bioinformatics. 2006;22: 2688–90.

50. Miller MA, Pfeiffer W, Schwartz T. Creating the CIPRES Science Gateway for inference of large phylogenetic trees. New Orleans: Proceedings of the Gateway Computing Environments Workshop (GCE); 2010. p. 1–8.

51. Posada D, Crandall KA. Modeltest: testing the model of DNA substitution. Bioinformatics. 1998;14:817–8.

52. Nylander JAA. MrModeltest v2. Program distributed by the author: Evolutionary Biology Centre, Uppsala University; 2004.

53. Ronquist F, Huelsenbeck JP. Mrbayes 3: Bayesian phylogenetic inference under mixed models. Bioinformatics. 2003;19:1572–4.

54. Felsenstein J. Confidence limits on phylogenetics: an approach using bootstrap. Evolution. 1985;39:783–91.

55. Berbee ML, Taylor JW. Dating the molecular clock in fungi – how close are we? Fungal Biol Rev. 2010;24:1–16.

56. Smith SY, Currah RS, Stockey RA. Cretaceous and Eocene poroid hymenophores from Vancouver Island, British Columbia. Mycologia. 2004;96:180–6.

57. Hibbett DS, Matheny PB. The relative ages of ectomycorrhizal mushrooms and their plant hosts estimated using Bayesian relaxed molecular clock analyses. BMC Biol. 2009;7:13.

58. James TY, Kauff F, Schoch CL, Matheny PB, Hofstetter V, Cox CJ, et al. Reconstructing the early evolution of fungi using a six-gene phylogeny. Nature. 2006;443:818–22.

59. Drummond AJ, Rambaut A. BEAST: Bayesian evolutionary analysis by sampling trees. BMC Evol Biol. 2007;7:214–21.

60. Molecular Evolution, Phylogenetics and epidemiology. 2007. http://tree.bio.ed.ac.uk/software/figtree/. Accessed 20 Sep 2016.

61. Ree RH, Smith SA. Maximum likelihood inference of geographic range evolution by dispersal, local extinction, and cladogenesis. Syst Biol. 2008;57:4–14.

62. Yu Y, Harris AJ, Blair C, He XJ. RASP (reconstruct ancestral state in phylogenies): a tool for historical biogeography. Mol Phylogenet Evol. 2015;87:46–9.

63. ArcGIS Platform. http://esri.com/arcgis. Accessed 18 Sep 2016.

14

Two new genera of songbirds represent endemic radiations from the Shola Sky Islands of the Western Ghats, India

V.V. Robin[1,2*], C. K. Vishnudas[1], Pooja Gupta[1], Frank E. Rheindt[3], Daniel M. Hooper[4], Uma Ramakrishnan[1] and Sushma Reddy[1,5*]

Abstract

Background: A long-standing view of Indian biodiversity is that while rich in species, there are few endemics or in-situ radiations within the subcontinent. One exception is the Western Ghats biodiversity hotspot, an isolated mountain range with many endemic species. Understanding the origins of the montane-restricted species is crucial to illuminate both taxonomic and environmental history.

Results: With evidence from genetic, morphometric, song, and plumage data, we show that two songbird lineages endemic to the Western Ghats montane forest each have diversified into multiple distinct species. Historically labeled as single species of widespread Asian genera, these two lineages are highly divergent and do not group with the taxa in which they were previously classified but rather are distinct early divergences in larger Asian clades of flycatchers and babblers. Here we designated two new genera, the Western Ghats shortwings as *Sholicola* and the laughingthrushes as *Montecincla*, and evaluated species-limits to reflect distinct units by revising six previously named taxa and describing one novel species. Divergence dating showed that both these montane groups split from their Himalayan relatives during the Miocene, which is coincident with a shift towards arid conditions that fragmented the previously contiguous humid forest across peninsular India and isolated these lineages in the Western Ghats. Furthermore, these two genera showed congruent patterns of diversification across the Western Ghats Sky Islands, coincident with other climatic changes.

Conclusion: Our study reveals the existence of two independent endemic radiations in the high montane Western Ghats or Shola Sky Islands with coincident divergence times, highlighting the role of climate in the diversification of these ancient lineages. The endemic and highly divergent nature of these previously unrecognized species underscores the dearth of knowledge about the biogeography of the Asian tropics, even for comparatively well-known groups such as birds. The substantial increase in the diversity of this region underscores the need for more rigorous systematic analysis to inform biodiversity studies and conservation efforts.

Keywords: Phylogenetics, Birds, Shola, Passerine, Montane, Sky-islands, Taxonomy, Tropics

* Correspondence: robinvijayan@gmail.com; sreddy6@luc.edu
[1]National Centre for Biological Sciences, TIFR, Bellary Road, Bangalore 560065, India
Full list of author information is available at the end of the article

Background

The Western Ghats (WG), an isolated coastal mountain chain in the southwest of India, is a global biodiversity hotspot [1]. Despite a long history of human occupation, knowledge of this biodiversity remains poor [2, 3]. Modern systematic analyses to assess species distinctiveness and their responses to past climatic events are urgently needed to inform conservation efforts in such montane tropical systems as the Western Ghats, where diversity and threat levels are high [1, 4, 5].

In most current avian taxonomic treatments, WG lineages are a subset of the diverse avifaunal groups in the Himalayas and Southeast Asia. WG endemics are usually circumscribed as single species of larger Asian groups with limited differentiation across the WG mountain range [6–8]. Contrary to this traditional view, the hitherto first and only phylogenetic investigation of a WG endemic songbird challenged both these ideas – it revealed considerable genetic divergence between populations across the mountain range as well as from its congeners, rendering the traditional genus non-monophyletic [9]. The Indian subcontinent underwent dramatic climatic changes during the Cenozoic, which may have influenced species dispersal to and diversification within the WG [10, 11]. The long-favored 'Satpura Hypothesis' [12–16] suggested a specific colonization route for species to disperse from the Himalayas to the WG through the Satpura Hills, a narrow band of wet forest across central India. In the Miocene, peninsular India was much more humid with near continuous forest cover and since this period, changing climate conditions and local tectonic events led to gradual drying of northwestern and central India and the establishment of the distinct patches of wet-zone forested regions on the highlands [15, 17]. Until now, support of this hypothesis was mainly from observations of avian species distributions [6] due to the dearth of phylogenetic analyses of Indian birds [3].

The peaks of the WG range host a unique form of tropical montane cloud forest known as Shola, a natural matrix of forests and grasslands [18]. The Shola habitat is restricted to the highest elevation zone and is characterized by high rainfall, humidity and low temperatures relative to lower elevations [18, 19]. A variety of endemic taxa are exclusively found in this habitat [18, 19]. These species often have disjunct distributions across the WG mountain-tops, which have been likened to 'islands' of specific habitat and microclimatic conditions in an 'ocean' of unsuitable habitat [20]. While there are several classic examples of speciation and adaptive radiations across oceanic islands, only a few studies have examined similar patterns across montane or sky islands [2].

Two endemic avian species groups in the Western Ghats sky islands, the Kerala or Black-chinned

Laughingthrushes (formerly placed in *Garrulax*, currently *Trochalopteron* [8, 21] or *Strophocincla* [7, 22]) and the Western Ghats Shortwing (alternatively placed in *Brachypteryx* [8, 21], *Myiomela* [7, 23] or *Callene* [24]) have been surrounded by taxonomic confusion, complicating studies of their diversity. The laughingthrushes exhibit striking plumage variation across the different isolated sky island peaks and have been alternatively considered as one [6, 25], two [7, 8, 21, 22], or four [26] species. The shortwings have been described as one [6, 21, 25] or two [7, 8] species. Previous phylogenetic analysis for some populations of WG Shortwings showed deep divergences across the sky island complex [9].

Elucidating the origins of montane species is crucial to illuminate the evolutionary and environmental history of this landscape [2]. We reconstructed the phylogenies of the WG laughingthrush and shortwing complexes to determine their evolutionary history and test hypotheses of diversification in relation to past climatic events. Furthermore, we examined discrete variation across populations of these two lineages using multiple types of data – genetic, song, plumage and morphometric – to determine species-limits and assess differentiation within the WG range.

Methods

Sample collection

From January 2012 to May 2013 we conducted expeditions to survey and collect samples across the entire distributional range of both species complexes (Additional file 1: Figure S1). We followed Robin et al. [9] for field sampling techniques to capture birds with multiple 12m * 2 m mist-nets and collect blood samples from the brachial vein in Queen's lysis buffer. For one location where field sampling proved difficult (Bababudan Hills) for the shortwings, we used two museum samples from the Natural History Museum (Tring, UK; NHMUK) for DNA analysis.

DNA sequencing

We extracted DNA (using the Qiagen Blood and Tissue Extraction Kit) and used standard procedures [9, 27] to sequence 26 and 31 individuals of the laughingthrush and the shortwing complexes, respectively, across their entire distributions (Additional file 1: Tables S1 and S2). We generated sequence data from five and eight loci, respectively, for both groups to match published phylogenies of their relatives [27–29]. For the laughingthrushes, we sequenced 5 loci: cytochrome b (CYTB), NADH dehydrogenase subunit-3 (ND3), the fifth intron of nuclear b-fibrinogen (FIB5), the third intron of the muscle-specific kinase Gene (MUSK), and the fifth intron of transforming growth factor β2 (TGF) using standard primers [30] and standard PCR procedures (see [9, 30, 31]). For shortwings, we sequenced 8 loci: NADH dehydrogenase subunit-2 (ND2), ND3, CYTB, cytochrome c oxidase 1 (CO1), intron 2 of myoglobin (MYO), introns 6 and 7 of ornithine decarboxylase (ODC), intron 11 of the glyceraldehyde-3-phosphodehydrogenase (GAPDH), and intron 3 of lactate dehydrogenase (LDH). DNA sequences were assembled, annotated, and aligned using Geneious 6.1.4.

Phylogenetic analyses
WG laughingthrushes
To reconstruct the evolutionary relationships of the WG Laughingthrushes, we re-analyzed a larger clade encompassing laughingthrushes in general (Leiothrichidae) as per recent studies on sylvioid songbirds [27, 32, 33]. We assembled a matrix by incorporating representatives of key members of other laughingthrush species with our data from all lineages found in the Western Ghats. As outgroups, we included one representative of each of the other major clades of babblers [27]. We assembled two matrices of five loci each: to examine divergences within the WG Laughingthrush complex, we included all 26 WG individuals and several other babblers as outgroups; and to examine the relationships of the WG Laughingthrushes with other babblers, we assembled a matrix comprising 73 taxa (Additional file 1: Table S1). PartitionFinder v1.01 [34] determined that the best partitioning scheme of gene regions divided by rates of evolution was four partitions: 1st position CYTB + 1st position ND3, 2nd position CYTB + 2nd position ND3, 3rd position CYTB + 3rd position ND3, FIB5 + MUSK + TGF.

WG shortwings
Preliminary analyses found that shortwings belong within the flycatcher/chat complex. To examine variation within the WG Shortwings, we assembled a phylogenetic matrix of 34 taxa for four loci: ND2, ND3, CYTB, CO1 (Additional file 1: Table S2). To examine their placement in a broader phylogenetic context, we compiled data for species across a larger clade based on other published studies [28, 29, 35] and included all distinct WG Shortwing lineages for a total of 96 species for six loci – ND2, CYTB, MYO, ODC, GAPDH, and LDH (Additional file 1: Table S2). The best partitioning scheme for this dataset according to PartitionFinder was 8 partitions: 1st position CYTB, 2nd position CYTB, 3rd position CYTB, 1st position ND2, 2nd position ND2, 3rd position ND2, MYO, and GAPDH + LDH + ODC.

For both groups, we conducted maximum likelihood (ML) phylogenetic analyses using RAxML 8 [36] by partitioning genes according to similar evolutionary rates determined with PartitionFinder. We conducted 1000 random bootstrap replicates that were subsequently used to search for the best ML tree. We compared relationships from this analysis to those using different

optimality criteria such as maximum parsimony, using PAUP* [37], and Bayesian inference, using MrBayes 3.2 [38]. In PAUP*, we treated all characters with equal weights and ran heuristic searches of 1000 random addition replicates. In MrBayes, we used the same partition scheme as ML analyses and ran two Monte Carlo Markov Chain runs of four chains each for 20 million generations, sampling every 500th generation. We used default priors and unlinked parameters across partitions except for branch length calculations. We assessed convergence and stationarity of runs using Tracer v1.6 [39] and AWTY [40], discarding the first 5000 generations of each run as burn-in.

Divergence dating
We used BEAST v1.75 [41] to estimate the timing of lineage divergence. For both groups, we used one representative of each distinct lineage for divergence dating analysis. In each analysis, we unlinked substitution and clock rates, and linked tree models for each locus. To calibrate the laughingthrush/babbler phylogeny, we used a secondary calibration for the divergence time between the families Timaliidae, Pellorneidae + Leiotrichidae as 20.92 Ma (standard deviation of 2.11 Ma), taken from [29], a recent time-calibrated phylogeny of Asian passerines using 13 corroborated clade ages based on fossil and biogeographic calibrations (see Additional file 1 in [29]). To calibrate the shortwing/flycatcher tree, we used the same source to date the split between the families Muscicapidae and Turdidae as 21 Ma (standard deviation of 2 Ma) [29]. We understand that dates for the bird tree of life can be controversial but we chose this source [29] because it was the most densely-sampled, recent phylogeny for taxa relevant to this study. For each analysis, we used an uncorrelated lognormal relaxed clock model with a birth-death speciation tree prior. We conducted two runs of 20 million generations, sampling every 1000th and discarding the first 5000 as burn-in. We used Tracer v1.6 [39] to ensure stationarity of chains for all parameters (ESS values >200).

Ancestral area reconstruction
We used Lagrange v. 20130526 [42] to reconstruct ancestral areas for both groups. For laughingthrushes and relatives, we defined eight areas – Africa, Peninsular India, Himalayas, Southeast Asia, China, Sundaland, Philippines, and Assam Hills. For shortwings and relatives, we defined nine areas based on the geographic extent of these species: Africa + Southwest Asia, Peninsular India, Himalayas, Southeast Asia, China, Sundaland, Philippines, Eurasia, and Australasia/New World. In Lagrange, we assigned the ranges of each species based on their current geographical distributions. We allowed the ancestral area reconstructions to include any

combination of areas. We used the time-calibrated tree from our BEAST analysis to reconstruct the likelihood of ancestral changes in distribution.

Species limits

We conducted focused mist-net based sampling over 3 years (2011 – 2014) and several years (2000 – 2010) of mixed sampling effort (point counts, transects, *ad libitum* observations, and mist-net sampling). During this period, we captured and examined 359 WG laughingthrushes and 430 WG shortwings. We collected a variety of vocal, plumage, morphometric and genetic data, along with extensive natural history observations of these species.

To assess species limits, we compared all population clades to determine statistically discrete units of nucleotide substitutions as well as distinct differences in plumage characteristics, morphometric variation, and song features. We chose this method because it allows using all available evidence to infer absence of gene flow and to delineate evolutionarily distinct units as species. Our species delineations are consistent with multiple species concepts - phylogenetic species concept, biological species concept, general lineage concept, and integrative taxonomy.

Coalescent-based test of species delimitation

We utilized a coalescent-based approach to statistically test species delimitation using the program BPP v3.1 [43]. This method employs the multispecies coalescent to compare different models of species delimitation and phylogeny in a Bayesian framework while accounting for incomplete lineage sorting due to ancestral polymorphism and gene tree-species tree conflicts [43]. We used the joint species delimitation and species tree analysis in BPP to test if the number of distinct units/species as delineated by our character-based analysis (see below) was statistically significant. For each lineage, we used the genetic dataset assembled to examine within-group variation and designated individuals by population/clade. For population size parameters, we assigned the gamma prior G (2, 1000), with mean 2/2000 = 0.001. We ran each analysis at least twice to confirm consistency between runs. Each run was for 100,000 samples, sampling frequency was 5, and the burnin was set to 20,000.

Morphometrics

All captured birds were banded and measurements of the length of right tarsus, flattened chord right wing, bill (from base of skull) and tail were taken following SAFRING manual [44] and USDA [45] guidelines. Each measurement of a feature was repeated thrice to examine measurement error [46]. Tarsus and bill measurements were taken with Mitutoyo ABS Digimatic Caliper (Mitutoyo Corp Japan) with accuracy of 0.02mm. Wing and tail measurements were taken with a wing rule (WING15ECON Avinet Inc.) that had a flush stop and calibration from both directions.

We analyzed the four morphometric variables (bill length, tarsus length, wing length and tail length; Additional file 1: Tables S3, S4) using multivariate statistics. For each species complex, we considered each distinct population clade separately and built a linear discriminant function with a common covariance matrix for all groups (JMP version 8). This was visualized with a canonical plot including a 95% confidence ellipse of the mean of each group and biplot rays indicating the direction of variables in canonical space. The groups with non-overlapping ellipses are considered different in morphology with the direction of biplot rays indicating the variables contributing to the observed differences. Wing and tail lengths for WG Shortwings were found to be collinear, but no significant collinearity among morphometric variables was detected in WG Laughingthrushes. However, a repetition of the discriminant analyses with reduced data set of three principal components (comprising 89.76% of the variation for WG Shortwings, 85% for WG Laughingthrushes) did not alter the results, hence we retained the analyses with the original variables for easier interpretation.

Plumage

For plumage comparisons, we examined and noted features of each population by studying live birds (sampled in the field), photographs of sampled birds and museum specimens (Bombay Natural History Society; BNHS, India) to make side-by-side comparisons, and descriptions in the literature. We noted differences in coloration across discrete feather patches to assess whether populations had unique plumage patterns.

Song

Field recording of 38 individual singing males was carried out between 0700 to 1100 h following Robin et al. [47]. Each recording consisted of only one continuous song bout from a single individual recorded using a Sennheisser shotgun microphone (ME66-K6) and Marantz Digital Audio Recorder (PMD660). Recordings were converted into spectrograms in Raven Pro 1.3 [48] at a sampling rate of 48 kHz. For each song, we collected data on frequency and note length parameters for each note, with multiple songs per individual and multiple individuals per population. A detailed statistical analysis of songs and syllables across multiple populations is presented in Purushotham and Robin [49].

Results

Phylogenetic analyses

The larger, family-level phylogenetic analyses indicated that both WG lineages diverged early within their respective clades (Fig. 1). Each of these two lineages is a distinct clade not closely related to the genus historically or currently grouped within. Additionally, both were found to be discrete clades deeply divergent from their closest relatives. The WG Laughingthrushes did not group with any of the other traditional laughingthrush clades (including ones designated as *Trochalopteron* or *Strophocincla*), but instead were sister to a clade composed of *Heterophasia*, *Minla*, *Actinodura*, *Leiothrix*, *Liocichla* and *Crocias* (Fig. 1a). Similarly, we reconstructed the WG Shortwings as sister to a newly uncovered clade comprised of several genera of mainly Asian blue flycatchers, the Niltavinae [28, 35], and not closely related to traditional shortwings in the genera *Myiomela* or *Brachypteryx* (Fig. 1b).

Both WG lineages are not only divergent from their closest relatives, they are also not sister to any single genus as defined under current taxonomic treatments. Thus, they cannot easily be lumped or subsumed under an existing clade without an extensive reorganization of multiple genera. As a better alternative, we designated new genus names: Western Ghats Laughingthrushes as *Montecincla* and Western Ghats Shortwings as *Sholicola* (see below for descriptions), to recognize their distinctiveness and to avoid confusion with older names associated with other taxa that are not part of these new groups.

The ancestral area reconstruction showed dispersal into the WG for *Montecincla* as most likely from the Himalayas and for *Sholicola* from the Himalayas + Southeast Asia (Additional file 1: Figures. 1, S2, S3). The two WG lineages have a similar estimated timing of separation from their closest relatives (Table 1). These dates also coincide with climatic events that led to the drying of peninsular India (Table 1), perhaps leading to discontinuing gene flow and a differentiation between

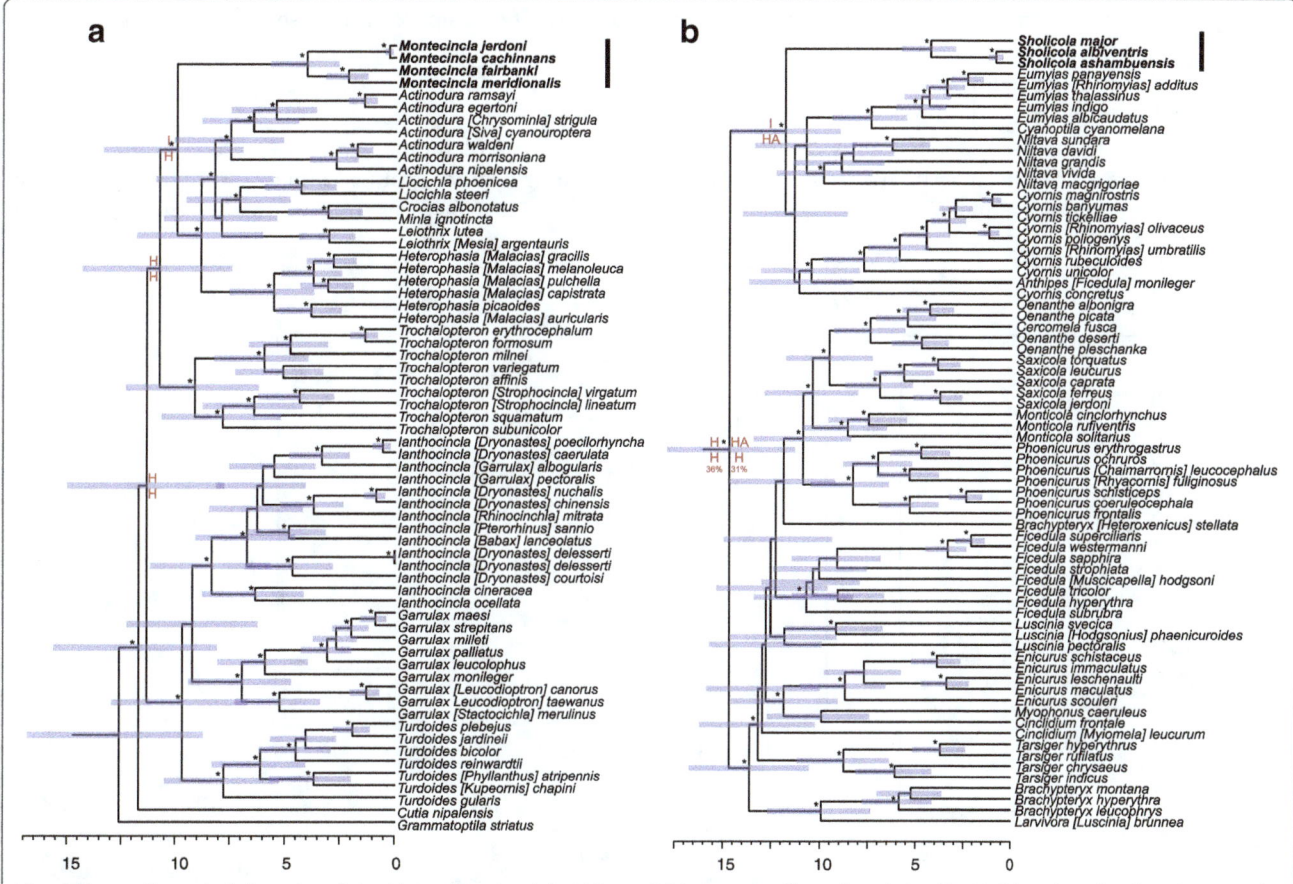

Fig. 1 Time-calibrated phylogenies of the (**a**) laughingthrush/babbler and (**b**) shortwing/flycatcher clades. The WG Laughingthrushes, *Montecincla*, did not group with any of the other laughingthrush clades and are sister to a clade of birds traditionally not included in laughingthrushes. The WG Shortwings, *Sholicola*, are not part of shortwings (*Brachypteryx* or *Myiomela*) but are sister to a clade of Asian blue flycatchers, the Niltavinae. Generic designations follow *Clements Checklist* [8]; *Handbook of the Birds of the World* [22] names are shown in brackets if different. Node bars show 95% HPD estimates of divergence dates and stars indicate ML bootstrap of 70% or higher and Bayesian posterior probability of 95% or higher. Letters (in red) show ancestral area reconstruction at relevant nodes (I = Peninsular India; H = Himalayas; A = Southeast Asia)

Table 1 Estimates of divergences dates and geological events

Event	Timing estimates (Ma)
Montecincla vs. *sister group* - split WG/Himalayas	11.57 (14.65 - 8.70)
Sholicola vs. *sister group* - split WG/Himalayas	11.83 (14.59 - 9.00)
within *Montecincla* - split across Palghat Gap [AB,CD]	4.7 (6.34 - 3.19)
within *Sholicola* - split across Palghat Gap [AB,CD]	4.33 (5.79 - 2.98)
within *Montecincla* - split across Shencottah Gap [C,D]	2.51 (3.60 - 1.60)
within *Sholicola* - split across Shencottah Gap [C,D]	0.86 (1.24 - 0.49)
within *Montecincla* - split across Chaliyar Valley [A,B]	0.33 (0.55 - 0.15)
Himalayas - peak constructional phase [55]	15 - 10.5
Tibetan Plateau - significantly uplift [55, 66]	10 - 8
Climate - enhanced aridity of Asia; onset of Indian monsoons [56–58, 66]	9 - 8
Vegetation - C4 plants start to replace C3 [56, 57]	6
Climate - major global shift; monsoons weakened [58]	2.6

peninsular forms from Himalayan / Southeast Asian forms. Furthermore, the divergence of populations within both lineages across the sky islands exhibits similar timing of diversification in response to the same biogeographic barriers (see below).

Species limits

Phylogenetic analysis of both groups showed clades corresponding to four major biogeographic regions in the WG, north to south: A- Bababudan & Banasura hills; B- Nilgiri hills; C- Anamalai, Palani, and Meghamalai hills; D- Ashambu hills (Fig. 2). *Montecincla* was comprised of four reciprocally monophyletic clades (Fig. 2b). Bayesian species delimitation using BPP also showed that these four clades were distinct species. Elevating former subspecies names, these are: *Montecincla jerdoni* (range A), *M. cachinnans* (B), *M. fairbanki* (C), *M. meridionalis* (D). Similarly, analysis of *Sholicola* revealed three reciprocally monophyletic clades (Fig. 2c) that were distinct species according to BPP. While two existing subspecies names that match these units can be elevated, namely *Sholicola major* (range A, B) and *S. albiventris* (C), the third (D) is a new species we designate below. Analyses of fixed characters of plumage, morphometrics and song data (below) match the distinct units uncovered in the phylogenetic and species delimitation analyses. Therefore multiple lines of data support four species in the *Montecincla* complex and three species in the *Sholicola* complex.

Plumage

Our analysis of plumage variation for *Montecincla* supports four distinct groups (Table 2). All four species have distinguishing features in multiple color patches that

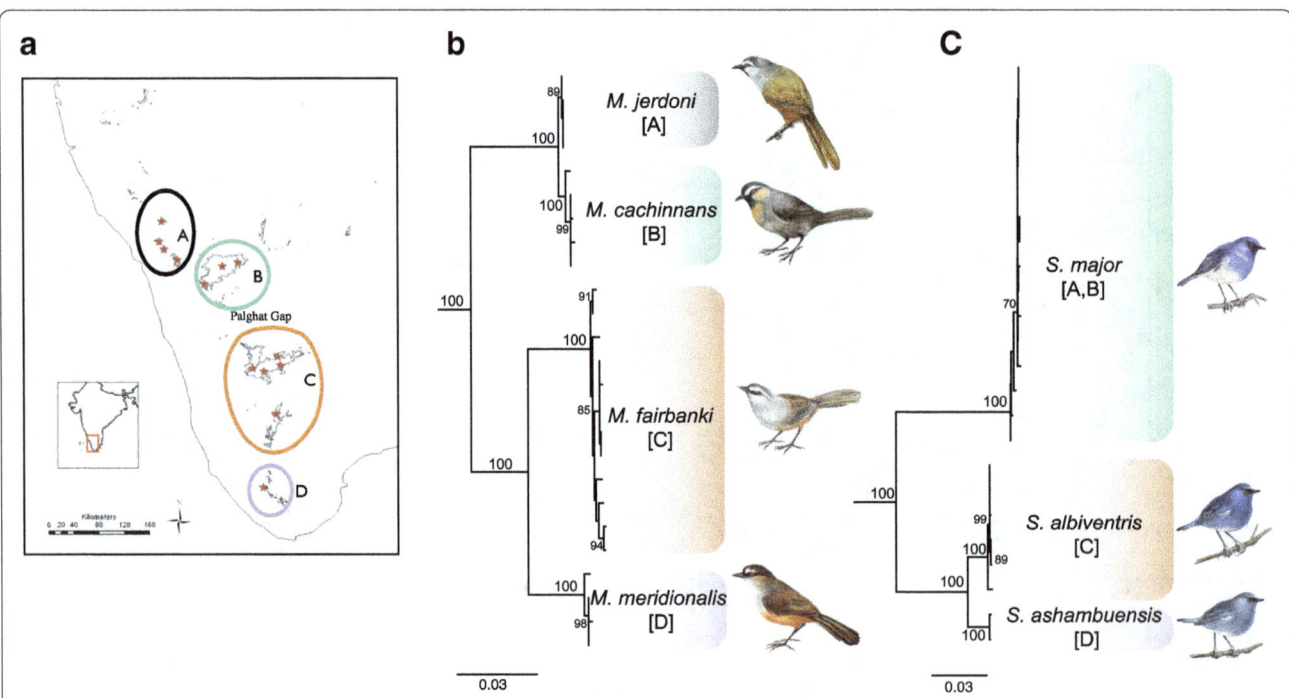

Fig. 2 Ranges (**a**) and phylogenetic relationships of *Montecincla* (**b**) and *Sholicola* (**c**) species. **a** Inset map of the Western Ghats shows sampling localities (stars), and divisions of the sky islands based on differentiated taxa. **b & c** ML bootstrap values are shown at nodes. Bird illustrations by Maya Ramaswamy

Table 2 Plumage differences between species of *Montecincla*

Plumage feature	M. jerdoni	M. cachinnans	M. fairbanki	M. meridionale
Crown	Slaty brown	Slaty brown	Dark brown almost black	Grey brown
Chin	Black	Black	Grey	Pale grey
Ear coverts	Greyish white	Pale rufous	Pale grey	Brownish grey
Supercilium	White, long, reaching behind eye; black eye strip from lore	White, long reaching behind eye; black eye strip from lore	White, Long, extending behind eye; black stripe below lores through eyes	White, short, not extending behind eye
Nape	Slaty brown	Ashy brown	Brownish grey	Pale brown
Upper parts	Olive brown up to tail	Olive brown	Olive brown	Dull grey at nape brownish towards rump
Breast	Grey, with faint streaks	Bright rufous	Pale grey with faint streaks	Whitish grey with prominent dark streaks
Belly	Olive brown but centre of belly pale rufous	Ochraceous	Rufous to chestnut	White at centre with mild dark streaks, sides deep chestnut
Centre of belly	Pale rufous	Rufous	Rufous to chestnut	White centre
Flanks	Olive brown	Olive brown	Rufous to chestnut	Reddish chestnut

separate these evolutionary units. Within-species variation is low and populations with similar features occur in geographically clustered sky islands (shown in Fig. 2). In *Sholicola*, *S. major* (A, B) showed striking differences from the other two species south of the Palghat Gap. The differences in plumage of between *S. albiventris* (C) and *Sholicola* sp. nov. (D; see below) are mainly in terms of the extent of white coloration on the belly. The newly described species has a considerably larger white belly patch (~24mm in length) than *S. albiventris* (~19mm) (Table 3).

Table 3 Plumage differences between species of *Sholicola*

Plumage feature	S. major	S. albiventris	S. ashambuensis
Crown	Slaty blue	Slaty blue	Slaty blue
Supraloral stripe	Faint blue	Whitish blue	Faint blue
Throat	Slaty blue	Slaty blue	Greyish blue
Breast	Slaty blue	Slaty blue	Greyish blue
Belly	Broad white patch	Narrow white patch from centre to vent	Narrow white patch extending anteriorly from vent to breast
Flanks	Pale rufous	Greyish blue	Greyish blue
Undertails coverts	Pale rufous	White	White
Upper parts	Slaty blue	Deep blue	Pale, slaty blue

Morphometrics

The four species of *Montecincla* and the three species of *Sholicola* differed significantly from each other in morphospace occupied (Fig. 3). This differentiation within both *Montecincla* and *Sholicola* was largely driven by tarsus length, although the direction of increase was reversed – *Montecincla* in the southernmost region of Ashambu Hills (D) is larger than the other species further north, while in *Sholicola* the species in the southern hills were the smallest (Additional file 1: Tables S3, S4).

Song

Spectrograms reveal the presence of unique song types in all four species of *Montecincla* and all three *Sholicola* species (Fig. 4). All four species of *Montecincla* show distinct features in song. *Montecincla fairbanki* sings at a higher bandwidth and song rate than the other populations, whereas *M. meridionalis* sings at a lower bandwidth than other species, while *M. jerdoni* has a higher song complexity with longer phrases than the other populations (Additional file 1: Table S5). Songs of all *Sholicola* species are also quantitatively distinct, with differences in song length and frequency across different species (more details in Purushotham and Robin [49]). The song of the newly described species of *Sholicola* is shorter in length and higher in frequency from *S. albiventris* (Additional file 1: Table S6).

Naming new taxa

As outlined above, the results of our analysis warrant new names for two genera and one species, in addition to elevating six subspecies to species level.

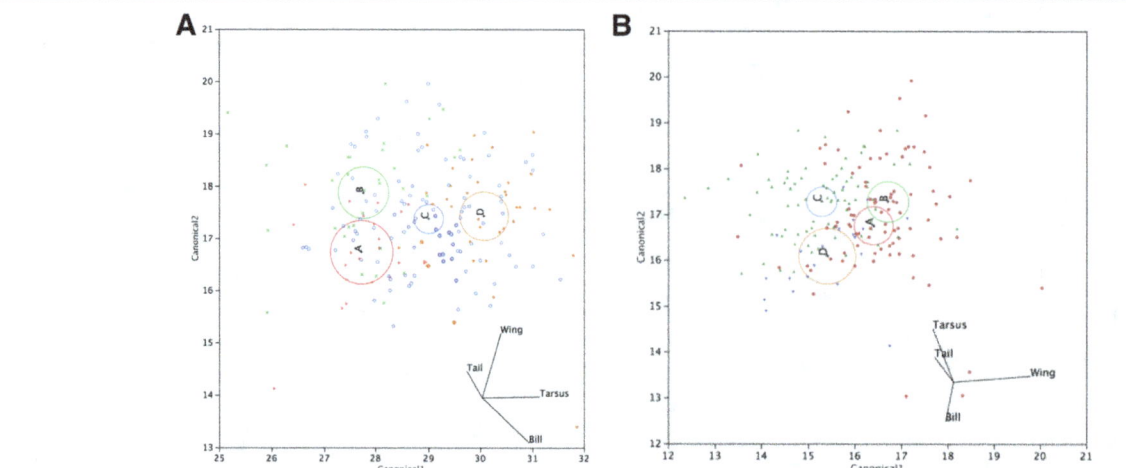

Fig. 3 Discriminant Function Analysis showing morphometric differentiation between **a**) *Montecincla* and **b**) *Sholicola* populations/species. A-D labels refer to geographic regions (as in Fig. 2). Circles represent 95% confidence ellipses of the mean of each group, which are significantly different when not overlapping. The direction of the biplot rays show how variables contributed to the observed differences

Western Ghats laughingthrushes

The laughingthrushes of the Western Ghats have been treated as part of the genus *Garrulax* Lesson, 1831 by Ali and Ripley [6] and as *Strophocincla* Wolters, 1981 by Rasmussen and Anderton [50], which was followed by others [22, 26]. However, in most global avian checklists, they are currently placed in *Trochalopteron* [8, 21] following phylogenetic insights [27]. Genus-level treatments in laughingthrushes have clearly been very unstable given extensive morphological variation and labile trait evolution in these birds. Our phylogenetic results reveal that all previous genus designations are inappropriate for Western Ghats Laughingthrushes and here we erect a new genus to recognize their distinctiveness.

Furthermore, we propose elevation of the four subspecies to full species level. This fits assertions by others: Rasmussen and Anderton [50] tentatively delimited two species (one north and the other south of the Palghat Gap with two subspecies each) but suggested that the four allopatric populations may each warrant species-level status with more evidence. Additionally, Praveen and Nameer [26] conducted an analysis of plumage differences and suggested that this complex was best represented as four distinct species. In line with our results of multiple character systems, we hereby propose elevation of these subspecies as four full allopatric species (Additional file 1: Figure S4). This is essentially a revival of the taxonomic treatment at the time of the original descriptions of these taxa.

Montecincla genus novum

Order Passeriformes: Family Leiothrichidae

Suggested common name: Chilappan

Type species: *C. [=Crateropus] cachinnaus* [sic] Jerdon, 1839 = *Strophocincla cachinnans* = *Montecincla cachinnans* comb. nov.

Additional species included: *Garrulax* (?) *Jerdoni* [sic] Blyth, 1851 = *Strophocincla cachinnans jerdoni* = *Montecincla jerdoni* comb. nov.; *Trochalopteron Fairbanki* Blanford, 1869 = *Strophocincla fairbanki* = *Montecincla fairbanki* comb. nov.; and *Trochalopterum meridionale* Blanford, 1880 = *Strophocincla fairbanki meridionalis* = *Montecincla meridionalis* comb. nov.

ZooBank Registry: urn:lsid:zoobank.org:act:88B21D13-2639-4BFD-8729-74308F609E4E

Diagnosis: *Montecincla* gen. nov. can be differentiated from other genera of traditional laughingthrushes - *Garrulax*; *Ianthocincla* Gould, 1835; *Trochalopteron*; and from *Turdoides* babblers by the combination of a prominent white supercilium, olive brown upperparts, small size, brown wings, rufous flanks and dark bill. *Montecincla* lacks the complex upperwing colouration (consisting of strong barring or conspicuous wing panels) of *Actinodura* Gould, 1836, and of *Liocichla* Swinhoe, 1877. *Montecincla* also lacks the slender greyish body and long tail of sibias of the genus *Heterophasia* Blyth, 1842. *Laniellus* Swainson, 1832 (=*Crocias* Temminck, 1836) has a diagnostic whitish underside with bold, streaked flanks that are absent in *Montecincla*. *Minla* Hodgson, 1837 is a genus of small birds (14cm) with a colourful reddish tail or rump, while *Montecincla* is larger (20cm) with a dull olive-brown tail and rump. *Leiothrix* Swainson, 1832, is characterized by a colourful bill and a yellow throat, contrasting with the black bill and greyish-white or black throats of *Montecincla*.

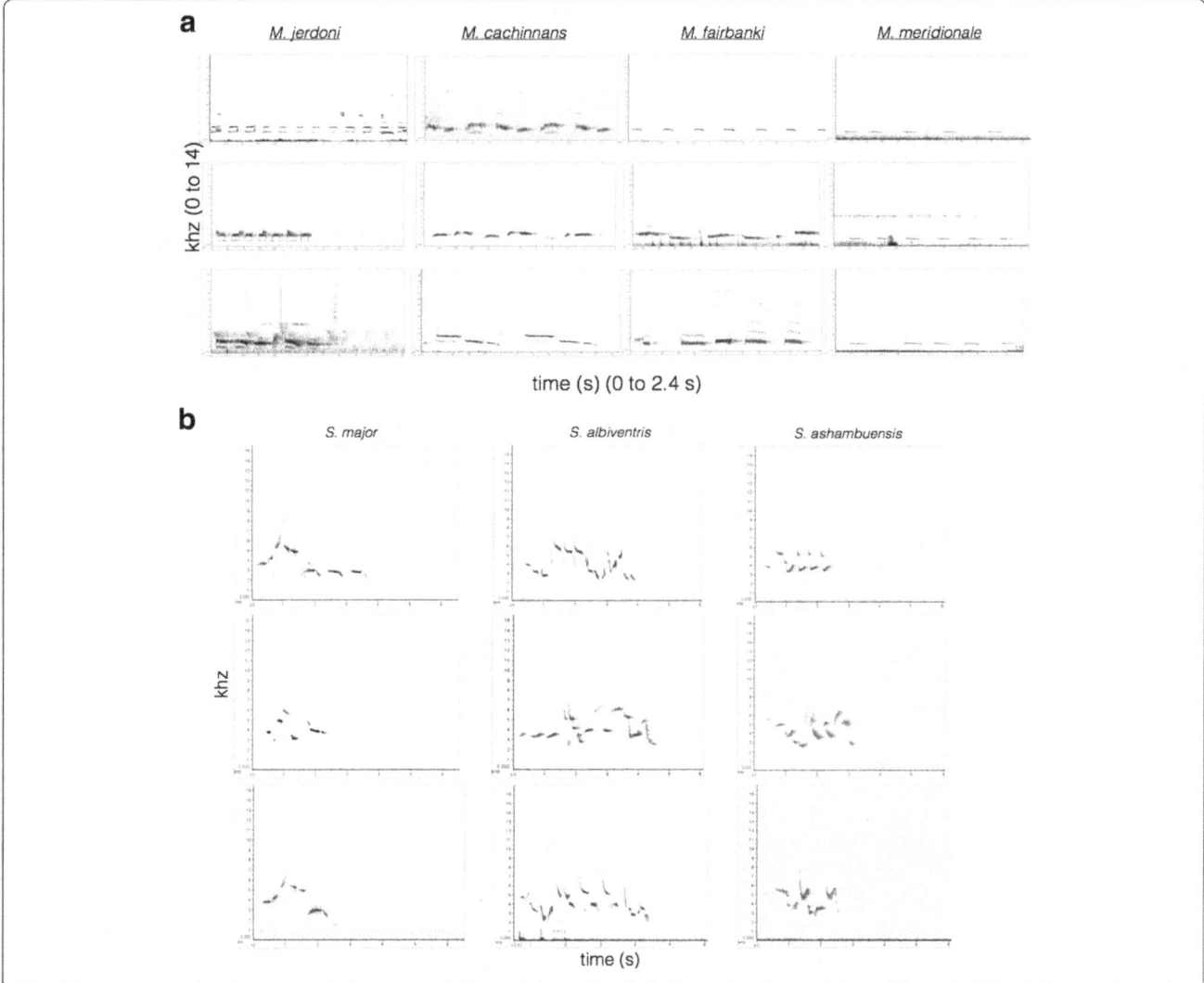

Fig. 4 Spectrograms showing song variation across **a)** *Montecincla* and **b)** *Sholicola* species. Songs of three different individuals from each species are shown

Description: *Montecincla* chilappans are medium-sized (~20cm) songbirds with rounded tails. Their upperparts are olive grey and they have a prominent white supercilium. The breast is greyish-white in three species and rufous in one (Table 2).

Etymology: *Montecincla*, of feminine gender, is a combination of Latin "mons" (gen. "montis"), with the meaning "mountain", and Greek "kinklos", denoting an unidentified type of songbird (often assumed to be a thrush). We chose this appropriate moniker because the genus is confined to the higher mountains of the Western Ghats. Our suggested vernacular name "Chilappan" stems from the local name of this genus (in Malayalam language), denoting their joyful cackling calls.

Comments: We suggest Banasura Chilappan as a common name for *Montecincla jerdoni* in recognition of Mt. Banasura, one of its strongholds and its type locality.

Similarly, we suggest the common names Nilgiri Chilappan for *Montecincla cachinnans*, Palani Chilappan for *Montecincla fairbanki*, and Ashambu Chilappan for *Montecincla meridionalis*.

Western Ghats blue robins or shortwings

The species were historically placed in the genus *Phoenicura* Swainson, 1831, later in *Callene* Blyth, 1847, then in *Brachypteryx* Horsfield, 1821 and more recently in *Myiomela*. None of these generic names are appropriate given our phylogenetic analysis. We thus erect a new genus to recognize the distinctness of Western Ghats Blue Robins (also known as Western Ghats Shortwings). In addition, we recognize three distinct species in this genus, including one new, previously unrecognized species.

Sholicola genus novum

Order Passeriformes: family Muscicapidae

Suggested common name: Sholakili

Type species: *Phoenicura major* Jerdon, 1841 = *Myiomela major* = *Sholicola major* comb. nov.

Additional species included: *Callene albiventris* Blanford, 1868 = *Myiomela albiventris* = *Sholicola albiventris* comb. nov.; *Sholicola ashambuensis* Robin, Vishnudas, Rheindt, Gupta, Hooper, Ramakrishnan & Reddy (see below).

ZooBank Registry: urn:lsid:zoobank.org:act:66837381-635E-476C-9454-092BB03B0A91

Diagnosis: *Sholicola* is sexually monomorphic, thus differing from the following closely related and sexually dimorphic genera: *Cyornis* Blyth, 1843; *Eumyias* Cabanis, 1850; *Niltava* Hodgson, 1837; and *Cyanoptila* Blyth, 1847. Being largely blue, it differs from *Anthipes* Blyth, 1847, which comprises brown birds with distinctive white throats. Members of *Sholicola* are resident birds with no known seasonal migration as in members of *Niltava* and *Cyanoptila*. The three species of *Sholicola* are phylogenetically distant from the genera they were previously placed under (*Brachypteryx, Callene, Phoenicura, Myiomela*). They also differ from these genera in having a sexually monomorphic plumage dominated by blue colouration.

Description: *Sholicola* is a genus of terrestrial blue flycatchers confined to high altitude (above 1200m) forests in the Shola-grassland complexes of the Western Ghats. Their general plumage is dominated by blue. They have a bluish-white band above the black lores, a slightly curved bill tip, well-developed rictal bristles, a short, nearly square tail, and long tarsi. The birds are mainly restricted to the understory, rarely venturing above three meters and feeding on insects singly or in pairs. There is no known sexual dimorphism in plumage though morphometric differences in wing length are recognized [10].

Etymology: *Sholicola*, of masculine gender, is a combination of Shola (the local name for montane forests in the Western Ghats) and the suffix –cola (from the Latin verb "colere"), meaning "dweller". We also suggest the common English name Sholakili, where -kili is the local name for "bird".

Description of a new species in the genus Sholicola:

Sholicola ashambuensis, species nova

English name: Ashambu Sholakili

Holotype: Trivandrum Museum of Natural History (TMNH) No. 725, collected by H.S. Fergusson on 3 May 1903 in the Chemunji Hills, Travencure (Travancore).

Etymology: The adjective *ashambuensis* denotes the species' geographical locality, the Ashambu Hills of southern India.

ZooBank Registry: urn:lsid:zoobank.org:act:CB9F32BC-750B-48AA-8F87-C9727144019D

Diagnosis: The following characters can be used to diagnose *Sholicola ashambuensis*.

1. Despite being smaller in general body size, *S. ashambuensis* has a considerably larger white belly patch (~24mm in length) than *S. albiventris* (~19mm). Only one *S. ashambuensis* specimen was available for measurements.
2. *S. ashambuensis* is a smaller bird (Additional file 1: Figure S5) with a shorter tarsus and longer bill (Additional file 1: Table S7; Additional file 1: Figure S6) than *S. albiventris* based on an examination of 76 *S. albiventris* and 21 *S. ashambuensis*.
3. *S. ashambuensis* has a distinct song (Additional file 1: Figure S7) with a higher mean maximum frequency, but narrower song bandwidth, shorter notes and shorter song bouts than *S. albiventris*, based on 119 *S. ashambuensis* songs and 203 *S. albiventris* songs.
4. When compared side-by-side, *S. ashambuensis* is paler blue than *S. albiventris*.

In addition to these phenotypic diagnostic characters, *S. ashambuensis* forms a reciprocally monophyletic clade based on DNA data from about 25 individuals. Further population genetic data with 15 microsatellites and 218 individuals (including 17 of *S. ashambuensis*) also support significant genetic differentiation [51].

Description of holotype: A small, overall dark-blue flycatcher with a large white belly patch reaching to the vent, black lores, and a bluish-white band above the lores. The holotype measures: bill – 16mm, wing– 81mm, tail – 61mm, tarsus – 21mm.

Distribution: Ashambu hills south of Shenkottah gap, southern India, mostly above 1200m elevation (Additional file 1: Figure S4).

Comments: The subtle plumage differentiation and limited fieldwork in the Ashambu Hills, including an absence of systematic capture-based studies, are perhaps the reason why *S. ashambuensis* has gone taxonomically unrecognized until now. The only museum specimen to our knowledge, the holotype (described above), was rediscovered (by CKV) after being locked up for about 120 years in a drawer of the Travancore Museum. Other specimens thought to be from the southern range are missing (and perhaps destroyed). There are possibly two more specimens in NHMUK (Tring) based on geography, but these were not examined by the authors.

Discussion

Using a modern systematic approach, we uncovered two deeply divergent lineages in the Indian avifauna and provided the first evidence of in-situ avian radiations within the Indian subcontinent. These new genera are not only phylogenetically distant from the clades with which they were previously classified, but also constitute lineages that diverged early from their closest relatives in larger Asian clades, many of which were only recently identified themselves. Our study underscores the importance of continued systematic studies in untangling taxonomic confusions to better understand local, regional, and global patterns of diversification.

The two WG lineages we examined, previously known as laughingthrushes and shortwings, do not belong to either of these two groups. Our results corroborate previous studies [27, 32, 33] that showed laughingthrushes to be polyphyletic. The WG Laughingthrushes do not group with any of the other traditional laughingthrush clades including species placed within *Trochalopteron* or *Strophocincla*. Rather, they are in a distinct clade of their own that is sister to several genera traditionally not considered laughingthrushes. We introduced a new genus name, *Montecincla*, to highlight the phylogenetic and biological distinction of the Western Ghats Laughingthrushes as well as to avoid confusion with alternative names. The taxonomic history of the WG Shortwings has gone through a similar level of confusion. Due to behavioral and plumage traits, they were traditionally thought to be shortwings (genus *Brachypteryx*). The first phylogenetic analysis of the WG species found that they were not closely related to *Brachypteryx* [9]. Our analysis conclusively shows that WG Shortwings are actually flycatchers and sister to a newly discovered clade of several genera of Asian blue flycatchers, the Niltavinae.

We present the first evidence of in-situ bird radiations in the Western Ghats mountains. Our examination of various character data – genetic, song, plumage and morphometric – all point to considerable differentiation among populations that we propose to comprise four species in the genus *Montecincla* and three species in the genus *Sholicola*. The complete distributions of both species complexes lie along a 400 km latitudinal gradient comprising the highest elevations, the sky islands of the Western Ghats. Further intensive examinations in this region may reveal patterns of similar endemic radiations in other taxa.

The diversification of *Montecincla* and *Sholicola* from their respective sister taxa supports the classic model of species colonizing the WG from the Himalayas. The estimated divergence dates for corresponding species on either side of common barriers are similar across these two groups, which is compelling evidence for a vicariance model. We interpret our estimated dates using the best calibration currently available and in light of the corresponding known climatic events that may have led to these biogeographic patterns. Given that there is much disagreement about the age of bird diversification (see [52–54]), we acknowledge that different divergence

dating methods and calibrations may provide alternative estimates of these dates. However, the relative timings of these divergences will remain the same. Our analyses provides a compelling demonstration that these two lineages have nearly identical divergence times, regardless of the exact ages, implying a common mechanism.

The coincidence in timing of colonization of the WG by *Montecincla* (11.57 Ma) and *Sholicola* (11.83 Ma) supports a scenario of initial range expansion during a wet, cool period in the mid-to-late Miocene [15, 55, 56] that allowed these cool-adapted lineages to disperse across the moist forests that covered the Indian subcontinent followed by subsequent vicariance as the subcontinent became drier and more seasonal in the late Miocene [56–58], which likely led to local extinction in the central region and current isolation in the WG. Intriguingly, another species of laughingthrush found in the Western Ghats, *Ianthocincla* [*Dryonastes*] *delesserti*, represents a more recent colonization into this region (see Fig. 1a).

Diversification within the WG in the Pliocene and Pleistocene was likely driven by climatic events that expanded and contracted the cool montane forest in southwestern India. Within the WG, both lineages show similar divergence dates (4.7 and 4.33 Ma) across the Palghat Gap. This break in the mountain chain is a major habitat barrier for many taxa [59–61]. It is the widest and deepest valley in the range and is thought to be the result of an ancient geological fissure (~500 Myr old) [9]. The tight correspondence of divergence dates in *Montecincla* and *Sholicola* across this now dry gap indicates that suitable wet, cool habitat, which was needed for these taxa to disperse, likely only lasted for a brief period in the Pliocene. Divergences across narrower gaps are not consistent across these two lineages, or several other bird species [60]. Climatic fluctuations in the Pleistocene were more numerous and variable in strength and duration [58, 62], perhaps leading to more individual responses by species across the narrow valleys within the WG. A more thorough survey of endemic taxa and their diversification history is crucial for a better understanding of the evolutionary history and assembly of this unique ecosystem.

Conclusions

The discovery of two independent endemic bird radiations in the Shola Sky Islands highlights the evolutionary role that such habitats can play in the diversification of lineages and the need for additional systematic studies that can potentially find other taxa with similar patterns. An important result from our findings is the dramatic increase in the biodiversity inventory of the WG, from the previously recognized three species to seven endemic species, each with much narrower ranges. Tropical sky island species are thought to be the most susceptible to anthropogenic climate change [4], and with 50% of Shola habitat in the WG already lost [5], these findings provide a much needed impetus for conservation [63]. With the Western Ghats continuing to lose between 0.57%–0.91% [64, 65] of their forest habitats each year, endemic species face extraordinary conservation challenges. Studies such as this not only clarify the taxonomic and phylogenetic information needed to quantify biodiversity but also urge the need to assess the possible responses of these species to anthropogenic climate change.

Additional file

Additional file 1: Figure S1. Sample collection sites across the sky islands of the Western Ghats. **Figure S2.** Ancestral area analysis using Lagrange for *Montecincla*. Area codes for terminal taxa are A-Africa, B-peninsular India, C-Himalayas, D-Southeast Asia, E-China, F-Sundaland, G-Philippines, and H-Assam. The most likely ancestral range reconstruction for relevant nodes (in red) are shown. **Figure S3.** Ancestral area analysis using Lagrange for *Sholicola*. Area codes for terminal taxa are I-Peninsular India, H-Himalayas, A-Southeast Asia, D-Eurasia, E-Australasia F- Africa + Southwest Asia, C-China, S-Sundaland, P-Philippines, E-Eurasia, and N-Australasia/New World. The most likely ancestral range reconstruction for relevant nodes (in red) are shown with their relative probabilities. **Figure S4.** Map showing the species distributions of *Montecincla* chilappans and *Sholicola* sholakillis. **Figure S5.** Photographs of *Sholicola ashambuensis sp. nov.* a) holotype in the Trivandrum Museum of Natural History; b) in comparison with other members of *Sholicola*: *S. ashambuensis* (right), with larger white belly contrasted with *S. albiventris* (middle), *S. major* (left). **Figure S6.** a) Discriminant Function Analysis of morphometric differences between *Sholicola ashambuensis* and *S. albiventris* based on 97 individuals shows 79.4% accurate classification and significant differences in the means (non overlapping 95% confidence interval circles); b) Box plots showing morphometric differences in bill and tarsus lengths in *Sholicola ashambuensis* and *S. albiventris*. **Figure S7.** Discriminant Function Analysis of *Sholicola ashambuensis* and *S. albiventris* songs based on 217 song recordings shows 80% accurate classification and significant differences in the means (non overlapping 95% confidence interval circles). **Table S1:** Samples used to reconstruct *Montecincla* phylogeny. **Table S2:** Samples used to reconstruct *Sholicola* phylogeny. **Table S3:** Morphometric variation in *Montecincla* species. **Table S4:** Morphometric variation in *Sholicola* species. **Table S5:** Song variation in *Montecincla* species in the Western Ghats Sky Islands. **Table S6:** Song variation in *Sholicola* species in the Western Ghats Sky Islands. **Table S7:** Comparison of morphometric measurements of *Sholicola ashambuensis* and *S. albiventris*. (PDF 3728 kb)

Acknowledgements

We thank Forest Departments of Kerala and Tamil Nadu; for exceptional support, we specifically thank - Chief Wildlife Wardens of Kerala - R. Rajaraja Varma, V. Gopinathan, G. Harikumar, PCCF Kerala T.M. Manoharan, Bransdon Corrie, former PCCF Tamil Nadu R. Sunder Raju, DCF K.I. Pradeep Kumar, ACF Santhosh Kumar, Senior Wildlife Assistant Reney R. Pillai; Abhilash Babu, Sahas Barve, Ravi Kiran, Chetana Purushotham, Anusha Shankar, Sriranjini Swaminathan for field assistance; Chetana Purushotham and Suma M. Shirley for song analyses; Joli Borah, Himanshu Chattani, Aleeson E., Shilpa M., Jyothi Nair, Abhinav Sur, Nelum Wickramasinghe for lab assistance; Tanya Balcar, Sumin George, Reney R. Pillai, Suhel Quader, Prathim Roy, Robert Stewart, for various support; H.S. Sudhira and Gubbi Labs for help with maps; Robin Abraham, Edward Dickinson, K.P. Dinesh, Praveen J., Rajah Jayapal, Muhammed Jafer Palot, Aasheesh Pittie, Pamela Rasmussen, C. Sasikumar, Anindya Sinha, Richard Schodde, for discussion on nomenclature and taxonomy; Ishan Aggarwal, John Bates, Balaji Chattopadhyay, Nishma Dahal, K.P. Dinesh, Varad Giri, Peter Makovicky, Nandini Rajamani, Vivek Ramachandran, Krishnapriya Tamma and S.P. Vijaykumar for comments on the manuscript; Prasenjeet Yadav for photographs of the museum birds;

Maya Ramaswamy for bird illustrations; S. Abu at Trivandrum Museum of Natural History; M. Adams at NHMUK; Vithoba Hegde, Rahul Khot, Deepak Apte at BNHS. This manuscript also greatly benefited from submission to Axios Review including comments from Dieter Thomas Tietze, George Sangster, Jon Fjeldså, Jason Weir, and an anonymous reviewer. We used color-blind safe colors from www.colorbrewer2.org where possible.

Funding

The project was supported by a National Geographic Society Research and Exploration Grant to VVR; Indian DAE, NCBS, Ramanujan Fellowship to UR; U.S. National Science Foundation (DEB-0962078; DEB-1457624) to SR. FER was supported by Singaporean Ministry of Education Tier I grants (WBS R-154-000-570-133; WBS R-154-000-658-112).

Authors' contributions

VVR, CKV, SR conceived of the study; VVR, PG carried out the molecular lab work; VVR, CKV, PG, DH, FER, SR analyzed the data; VVR, CKV, SR, UR coordinated the study; VVR, CKV, FER, SR drafted the manuscript; CKV, VVR carried out the field work. All authors gave final approval for publication.

Competing interests

The authors declare that they have no competing interests.

Author details

[1]National Centre for Biological Sciences, TIFR, Bellary Road, Bangalore 560065, India. [2]Present address – Indian Institute of Science Education and Research Tirupati, Mangalam, Tirupati 517507, India. [3]Avian Evolution Lab, Department of Biological Sciences, Faculty of Science, National University of Singapore, Singapore 117543, Singapore. [4]Committee on Evolutionary Biology, University of Chicago, Chicago, IL 60637, USA. [5]Biology Department, Loyola University Chicago, Chicago, IL 60660, USA.

References

1. Myers N, Mittermeier RA, Mittermeier CG, da Fonseca G, Kent J. Biodiversity hotspots for conservation priorities. Nature. 2000;403:853–8.
2. Fjeldså J, Bowie RCK, Rahbek C. The role of mountain ranges in the diversification of birds. Annu Rev Ecol Evol Syst. 2012;43:249–65.
3. Reddy S. What's missing from avian global diversification analyses? Mol Phylogenet Evol. 2014;77:159–65.
4. Freeman BG, Class Freeman AM. Rapid upslope shifts in New Guinean birds illustrate strong distributional responses of tropical montane species to global warming. Proc Natl Acad Sci U S A. 2014;111:4490–4.
5. Sukumar R, Suresh HS, Ramesh R. Climate change and its impact on tropical montane ecosystems in southern India. J Biogeogr. 1995;22:533–6.
6. Ali S, Ripley SD. Handbook of the Birds of India and Pakistan, Together with those of Nepal, Sikkim, Bhutan and Ceylon, vol. 1–10. Bombay: Oxford University Press; 1978.
7. Rasmussen P, Anderton JC. Birds of South Asia: The Ripley Guide. 2nd Edition. 2 vols. Barcelona: Lynx Edicions; 2012.
8. Clements JF, Schulenberg TS, Iliff MJ, Robertson D, Fredericks TA, Sullivan BL, et al. The eBird/Clements checklist of birds of the world: v2015. 2015. Downloaded from http://www.birds.cornell.edu/clementschecklist/. Accessed 10 Aug 2015.
9. Robin VV, Sinha A, Ramakrishnan U. Ancient geographical gaps and paleoclimate shape the phylogeography of an endemic bird in the sky islands of southern India. PLoS ONE. 2010;5:e13321.
10. Widdowson M, Cox K. Uplift and erosional history of the Deccan Traps, India: Evidence from laterites and drainage patterns of the Western Ghats and Konkan Coast. Earth Planet Sci Lett. 1996;137:57–69.
11. Gunnell Y, Gallagher K, Carter A, Widdowson M, Hurford AJ. Denudation history of the continental margin of western peninsular India since the early Mesozoic – reconciling apatite fission-track data with geomorphology. Earth Planet Sci Lett. 2003;215:187–201.
12. Hora SL. Satpura hypothesis of the distribution of the Malayan fauna and flora to peninsular India. Proc Nat Instit Sci India. 1949;15:309–14.

13. Islam MA. Satpura Hypothesis and the distribution of laughing thrushes *Garrulax* Lesson of India. J Bomb Nat Hist Soc. 1990;86:318–22.

14. Ali S. The Satpura trend as an ornithogeographical highway. Proc Nat Instit Sci India. 1949;15:379–86.

15. Karanth K. Evolution of disjunct distributions among wet-zone species of the Indian subcontinent: Testing various hypotheses using a phylogenetic approach. Curr Sci. 2003;85:1276–83.

16. Ripley SD, Beehler BM. Patterns of speciation in Indian birds. J Biogeogr. 1990;17:639–48.

17. Patnaik R. Fossil murine rodents as ancient monsoon indicators of the Indian subcontinent. Quatern Int. 2011;229:94–104.

18. Robin VV, Nandini R. Shola habitats on sky islands: status of research on montane forests and grasslands in southern India. Curr Sci. 2012;103:1427–37.

19. Daniels RJR, Joshi NV, Gadgil M. On the relationship between bird and woody plant species diversity in the Uttara Kannada district of south India. Proc Natl Acad Sci U S A. 1992;89:5311–5.

20. Ripley SD. Avian relicts and double invasions in peninsular India and Ceylon. Evolution. 1949;3:150–9.

21. Dickinson EC, Christidis L. The Howard and Moore Complete Checklist of the Birds of the World Fourth Edition, Volume 2: Passerines. Dickinson EC, Christidis L, editors. Eastbourne, UK: Aves Press; 2014.

22. Collar NJ, Robson C. Family Timaliidae (Babblers). In: del Hoyo J, Elliott A, Christie DA, editors. Handbook of the Birds of the World, Vol. 12. Barcelona: Lynx Edicions; 2007. p. 70–291.

23. Collar N, de Juana E, Sharpe CJ. White-bellied Blue Robin (*Myiomela albiventris*). In: del Hoyo J, Elliott A, Sargatal J, Christie DA, de Juana E, editors. Handbook of the Birds of the World Alive. Barcelona: Lynx Edicions. 2016. (retrieved from http://www.hbw.com/node/58510 on 7 March 2016).

24. Rasmussen P. Biogeographic and conservation implications of revised species limits and distributions of South Asian birds. Zool Med Leiden. 2005;79-3:137–46.

25. Sibley CG, Monroe BJ. Distribution and Taxonomy of Birds of the World. New Haven: Yale University Press; 1991.

26. Praveen J, Nameer PO. *Strophocincla* Laughingthrushes of South India: a case for allopatric speciation and impact on their conservation. J Bomb Nat Hist Soc. 2012;109:46–52.

27. Moyle RG, Andersen MJ, Oliveros CH, Steinheimer FD, Reddy S. Phylogeny and biogeography of the core babblers (Aves: Timaliidae). Syst Biol. 2012;61:631–51.

28. Sangster G, Alström P, Forsmark E, Olsson U. Multi-locus phylogenetic analysis of Old World chats and flycatchers reveals extensive paraphyly at family, subfamily and genus level (Aves: Muscicapidae). Mol Phylogenet Evol. 2010;57:380–92.

29. Price TD, Hooper DM, Buchanan CD, Johansson US, Tietze DT, Alström P, et al. Niche filling slows the diversification of Himalayan songbirds. Nature. 2014;509:222–5.

30. Kimball R, Braun EL, Barker FK, Bowie RCK, Braun MJ, Chojnowski JL, et al. A well-tested set of primers to amplify regions spread across the avian genome. Mol Phylogenet Evol. 2009;50:654–60.

31. Reddy S, Driskell AC, Rabosky DL, Hackett SJ, Schulenberg TS. Diversification and the adaptive radiation of the vangas of Madagascar. Proc R Soc B. 2012; 279:2062–71.

32. Luo X, Qu YH, Han LX, Li SH, Lei FM. A phylogenetic analysis of laughingthrushes (Timaliidae: *Garrulax*) and allies based on mitochondrial and nuclear DNA sequences. Zool Scripta. 2009;38:9–22.

33. Gelang M, Cibois A, Pasquet E, Olsson U, Alström P, Ericson PGP. Phylogeny of babblers (Aves, Passeriformes): major lineages, family limits and classification. Zool Scripta. 2009;38:225–36.

34. Lanfear R, Calcott B, Ho SYW, Guindon S. PartitionFinder: Combined selection of partitioning schemes and substitution models for phylogenetic analyses. Mol Biol Evol. 2012;29:1695–701.

35. Zuccon D, Ericson PGP. A multi-gene phylogeny disentangles the chat-flycatcher complex (Aves: Muscicapidae). Zool Scripta. 2010;39:213–24.

36. Stamatakis A. RAxML-VI-HPC: maximum likelihood-based phylogenetic analyses with thousands of taxa and mixed models. Bioinformatics. 2006;22: 2688–90.

37. Swofford DL. PAUP*. Phylogenetic Analysis Using Parsimony (*and Other Methods). Sunderland, Massachusetts: Version 4. Sinauer Associates; 2003.

38. Ronquist F, Teslenko M, van der Mark P, Ayres DL, Darling A, Hohna S, et al. MrBayes 3.2: Efficient Bayesian phylogenetic inference and model choice across a large model space. Syst. Biol. 2012;61:539–42.

39. Rambaut A, Suchard MA, Xie D, Drummond AJ. Tracer v1.6. 2014. Available from: http://tree.bio.ed.ac.uk/software/tracer/.

40. Wilgenbusch JC, Warren DL, Swofford DL. AWTY: a system for graphical exploration of MCMC convergence in Bayesian phylogenetic inference. 2004. Available from: http://ceb.csit.fsu.edu/awty.

41. Drummond AJ, Suchard MA, Xie D, Rambaut A. Bayesian Phylogenetics with BEAUti and the BEAST 1.7. Mol. Biol. Evol. 2012;29:1969–73.

42. Ree RH, Smith SA. Maximum likelihood inference of geographic range evolution by dispersal, local extinction, and cladogenesis. Syst Biol. 2008;57:4–14.

43. Yang Z. The BPP, program for species tree estimation and species delimitation. Curr Zool. 2015;61:854–65.

44. De Beer S, Lockwood G, Raijmakers JHFA, Raijmakers JMH, Scott WA, Oschadleus HD, et al. SAFRING bird ringing manual. Cape Town, South Africa: University of Cape Town; 2000.

45. Ralph CJ, Geupel GR, Pyle P, Martin TE, DeSante DF. Handbook of Field Methods for Monitoring Landbirds. Gen. Tech. Rep. PSW-GTR-144-www. Albany, CA: Pacific Southwest Research Station, Forest Service, U.S. Department of Agriculture; 1993.

46. Lougheed SC, Arnold TW, Bailey RC. Measurement error of external and skeletal variables in birds and its effect on principal components. Auk. 1991; 108:432–6.

47. Robin VV, Katti M, Purushotham C, Sancheti A, Sinha A. Singing in the sky: song variation in an endemic bird on the sky islands of southern India. Anim Behav. 2011;82:513–20.

48. Bioacoustics Research Program. Raven Pro: Interactive Sound Analysis Software (Version 1.3) [Computer software]. Ithaca, NY: The Cornell Lab of Ornithology. 2008. Available from http://www.birds.cornell.edu/raven.

49. Purushotham CB, Robin VV. Sky island bird populations isolated by ancient genetic barriers are characterized by different song traits than those isolated by recent deforestation. Ecol Evol. 2016;20:7334–43.

50. Rasmussen PC, Anderton JC. Birds of South Asia - The Ripley Guide. Lynx Edicions: Barcelona; 2005.

51. Robin VV, Gupta P, Thatte P, Ramakrishnan U. Islands within islands: two montane palaeo-endemic birds impacted by recent anthropogenic fragmentation. Mol Ecol. 2015;24:3572–84.

52. Jarvis ED, Mirarab S, Aberer AJ, Li B, Houde P, Li C. Whole-genome analyses resolve early branches in the tree of life of modern birds. Science. 2014;346: 1320–31.

53. Claramunt S, Cracraft J. A new time tree reveals Earth history's imprint on the evolution of modern birds. Sci Adv. 2015;1:e1501005–5.

54. Prum RO, Berv JS, Dornburg A, Field DJ, Townsend JP, Lemmon EM, et al. A comprehensive phylogeny of birds (Aves) using targeted next-generation DNA sequencing. Nature. 2015;526:569–73.

55. Potter PE, Szatmari P. Global Miocene tectonics and the modern world. Earth-Sci Rev. 2009;96:279–95.

56. Pound MJ, Haywood AM, Salzmann U, Riding JB. Global vegetation dynamics and latitudinal temperature gradients during the mid to late Miocene (15.97–5.33 Ma). Earth Sci Rev. 2012;112:1–22.

57. Sanyal P, Bhattacharya SK, Kumar R, Ghosh SK, Sangode SJ. Mio–Pliocene monsoonal record from Himalayan foreland basin (Indian Siwalik) and its relation to vegetational change. Palaeogeogr Palaeoclimatol Palaeoecol. 2004;205:23–41.

58. Zhisheng A, Guoxiong W, Jianping L, Youbin S, Yimin L, Weijian Z, et al. Global Monsoon Dynamics and Climate Change. Annu Rev Earth Planet Sci. 2015;43:29–77.

59. Vidya TNC, Fernando P, Melnick DJ, Sukumar R. Population differentiation within and among Asian elephant (*Elephas maximus*) populations in southern India. Heredity. 2004;94:71–80.

60. Robin VV, Vishnudas CK, Gupta P, Ramakrishnan U. Deep and wide valleys drive nested phylogenetic patterns across a montane bird community. Proc R Soc B. 2015;282:20150861–7.

61. Van Bocxlaer I, Biju SD, Willaert B, Giri VB, Shouche YS, Bossuyt F. Mountain-associated clade endemism in an ancient frog family (Nyctibatrachidae) on the Indian subcontinent. Mol Phylogenet Evol. 2012;62:839–47.

62. Clift PD, Hodges KV, Heslop D, Hannigan R, Van Long H, Calves G. Correlation of Himalayan exhumation rates and Asian monsoon intensity. Nat Geosci. 2008;1:875–80.

63. Padma TV. India faces uphill battle on biodiversity. Nature. 2013;504(7479):200.

64. Menon S, Bawa KS. Applications of geographic information systems, remote-sensing, and a landscape ecology approach to biodiversity conservation in the Western Ghats. Curr Sci. 1997;73:134–45.

65. Prasad SN, Vijayan L, Balachandran S. Conservation planning for the Western Ghats of Kerala: I. A GIS approach for location of biodiversity hot spots. Curr Sci. 1998;75:211–9.

PERMISSIONS

LIST OF CONTRIBUTORS

Matthew P. Heinicke
Department of Natural Sciences, University of Michigan-Dearborn, 4901 Evergreen Rd., Dearborn, MI 48128, USA

Todd R. Jackman and Aaron M. Bauer
Department of Biology, Villanova University, 800 Lancaster Avenue, Villanova, PA 19085, USA

Yuyini Licona-Vera and Juan Francisco Ornelas
Departamento de Biología Evolutiva, Instituto de Ecología, A.C., Carretera Antigua a Coatepec No. 351, El Haya, Xalapa, 91070 Veracruz, Mexico

V. Valcárcel
Department of Biology (Botany), Universidad Autónoma de Madrid, Madrid, Spain

B. Guzmán and P. Vargas
Department of Biodiversity and Conservation, Real Jardín Botánico, CSIC, Madrid, Spain

N. G. Medina
Department of Botany, Faculty of Science, University of South Bohemia, Ceske Budejovice, Czech Republic

J. Wen
Department of Botany/MRC 166, Smithsonian Institution, Washington, DC, USA

Nadine Bernhardt and Jonathan Brassac
Leibniz Institute of Plant Genetics and Crop Plant Research (IPK), Gatersleben, Germany

Benjamin Kilian
Leibniz Institute of Plant Genetics and Crop Plant Research (IPK), Gatersleben, Germany

Frank R. Blattner
Leibniz Institute of Plant Genetics and Crop Plant Research (IPK), Gatersleben, Germany
German Centre for Integrative Biodiversity Research (iDiv) Halle-Jena-Leipzig, Leipzig, Germany

Marta Vidal-García and J. Scott Keogh
Research School of Biology, The Australian National University, Canberra, Australia

Bernhard Misof
Centre for Molecular Biodiversity Research (ZMB), Museum Alexander Koenig, Adenauerallee, 53113 Bonn, Germany

Adam Ślipiński
Australian National Insect Collection, CSIRO, Canberra ACT 2601, Australia

Hermes E. Escalona
Centre for Molecular Biodiversity Research (ZMB), Museum Alexander Koenig, Adenauerallee, 53113 Bonn, Germany
Australian National Insect Collection, CSIRO, Canberra ACT 2601, Australia

Andreas Zwick and Diana Hartley
Australian National Insect Collection, CSIRO, Canberra ACT 2601, Australia

Hong Pang and Hao-Sen Li
State Key Laboratory of Biocontrol, Key Laboratory of Biodiversity Dynamics and Conservation of Guangdong Higher Education Institute, College of Ecology and Evolution, Sun Yat-Sen University, Guangzhou 510275, China

Jiahui Li
College of Environment and Plant Protection, Hainan University, No. 58 Renmin Avenue, Haikou 570228, China

Xingmin Wang
Key Laboratory of Bio-Pesticide Innovation and Application, Guangdong Province, Guangzhou, China

Lars S. Jermiin
Centre for Biodiversity Analysis, Australian National University, ACT, Acton 2601, Australia

Oldřich Nedvěd
Institute of Entomology, Biology Centre, Branišovská 31, -37005 České Budějovice, CZ, Czech Republic.
University of South Bohemia, Branišovská, 31 České Budějovice, Czech Republic

Oliver Niehuis
Department of Evolutionary Biology and Ecology, Institute of Biology I (Zoology) Albert Ludwig University of Freiburg, Hauptstr. 1, 79104 Freiburg, Germany

Wioletta Tomaszewska
Museum and Institute of Zoology, Polish Academy of Sciences, Wilcza 64, 00-679 Warszawa, Poland

Adam Dawid Urantowka
Department of Genetics, Wroclaw University of Environmental and Life Sciences, ul. Kożuchowska7, 51-631, Wroclaw, Poland

Aleksandra Kroczak and Paweł Mackiewicz
Department of Genomics, Faculty of Biotechnology, University of Wrocław, ul. Fryderyka Joliot-Curie 14a, 50-383 Wrocław, Poland

Juan Carlos Aledo
Departamento de Biología Molecular y Bioquímica, Facultad de Ciencias, Universidad de Málaga, 29071 Málaga, Spain

Anna Drews, Maria Strandh, Lars Råberg and Helena Westerdahl
Department of Biology, Lund University, Ecology Building, 223 62 Lund, Sweden

Samuel Abalde, Juan E. Uribe, Ana M. Echeverry and Rafael Zardoya
Museo Nacional de Ciencias Naturales (MNCN-CSIC), José Gutiérrez Abascal 2, 28006 Madrid, Spain

Manuel J. Tenorio
Departamento CMIM y Q. Inorgánica-INBIO, Facultad de Ciencias, Universidad de Cádiz, 11510 Puerto Real, Cádiz, Spain

Carlos M. L. Afonso
Centre of Marine Sciences (CCMAR), Universidade do Algarve, Campus de Gambelas, 8005 - 139 Faro, Portugal

Ian J. Pepper, Robert E. Van Sciver and Amy H. Tang
Department of Microbiology and Molecular Cell Biology, Eastern Virginia Medical School, Leroy T. Canoles Jr. Cancer Research Center, Harry T. Lester Hall, Room 454-457, 651 Colley Avenue, Norfolk, VA 23501, USA

Seong-il Eyun
Center for Biotechnology, University of Nebraska-Lincoln, Lincoln, NE 68588, USA

Jie Song and Bao-Kai Cui
Institute of Microbiology, Beijing Forestry University, Qinghua East Road, Haidian District, Beijing 100083, People's Republic of China

C. K. Vishnudas, Pooja Gupta and Uma Ramakrishnan
National Centre for Biological Sciences, TIFR, Bellary Road, Bangalore 560065, India

V.V. Robin
National Centre for Biological Sciences, TIFR, Bellary Road, Bangalore 560065, India
Indian Institute of Science Education and Research Tirupati, Mangalam, Tirupati 517507, India

Frank E. Rheindt
Avian Evolution Lab, Department of Biological Sciences, Faculty of Science, National University of Singapore, Singapore 117543, Singapore

Daniel M. Hooper
Committee on Evolutionary Biology, University of Chicago, Chicago, IL 60637, USA

Sushma Reddy
National Centre for Biological Sciences, TIFR, Bellary Road, Bangalore 560065, India
Biology Department, Loyola University Chicago, Chicago, IL 60660, USA

Index